MAN-MADE WONDERS OF THE WORLD

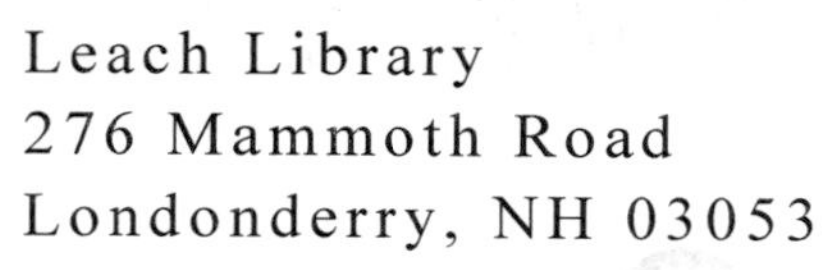

SMITHSONIAN

MAN-MADE WONDERS OF THE WORLD

FOREWORD BY DAN CRUICKSHANK

CONTENTS

DK LONDON

Senior Editor Peter Frances
Editors David Summers, Hannah Westlake
US Editor Kayla Dugger
Indexer Helen Peters
Jacket Designer Surabhi Wadhwa-Gandhi
Jacket Design Development Manager Sophia MTT
Managing Editor Angeles Gavira Guerrero
Associate Publisher Liz Wheeler
Publishing Director Jonathan Metcalf

Senior Art Editor Sharon Spencer
Project Art Editor Francis Wong
Senior Graphics Co-ordinator Sharon Spencer
Illustrators Phil Gamble, Mike Garland
Producer (Pre-production) Rob Dunn
Senior Producer Meskerem Berhane
Managing Art Editor Michael Duffy
Art Director Karen Self
Design Director Phil Ormerod

DK INDIA

Senior Editor Dharini Ganesh
Editors Aishvarya Misra, Priyanjali Narain
Assistant Editors Aashirwad Jain, Ishita Jha
Managing Editor Rohan Sinha
Picture Researcher Deepak Negi
Picture Research Manager Taiyaba Khatoon

Lead Senior Art Editor Mahua Sharma
Senior Art Editor Vaibhav Rastogi
Project Art Editor Anjali Sachar
Art Editor Sonali Sharma
Assistant Art Editor Garima Agarwal
Managing Art Editor Sudakshina Basu
DTP Designers Nand Kishor Acharya, Anita Yadav
Pre-production Manager Balwant Singh Pankaj Sharma

COBALT ID

Editors Marek Walisiewicz, Johnny Murray, Iain Bowden

Picture Editor Paul Reid

First American Edition, 2019
Published in the United States by DK Publishing
1450 Broadway, Suite 801, New York, NY 10018

19 20 21 22 23 10 9 8 7 6 5 4 3 2 1
001-310225-October/2019

Published in Great Britain by Dorling Kindersley Limited

A catalog record for this book
is available from the Library of Congress.
ISBN 978-1-4654-8252-5

DK books are available at special discounts when purchased in bulk for sales promotions, premiums, fund-raising, or educational use. For details, contact: DK Publishing Special Markets, 1450 Broadway, Suite 801, New York, NY 10018
SpecialSales@dk.com

Printed in China

A WORLD OF IDEAS:
SEE ALL THERE IS TO KNOW

www.dk.com

Smithsonian

Established in 1846, the Smithsonian—the world's largest museum and research complex—includes 19 museums and galleries and the National Zoological Park. The total number of artifacts, works of art, and specimens in the Smithsonian's collections is estimated at 154 million, the bulk of which is contained in the National Museum of Natural History, which holds more than 126 million specimens and objects. The Smithsonian is a renowned research center, dedicated to public education, national service, and scholarship in the arts, sciences, and natural history.

Contributors

Simon Adams, Alexandra Black, Thomas Cussans, Dr. Kay Celtel, Reg Grant, Owen Hopkins, Andrew Humphreys, Dr. Diana Loxley, Ellie Stathaki, Marcus Weeks, Iain Zaczek

Consultants

Dr. Andrew Law, Professor Ola Uduku

Smithsonian Consultants

Smithsonian's Office of Planning, Design & Construction Architectural History and Historic Preservation

Carly Bond Historic Preservation Specialist
Sharon Park, FAIA Associate Director for Architectural History and Historic Preservation

Half-title page: Lotus Temple, New Delhi
Title page: Hampi, India
Foreword: Leshan Giant Buddha, China

FOREWORD

The story of the world's man-made wonders—of architecture, engineered structures, and sublime works of sculpture—is also the story of mankind. These creations commemorate breathtaking tangible achievements, but they go beyond just the material world because they also reveal mankind's dreams and aspirations. And when the wonders are chosen through time and space—as in this book—the cumulative power is extraordinary.

The subjects raised by the exploration of these wonders cover all aspects of life because architecture, art, and engineering are the epitome of mankind's creative spirit. They touch upon the arcane and subjective aspects of aesthetics but also upon the more objective disciplines of science and physics because structures, by their nature, are often obliged to emulate and echo the stupendous power and laws of nature. This functional—almost elemental—approach is expressed best perhaps by the design of heroic bridges, such as the Akashi Kaikyo Bridge in Japan.

But the story of man-made wonders also includes the more pragmatic cut-and-thrust of politics, the exigencies of economics, and the evolution and refinement of building technologies and techniques. Among the most exciting moments in world architecture come when sudden leaps are made, often through audacious, almost revelatory, experimentation. This happened in Europe in the very early 12th century, when the theories of Gothic construction—embracing the notion of a skeletal frame formed by ribs, piers, and buttresses balanced by pinnacles and utilizing the structural power of the pointed arch—emerged with dramatic speed. This offered revolutionary possibilities to traditional stone construction by imbuing it with added strength realized through finely calculated design.

Even theology plays a key role in this story because so many of mankind's most memorable structures are not to do with structural innovation or the practical issues of providing shelter or security for the living, but with the more poetic challenge of honoring the gods and sustaining the dead. And some—like the Karnak temple complex in Luxor, Egypt—suggest answers to the great riddles that have enthralled humanity across the world and through the ages. Does anything survive death? Can the living commune with the dead? Can the dead intercede for the living? And can there be reincarnation—or even resurrection?

This spiritual quest is usually couched in the most complex of architecture, defined by geometry that simulates the apparent order of the natural world or of Divine creation and by ornament that is both symbolic and emblematic. Aachen Cathedral and the Dome of the Rock in Jerusalem are powerful examples.

Many other themes—tantalizing and inspiring, informative and entertaining—arise when contemplating the wonders in this book. One is the thrilling role of technology—not at least a dry topic. From the early 19th century onward, architecture of the Western world was transformed in majestic manner by the application of new building materials, primarily cast iron and wrought iron and then steel and reinforced concrete, to make unheralded building types possible—notably the skyscraper with a structural frame and heavily glazed curtain walls, including the Shard in London and the Burj Khalifa in Dubai. And then there is the integration of art and architecture and, in certain parts of the world, the reversal of common practice with buildings and volumes excavated rather than constructed—such as the rock-cut churches of Lalibela in Ethiopia—that make architecture, in a sense, giant carved pieces of sculpture.

And it is intriguing to see the principles of sustainable architecture—the use of cheap and available materials, such as unfired clay or mud, that involves little or no environmental damage or pollution, and techniques of natural insulation and ventilation—being applied creatively in traditional vernacular building. The outstanding example in this book is the Great Mosque at Djenné, Mali, built with sun-dried waferlike bricks of mud, dung, and straw covered with render.

Reading this book is like taking a journey through the world—not only the world of the present, but also of the past because the roots of many of the wonders described lie in antiquity. It is also a journey that is increasingly difficult to make in reality as the world becomes more hostile and more volatile, with key sites becoming practically inaccessible. And, in a few cases, the journey is impossible because some of the wonders shown here—notably Palmyra in Syria—have been, in an unthinkable manner, badly damaged and in part destroyed in very recent times. This gives the book another dimension, and a very important one. It not only displays wonders in telling detail that are now hard to see, but also documents wonders that are no more. In this respect, it is a solemn reminder. Never take what we love for granted, and be prepared to fight—if necessary—for our man-made wonders that do much to tell the story of mankind and that make the world such an exciting place.

DAN CRUICKSHANK
MARCH 2019

Long story
There has been a settlement on the site of Barcelona since well before Roman times, and the modern city is made up of buildings from many different periods, built with materials ranging from timber, brick, and stone to concrete and glass.

Introduction

EARTH AND TIMBER

As human societies developed agriculture and former nomads began to settle, temporary encampments gave way to permanent structures, perhaps as far back as 10000 BCE. The earliest building material was the earth itself, combined with other materials, such as straw, to add strength, and timber to support roofs.

Building with earth

Earthen structures are highly susceptible to erosion and weathering, and little evidence of early human earthen dwellings remains today. The structures that do survive have done so largely because of their massive size. These tend to be barrows, mounds, and tumuli; the names tend to be interchangeable, but they are all huge, man-made hills of compacted earth of a size to defy the elements. Such mounds are found across northern and central Europe and on the steppes of Russia and Central Asia. Many date from the Bronze Age (c.3500–300 BCE). Their function is almost always as a tomb, and archaeologists have discovered skeletons and funerary items in many of them. One of the largest mounds in Europe is the Leeberg near Großmugl in Austria, which dates from c.600–500 BCE. This is dwarfed by the burial mound at Bin Tepe in Lydia, Turkey, built in around 560 BCE; with a height of 196 ft (60 m), it is one of the largest tumuli known.

Mud and mud-mixture structures

Some of the earliest habitable buildings were made from mixtures of mud or dung strengthened with straw or horse hair, sometimes applied to a frame of branches or woven reeds. It does not seem a large step from building with mud and straw in bulk to building with bricks made of mud and straw, but the compressive strength and versatility of fired bricks allowed humans to build sizeable and sophisticated structures. The earliest large-scale buildings of mud-brick, or adobe, belong to the civilizations of ancient Mesopotamia (modern-day Iraq). Built in around 2100 BCE, the Ziggurat of Ur (see p.243) was a huge stepped pyramid that is estimated to have risen to over 98 ft (30 m). Later Mesopotamian civilizations developed glazed bricks, which they used to decorate their buildings, as at Babylon, their capital.

Earth remains one of the most abundant basic materials. It is low-tech, easily worked, durable, fire resistant, and cheap, so mud-brick is still popular for building—although not in regions facing regular rainfall.

The laminated **timber** used in modern high-rise buildings is similar in **strength** to **steel** but much **lighter**

EARLY STRUCTURES
Formed from a core of mud-bricks with a further mud covering, these beehive houses can still be found in southeast Turkey and parts of central Syria.

▷ **MAN-MADE HILL**
One of the largest artificial mounds in Europe, Silbury Hill dates from around 2400 BCE. It is composed of clay and chalk and apparently contains no burial, which makes its purpose something of a mystery.

Timber

There is archaeological evidence that wood was used to build dwellings in Europe as early as 6000 or 5000 BCE. Numerous examples of longhouses dating from this time have been found, particularly in the region of Transdanubia, comprising northwestern Hungary, northern Austria, and southern Germany. These long, rectangular, single-room structures with oak posts supporting the roof are thought to have provided living quarters for up to 20–30 people, with multiple longhouses forming villages. Later, the discovery of metals such as bronze and steel, which could be fashioned into tools for cutting and shaping, advanced the ways in which wood could be used in construction.

The ancient Egyptians, Greeks, and Romans all made use of cut and carved timber, particularly in roof construction. It was the Romans who developed timber framing—the construction of wooden skeletons to support a building. This technique reached its peak during the Middle Ages, when carpenters created magnificently elaborate, ingenious, and frequently beautiful arrangements of timber to support the roofs of halls, churches, and cathedrals. These include the hammer-beam roof of Westminster Hall in London, which is the largest medieval timber roof in northern Europe. Scandinavia had its own spectacular take on timber construction in the form of tumbling stave churches. Around the same time, the Chinese were constructing timber buildings of great complexity and refinement, exemplified by the Pagoda of Fogong Temple, built in 1056 and reaching over 220 ft (67 m) high, without a single nail, screw, or bolt.

Timber continues to be used in construction because it is durable, often widely available, and renewable, and relatively little energy is needed to turn trees into usable timber. Architects are experimenting with new ways of using timber, exploiting its light weight and flexibility, including, most ambitiously, wooden skyscrapers, or "plyscrapers."

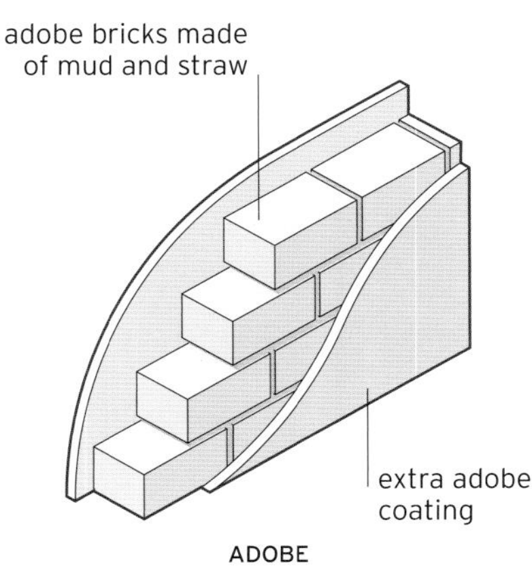

ADOBE

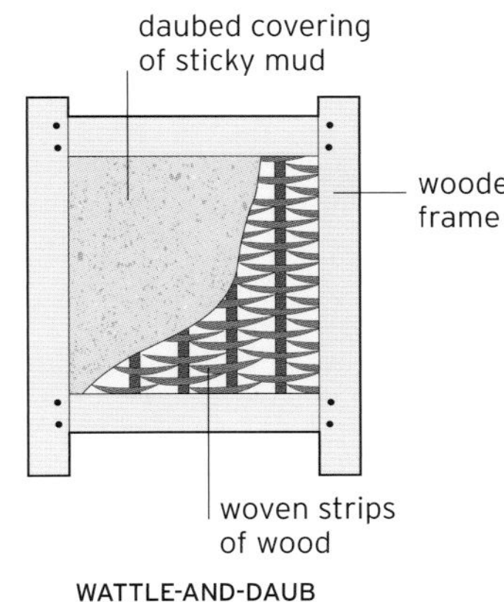

WATTLE-AND-DAUB

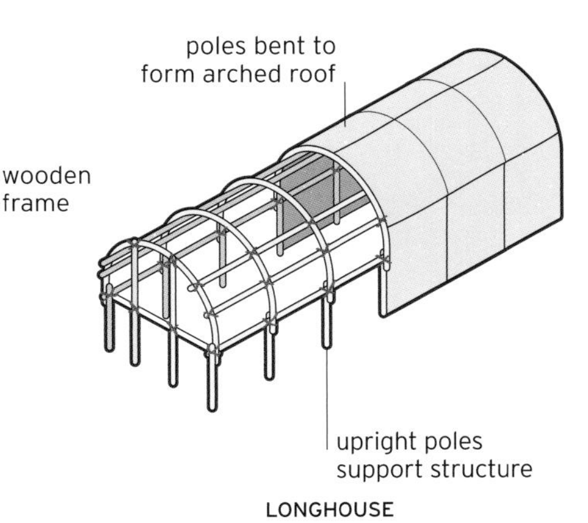

LONGHOUSE

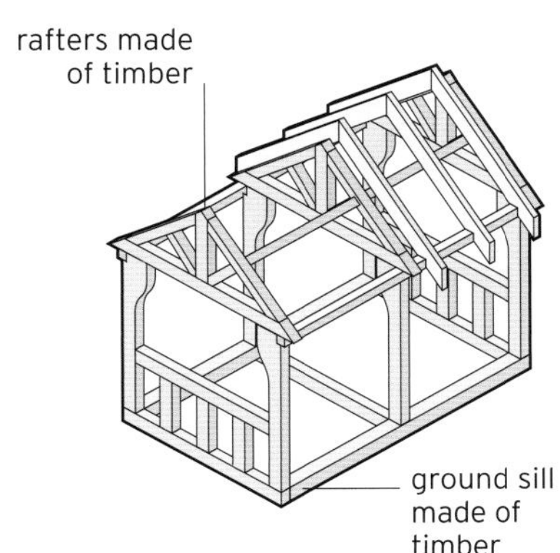

TIMBER-FRAME

△ **EARLY ADOBE CONSTRUCTION**
Mud mixed with straw (adobe) is compressed into molds and baked dry to form regular-sized bricks. Once laid, the bricks are faced with an additional coating of adobe, as seen above in Taos Pueblo, New Mexico.

△ **WATTLE-AND-DAUB CONSTRUCTION**
In this age-old construction technique, a woven lattice of wooden strips or branches (wattle), held between upright stakes, is daubed with a mixture of sticky mud and straw. This Tudor-style house in England employs this technique.

△ **LONGHOUSE CONSTRUCTION**
Ancient longhouses survived into the 19th century in North America, built and inhabited by the native tribes—notably the Iroquois of the northeastern US. This longhouse at the Ganondagan State Historic Site, New York, is one such example.

△ **TIMBER-FRAME CONSTRUCTION**
In parts of the world with a plentiful supply of wood, timber-framing became the most common form of construction because of its cheapness and versatility. The Paul Revere House in Boston is famous for its timber construction.

BRICK AND STONE

First appearing around 3000 BCE, robust fired bricks facilitated the development of new forms of architecture such as arches, vaults, and domes. However, stone has greater compressive strength than brick—and, once tools had been invented to cut it, buildings began to soar.

Early stone structures

Evidence survives for early stone structures erected during the Neolithic Age (10000–3000 BCE), such as the Megalithic Temples of Malta (see p.94), dated about 3700 BCE, and the Cairn of Barnenez in France, dating from 4850 BCE. These are relatively crude structures of piled, undressed stones. The advanced civilizations in the Nile valley were the first to develop the ability to cut and shape stone, and thus to build structures of immense size and durability. Their first great monument was the Stepped Pyramid of Djoser (2667–2648 BCE)—the earliest known large-scale cut-stone construction. This was followed roughly 100 years later by the Great Pyramid of Giza (see pp.208–209), whose visual impact remains stunning. The Egyptians' building techniques advanced quickly, and by 2000 BCE, they were creating vast temple complexes with pylons, courtyards, colonnades, and huge roofed halls. The culmination was Karnak (in modern-day Luxor; see pp.210–211), the largest religious complex ever known, built on such a scale that its columns rose up to 80 ft (24 m) high.

Greek and Roman architecture

Across the Mediterranean, the Greeks adopted the same stone-frame temple construction form as the Egyptians—column-and-beam (or post-and-lintel). They brought to building in stone a new refinement and a set

◁ **INTRICATE DECORATION**
Around the same time that Gothic cathedrals were spreading across Europe, the Khmer empire (802–1431) in Southeast Asia was constructing temple complexes of great elaboration out of brick and sandstone.

of aesthetics and architectural orders that are still adhered to, notably Doric, Ionic, and Corinthian (see p.97). They also developed the bas-relief decoration of the Egyptians into full-blown sculpted friezes and introduced optical corrections to show their buildings to best advantage.

It was the Roman builders, however, who exploited masonry to its fullest advantage. Archaeological digs have revealed the first true arches in brick and the earliest forms of vaulting in the sewers and tombs of Mesopotamia. It is clear, then, that the Romans did not invent

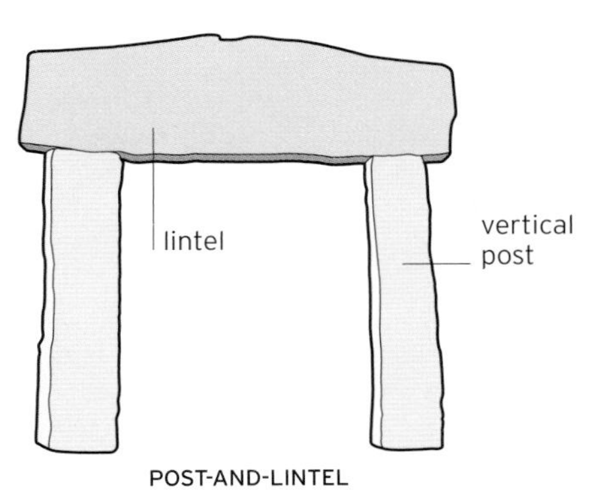

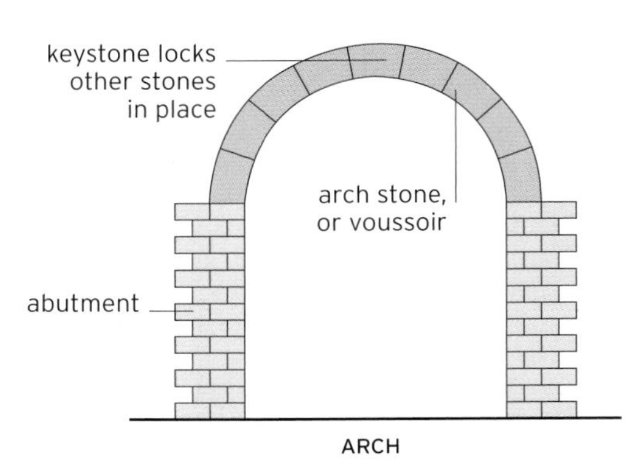

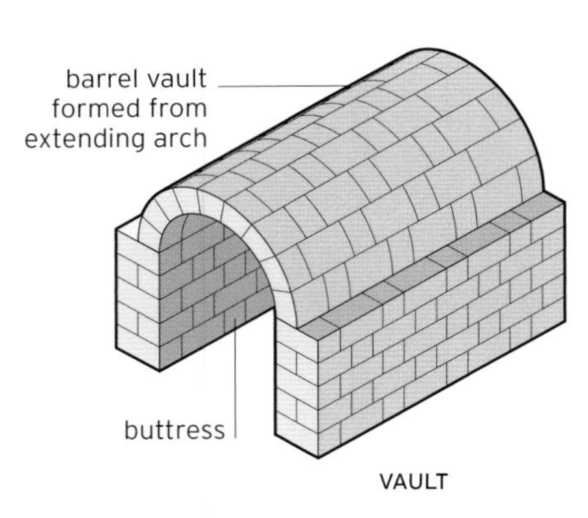

△ **POST-AND-LINTEL CONSTRUCTION**
Some of the earliest forms of stone architecture were post-and-lintel structures, with horizontal blocks supported on vertical posts. The Egyptians and Greeks refined this into columns, entablatures, and pediments, as seen in the Parthenon temple in Athens (see pp.98–99).

△ **ARCHES**
The invention of the arch enabled builders to span far larger openings than had been possible previously. The Romans were masters of arch building, as in this aqueduct, the Pont du Gard at Nîmes in France (see p.101).

△ **VAULTING**
From the arch came the basic barrel or tunnel vault. This could be used to create vast internal spaces, as at Ctesiphon in Iraq. Later, vaulting developed into ever more elaborate forms, such as the intersecting groin vault.

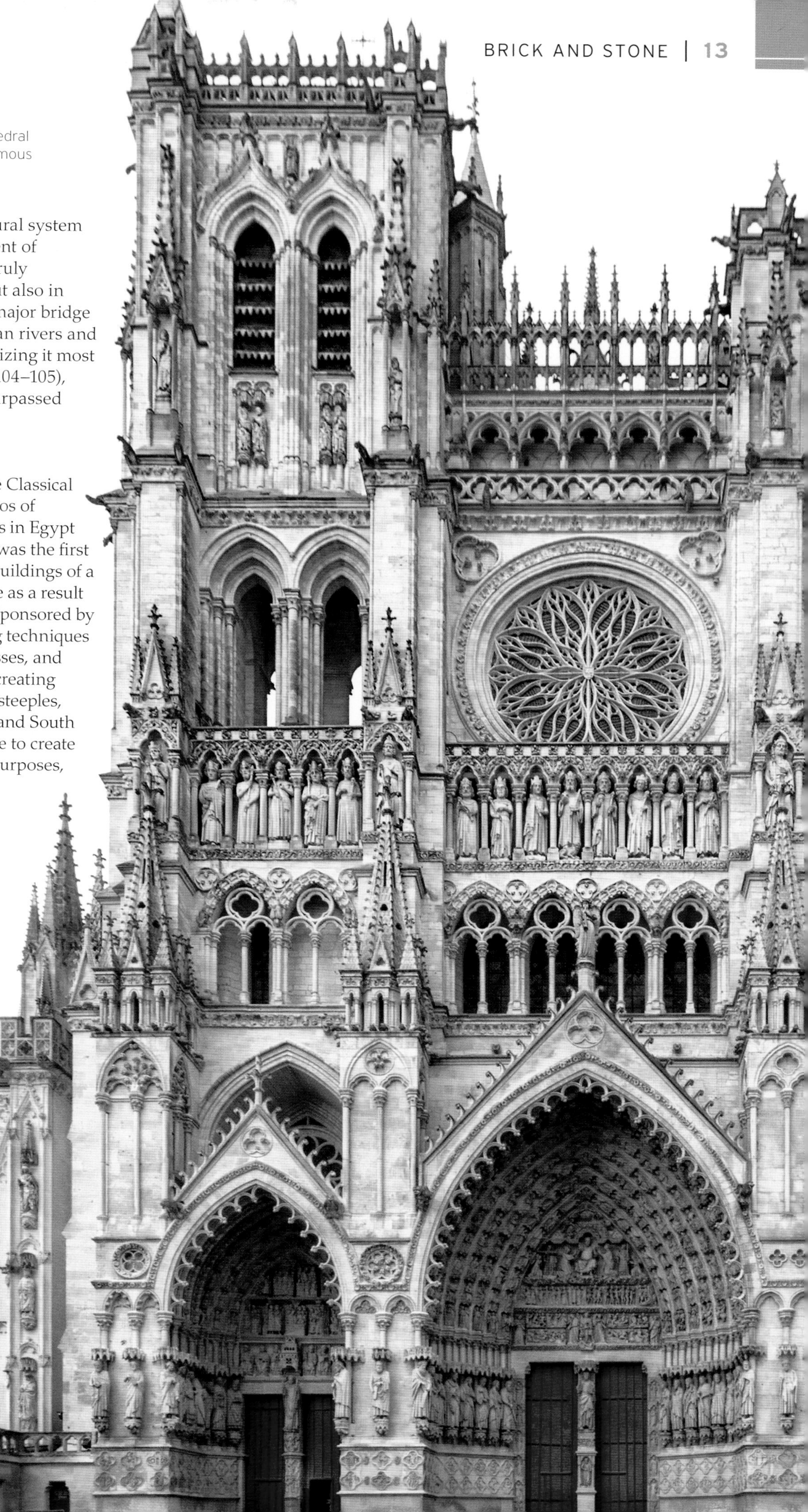

▷ TOWERING CATHEDRAL
Built between 1220 and 1270, Amiens is the tallest complete cathedral in France and has the greatest interior volume. The building is famous for its highly ornate exterior, built using locally quarried stone.

these forms, but they developed them to create a structural system that was not substantially improved upon until the advent of construction in iron and steel. It was the Romans who truly developed the arch, not only in monumental fashion, but also in revolutionary public works. The Romans were the first major bridge and aqueduct builders, employing multiple arches to span rivers and valleys. They were also early proponents of concrete, utilizing it most famously in the glorious dome of the Pantheon (see pp.104–105), which, with a diameter of 142 ft (43.2 m), remained unsurpassed until the 19th century.

Attaining height

Perhaps the most spectacular building achievement of the Classical age—if the descriptions are to be believed—was the Pharos of Alexandria, a great lighthouse built by the Greco–Romans in Egypt in the 3rd century BCE. Perhaps as tall as 330 ft (100 m), it was the first high-rise building. It was not until the 14th century that buildings of a comparable height were rising in Europe again. This came as a result of a boom in cathedral building—a "cathedral crusade." Sponsored by church and state, medieval masons pushed stone building techniques to even greater heights. The use of trunklike piers, buttresses, and Gothic arches and vaults raised the roofs to the heavens, creating soaring internal spaces, echoed on the outside by towers, steeples, and spires that strained skyward. In what is now Central and South America, pre-Columbian civilizations also employed stone to create sophisticated architectural wonders used for a variety of purposes, including their own version of the stepped pyramid.

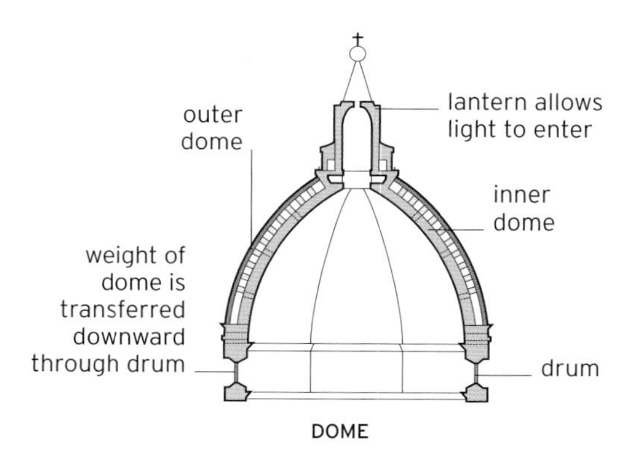

△ DOMES
The arch was swiveled through 360 degrees to give us the dome. Developed by the Romans, it reached its apex with Filippo Brunelleschi's dome for Florence Cathedral (see pp.140–141), completed in 1461 and the largest brick dome ever built.

IRON, STEEL, AND BEYOND

For over 10,000 years, the basic materials for construction were earth, timber, bricks, and stone. In the last 200 years, however, there has been radical change. Our architecture is now given new and ever greater freedoms of form through materials such as iron, steel, and concrete.

Advent of new materials

Not only did the Industrial Revolution in the 18th–19th centuries herald the advent of industry and new forms of transportation, but it changed the way buildings were made. Engineers were quick to see the possibilities in the large-scale production of iron. The metal was fire-resistant and could be molded into shapes that were impossible to make with stone or brick. It was first tested out in 1779, with the Iron Bridge over the River Severn in England, and was then adapted for columns, beams, and full iron-frame buildings elsewhere. Ditherington Mill, built in 1796 in Shrewsbury, England, is considered the first iron-framed building in the world. In the late 19th century, iron was superseded by steel, which was much tougher and less brittle. Its qualities were shown off to best advantage in 1889, when steel was the main building material for the revolutionary Eiffel Tower (see pp.180–181).

△ **MAIDEN USE**
One of the earliest uses of the full iron frame was in the building of greenhouses, such as the Palm House at Kew Gardens, England.

▷ **FREE-FLOWING STRUCTURE**
Supported by 16 inward-curving columns, Oscar Niemeyer's Cathedral of Brasilia (see pp.84–85) is an example of the freedom of structure offered to architecture by the innovation of prestressed reinforced concrete.

Further innovations

Steel, along with another piece of new technology, the elevator, was key to the most significant innovation in modern architecture: the skyscraper. It began in Chicago, with the 10-story Home Insurance Company Building of 1885, before the momentum shifted to New York, where the Manhattan Life Building hit 26 stories in 1889, followed by the Singer Building of 47 stories in 1907, and topping out with the 102-story Empire State Building (see p.43) in 1931.

As with height for buildings, steel facilitated longer spans for bridges. In 1874, the Eads Bridge over the Mississippi River at St. Louis was completed. A distance of 520 ft (158 m) between the piers made it the longest rigid span ever built at the time, which was achieved through the use of a steel arch. Nine years later, the monumental Brooklyn Bridge (see pp.26–27) opened in New York as the world's first major suspension bridge with steel cables, suspenders, and a deck. It was the first step toward future grand spans, such as the Golden Gate Bridge in San Francisco (see pp.46–47), with its main span of 4,200 ft (1,280 m), and the Akashi Kaikyo Bridge (see p.308) in Japan—the world's longest—with a main span of 6,532 ft (1,991 m).

Due to environmental concerns, **100 percent recyclable steel** is now being used to build many structures

Redefining architecture

The industrial age also saw the reintroduction of concrete, a material used in Roman times but now revived in reinforced form. Engineers were now able to create a dome of a size that surpassed ancient Rome's Pantheon (see pp.104–105), with the Centennial Hall at Breslau, Germany (now Wroclaw, Poland), in 1913. The facility to precast concrete in almost any shape freed architects' minds, resulting in the visionary church Notre Dame-du-Haut by Le Corbusier, the organic houses of Frank Lloyd Wright, and Jørn Utzon's sail-like Sydney Opera House (see pp.304–305). The trend continues, but not just with concrete. It is aided by computer modeling of new buildings, which now lean, twist, loom, hover, and of course climb even higher.

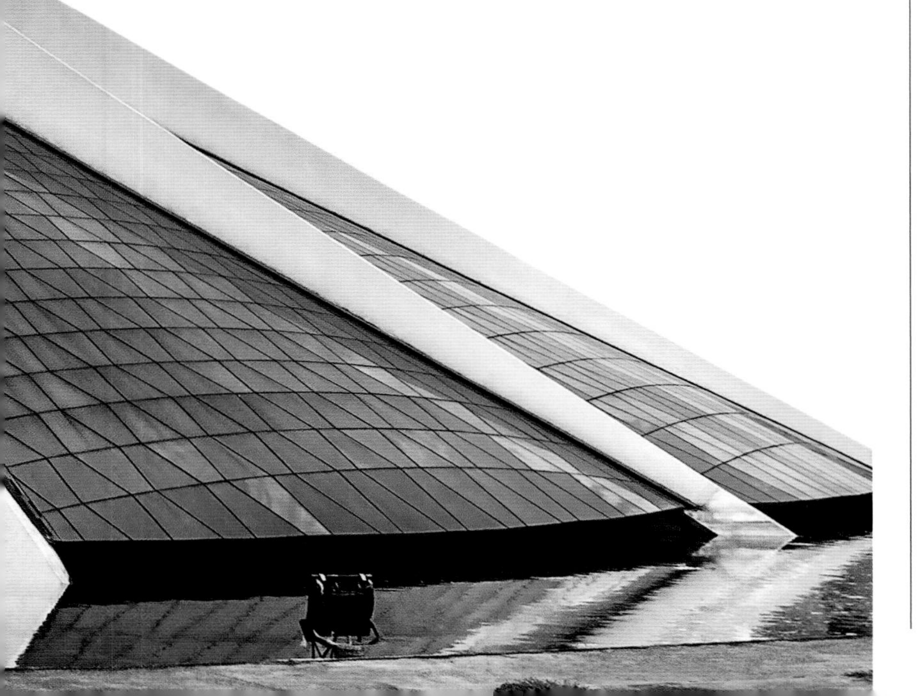

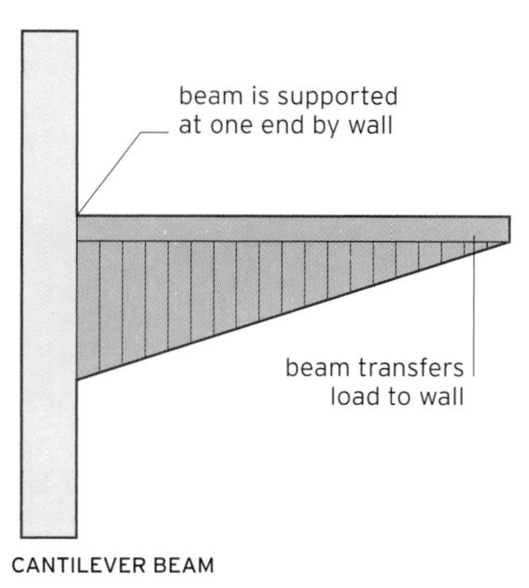

△ **CANTILEVER**
The cantilever, in which only one end of a structure is supported, has often been used in bridge structures. However, in the China Central Television (CCTV) Building (shown above), built in 2015 in Beijing, the whole upper portion of the tower is cantilevered for 245 ft (75 m).

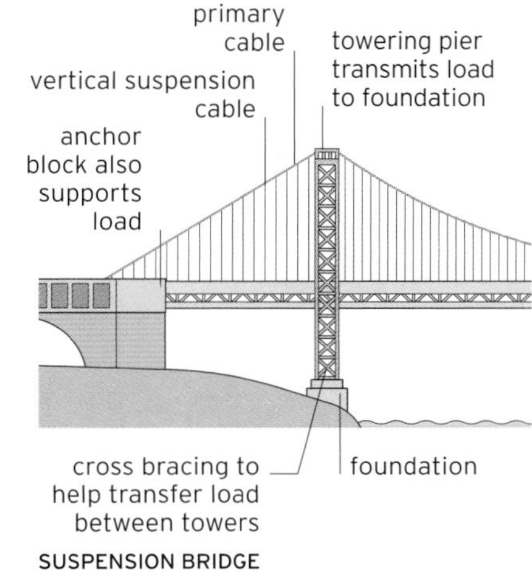

△ **SUSPENSION**
As a material, steel works well in compression but also in tension, which is exploited in suspension bridges such as New York's George Washington Bridge. In such structures, the deck of the bridge is suspended from giant cables that are hung off towering piers.

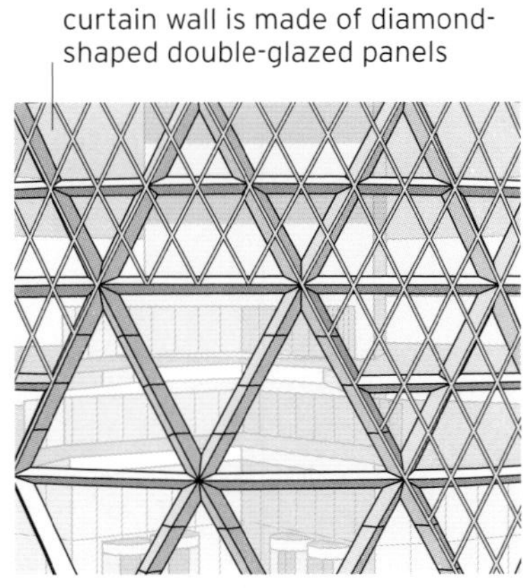

△ **NEW MATERIALS AND DESIGN TOOLS**
By taking advantage of new materials and manufacturing techniques, architects can now design buildings with sinuous curves and irregular outlines, such as Frank Gehry's Weisman Art Museum in Minnesota. Many modern buildings are enclosed by an outer curtain wall, often made of metal and glass.

Reaching up
At the end of the 19th century, North America gave birth to the skyscraper when iron framing and the invention of the elevator made it possible to build above 10 stories. Today, skyscrapers dominate the skylines of New York and many other US cities.

North America

PUEBLOS TO SKYSCRAPERS

North America

Ambitious ideas, expansive spaces, and a strong appreciation of regional identity have underpinned some of North America's most remarkable structures. The Pueblo architecture of the 1st century CE was revived and reinterpreted in the 20th century as the Santa Fe style. Meanwhile, in the Deep South, the ancient construction of huge platform mounds was still being practiced when Europeans arrived in the 17th century. The colonizers evolved a distinctive American style, supported by Classical ideals and concepts of freedom and democracy. Reflecting the vastness of the continent and the potential to dream large, man-made structures were supersized, an approach that continued through the Industrial Revolution and into the era of skyscrapers.

KEY SITES

1. Chaco Canyon
2. Cliff Palace
3. Serpent Mound
4. Monticello
5. The White House
6. The US Capitol
7. Brooklyn Bridge
8. Washington Monument
9. Flatiron Building
10. Château Frontenac
11. Biltmore House
12. The Statue of Liberty
13. Washington National Cathedral
14. Grand Central Terminal
15. Lincoln Memorial
16. Jefferson Memorial
17. Mount Rushmore
18. Chrysler Building
19. Empire State Building
20. Hoover Dam
21. Golden Gate Bridge
22. The Pentagon
23. TWA Flight Center, JFK Airport
24. Guggenheim Museum
25. Space Needle
26. Gateway Arch
27. Habitat 67
28. Vietnam Veterans Memorial
29. CN Tower
30. Walt Disney Concert Hall
31. National Museum of African American History and Culture

COLONIZATION LEAVES ITS MARK

What remains of indigenous American structures consists largely of prehistoric effigy mounds in the east and Pueblo architecture, made of adobe, in the southwest. Colonization of the eastern and western coasts from the 17th to 19th centuries introduced different European styles in more durable materials.

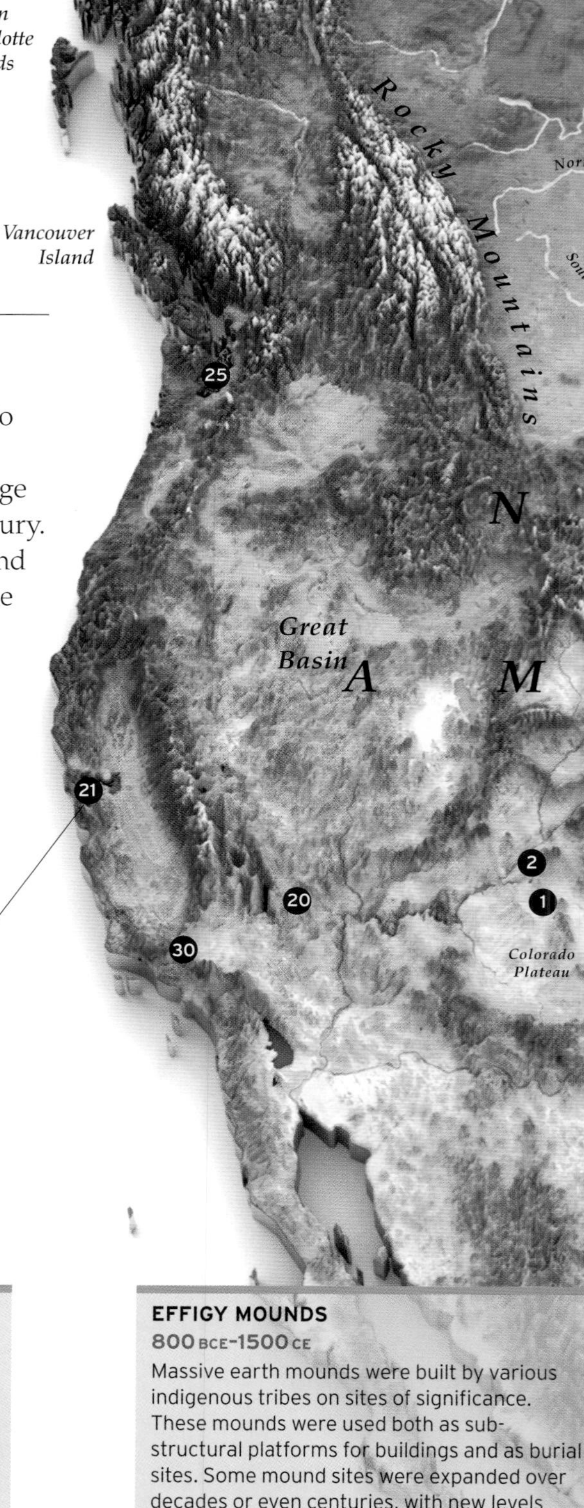

when it was built in 1937, the Golden Gate Bridge was the tallest and longest suspension bridge in the world

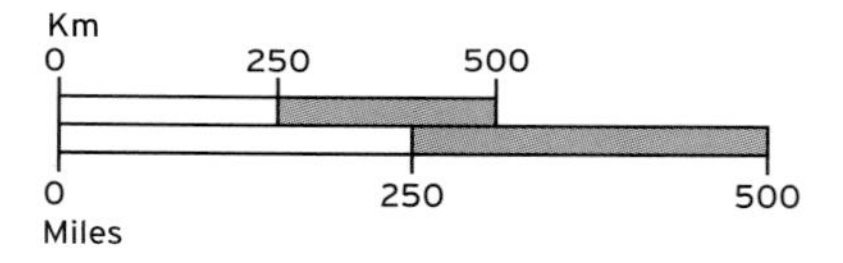

PUEBLO BUILDERS

700–1200 CE

Native American people living in the southwest of what is now the US developed a tradition of building stone and adobe pueblos. Using limestone blocks and bricks, made from clay and water, they designed villages with tiered, multistory terraces, with each floor set back slightly from the one below.

1 CHACO CANYON

EFFIGY MOUNDS

800 BCE–1500 CE

Massive earth mounds were built by various indigenous tribes on sites of significance. These mounds were used both as sub-structural platforms for buildings and as burial sites. Some mound sites were expanded over decades or even centuries, with new levels added and layers applied to prevent slumping.

3 SERPENT MOUND

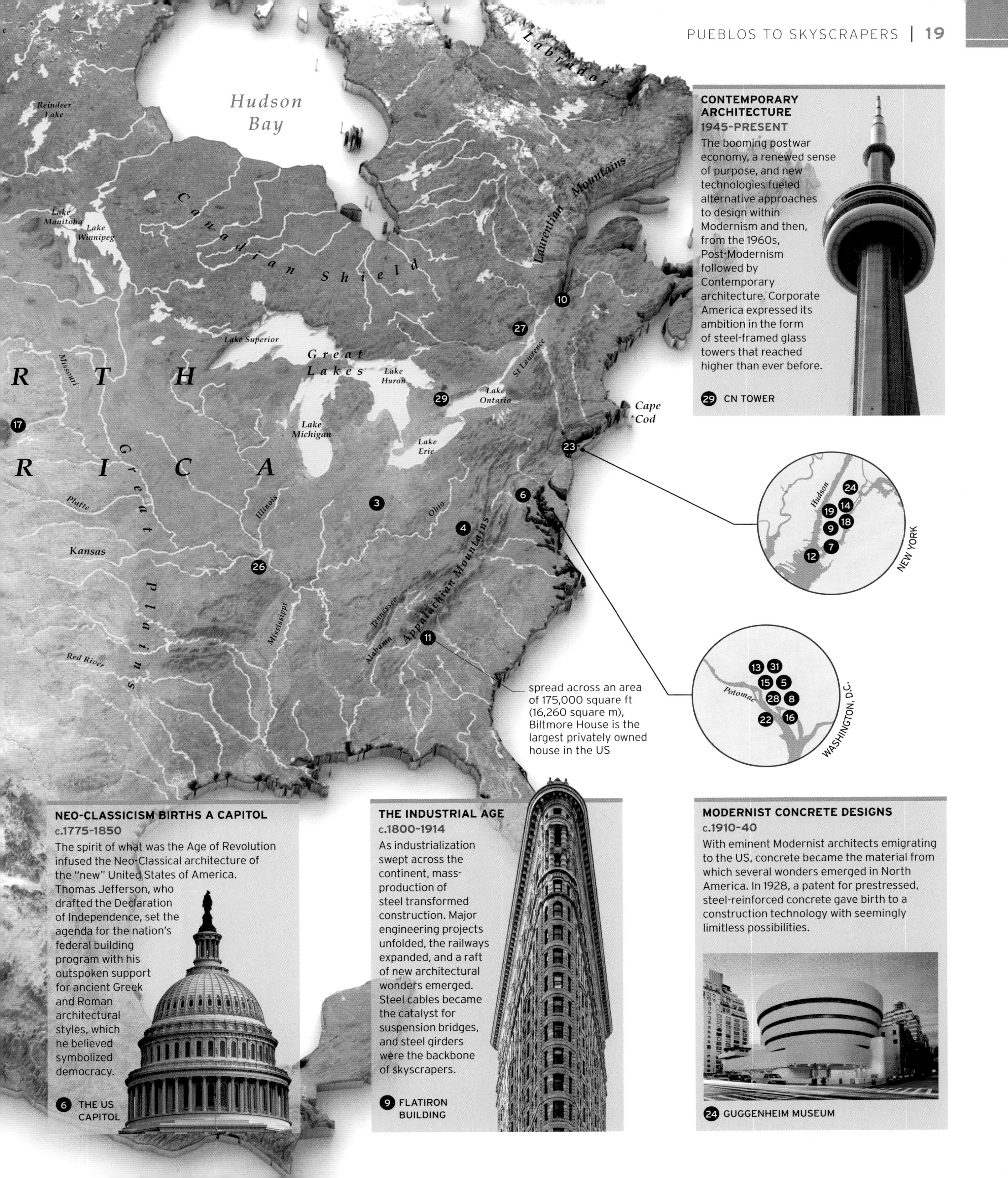

spread across an area of 175,000 square ft (16,260 square m), Biltmore House is the largest privately owned house in the US

CONTEMPORARY ARCHITECTURE

1945–PRESENT

The booming postwar economy, a renewed sense of purpose, and new technologies fueled alternative approaches to design within Modernism and then, from the 1960s, Post-Modernism followed by Contemporary architecture. Corporate America expressed its ambition in the form of steel-framed glass towers that reached higher than ever before.

29 CN TOWER

NEO-CLASSICISM BIRTHS A CAPITOL

c.1775–1850

The spirit of what was the Age of Revolution infused the Neo-Classical architecture of the "new" United States of America. Thomas Jefferson, who drafted the Declaration of Independence, set the agenda for the nation's federal building program with his outspoken support for ancient Greek and Roman architectural styles, which he believed symbolized democracy.

6 THE US CAPITOL

THE INDUSTRIAL AGE

c.1800–1914

As industrialization swept across the continent, mass-production of steel transformed construction. Major engineering projects unfolded, the railways expanded, and a raft of new architectural wonders emerged. Steel cables became the catalyst for suspension bridges, and steel girders were the backbone of skyscrapers.

9 FLATIRON BUILDING

MODERNIST CONCRETE DESIGNS

c.1910–40

With eminent Modernist architects emigrating to the US, concrete became the material from which several wonders emerged in North America. In 1928, a patent for prestressed, steel-reinforced concrete gave birth to a construction technology with seemingly limitless possibilities.

24 GUGGENHEIM MUSEUM

PUEBLOAN JEWEL
The large-scale structures at Chaco Canyon were highly organized. The multistory construction and sophisticated masonry used by Ancestral Puebloans illustrate the complexity of their social structure.

Chaco Canyon

A remote canyon that holds the huge structures built by the Ancestral Puebloans more than 1,000 years ago

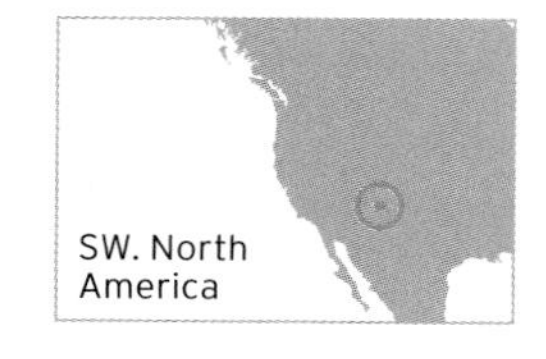

The Ancestral Puebloans were a Native American people who thrived in the southwestern US from around 750 to 1350 CE. They were great builders, constructing around 125 towns connected by an impressive system of roads. The finest of these towns lie in the Chaco Canyon in a remote area of northwest New Mexico. Most notable are the 15 large, D-shaped complexes—the largest buildings in North America until the late 1800s. These apartment-style structures were made from stone, adobe mud, and other materials, with timber often hauled to the site from up to 70 miles (110 km) away.

Pueblo Bonito

The most famous Ancestral Puebloan town, Pueblo Bonito, built around 1050 CE, spreads over more than 110,000 square ft (10,000 square m) and contains at least 650 rooms. It supported a population of more than 1,200 people. Pueblo Bonito and the other towns acted both as ceremonial centers—many of their buildings align with key stages of the solar and lunar cycles—and trading centers for food and luxury goods, such as turquoise.

△ **CHOCOLATE MUGS**
It is now believed that the cylindrical jars found in Pueblo Bonito were used for drinking chocolate, made using cacao beans brought from Mexico around 1,200 miles (1,900 km) away.

THE *KIVA*

The *kiva*—a room used for rituals and meetings—attests to the organized society of the ancient Puebloans. Partly underground and roughly circular in shape, it had benches around the edge and a firepit in the center. A *sipapu*, or hole, dug in the north part of the chamber represented the Ancestral Puebloans' place of emergence from the underworld.

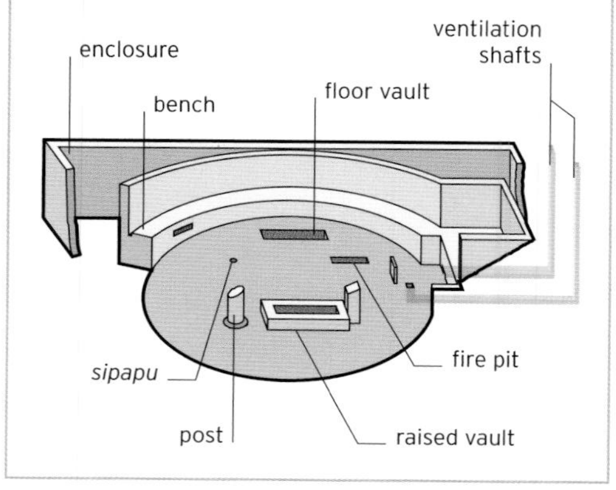

Chocolate was drunk for the **first time** in North America in Chaco Canyon about **1,000 years ago**

Cliff Palace

A palace of 150 rooms built by the Ancestral Puebloans into the side of a steep and inaccessible cliff

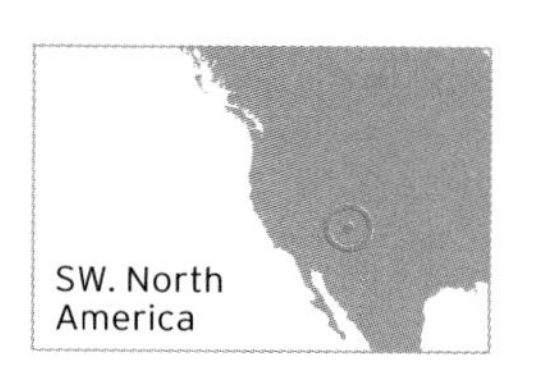

While some Puebloan towns lay on open ground, many more were built into the sides of steep cliffs. Such protected locations were easily defensible, suggesting that they were built at times of increasing competition among local tribes for scarce resources.

Primeval palace

The most impressive of these dwellings is Cliff Palace, found in a canyon in the Mesa Verde plateau of southwest Colorado. Built between 1190 and 1260, it was mainly constructed of sandstone blocks and wooden beams held together with a mortar of soil, ash, and water.

The Palace contains around 150 rooms and 23 sunken *kivas* (see opposite); the high number of *kivas* suggests that Cliff Palace was the center of a widespread local community. However, the site was abandoned by 1300, probably in response to extreme drought.

CAVE PROTECTION

Cliff Palace was built in a large cave under the overhang of a sandstone cliff that had been eroded by wind and water. The cave is 89 ft (27 m) deep, 59 ft (18 m) high, and 324 ft (99 m) long. Some of the dwellings within have four stories, and their walls are decorated with colored plaster.

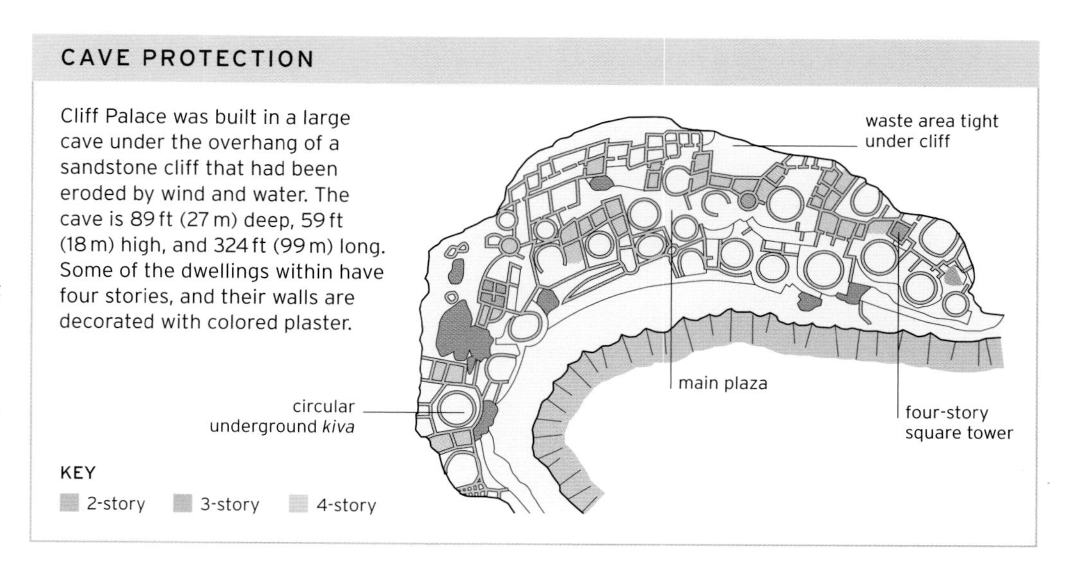

Cliff Palace was rediscovered in 1888 by ranchers searching for their cattle

▽ PALACE RUINS
In Cliff Palace, the cramped buildings rise up above each other on terraces, many of them excavated to produce subterranean *kivas*.

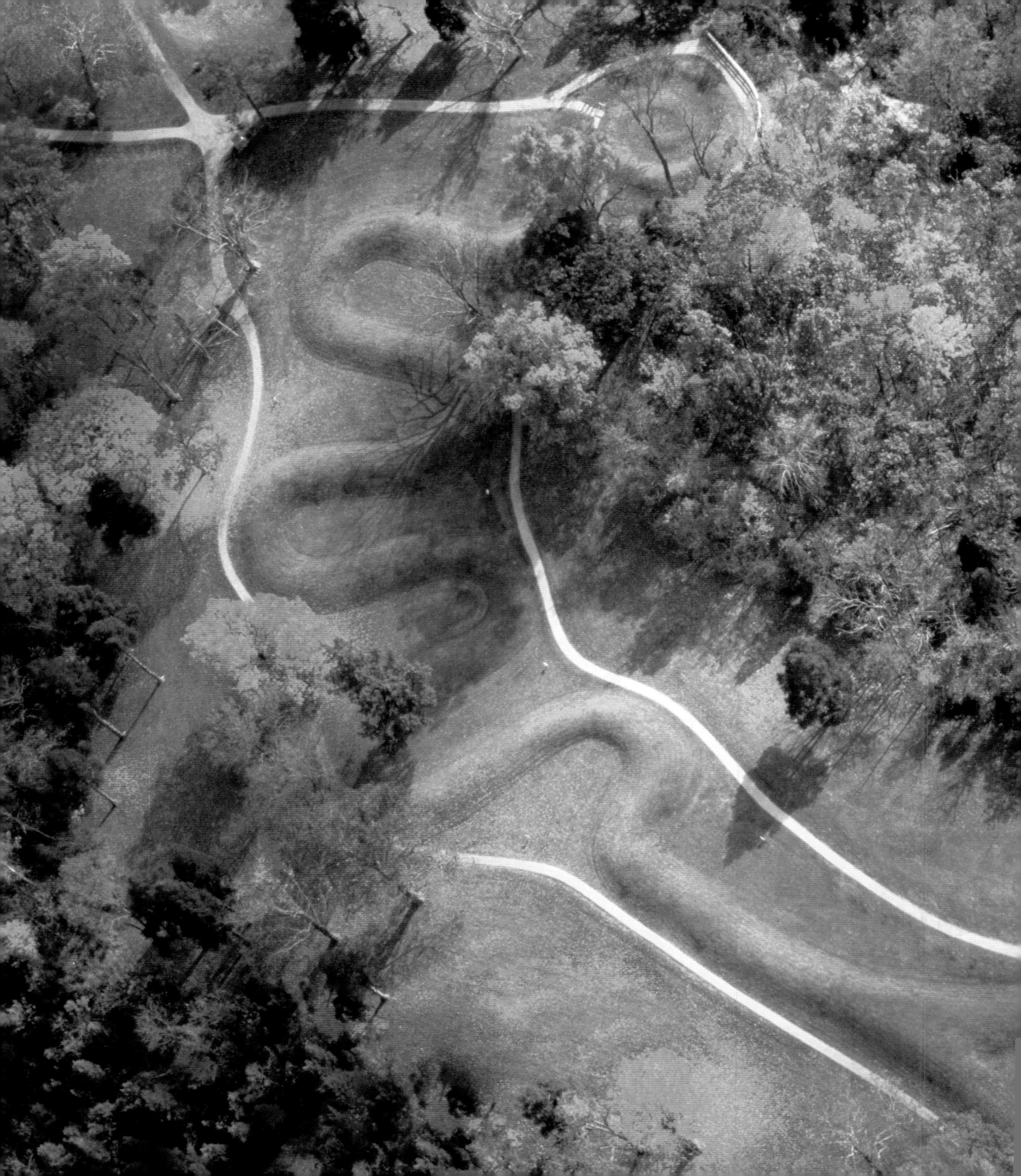

Serpent Mound

A giant snake-shaped mound in Ohio that continues to puzzle archaeologists

The Serpent Mound winds for 1,348 ft (411 m) along a plateau beside the Ohio Brush Creek in southern Ohio. It curves comfortably around the land, its head close to a cliff above the creek and its seven-coiled body ending in a triple-coiled tail. It lies on the now-concealed site of an ancient crater formed by the impact of a meteorite millions of years ago. It is not known if this event affected its placement or design. The earthwork was one of numerous mounds created by the Native American cultures that cultivated the fertile river valleys in Ohio. Most have been destroyed by modern agricultural practices.

The mouth of the snake

The snake is made of a layer of yellowish clay and ash reinforced with a layer of rocks and covered with soil. Its open mouth extends around the end of a 120-ft- (37-m-) long oval hollow that might represent the snake eating an egg, although the oval could also symbolize the Sun or a frog, or merely be the remnant of a platform.

Dating difficulties

Originally thought to be the work of the Adena people, who lived between 1000 BCE and 200 BCE, the serpent was later attributed to the Fort Ancient culture from around 1070 CE, on the basis of carbon dating undertaken in 1996. However, more recent dating carried out in 2014 reallocates it to around 320 BCE, apparently reaffirming its Adena origins. The purpose of the mound remains obscure. Its head does align with the summer solstice sunset, indicating some calendrical or ceremonial function, but it is more likely that the mound had a place in funerary rites, perhaps directing spirits of the dead from nearby burial mounds.

The Serpent Mound is by far the **largest snake effigy** in the world

△ SERPENTINE LANDSCAPE
The Serpent Mound varies greatly in height, from less than 1 ft (30 cm) to more than 3 ft (1 m), and has a width of 21–26 ft (6.5–8 m).

◁ SNAKING ACROSS
While the coils of the snake effigy seem fairly uniform from above, their undulating construction makes the creature appear to slither as it winds across Ohio Brush Creek.

ASTRONOMICAL PURPOSES

There has been a lot of speculation about the purpose of the Serpent Mound. The mound does align with some important astronomical events, notably those concerning the summer and winter solstices and the two equinoxes, suggesting that these dates were of great significance to an agricultural people dependent on the production of corn to eat.

head aligns with direction of summer solstice sunset

main body of snake

Ohio Brush Creek

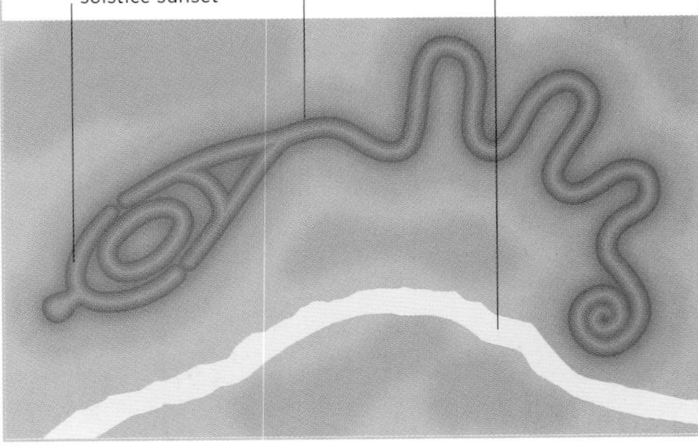

Monticello

A hilltop Palladian villa, designed according to Classical proportions by a future president of the United States

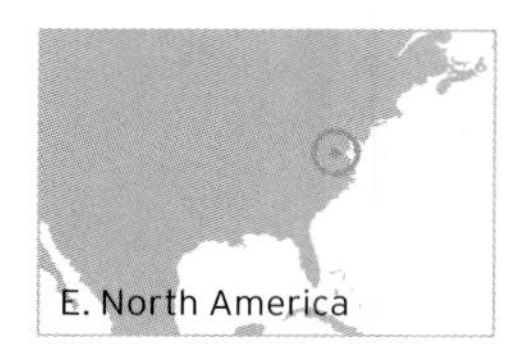

In 1768, aged just 26, the lawyer and politician Thomas Jefferson (1743–1826) inherited 8 square miles (20 square km) of land just outside Charlottesville, Virginia, from his father. He farmed the land (using slave labor) and, indulging his interest in architecture, drew the blueprints for a plantation house. He based his design on the principles of the Italian Renaissance architect Andrea Palladio (1508–1580).

Little mount

Jefferson lived in the house—named Monticello, from the Italian for "little mount"—from 1770; following the death of his wife Martha, he traveled to France in 1784, and in 1785 took the post of Minister to France. Inspired by the architecture he saw in Paris, he returned to the US with new plans for Monticello, which he remodeled and enlarged, adding a central octagonal dome and transforming the eight-room villa into a 21-room house in the Neo-Classical style. He lived in the house until his death in 1826, moving to Washington during his presidency (1801–1809).

▽ **IMPRESSIVE ENTRANCE**
Monticello has two main entrances. Visitors enter through one dominated by a columned portico, which leads into a domed reception room.

The White House

The official residence and workplace of the president of the United States

The Neo-Classical building at 1600 Pennsylvania Avenue in Washington, D.C., is perhaps the most famous structure in the world. Every president since John Adams in 1800 has lived in the White House, as it is popularly known.

President's residence

The residence was designed by James Hoban (1755–1831), who modeled it on Leinster House, Dublin. It was built between 1792 and 1800. The West Wing was added in 1901, and the famous Oval Office was created in 1909.

A myth has arisen that during the rebuilding of the mansion after the British sacked and burned Washington in 1814, white paint was used to mask the damage, hence its modern name. But the exterior had been coated with a lime-based whitewash to protect it from moisture and frost from as early as 1798.

▷ **LOOKING DOWN**
This photograph, taken from the top of the nearby Washington Monument, shows the modest size of the White House—the building is 168 ft (51 m) long and 85 ft (26 m) wide.

US Capitol

A new legislative building in the new capital of a new state

△ **BUILT AND REBUILT**
The US Capitol's first dome was made of wood and copper. A new cast-iron dome—featuring columns, brackets, several windows, and a crowning statue—was added in 1855–1866.

When the US became an independent nation in 1783, it had neither a capital city nor a congress hall. In 1790, the Residence Act settled on a site on the Potomac River in Maryland to be the nation's new capital. It was named Washington in honor of the first president, George Washington (1732–1799). Pierre Charles L'Enfant (1754–1825), a French–American military engineer, drew up a basic plan for the city, placing the legislative building on what is now Pennsylvania Avenue, linking it to the President's House. He named the new building the Congress House, but Thomas Jefferson insisted it be named the Capitol, a Latin word associated with the Capitoline Hill, one of the seven hills of ancient Rome.

Creating the Capitol

In 1792, Jefferson proposed a design competition for the new building. Amateur architect William Thornton (1759–1828) submitted a design inspired by the east front of the Louvre Palace in Paris in January 1793. The cornerstone was laid by Washington on September 18, 1793, and the building was completed by 1811.

EXPANDING THE CAPITOL

By 1850, the increasing number of legislators from newly admitted states led to the building of two new wings to house the growing Senate and House of Representatives. A new central dome was added in 1863, the east front was rebuilt in 1904, and the East Portico extended in 1958.

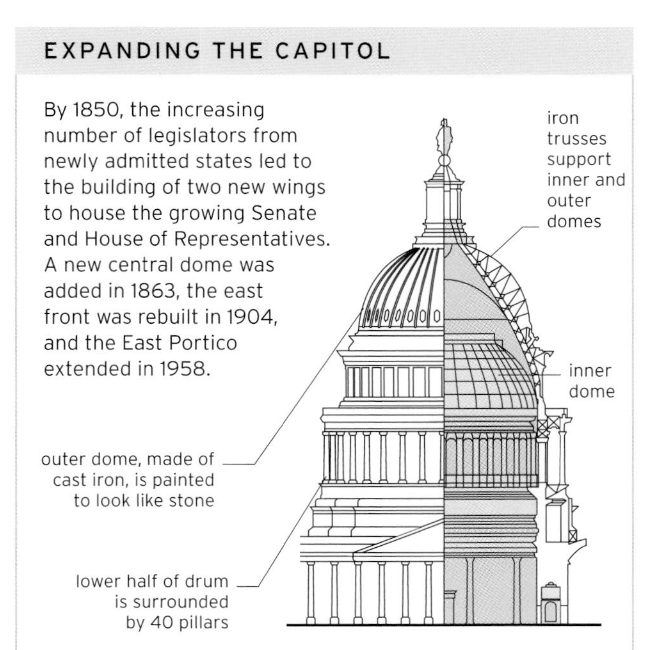

In the **early 1800s**, the Capitol was used for **Sunday religious services**

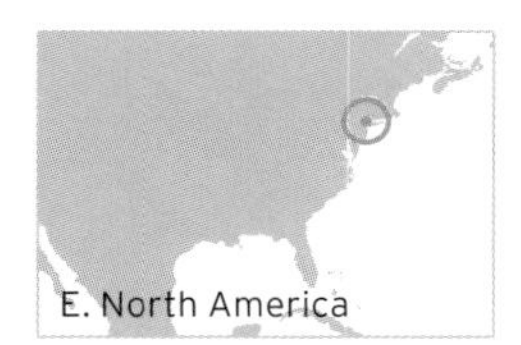

Brooklyn Bridge

A majestic bridge that was an engineering marvel in its time and steadily serves the commuters of New York

Brooklyn Bridge is by no means the longest or the tallest bridge in the US—although it was the longest, and first steel-wire, suspension bridge in the world when it was built—nor is it the most advanced. It is not even the only bridge that crosses New York's East River, but it can claim to be one of the most famous bridges in the world, as much an icon of its home city as the Statue of Liberty or the Empire State Building. The 5,989-ft- (1,825-m-) long bridge links Manhattan with its neighboring borough of Brooklyn.

Spanning the river

The idea of a bridge across the East River was conceived in 1852 by John Augustus Roebling (1806–1869), a German immigrant. Work began in 1869 under his son, Washington (1837–1926). While the finished bridge was technically a suspension bridge—one in which the deck is hung below suspension cables on vertical suspenders—it employed a hybrid cable-stayed design in which a fanlike pattern of cables also ran directly down from the two towers to support the main deck.

On May 24, 1883, the bridge was officially opened by President Chester A. Arthur, who crossed it together with New York's mayor Franklin Edison. The bridge was initially designed to carry horse-drawn and rail traffic, with a separate, elevated walkway in the center for pedestrians and cyclists. The last trains ran in 1944, and in 1950, the streetcars that shared the roadway also stopped. The bridge was then reconfigured to carry six lanes of cars. Height and weight restrictions keep commercial vehicles and buses off the bridge.

△ **CABLE POWER**
The four main cables that support the roadway are each 3,578 ft (1,090 m) long and 15½ in (40 cm) wide, with each made up of 21,000 individual wires wrapped together.

▷ **HAZARDOUS WORK**
High up on a catwalk, two men survey the unfinished Brooklyn Bridge. Building the bridge was dangerous, with at least 20 workers killed during construction.

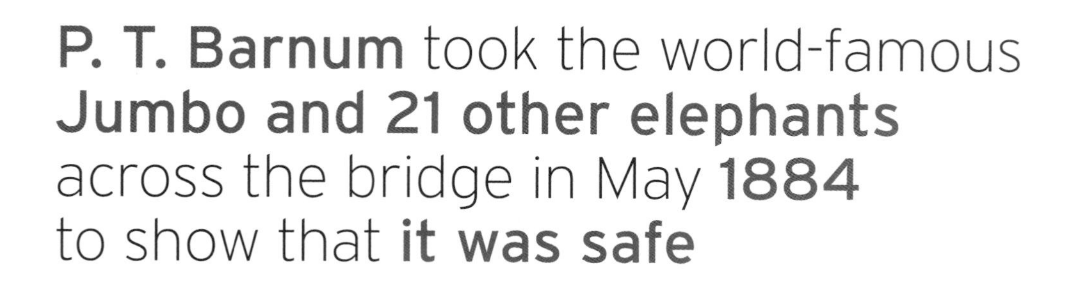

△ **PEDESTRIAN WALKWAY**
The bridge's cables form an intriguing weblike pattern that is best seen from the shared pedestrian and cycle path.

BUILDING THE TOWERS

The twin Neo-Gothic towers of the Brooklyn Bridge stand 277 ft (84.3 m) above the water and are built of limestone, granite, and Rosendale cement, a natural cement produced in New York since 1825. They were built by floating two caissons—huge upside-down pine boxes—into position and then building the stone towers on top of them until they sank to the bottom. Compressed air was then pumped into the caissons to allow workers to dig out the sediment until the caissons sank down to the bedrock, where they were filled in with brick piers and concrete.

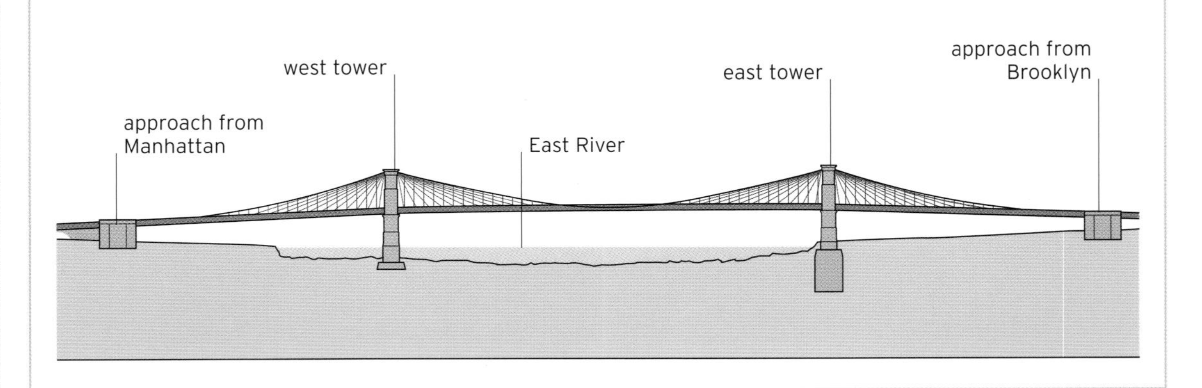

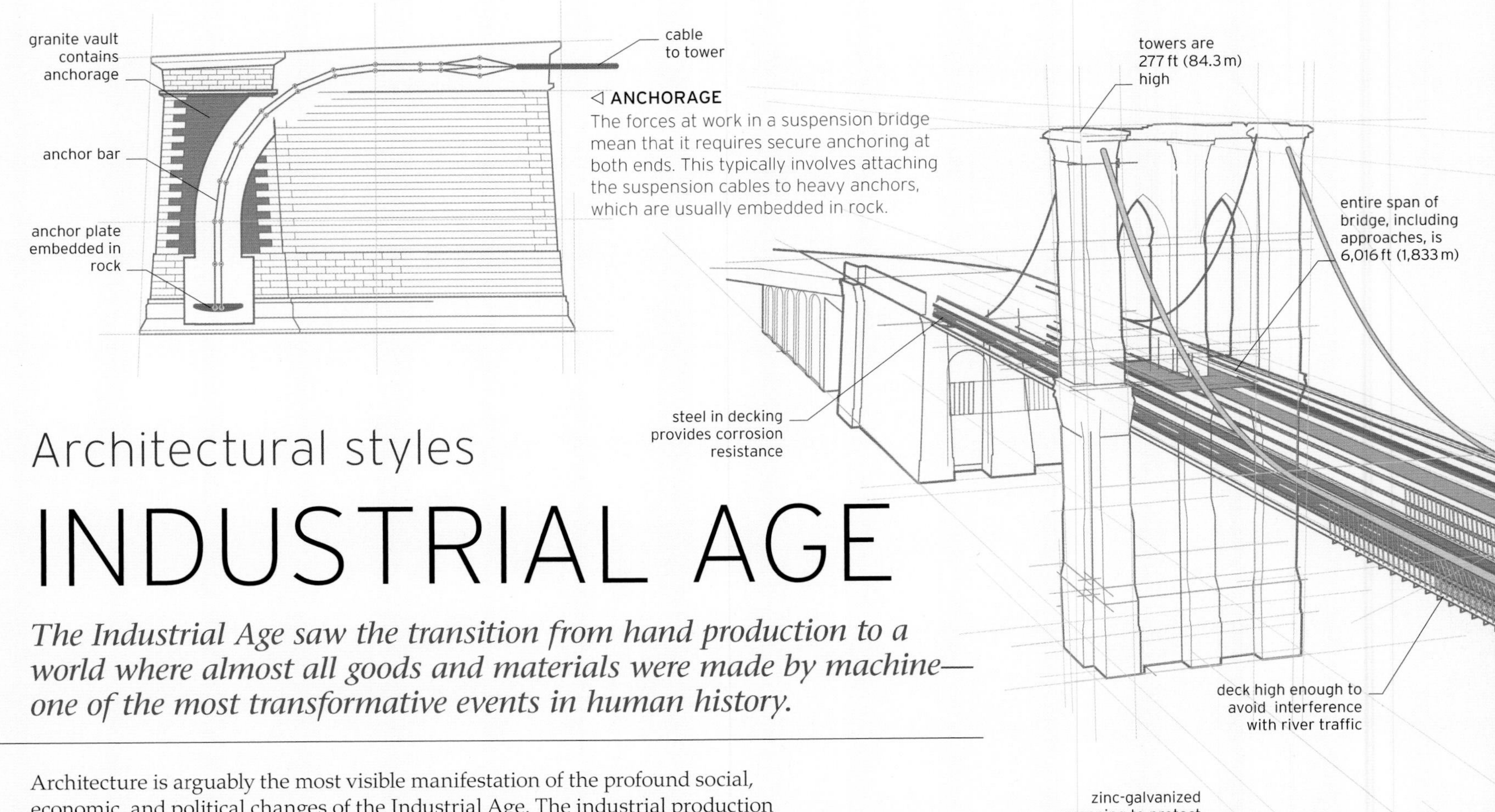

◁ **ANCHORAGE**
The forces at work in a suspension bridge mean that it requires secure anchoring at both ends. This typically involves attaching the suspension cables to heavy anchors, which are usually embedded in rock.

Architectural styles

INDUSTRIAL AGE

The Industrial Age saw the transition from hand production to a world where almost all goods and materials were made by machine—one of the most transformative events in human history.

Architecture is arguably the most visible manifestation of the profound social, economic, and political changes of the Industrial Age. The industrial production of materials such as iron and glass allowed the creation of buildings and structures that simply would not have been possible before. More important, however, were the new types of building and infrastructure that these new materials allowed and which industry demanded. With the new wealth that industry created came new civic and municipal buildings. New bridges, train stations, sewage systems, factories, industrial buildings, and workers' housing all fundamentally changed the cities and landscapes in which they were built.

In terms of appearance, the stern Neo-Classicism of the late 18th and early 20th centuries gave way to a range of styles. Of these, Neo-Gothic was the most prevalent. It quickly became a feature of a range of structures, from hotels to government buildings.

zinc-galvanized wrapping to protect from corrosion
each cable contains 3,515 miles (5,657 km) of wire
one central strand with 18 strands compressed around it
cable bound with a close spiral wrapping of wire

△ **SUSPENSION CABLE**
The development of wire rope in the mid-19th century was vital in making suspension bridges of the scale of the Brooklyn Bridge. Formed from multiple strands of metal twisted together, wire rope was far stronger and much more durable than the metal chain links used previously.

GLASS, STEEL, AND STONE

wrought-iron arches
glass follows smooth curves

△ **CURVED GLASS**
The 19th century saw significant advances in glass making. The most dramatic use of large panes was seen in the advent of structures where glass was inserted into iron frames.

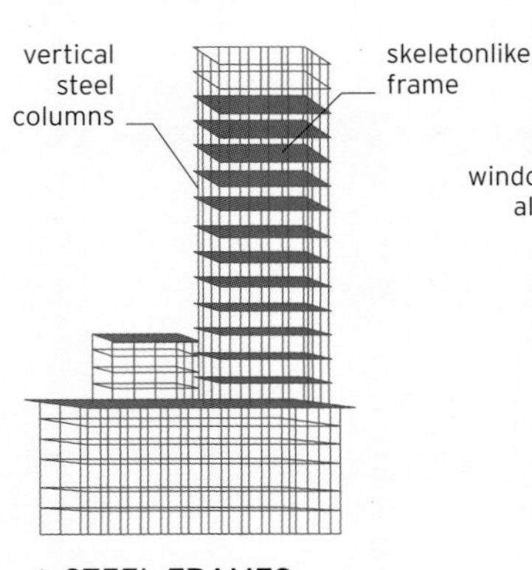

△ **STEEL FRAMES**
In the late 19th century, steel became available as a building material. Buildings with a structural steel frame no longer required walls for the purpose of load-bearing.

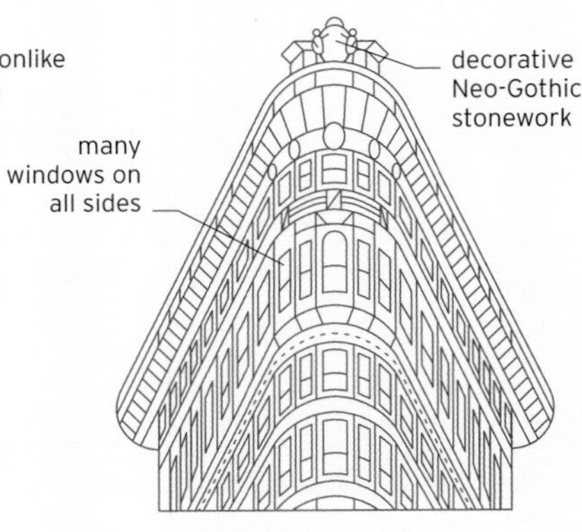

△ **STONEWORK**
In a steel-framed building, walls function essentially as cladding. This allows for more windows and for stonework to become decorative rather than structural.

Although now nearly **150 years old**, the Brooklyn Bridge still carries over **100,000 vehicles every day**

▼ BROOKLYN BRIDGE

Begun in 1869, Brooklyn Bridge (see pp.26–27) is a cable-stayed suspension bridge and the first to use steel rather than iron wires. The central span of 1,595 ft (486 m) is supported by cables attached from the two stone towers sunk deep into the bed of New York's East River.

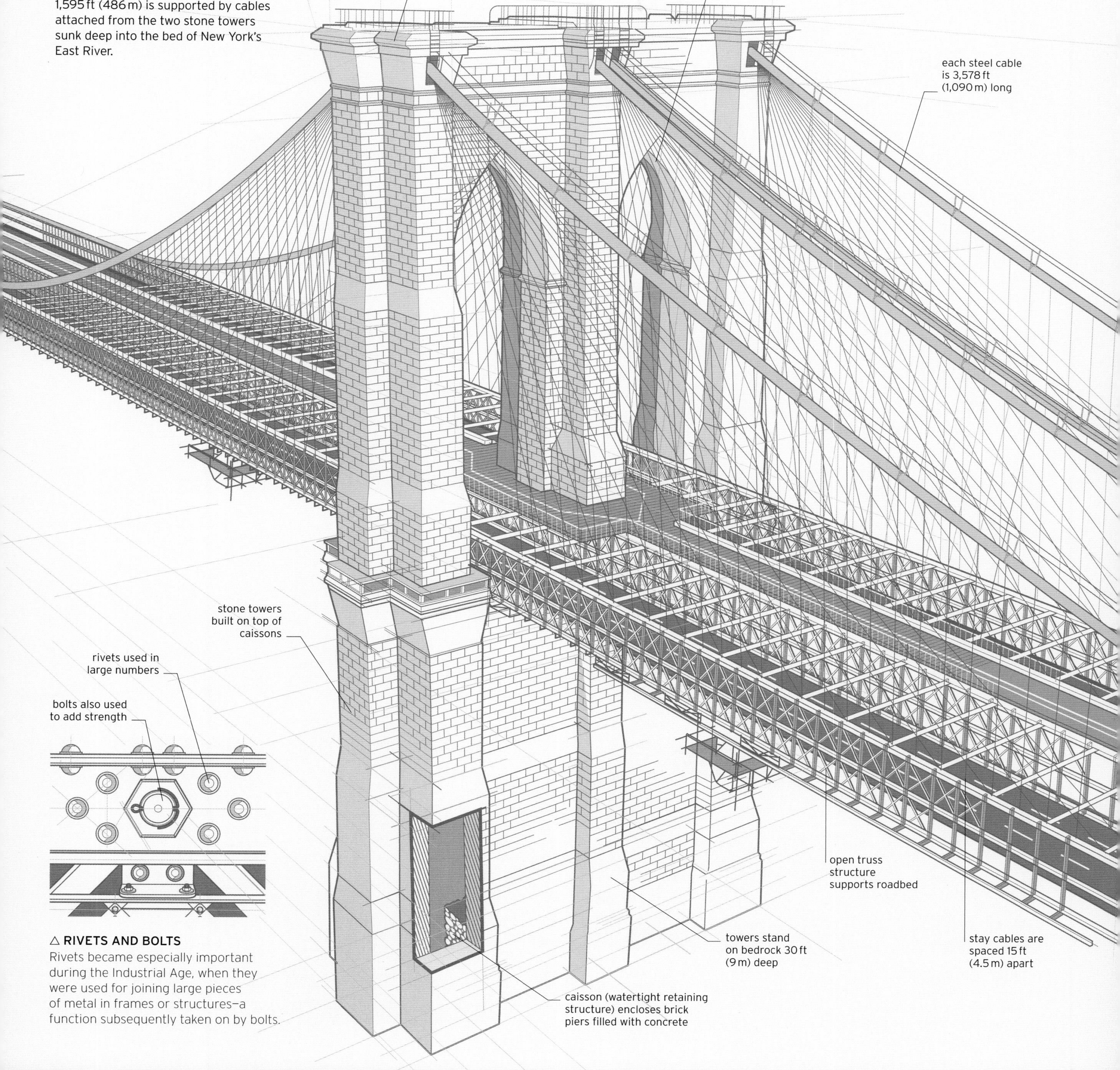

towers made of granite, limestone, and Rosendale cement

Neo-Gothic double arch

each steel cable is 3,578 ft (1,090 m) long

stone towers built on top of caissons

rivets used in large numbers

bolts also used to add strength

open truss structure supports roadbed

towers stand on bedrock 30 ft (9 m) deep

stay cables are spaced 15 ft (4.5 m) apart

caisson (watertight retaining structure) encloses brick piers filled with concrete

△ RIVETS AND BOLTS

Rivets became especially important during the Industrial Age, when they were used for joining large pieces of metal in frames or structures—a function subsequently taken on by bolts.

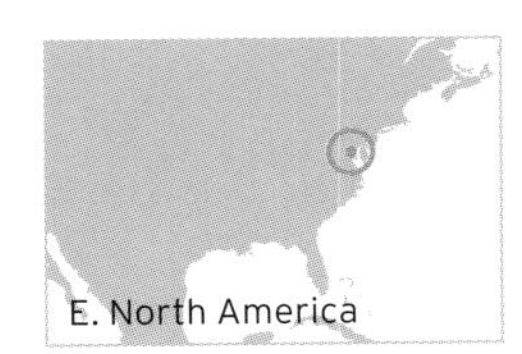

Washington Monument

An obelisk evoking the power of an ancient civilization, built to honor a new nation's founding father

Given its importance as a national landmark, the Washington Monument in the US capital had a perilous gestation. Built to honor George Washington (1732–1799), the first president of the US, the monument was started in 1848, but construction work was halted from 1854 to 1876 due to lack of funds and the intervention of the Civil War. It was eventually completed in 1884.

The original design for the monument by Robert Mills imagined a 600-ft- (183-m-) high obelisk surrounded by 30 pillars that were 100 ft (30 m) high, but the plans were scaled down when building resumed in 1876. The monument is a hollow obelisk made of marble, granite, and bluestone gneiss that stands 555 ft (169 m) high, capped by a 55-ft- (16.8-m-) high pyramidion.

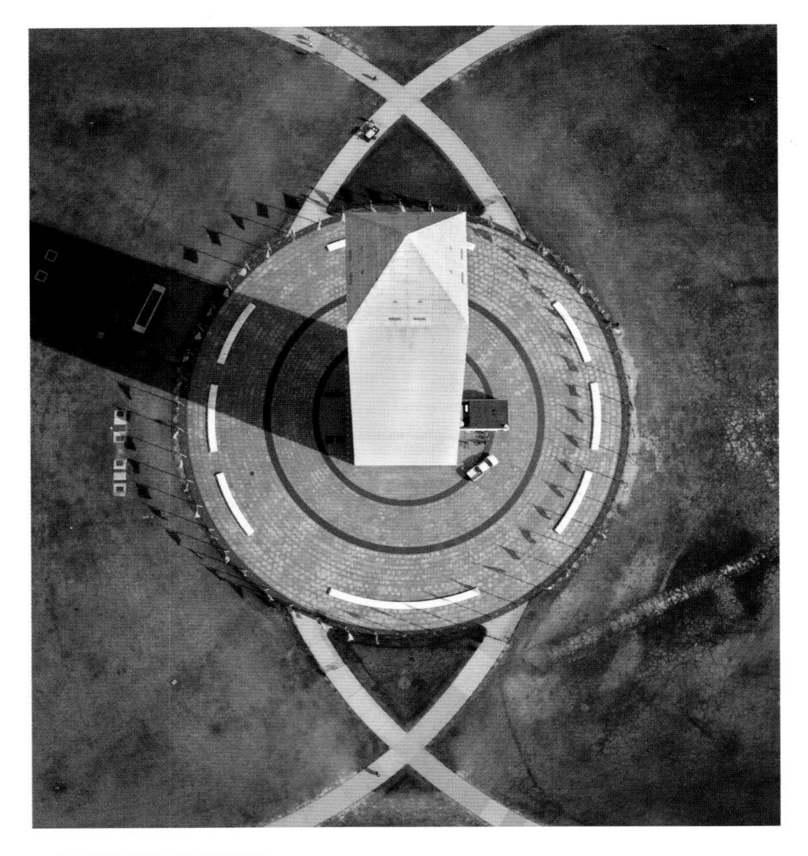

△ **VIEW FROM THE TOP**
The obelisk is capped by a hollow marble pyramidion (a small pyramid), at the apex of which is a smaller aluminum pyramid, part of the monument's lightning conductor.

Flatiron Building

A wedge-shaped skyscraper that came to symbolize New York's confidence and ambition

This dramatic Manhattan skyscraper owes its aggressive wedge-shape to a desire to make best use of expensive New York real estate. The site itself was known in the 1850s as Eno's flatiron, after the owner Amos Eno and its clothes-iron-like shape. The land was sold in 1901 to an investment partnership created by Harry S. Black of the George A. Fuller Company. Their new building was to be named the Fuller Building, but locals persisted in calling it the Flatiron.

Construction started in June 1901 and proceeded at a fast pace, as the steel used for the building's skeleton was precut, enabling the skyscraper to rise by a story a week. The 22-story building was finished in June 1902. The design by Daniel Burnham takes the form of a vertical Renaissance palazzo with Beaux Arts styling. The overall form of the Flatiron is based on a Classical Greek column, with a limestone base and a glazed terracotta shaft and capitol on top.

THE RIGHT TRIANGLE

The building is in the shape of a right-angled triangle. Each floor consists of a central lobby and corridor with 23 rooms around it, all but three of them with external windows. Access to each floor was originally provided by a water-hydraulic elevator powered by pressure. It was very slow and took 10 minutes to ascend the building.

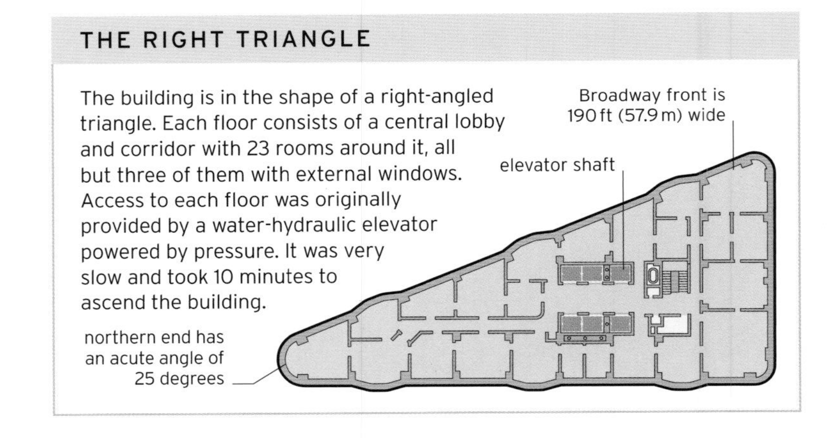

◁ **STEEL SKELETON**
In 1892 New York City removed its requirement that masonry had to be used for fireproofing a building, allowing new buildings, such as the Flatiron, to be constructed using a steel skeleton.

▷ **BEAUX ARTS STYLE**
The glazed terracotta tiles that cover the Flatiron Building are purely decorative; they have no structural role but merely cover up the steel skeleton that holds and supports the building.

Château Frontenac

A château-style Canadian hotel built in an age when railway travel was the height of luxury

As the railway network expanded across Canada, the rail companies developed a series of grand hotels to serve their passengers in style. Almost all of them were châteaulike in design, with towers, turrets, and other Scottish baronial and French château elements. Of these grand hotels, the most famous is Château Frontenac in Québec City, built by the Canadian Pacific Railway. This imposing building stands at the eastern edge of Old Québec's Upper Town high on a hill above the St. Lawrence River.

A French-Gothic extravaganza

Château Frontenac was designed by Bruce Price (1845–1903), one of the main architects employed by the Canadian Pacific Railway to design their hotels. Built in 1892–1893, the hotel is modeled on the châteaux found in the Loire Valley in France but with additional Gothic and high Victorian elements. Presenting an asymmetrical profile to the city beneath, its most commanding features are its steeply pitched roofs, massive towers and turrets, and tall chimneys. The hotel stands on a gray stone ashlar base and is faced with Glenboig fireclay bricks made in Lanarkshire, Scotland. Its interior is a wealth of marble staircases, mahogany panels, wrought iron, and carved stone. The hotel has 611 guest rooms and numerous reception rooms, bars, and other facilities. On its roof are four beehives, their 70,000 bees producing around 650 lb (295 kg) of honey a year. Among the hotel's executive suites are the Trudeau-Trudeau Suite, named after the father and son Canadian prime ministers.

THE HORSESHOE

Château Frontenac is basically horseshoe-shaped with four wings of uneven length. Initially meant to be a square building, the construction of the Terrasse Dufferin, which wraps around the eastern side of the hotel, resulted in a more complex structure.

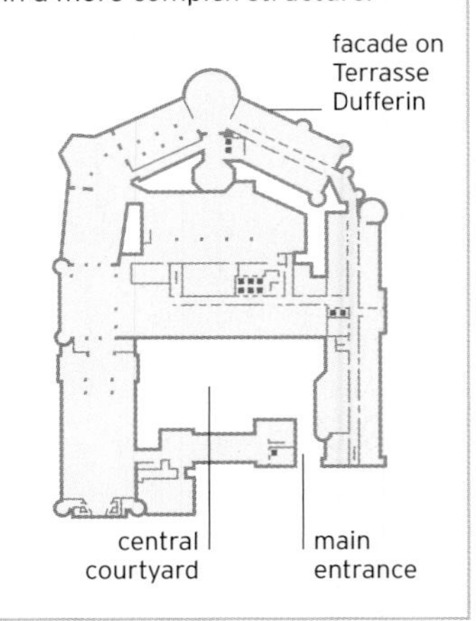

△ **OPULENT INTERIOR**
The grand double staircase within Château Frontenac is made of marble and wrought-iron banisters. The surrounding walls are painted a pale yellow-green color called eau de nil after its resemblance to the waters of the River Nile.

◁ **COPPER ROOF**
Château Frontenac's steeply pitched roofs and small dormer roofs are covered with copper, which has weathered to give it a distinctive green shading.

DOMINATING THE SKYLINE
The steel-framed, fortresslike tower of Château Frontenac provides exceptional views over the St. Lawrence River and is a part of the Historic District of Old Québec, a designated World Heritage Site.

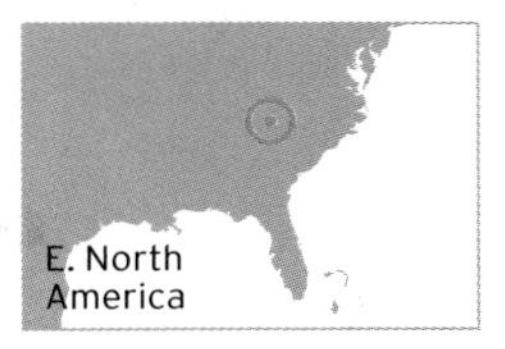

Biltmore House

A French Renaissance-style mansion that, when built, was the world's largest residence

The Gilded Age (1870–1900) was a period of rapid economic growth and conspicuous wealth in the US. None came wealthier than George Washington Vanderbilt II, whose childhood visits to Asheville, North Carolina, inspired him to build his own summer estate there. Designed by New York architect Richard Morris Hunt, Biltmore is modeled on the French châteaux of the Loire Valley, as well as the Rothschild-owned Waddesdon Manor in England. With 179,000 square ft (16,600 square m) of floor space, the 250-room, four-story house has steeply pitched roofs, turrets and towers, and sculptural ornamentation. Construction began in 1889 and continued into 1896. No expense was spared, as Biltmore was designed for an opulent, gilded age.

△ **WINTER LUXURY**
The octagonal Winter Garden is surrounded by stone archways and covered with a ceiling of sculpted wood and glass. The centerpiece is a sculpture and fountain entitled *Boy Stealing Geese*, created by the Austrian-born Karl Bitter.

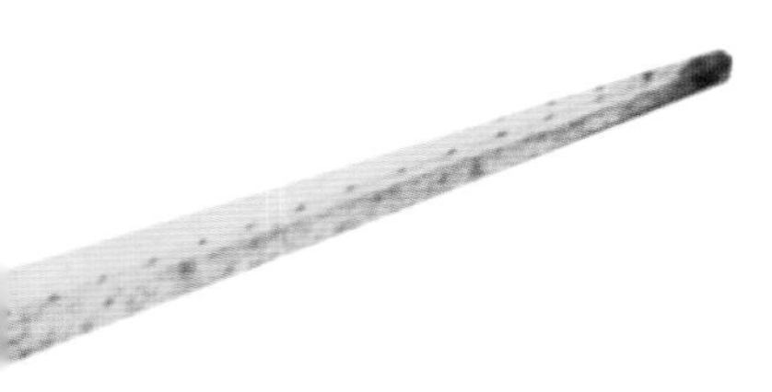

◁ **COPPER COVERING**
The sculptor Frédéric Bartholdi chose copper sheeting $^{3}/_{32}$ in (2.3 mm) thick to cover the statue, as cast bronze sheets or stone cladding would have been too expensive and too heavy to transport.

Statue of Liberty

A gift from the people of France to the people of the United States that symbolizes freedom and enlightenment

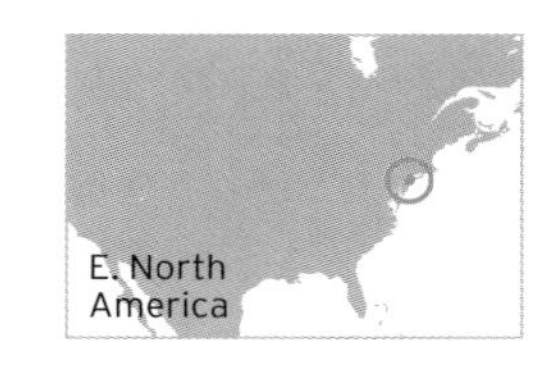

Standing tall on Liberty Island in New York Harbor, the Statue of Liberty has welcomed immigrants and visitors to the US ever since it was dedicated on October 28, 1886. The 151-ft- (46-m-) tall copper statue is a figure of Libertas, the Roman goddess of liberty. Holding a torch in her right hand to enlighten the world and a tablet in her left hand inscribed with the Roman numerals for the date of the US Declaration of Independence from Great Britain—July 4, 1776—the statue stands proudly free, the broken chains of bondage at her feet.

French origins, American finance

The idea of the statue arose in France when law professor and politician Edouard René de Laboulaye remarked to the sculptor Frédéric Auguste Bartholdi that any monument raised to commemorate American independence should be a joint project between the American and French people, given their revolutionary ties. Bartholdi completed the head and the torch-bearing arm before he had designed the rest of the statue. He exhibited the arm at the Centennial Exposition in Philadelphia in 1876, and later in New York. Fundraising proved a problem until Joseph Pulitzer, publisher of the *New York World*, appealed for funds, which then poured in from more than 120,000 contributors.

The statue itself was built in France by Gustave Eiffel, later famous for his eponymous tower in Paris (see pp.180–181), and shipped to the US. Its completion was marked by New York's first ticker-tape parade, its dedication presided over by President Grover Cleveland.

▷ **HOLDING THE TORCH**
Like the rest of the statue, the hand that holds the torch was made in Paris, then dismantled and shipped across the Atlantic in 200 crates. It was reassembled in what was then called Bedloe's Island, New York.

The torch alone is **16 ft (5 m)** tall and weighs **1$^{3}/_{4}$ tons (1.6 tonnes)**

IRON FRAMEWORK

The statue was designed around an iron frame. A secondary framework was attached to the central pillar to enable the statue to move slightly in the winds of the harbor. Then, sections of outer copper skin were connected to it using metal straps known as saddles. The skin was insulated with asbestos impregnated with shellac to prevent corrosion between the copper skin and the iron supports.

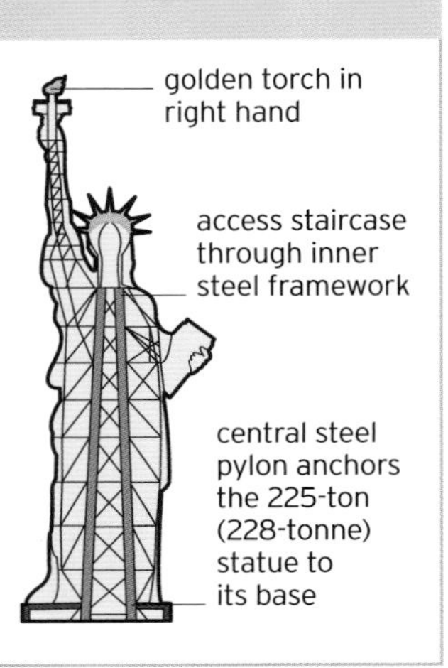

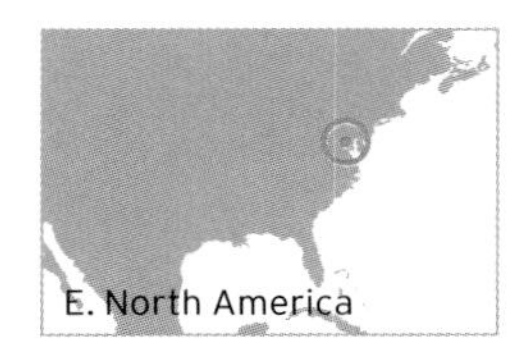

Washington National Cathedral

The spiritual home of the American people at the heart of the US capital

The US does not have an official or state religion, but the nearest it has to a spiritual home is the Washington National Cathedral. Technically called the Cathedral Church of St. Peter and St. Paul in the City and Diocese of Washington, this cathedral of the Episcopal Church has, from its earliest days, been a national shrine, a "house of prayer for all people," and a place where major state events and funerals could take place.

△ **GROTESQUE CARVINGS**
The Washington National Cathedral has hundreds of imaginative chimeras carved in the traditional style of cathedral grotesques.

Several years in the making

The building itself is in a Neo-Gothic style closely modeled on the English Gothic style of the late 14th century. Construction began in 1907, when President Theodore Roosevelt laid the foundation stone, and continued until 1990, when the final finial was placed in the presence of President George H. W. Bush, although decorative works continue to this day.

State funerals for **four** postwar **US presidents** have been held here

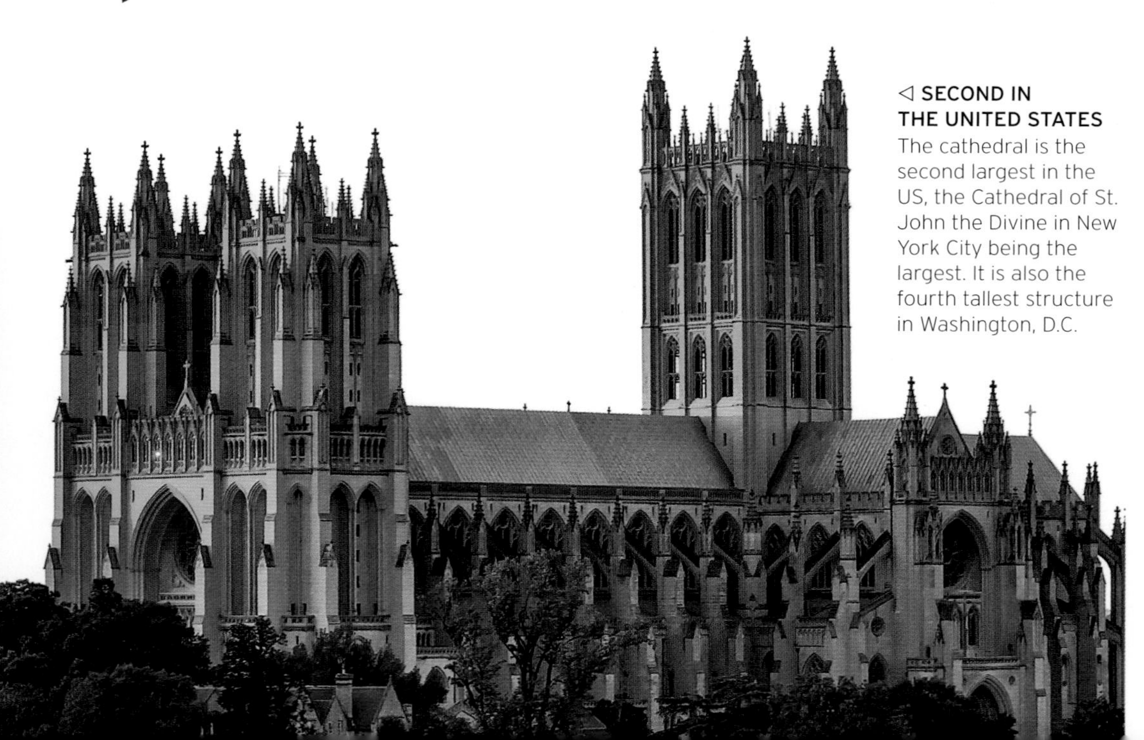

◁ **SECOND IN THE UNITED STATES**
The cathedral is the second largest in the US, the Cathedral of St. John the Divine in New York City being the largest. It is also the fourth tallest structure in Washington, D.C.

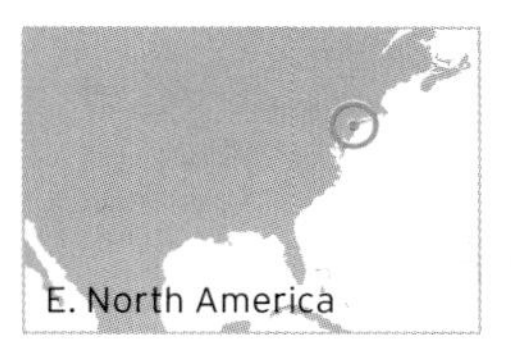

Grand Central Terminal

A majestic Beaux Arts railway terminus in New York with more than 750,000 passengers every weekday

Commuters and travelers from upstate New York and Connecticut flood into New York City via Grand Central Terminal—the largest working train station in the US and the third busiest in North America. Its 44 platforms are all underground, serving 56 different passenger tracks on two different levels with an additional 11 sidings. Two even deeper levels are currently under construction. The terminal is among the world's top 10 most visited tourist destinations, with numbers easily exceeding 22 million people each year.

The Main Concourse

The current station—the third on the site—was built in 1903–1913. Its Main Concourse, originally known as the Express Concourse because of the departing intercity trains, is deliberately vast—a huge hall 275 ft (84 m) long, 120 ft (37 m) wide, and 125 ft (38 m) high—and is intended to underline the terminal's status. Its elliptical, barrel-vaulted ceiling displays a highly decorated mural of the constellations; 10 globe-shaped chandeliers shed light on the crowds below. An 18-sided information booth in the center of the concourse is topped by a four-sided brass clock, probably the most iconic feature of the terminal.

◁ **MEETING POINT**
The geographical center of the station, the Main Concourse is lit by 10 chandeliers and large, arched windows. It also contains two working fountains.

△ **GRAND SCULPTURE**
The 66-ft- (20-m-) wide *Glory of Commerce*, sculpted by French artist Jules-Félix Coutan, on the south facade of the station, features the Roman gods Minerva, Hercules, and Mercury.

TRACKS AND TERMINALS

Trains arrive at the terminal every 58 seconds, rising up a gentle slope to slow down. The train tracks are arranged on two levels, with 30 on the top level and 26 on the lower. North of the station, the tracks converge to pass under the length of Park Avenue, finally emerging at 97th Street on the Upper East Side.

Track 61 was reserved for the private train of President Franklin D. Roosevelt

KEY

Main tracks | Sidings

UPPER LEVEL

LOWER LEVEL

Lincoln Memorial

A monument built to honor the 16th President and to help repair the divisions of the Civil War

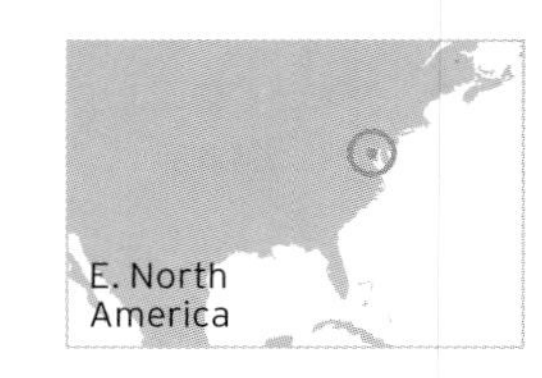

Just as the turmoil of the Civil War was ending, President Abraham Lincoln (1809–1865) was assassinated in Washington, D.C., by a Confederate sympathizer. A simple statue was erected in his honor in 1868, but in response to public demand, the US Senate approved a new, grander memorial in 1910. Designed by Henry Bacon (1866–1924) in the form of a Greek Doric temple, it was built between 1914 and 1922; its construction was delayed by US entry into World War I and wartime shortages of materials.

Immortalized in stone

Within the memorial is a monumental seated statue of Lincoln, made from Georgia white marble, by sculptor Daniel Chester French. Behind the statue are carved inscriptions of two of Lincoln's famous speeches: the Gettysburg Address of 1863 and his second inaugural address of 1865. Standing at the western end of the National Mall, it has been the site of many historic events; most notably, Martin Luther King Jr.'s "I Have A Dream" speech was delivered here on August 28, 1963, at the end of the March on Washington.

DORIC DESIGN

The memorial stands 98 ft (30 m) high and measures 189 ft by 118 ft (58 m by 36 m). It is surrounded by 36 columns: one for each of the states in the Union at the time of Lincoln's death. The stone used in its construction was sourced from different parts of the US, symbolizing unity.

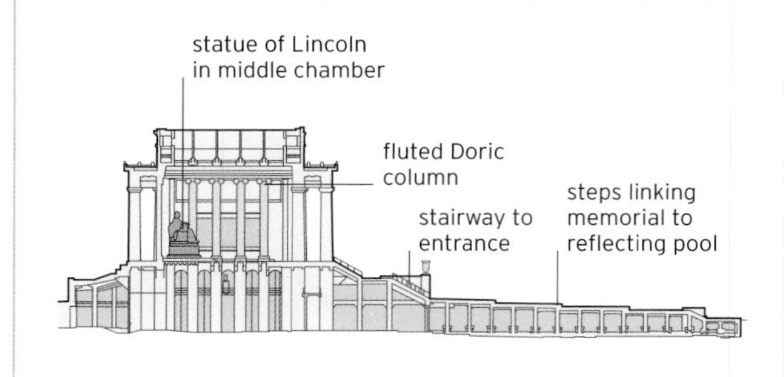

▷ **STATELY AND DIGNIFIED**
The seated statue of Lincoln was originally intended to be only 10 ft (3 m) tall, but it was enlarged to almost double that size. Under the supervision of the sculptor Daniel Chester French, the sculpture took 4 years to complete.

Jefferson Memorial

A Neo-Classical white marble monument to a great US statesman, thinker, and architect

Thomas Jefferson (1743–1826) is one of the towering figures of American history: the principal author of the Declaration of Independence, written after separation from Great Britain in 1776; the new nation's first ever Secretary of State, under President George Washington from 1790–1793; and its third president, from 1801 to 1809. After retiring from public office, he founded the University of Virginia. Not that he is without reproach, for despite enshrining in the preamble to the Declaration of Independence that "all men are created equal," he was also the owner of many slaves who worked on his Virginia plantation.

A Neo-Classical tribute

Jefferson was an architect as well, designing his own Neo-Classical house at Monticello in Virginia (see pp.24–25), which makes the Neo-Classical memorial to him in Washington, D.C., all the more apt. Designed by John Russell Pope and begun after his death in 1938, the memorial was officially dedicated by President Franklin D. Roosevelt on April 13, 1943, the 200th anniversary of Jefferson's birth. A 19-ft- (5.8-m-) tall bronze statue of Jefferson by the sculptor Rudulph Evans was placed inside the memorial in 1947, which replaced a bronze-painted plaster statue that was initially installed due to material shortages during World War II. The pediment on top of the main portico features a sculpture depicting the Committee of Five, the five members of the drafting committee of the Declaration of Independence. Excerpts from that declaration and the 1777 Virginia Statute for Religious Freedom—drafted by Jefferson when he was a member of the Virginia General Assembly—are carved on panels on the inside walls.

CIRCULAR STRUCTURE

The Jefferson Memorial consists of a portico with a triangular pediment and a circular colonnade of Ionic columns enclosed by a shallow dome. It is constructed of white Imperial Danby marble from Vermont and is flanked by granite and marble stairs and platforms.

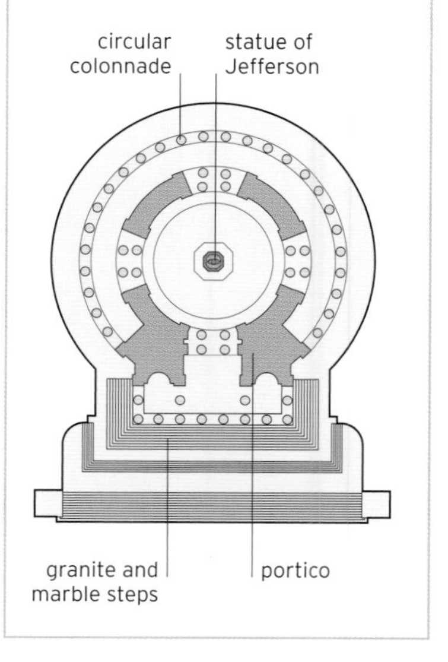

▽ **INSPIRED BY THE PAST**
Modeled on the Pantheon in Rome (see p.104), the circular colonnaded memorial was built in the Neo-Classical style that Thomas Jefferson himself helped introduce in the US.

▷ **AGAINST TYRANNY**
Inscribed in a frieze below the dome are Jefferson's words defending the US Constitution's refusal to recognize a state religion: "I have sworn ... hostility against every form of tyranny ...."

The **cherry trees** around the monument were **a gift** from the people of **Japan**

Each president was to be shown from **head to waist**, until a lack of funding **cut them off** at the neck in **1941**

▷ **UNDER CONSTRUCTION**
Workers used climbing equipment, cradles, and ski lifts to access the cliff-face on which they were working.

▽ **THE FOUR PRESIDENTS**
More than 450,000 tons (410,000 tonnes) of granite were blasted off the side of Mount Rushmore to carve the faces of the four presidents.

Mount Rushmore

A project aimed at attracting tourists to South Dakota in the 1920s that now draws 3 million people a year to admire the faces of four American presidents

The huge sculptures of Mount Rushmore in the Black Hills of South Dakota were conceived to promote tourism. Doane Robinson, a local historian who read about plans for the massive Confederate Memorial on Stone Mountain in Georgia, decided in 1923 to create a similar site to attract visitors to South Dakota.

Sculpting the mountain

The artist Gutzon Borglum suggested Mount Rushmore as a possible site. In 1929, President Calvin Coolidge signed a bill creating a commission to oversee the project. Construction began in 1927, and the four faces, each 60 ft (18 m) high, were completed between 1930 and 1939. The presidents—George Washington, Thomas Jefferson, Theodore Roosevelt, and Abraham Lincoln—were chosen for their role in founding and preserving the Union and expanding its territory.

The bulk of the mountain was carved using dynamite until only about 3–6 in (7.5–15 cm) of rock remained. The drillers and carvers then drilled small holes into the granite in a process called honeycombing. Finally, the workers evened up the granite, creating a surface as smooth as pavement. The construction was ultimately completed in 1941.

Chrysler Building

The Art Deco masterpiece that stylishly adorns the New York skyline and the first man-made structure to be taller than 1,000 ft (300 m)

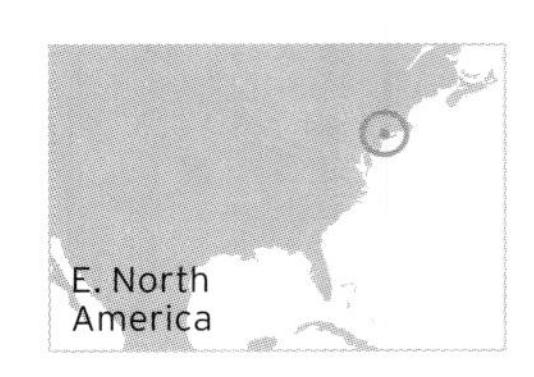

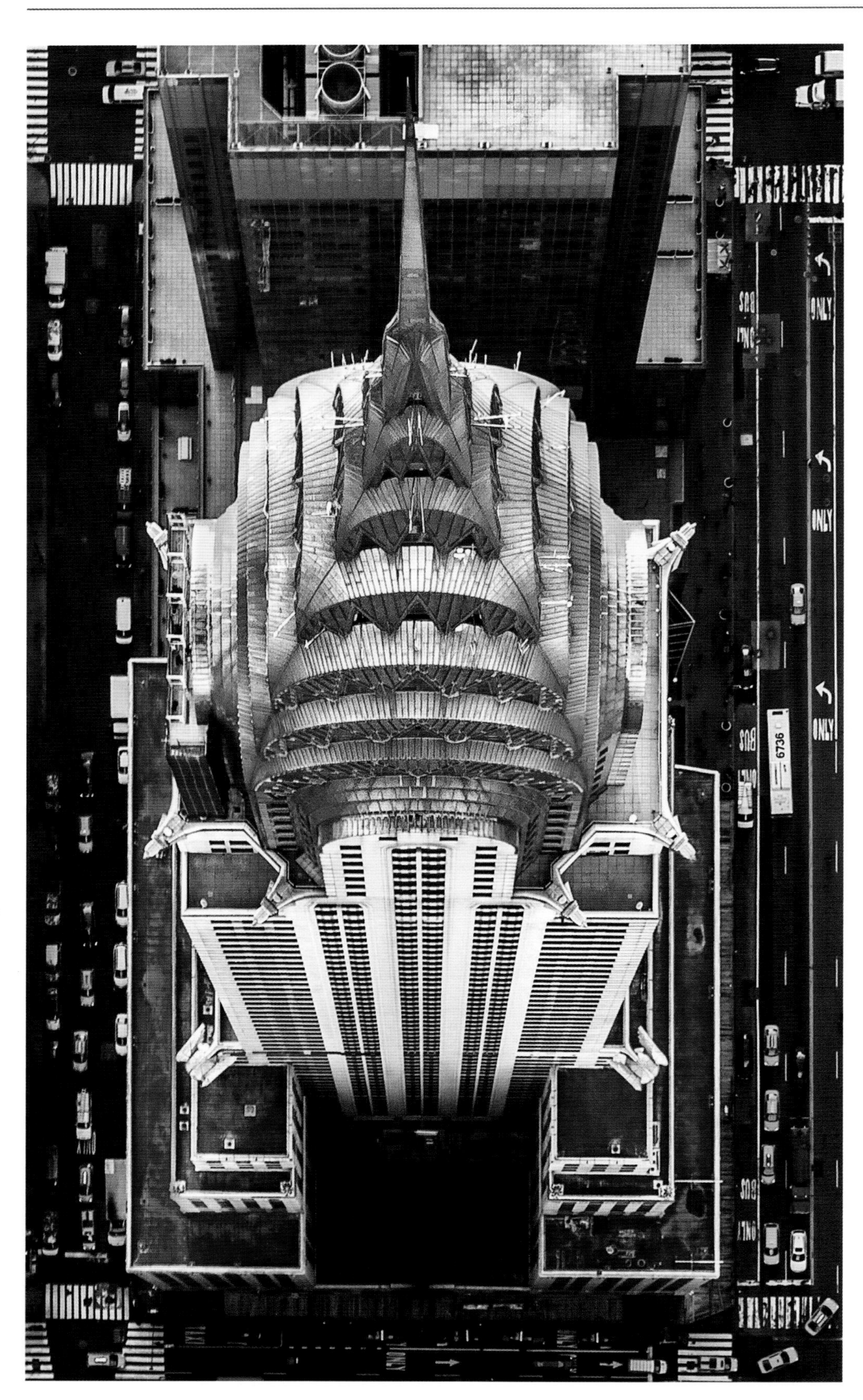

In a city famed for its skyscrapers, the Chrysler Building on the East Side of Manhattan in New York City is the most renowned of them all. Whereas most tall buildings are faceless, solid slabs, the Chrysler Building is an Art Deco masterpiece of high style and subtle design, its top a metallic crown that lights up the skyline.

Although the building was the headquarters of the Chrysler Corporation of carmakers, the company did not own the building. It was built by Walter Chrysler (1875–1940), the company's founder, and served as the corporation's headquarters from 1930 until the mid-1950s. Walter liked the building so much that he decided to pay for it and own it himself so that his children could inherit it. Designed by William Van Alen, at 1,046 ft (318.9 m) high, it aimed to be the tallest building in the world. During construction, it overtook its nearest rival—40 Wall Street—and when it opened on May 27, 1930, it claimed that towering prize.

◁ **SECRET SPIRE**
The Chrysler Building's architect secretly built a spire 125 ft (38 m) long and hoisted it on top of the tower, making it 117 3/4 ft (35.9 m) taller than 40 Wall Street.

△ **ART DECO STYLE**
The gargoyles on the 31st floor are designed to look like a car's mascot, and replicas of Chrysler cars' hubcaps and fenders adorn the walls.

TALLER AND TALLER

When the Chrysler Building opened in May 1930, it was the world's tallest building. On May 1, 1931, though, it lost the crown to the Empire State Building, and it has twice been overtaken in New York since then by the two World Trade Center buildings.

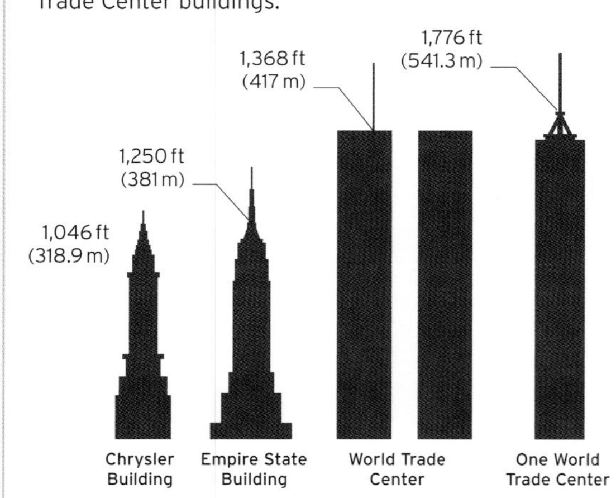

EMBLEM OF MANHATTAN
Dominating the city's skyline, the Empire State Building stands on Fifth Avenue in the Midtown South district of Manhattan.

Empire State Building

A New York icon, built in just 1 year and 45 days in a race to the sky

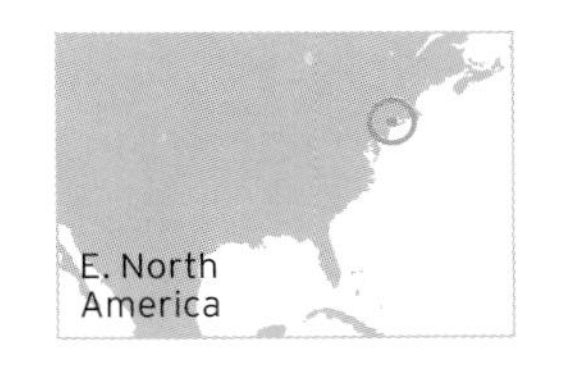

The Chrysler Building (see opposite) might win prizes for its beauty, but it is the Empire State Building that can claim to be New York's most iconic building. Ever since it opened on May 1, 1931—just 20 months after contracts had been signed with its architects, Shreve, Lamb, and Harmon Associates—the Empire State has defined the city skyline, featuring in countless movies and photographs. At 1,250 ft (381 m) high, it was the tallest building in the world until it was overtaken by the World Trade Center in 1970. Yet it has not always been so popular. In its first year of opening, which coincided with the worst effects of the Great Depression, only 23 percent of its available space was occupied, which gave it the nickname "the Empty State Building."

Deco exterior and details

Structurally, the 102-story building is a steel frame covered with 10 million bricks and 730 tons (660 tonnes) of aluminum and stainless steel. Its stepped, uncompromisingly modern Art Deco form is complemented by interior details, including the ceiling murals in the lobby that depict the mechanical wonders of the modern age in gold and aluminum leaf. The exterior of the tower comes to life when it is floodlit every evening.

◁ **LOBBY MURAL**
An aluminum mural of the building, without its antenna but with rays of light beaming from its pinnacle, decorates a wall in the entrance lobby.

DOCKING STATION

The top 200 ft (61 m) of the Empire State Building was originally intended to serve as a mooring mast for airships. It was planned that elevators would take passengers up to the 102nd floor, where they would board the airships via a gangplank. The scheme proved completely fanciful—no airship was ever able to moor safely—and was abandoned, after which a radio antenna was fixed to the peak of the docking station.

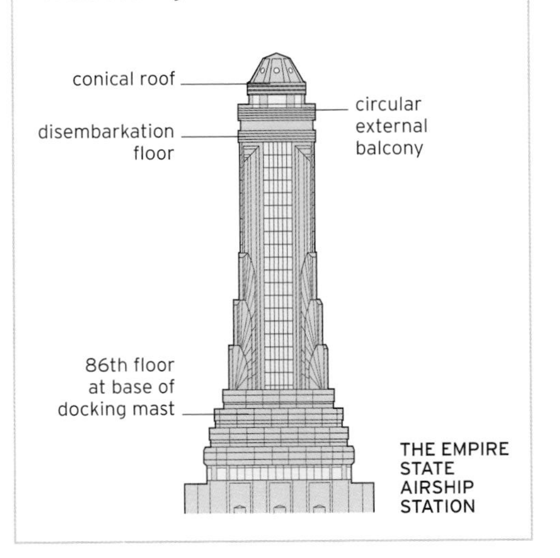

THE EMPIRE STATE AIRSHIP STATION

WATERING THE AMERICAN WEST
The dam is as high as a 60-story building, and its base is the length of two football fields. It stores the water that irrigates more than 31,000 square miles (80,000 square km) of land in California and Arizona.

Hoover Dam

A massive structure that powered the industrial and agricultural growth of the southwestern US

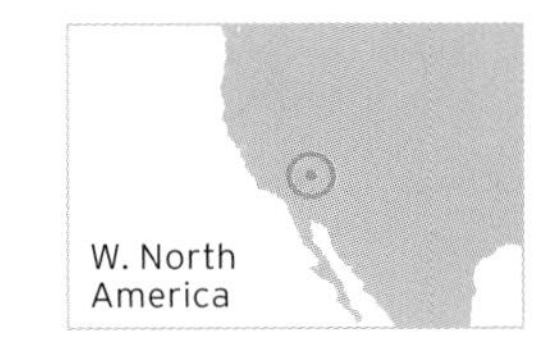

With the expansion of the southwestern economy in the 1800s and 1900s, planners began to look to the mighty Colorado River as a source of power and water for a growing population. In December 1928, President Calvin Coolidge authorized the construction of a giant dam on the borders of Nevada and Arizona.

Building the dam

To allow construction to begin, the flow of the Colorado was diverted through tunnels driven into the canyon walls. Watertight enclosures called coffer dams were built to protect the site from flooding, and the ground was cleared. The first concrete was poured in 1933, and by the time the dam was completed, it was the largest concrete structure ever made.

The challenge for the engineers was to contain 8.44 cubic miles (35.2 cubic km) of water behind the dam. Part of the resistance was provided by the weight of the concrete itself; additionally, with the point of the arch facing upstream, water was directed against the canyon's walls, serving to compress and strengthen the structure. The partially built dam—then known as the Boulder Dam—was opened by President Franklin D. Roosevelt in 1935. It was later renamed the Hoover Dam after President Herbert Hoover, who oversaw its initial construction.

△ **ASSEMBLING THE SLABS**
The dam was built of columns of rectangular concrete slabs (see panel) containing steel pipes through which cool river water, and then ice-cold water from a refrigeration plant, was poured to cure the concrete.

COOLING THE CONCRETE

When planning the dam, engineers calculated that if the concrete for the dam was poured continuously, it would take 125 years to cool. It was therefore poured in separate slabs, each about 50 ft (15 m) square and 5 ft (1.5 m) high.

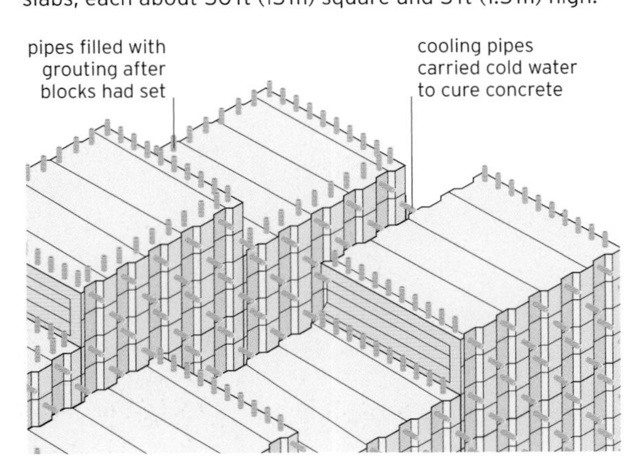

△ **PUBLIC APPEARANCE**
This posed publicity shot taken while the dam was under construction belies the fact that 112 people lost their lives building the structure, and a further 42 died from carbon monoxide poisoning while working in the diversion tunnels.

▽ **IN THE MIST**
The Golden Gate Bridge is often shrouded in fog, particularly during the summer months. There are two foghorns on the bridge, each with a different pitch, and multiple 360-degree flashing red beacons that serve as warning lights for air and sea travelers.

Golden Gate Bridge

An elegant suspension bridge that spans the strait of water that connects San Francisco Bay with the Pacific Ocean

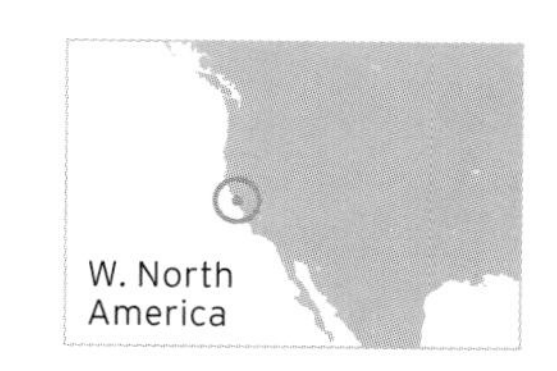

Until the mid-20th century, San Francisco was the largest US city still served by ferryboats. Access across the Golden Gate—the 1-mile- (1.6-km-) wide strait separating the northern tip of the San Francisco peninsula from Marin County to its north—was by a 20-minute ferry journey that cost $1 a vehicle.

Overcoming objections

Building a bridge across the strait had often been considered, but it was opposed by the US Department of War, as it could interfere with shipping. The Southern Pacific Railroad, a powerful business in California, also objected, as the bridge would provide competition for its ferry services. It filed a lawsuit against the project, prompting a mass boycott of its ferry services. After much argument, in 1928, the California legislature set up the Golden Gate Bridge and Highway District as the official body to design, construct, and finance the bridge. The Wall Street Crash of 1929 meant the district was unable to finance the project, but after bonds were raised locally, construction finally began on January 5, 1933.

The bridge itself is a simple suspension bridge designed by various structural engineers and architects. In order to cope with the high winds in the strait, the roadway was built to be thin and flexible enough to twist in the wind. The bridge is not golden—"Golden Gate" refers just to the strait below—but is painted in a special color known as international orange, which was originally used as a sealant for the bridge. The US Navy, however, had wanted the bridge to be painted with black and yellow stripes to ensure it was visible to passing ships.

The day before it opened on May 27, 1937, 200,000 local people crossed the bridge on foot and roller skates, but since then its main traffic has been motor vehicles, using both US Route 101 and California State Route 1, which merge to cross the bridge.

The **two towers** of the bridge are **held together** by **1.2 million steel rivets**

SPANNING THE GOLDEN GATE

The cables holding up the Golden Gate Bridge are held by anchorage blocks inside the two huge abutments on either side of the water. At the time of its opening in 1937, the bridge was both the longest and the tallest suspension bridge in the world, although 14 bridges are now longer and 20 are taller. The two towers each stand 746 ft (227 m) tall and are 4,198 ft (1,280 m) apart. The total length of the bridge from abutment to abutment is 8,981 ft (2,737 m), the roadway clearing the strait below by an average of 220 ft (67 m).

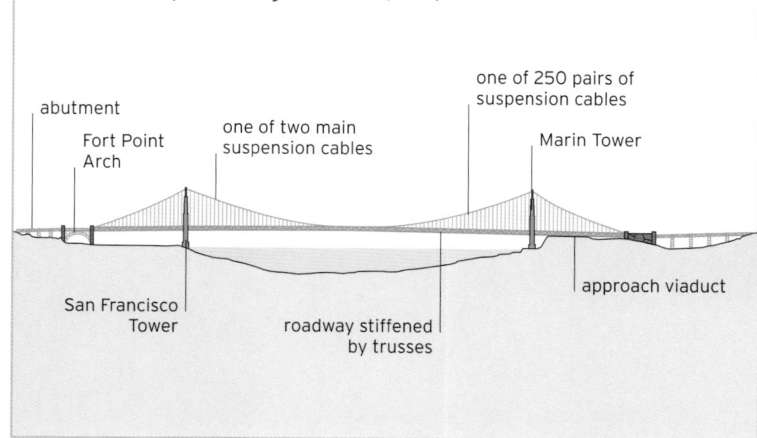

△ **SUSPENDED ROADWAYS**
Each of the 250 pairs of vertical suspender cables that support the roadway is made of 27,572 strands of galvanized steel wire, their total length estimated to be around 80,000 miles (130,000 km).

The Pentagon

The world's largest office building and an embodiment of US military power

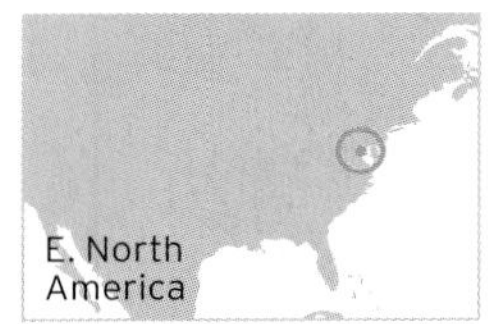

Few visitors to the Pentagon—located just across the Potomac River in Virginia from Washington, D.C.—describe it as a beautiful building. However, as the headquarters of the US Department of Defense, it serves an important purpose, housing around 25,000 military and civilian staff. The building was designed by American architect George Bergstrom and completed on January 15, 1943.

Prime target

The Pentagon has five even-length sides, with five floors above ground and two below it, providing more than 6,500,000 square ft (600,000 square m) of space and 17 miles (28 km) of corridors. At its center is a 215,000-square-ft (20,000-square-m) pentagonal plaza nicknamed "ground zero," on the presumption that if the Soviet Union had attacked the US in a nuclear conflict during the Cold War, the Pentagon would have been the prime target. Although it survived the Cold War intact, the Pentagon was attacked on 9/11, when hijackers flew American Airlines Flight 77 into the western side of the building, killing 189 people, including 125 Pentagon employees.

GIANT SCALE

The Pentagon is vast in scale, covering an area of 172,000 square ft (16,000 square m), including the central plaza. Each of its five sides is 918 ft (280 m) long, and the building is almost as wide as the Empire State building in New York is tall. The Pentagon's distinctive shape was determined by the need for it to fit between existing roads at the site, and its height was limited by the relative scarcity of steel at the time of its construction.

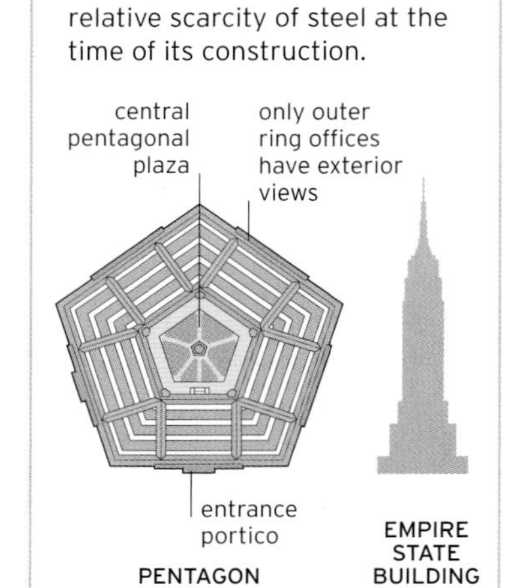

▽ **CAREFULLY PLANNED**
Constructed of steel and reinforced concrete with some limestone facing, the Pentagon has five floors. It comprises five concentric rings that are connected by 10 spokelike corridors.

It is possible for a person to **walk** between any **two rooms** in the Pentagon in less than **7 minutes**

△ **STREAMLINED DESIGN**
A symbol of rapid transformation in its time, the terminal's roof was deliberately designed to look like the wings of an aircraft. This also provided TWA with a new way of marketing its services. The roof's thin, reinforced concrete shell is lightly supported at its corners.

A NEW PURPOSE

After the closure of the TWA Flight Center in 2001, the building remained vacant but was used occasionally as a venue for art exhibitions and as a movie set. In 2008, a new terminal building capable of handling far higher volumes of passengers was built next to the old building, which was subsequently repurposed as a hotel.

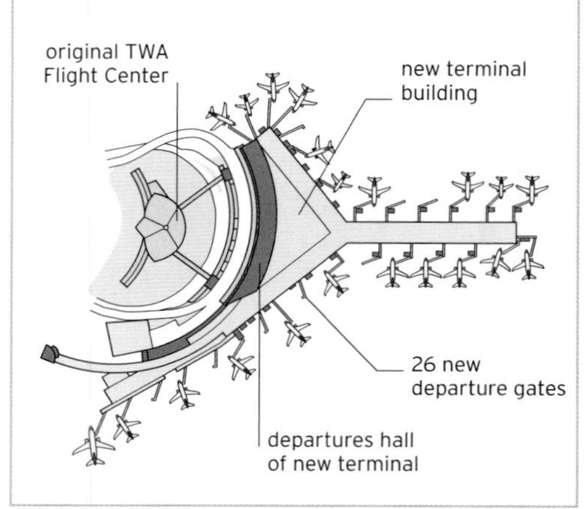

TWA Flight Center, JFK Airport

A space-age flight terminal, described as the "Grand Central" of the jet age

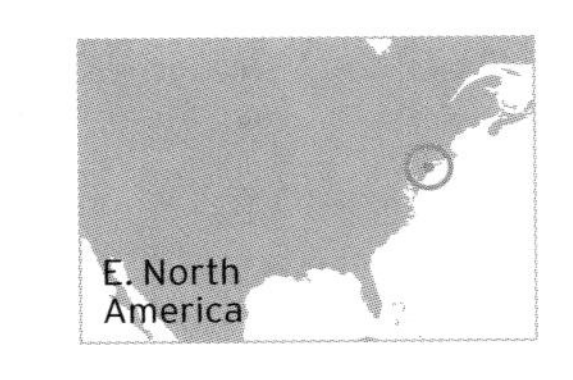

In 1955, the authorities at New York's Idlewild Airport—renamed John F. Kennedy Airport in December 1963 in honor of the late president—determined that each of the major airlines using the airport should build and operate their own terminal. Trans World Airlines (TWA) selected a bold, futuristic building by the Finnish American architect Eero Saarinen. Saarinen's terminal was a monument to aviation itself. It was covered with a prominent wing-shaped, white concrete roof, while tall windows allowed passengers to view the incoming and outgoing aircraft outside.

Passenger innovations

The new terminal opened on May 28, 1962, a year after Saarinen's premature death. It was one of the first to feature enclosed passenger jetways to allow passengers to enter and leave their aircraft safe from the elements outside. The terminal also sported baggage carousels, an electronic schedule board, and closed-circuit television (CCTV), all innovations at that time. In 1969, the terminal was expanded to handle more passengers. It closed in 2001, but was granted a new lease of life in 2015, when plans were announced to convert it into a hotel.

◁ **FLUID INTERIORS**
The inside of the terminal is remarkably coordinated and harmonious. It is free of obtruding pillars or walls, which gives an impression of openness inside.

Gateway Arch

A symbolic entrance to the western US, built on the banks of the mighty Mississippi

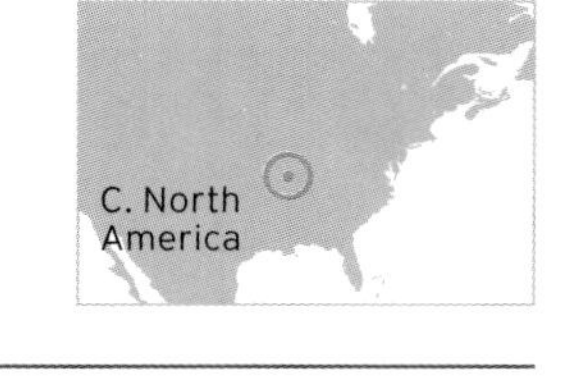

The Gateway Arch stands on the site where the inland port city of St. Louis was founded on the west bank of the Mississippi River by two French fur traders—Pierre Laclède and Auguste Chouteau—in 1764. St. Louis has long been considered the gateway for American expansion to the west, and it was here that Meriwether Lewis and William Clark set out on their famous expedition of 1804 to the Pacific Coast. The Gateway Arch was built as a monument to the city's pioneering spirit and officially dedicated to "the American people" on May 25, 1968.

Design and construction

The structure was designed in 1947 by the Finnish American architect Eero Saarinen (1910–1961), but construction was delayed because of funding and planning issues, so it did not start until almost 2 years after Saarinen's death. Built of carbon steel and concrete and coated with stainless steel, the arch is the tallest memorial in the US. A tram system inside it takes visitors up to the observation deck at the top.

▷ **GATEWAY TO THE WEST**
The width and the height of the Gateway Arch are the same, at 630 ft (192 m). The cross-section of each leg is an equilateral triangle that becomes smaller toward the top of the arch.

Guggenheim Museum

A Modernist triumph built to house an unrivaled collection of modern art

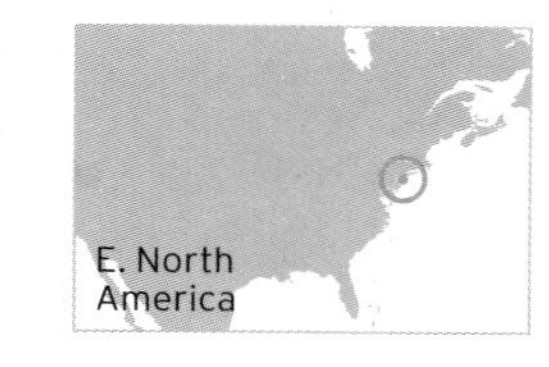

Solomon Robert Guggenheim (1861–1949) was born into wealth and increased his fortune with his gold mines and other investments. He began collecting art in the 1890s, and after World War I retired to devote himself full time to his collection. In 1939, his artistic foundation, set up to foster the appreciation of modern art, established a museum to house his growing collection. As the art overflowed its rented home, Guggenheim wrote in 1943 to Frank Lloyd Wright (1867–1959), America's preeminent modern architect, asking him to design a permanent home for his paintings.

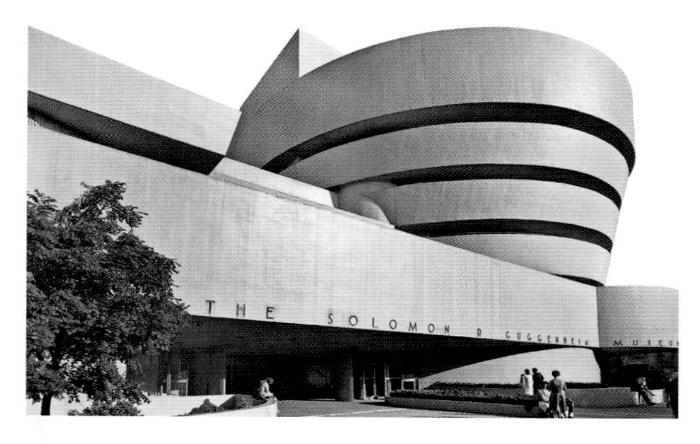

◁ **SPIRALING DOWN** Inside the museum, the continuous spiral gallery gently flows around the central atrium, which is lit from above by a central skylight.

△ **THE HONEY DIPPER** The museum has been nicknamed the "honey dipper," as it resembles the utensil used to scoop out honey from a jar.

A spiral of art

Wright responded with one of New York's most extraordinary buildings. Cylindrical in shape and wider at the top than the bottom, it features a unique ramp gallery that runs in a continuous spiral around the outer walls, reaching up to the ceiling skylight. The finished museum is an inverted circular ziggurat that took 7 years to build before it opened to the public in 1959. Inside and out, the building overwhelms.

THE SPIRAL RAMP

Wright designed the museum as an inverted ziggurat (an ancient tiered pyramid or mound), its spiral design similar to a nautilus shell. His plan was for visitors to take the elevator to the top, then slowly walk down the ramp, viewing the artworks before admiring the spectacular atrium at the bottom.

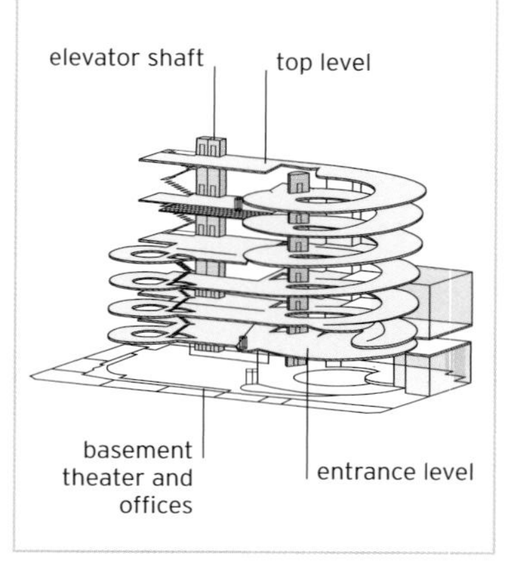

Guggenheim **never saw** his museum; he **died 10 years before** it opened

Space Needle

Seattle's saucer in the sky, built as a centerpiece of the 1962 World's Fair

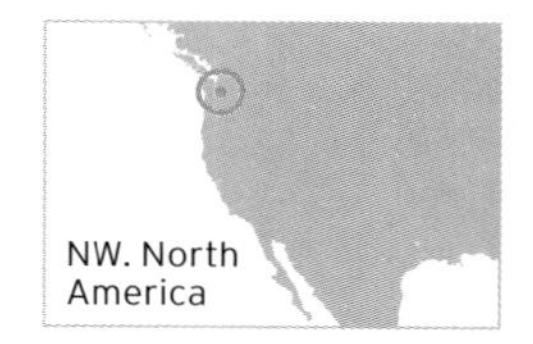

Seattle's landmark Space Needle, with its distinctive hourglass shape and elevated viewing platform, has been a symbol of the city since 1962. Standing 604 ft (184 m) tall, it commands views of downtown Seattle and its inlet, as well as the more distant Cascade Range and Olympic Mountains.

Strong and sturdy

Construction of the tower began on April 17, 1961, and was completed on December 8 the same year. Built to withstand earthquakes of up to magnitude 9.0 and winds of up to 199 mph (320 kph), as well as lightning strikes, the tower now features the world's first and only revolving glass floor, known as the Loupe. Recent modifications to remove the mullions that restrict the view from the observation deck have brought the tower closer to the design intended by the original architects.

▷ **INTO SPACE** The top of the Space Needle contains the rotating glass floor of the Loupe and an open-air observation deck above it.

Habitat 67

A revolutionary housing scheme built from prefabricated concrete modules to celebrate Canada's centennial

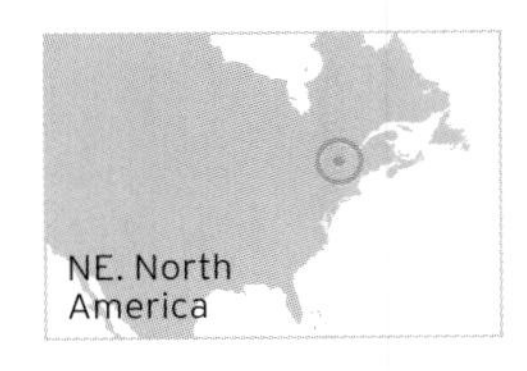

This extraordinary housing complex beside the banks of the St. Lawrence River in Montreal, Canada, began life as a master's thesis by the Israeli-Canadian architect Moshe Safdie (1938–). Safdie's academic advisor at McGill University later asked him to develop his plans for the forthcoming Expo 67, the world's fair being held to celebrate Canada's centennial as an independent nation.

Linked modules

Habitat 67 reaches up to 12 stories in height. It consists of 354 identical prefabricated concrete boxes arranged in various combinations to make 146 residences, each made up of one to eight linked boxes. The initial models of the project were built using LEGO® bricks in order to visualize how the finished building would look in three dimensions. Originally built as low-cost housing, Habitat has become a highly desirable residence. However, it failed to give rise to similar prefabricated buildings elsewhere, as its designer originally intended, nor did it revolutionize affordable housing for the masses.

▷ **LITTLE BOXES**
Concrete boxes are stacked in such a way that each box is a step back from its immediate neighbor. This allows each apartment a terrace of its own, as well as excellent ventilation.

Vietnam Veterans Memorial

A somber memorial to those Americans who gave their lives in the Vietnam War

History records that 58,320 men and women died serving with the US armed forces in the Vietnam War, which lasted from 1955 to 1975. For some years, they went unremembered, but in 1982, a new memorial was opened in Constitution Gardens, next to the National Mall in Washington, D.C.

A wall, a memorial, a statue

The memorial consists of two 247-ft- (75-m-) long, highly polished, black granite walls inscribed with the names of the dead in horizontal rows on 142 panels. Close to the wall is a memorial to the women who served in the war, most of them as nurses. There is also a bronze statue named the Three Soldiers, which includes the first representation of an African American on the National Mall. Together they form a poignant memorial to one of the US's lengthiest and most wasteful wars.

◁ **WALL OF REMEMBRANCE**
Designed by Maya Lin, the wall lists all those service members who were killed in action or otherwise declared dead, as well as those whose status is unknown and who went missing in action. Only their names are recorded, without rank, unit, or decorations.

△ **FINAL POUR**
This photograph was taken after the last concrete was poured on top of the tower on February 22, 1974. Since work began just over a year earlier, 1,532 people had worked to continually pour the concrete.

◁ **PANORAMIC VIEW**
Rising 1,135 ft (346 m) above the tallest neighboring buildings, the CN Tower affords uninterrupted views of Toronto and Lake Ontario. On a clear day, it is even possible to see Niagara Falls.

CN Tower

The most prominent landmark on the Toronto skyline, and in its day the tallest structure on the planet

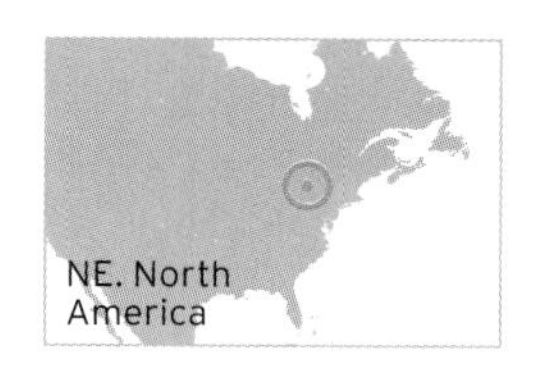

In 1968, the Canadian National (CN) railway built a communications and transmission tower to serve the Toronto area and as a symbol of its corporate vision. Work started in 1973 and was completed in 1976. The resulting tower measures 1,814 ft (553 m) from the ground to the tip of its antenna—tall enough to make it the world's highest tower until 2009, when it was overtaken by both the Burj Khalifa in Dubai (see p.310) and the Canton Tower in Guangzhou.

Up in the skies

The main column of the tower consists of a hollow, concrete, hexagonal pillar containing stairwells and services. The tower was built using a hydraulically raised slipform metal platform, which slowly moved upward at a rate of about 19 ft (6 m) a day, continually pouring new concrete as it rose above the just-set concrete below. In total, the tower contains 1.4 million cubic ft (40,500 cubic m) of concrete, all of which was mixed on site to ensure that it was of the correct consistency. To check that the tower was vertical, plumb lines were dropped from the slipform platform and observed by instruments on the ground.

On top of the column is a 335-ft- (102-m-) high metal broadcast antenna. Six glass-lined elevators located in the tower's outer supports take visitors to observation areas, although the glass floor and the outdoor observation deck are not for the faint-hearted.

The CN Tower's **revolving 360 Restaurant** makes one full rotation **every 72 minutes**

VIEWING AREAS

The public areas of the CN Tower are in two parts. The upper SkyPod sits just below the antenna at 1,465 ft (446.5 m). The lower LookOut Level, toward the tip of the main column, is at 1,122 ft (342 m); it has two observation decks and a revolving restaurant, as well as the tower's microwave receivers.

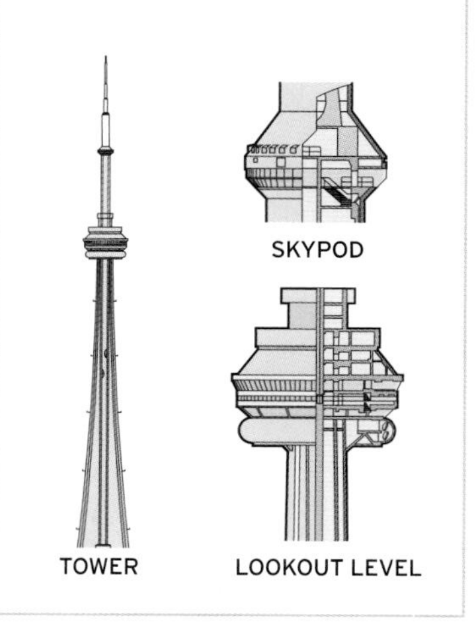

Architectural styles
CONTEMPORARY

There is no single style that defines contemporary architecture. Architects working today draw from a range of sources, ideas, influences, and technologies to create buildings that can be iconic or particular to their settings.

Contemporary architecture is defined by its plurality, with architects around the world working at a variety of scales, in a diversity of styles and formal modes, and utilizing a range of building materials. Computer-aided design and advanced engineering techniques offer architects today an extraordinary freedom to produce almost any conceivable shape—client and budget permitting. At the same time, many architects resist the temptations of form-making in favor of developing designs that emerge from an understanding of buildings' settings and contexts in terms of form, scale, or materials.

What connects the diversity of approaches is an awareness of a building's environmental footprint. Construction entails actually altering the physical environment and has significant environmental impacts on both a local and a global scale. Concrete, which is a key material of nearly every building, is responsible for around 8 percent of global CO_2 emissions. As a result, many contemporary buildings incorporate ecological features, such as solar panels, water recycling systems, and even passive ventilation systems to reduce the use of air conditioning. On a more fundamental level, sophisticated building modeling software allows a building's environmental performance to be measured—and subsequently improved—during the design phase.

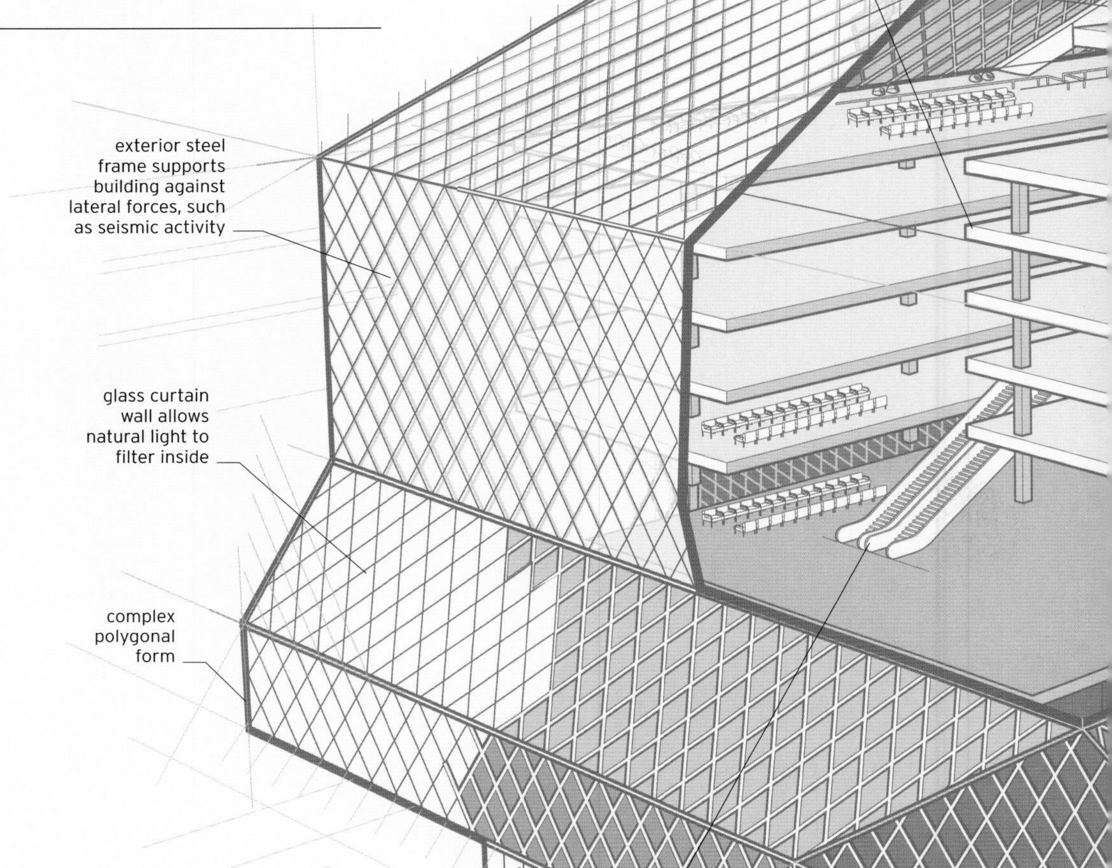

▲ SEATTLE CENTRAL LIBRARY
Opened in 2004, the 11-story building from architects Rem Koolhaas and Joshua Ramus consists of five interior platforms that appear to be unsupported. Encased in a curtain wall of steel and energy-efficient glass, the geometric shape provides adaptable space for the library's collection to grow and change.

CONTEMPORARY COMPLEXITIES

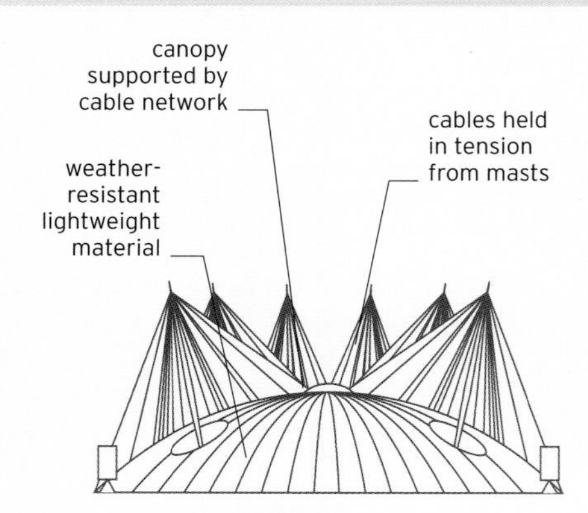

△ TENT ROOFS
Coming to the fore in the 1960s and 1970s, tent roofs or tensile structures allow the spanning of large spaces with a minimal amount of materials and lower budgets.

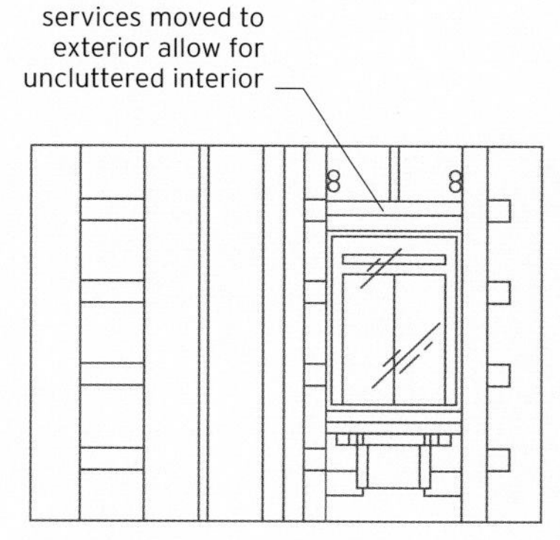

△ EXTERIOR ELEVATORS
The 1980s High-Tech movement aimed to create completely flexible internal spaces by moving structures, services, and sometimes circulation spaces to a building's exterior.

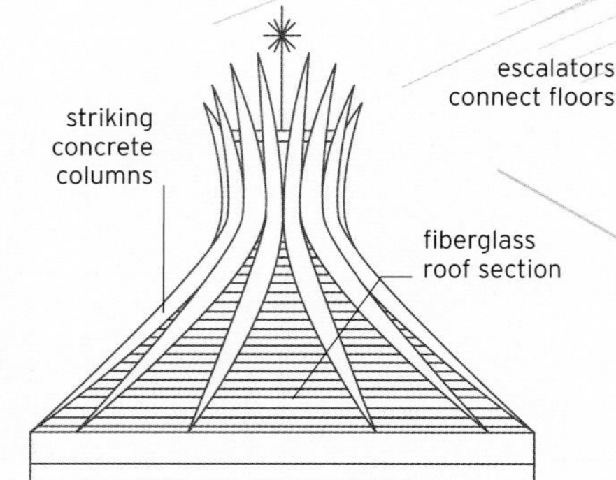

△ COMPLEX GEOMETRIES
Complex geometrical structures, such as the hyperboloid above, are a notable trend in contemporary architecture, utilized for their powerful formal presence and structural advantages.

Wind puts a greater **load** on many modern **high-rise buildings** than the actual **weight** of the structure

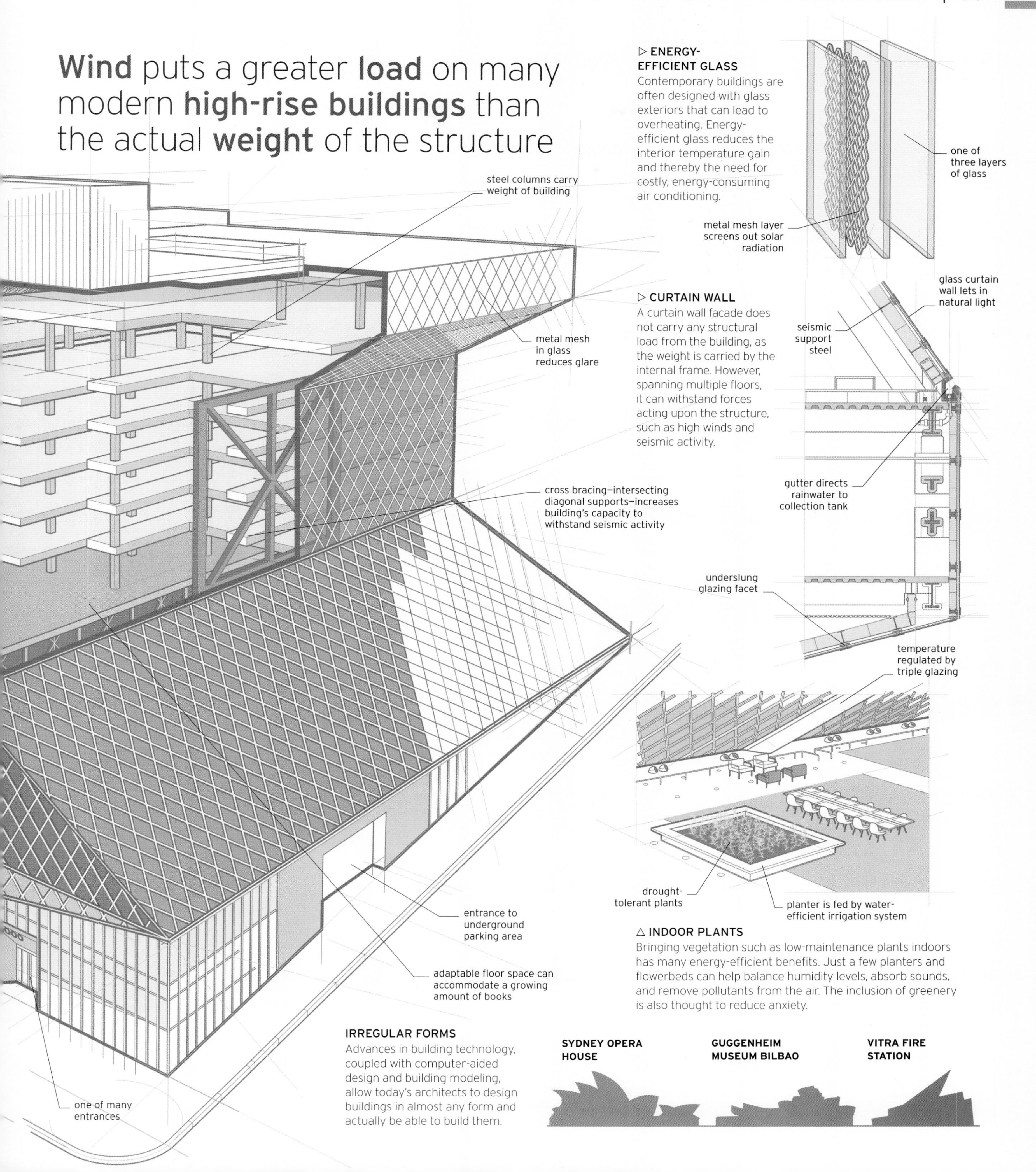

▷ **ENERGY-EFFICIENT GLASS**
Contemporary buildings are often designed with glass exteriors that can lead to overheating. Energy-efficient glass reduces the interior temperature gain and thereby the need for costly, energy-consuming air conditioning.

▷ **CURTAIN WALL**
A curtain wall facade does not carry any structural load from the building, as the weight is carried by the internal frame. However, spanning multiple floors, it can withstand forces acting upon the structure, such as high winds and seismic activity.

△ **INDOOR PLANTS**
Bringing vegetation such as low-maintenance plants indoors has many energy-efficient benefits. Just a few planters and flowerbeds can help balance humidity levels, absorb sounds, and remove pollutants from the air. The inclusion of greenery is also thought to reduce anxiety.

IRREGULAR FORMS
Advances in building technology, coupled with computer-aided design and building modeling, allow today's architects to design buildings in almost any form and actually be able to build them.

Walt Disney Concert Hall

A gleaming temple to the performing arts in Los Angeles, named after one of the movie industry's most famous sons

The long and successful association between film producer Walt Disney (1901–1966) and Hollywood and the city of Los Angeles was furthered in 1967 when his widow, Lillian, donated an initial $50 million to build a new concert hall. Her gift would serve not only to enrich the lives of people living in the city but also to stand as a memorial to her husband's close links to the arts.

The new hall, at 111 South Grand Avenue in downtown Los Angeles, was the work of Canadian-born architect Frank Gehry (1929–), whose radical design predated his titanium-clad Guggenheim Museum in Bilbao, Spain (see pp.200–201)—the building that made him famous. The hall's design was completed in 1991, but construction only began in earnest in 1999, due to a lack of funds, and was finished in 2003. The concert hall officially opened with a gala concert on October 24 of the same year.

Sailing walls

Gehry designed the auditorium to provide audiences with an intimate experience, wrapping them around the orchestra, and paid special attention to its acoustics, lining the walls and ceilings with Douglas fir and the floors with oak to enhance the sound. This functional interior is wrapped in a dramatic sculptural exterior inspired by the architect's love of sailing.

Gehry used a computer program more commonly put to work in the aerospace industry to design the stainless-steel sails that wrap around the building. Some of the concave panels had to be dulled by lightly sanding their surfaces to prevent glare—their reflected light overheating nearby condominiums and increasing the risk of traffic accidents—but once this problem was overcome, the hall took its place as one of Los Angeles's finest buildings.

A note **echoes** in the hall for **2 seconds after** it has been played

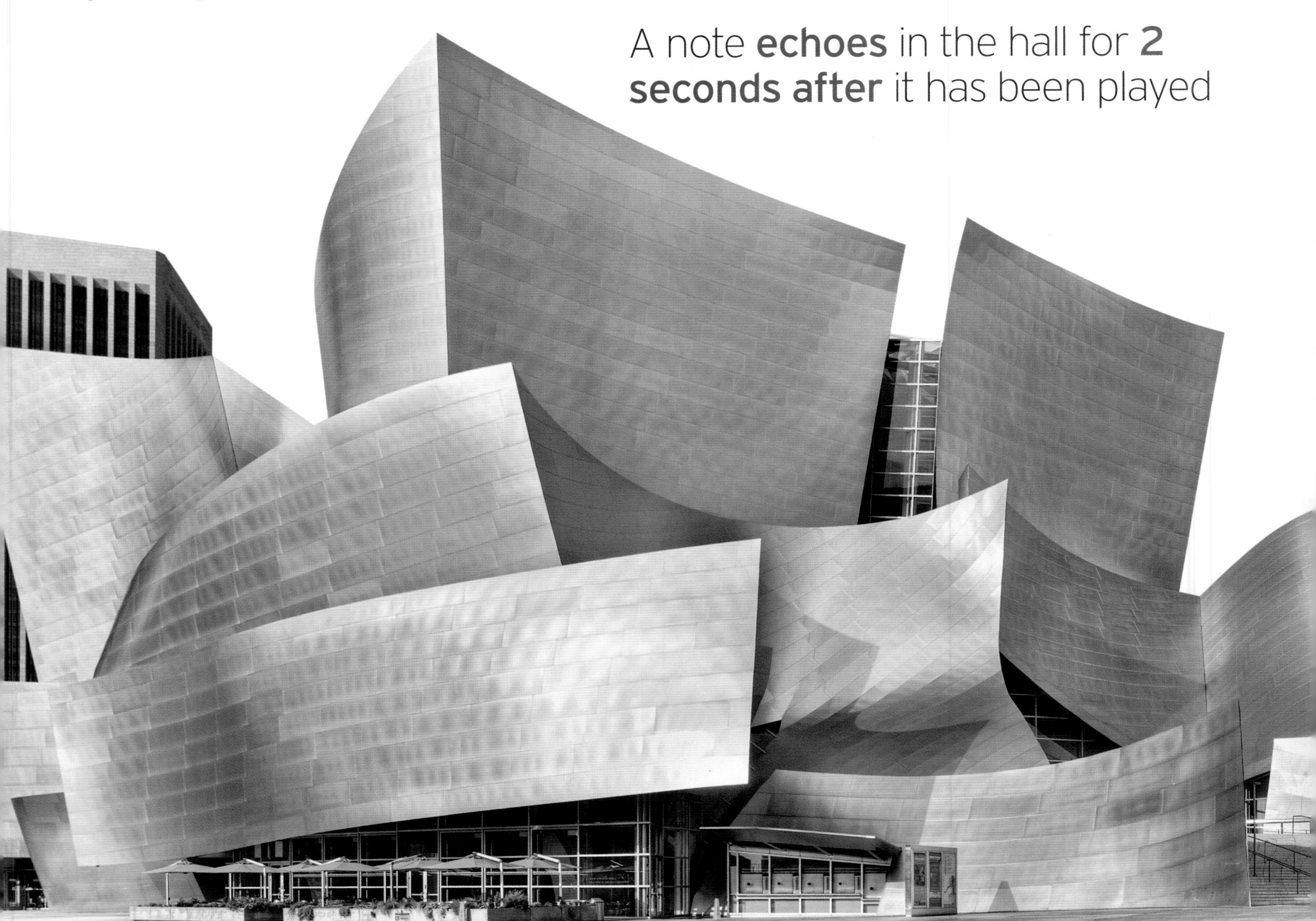

OUTER SHELL

More than 12,500 individual pieces of stainless steel were used to create the complex curves that wrap around the concert hall, and no two are the same. Glass and stone were used in the more conventional parts of the building.

tubular steel framework

anodized aluminum frame

galvanized and painted vertical steel strut

stainless-steel panel

◁ **SKIN OF STEEL**
The flamboyant stainless-steel exterior encloses a concert hall with seating for 2,265 people, as well as restaurants, an exhibition space, and other rooms.

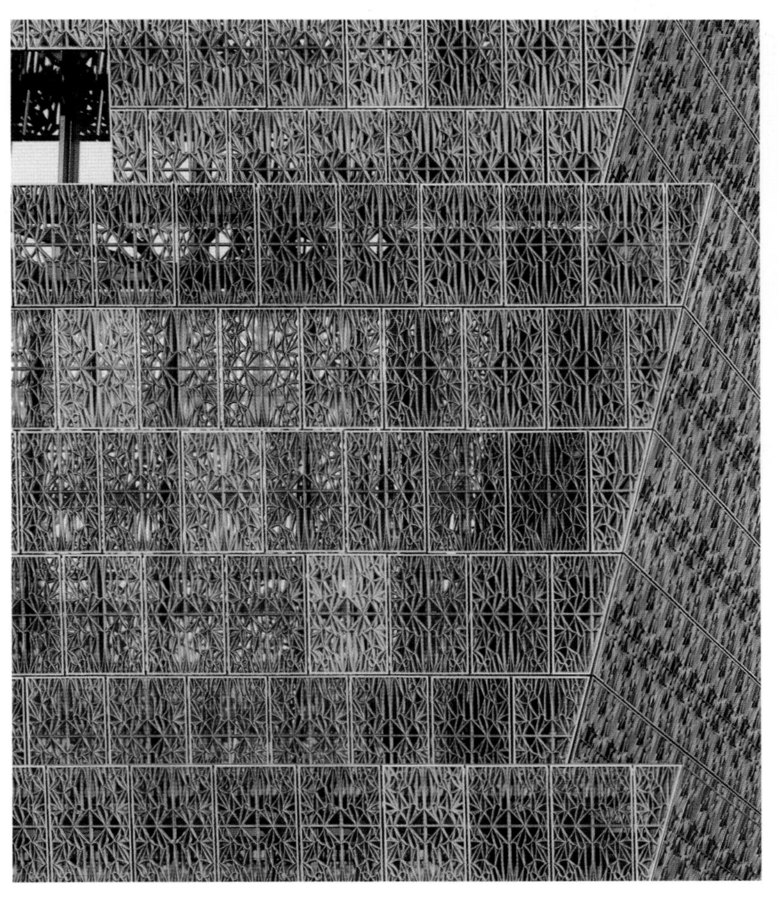

△ **THE CONTEMPLATIVE COURT**
The Contemplative Court is a space for visitors to reflect on the often painful exhibits in the museum. A cylindrical fountain rains down water into the square pool below.

◁ **SYMBOLIC PATTERNS**
The exterior of the building is covered with a thin, bronze-colored aluminum lattice—a homage to the skilled ironwork crafted by enslaved African Americans.

National Museum of African American History and Culture

A building that imaginatively brings together elements of cultural expression from Africa and the Americas

The Civil War (1861–1865) might have ended slavery, but its bitter legacy lived on for years. In 1915, a group of African American veterans of the Union Army that fought in that war met for a reunion, bemoaning the racial discrimination they still faced. They formed a committee to build a memorial to African American achievements, receiving approval in 1929 from President Hoover (1874–1964). Progress was slow, and opposition fierce, until the Smithsonian Institution's National Museum of African American History and Culture, situated proudly on the National Mall in Washington, D.C., was finally opened on September 24, 2016, by the nation's first black president, Barack Obama (1961–). The museum houses more than 37,000 exhibits celebrating African American history.

The building's dramatic shape—a three-tiered inverted pyramid—was inspired by the form of the crowns worn by Yoruban rulers in West Africa and mirrors the angles evident in the capstone of the nearby Washington Monument (see p.30). The exterior is covered by a corona made of 3,600 bronze-colored metal panels.

Sustainable future

Visitors enter from the National Mall through a sweeping porch into a huge central hall that offers expansive views to the upper and lower levels—60 percent of the museum's volume is below ground. The building was constructed from locally sourced or salvaged materials and incorporates the best practices of green design in order to reduce its power and water consumption.

Forest pyramids

The Maya of Central America built their own version of the stone pyramid. The El Castillo pyramid rises above the remains of the city of Chichén Itzá, its sides aligning with the position of the Sun at the summer and winter solstices.

Central and South America

LAND OF TEMPLES AND CITIES

Central and South America

Negotiating the challenging terrain of Mesoamerica and the Andean region without wheeled transportation, early civilizations of South and Central America produced an extraordinary number of impressive structural feats. Their engineering innovations included precision-cut, mortar-free masonry that was earthquake-proof, agricultural terracing that channeled water through every step, and pressurized water systems. Recurring themes were mastery of geometrical forms and extravagant decoration, such as the gold sheeting employed by the Incas. Colonization in the 16th century devastated the indigenous culture while transforming the built environment with elaborate edifices and ambitious schemes that in scale and grandeur were reminiscent of the Incan and Pre-Columbian civilizations. During the 20th century and beyond, the blending of indigenous and European elements evolved into novel modern architectural styles adapted to the local climate and materials.

KEY SITES

1. Calakmul
2. Tikal
3. Nazca Lines
4. Monte Albán
5. Teotihuacán
6. Chichén Itzá
7. Copán
8. Yaxchilán
9. Temple of the Inscriptions
10. Sacsayhuamán
11. Chan Chan
12. Easter Island Moai
13. Machu Picchu
14. Church of Santa Prisca
15. San Ignacio Miní
16. Citadelle Laferrière
17. Panama Canal
18. Christ the Redeemer
19. Cathedral of Brasília
20. Central University City Campus of UNAM
21. Salt Cathedral of Zipaquirá
22. Niterói Contemporary Art Museum
23. Itaipu Dam

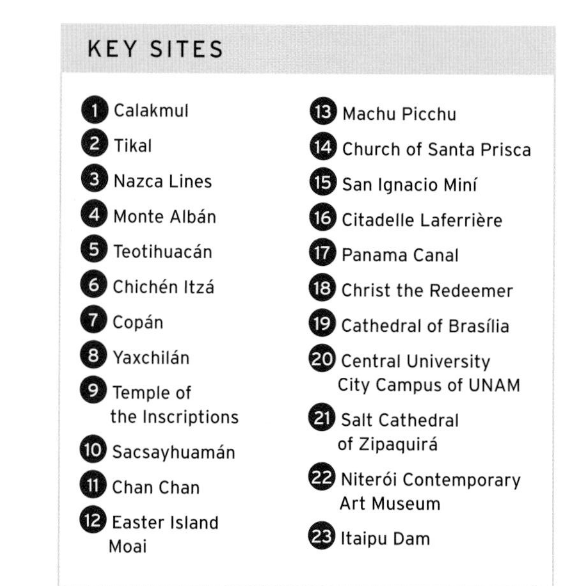

rising to prominence from c.600–1200 CE, Chichén Itzá was one of the largest cities built by the Maya civilization

a mountaintop fortress, Citadelle Laferrière was built in 1805–1820 by Henri Christophe, a leading figure in the Haitian Revolution

MOUNTAIN AND FOREST PIONEERS
Pre-Columbian cultures populated the rainforests and mountaintops of Central and South America with stone cities and temple complexes in astronomically significant locations. Colonization introduced new European styles, and the ensuing deforestation paved the way for the Industrial Age and modern architecture.

PRE-COLUMBIAN KINGDOMS
c.1200 BCE–1550 CE
Rising from the Central American jungle, the step pyramids of ancient Olmec culture gave way to the elaborate designs of the Maya, who masterminded the biggest cities in the world. Mesoamerican architects became renowned for their knowledge of astronomy, their engineering prowess, and their skillful incorporation of symbolic references.

2 TIKAL TEMPLE

MONOLITH BUILDERS
900–1600
From the highland plateaus of southern Colombia to the far-flung reaches of the southeastern Pacific, distinct indigenous cultures executed similar feats of stone construction and carving in the form of giant monoliths. Most notable are more than 800 *moai* erected by the Rapa Nui people of Easter Island.

12 EASTER ISLAND MOAI

INCA MOUNTAIN STRONGHOLDS

1438–1572

Faced with some of the roughest territory in the Andes, the engineers of the Inca civilization laid down an extensive network of roads, staircases, and suspension bridges to link the outposts of their empire. Without using mortar, they erected impressive stone citadels that were able to withstand earthquakes and remained impregnable for centuries.

13 MACHU PICCHU

created between 500 BCE and 500 CE, the Nazca Lines form mysterious images of animals, plants, and humanlike figures, as well as geometric designs

Rio de Janeiro's Art Deco statue of Christ the Redeemer is a symbol of the Christianity that European colonists brought to South America

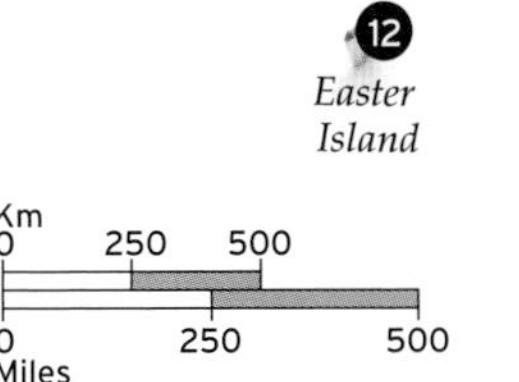

COLONIZERS EXPORT ARCHITECTURE

1521–1821

As colonial society was imposed on, and then absorbed into, the cultural landscape of Latin America, architectural hybrids evolved, marrying European tradition with local materials and artistic preferences. Among these new building styles were flamboyant Andean Baroque, expressive New Spanish Baroque, and elaborate Mexican Churrigueresque.

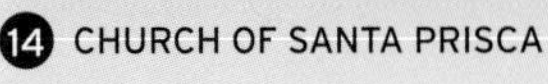

14 CHURCH OF SANTA PRISCA

INDUSTRY TAKES HOLD

1800–1914

The expansion of industry in Europe and the US propelled Latin America to ramp up its extraction of raw materials and its own manufacturing. At the same time, newly independent republics began to transform their cities. These two factors came together in the ambitious infrastructure and urban renewal projects that unfolded across the continent.

17 PANAMA CANAL

CONTEMPORARY ARCHITECTURE

1945–PRESENT

An explosion of creativity, experimentation, and innovation characterized the second half of the 20th century, as urban planners attempted to establish a forward-thinking architectural identity for the many nations of the continent—from the avant-garde Brazilian capital of Brasilia to the geodesic dome of the Poliedro de Caracas in Venezuela.

22 NITERÓI CONTEMPORARY ART MUSEUM

Calakmul

A lost Maya city deep in a tropical forest, dominated by two imposing pyramid temples

Once a thriving city of as many as 50,000 inhabitants, Calakmul, known by the Maya as Ox Te' Tuun ("Three Stones"), now stands empty, with much of its ruins overgrown by the surrounding jungle. Situated in the UNESCO-protected tropical forest of the Tierras Bajas in the Yucatán Peninsula, Mexico—the heartland of the Maya civilization—the settlement was established in the 1st millennium BCE. It became the seat of a powerful kingdom in the Mesoamerican Classic period (c.250–900 CE). Like several Maya cities, Calakmul was abandoned in the 9th century, but many of its structures are well preserved and have been reclaimed from the forest since the city's discovery in 1931. Thousands of structures—including residential buildings, tombs, and monuments—have been found, with a wealth of decorative murals, ceramic artifacts, and stone stelae. Dominating these buildings, however, are the two massive pyramid temples that give the city its modern name: Calakmul, "city of two adjacent pyramids."

◁ **CARVED STELA**
This is one of the 117 stelae found at Calakmul. The stelae featured portraits and inscriptions carved into the soft limestone, commemorating members of the ruling families.

American botanist **Cyrus Lundell discovered the ruins** of Calakmul **by chance** while working in the rainforest

△ **THE GREAT PYRAMID**
The largest of the pyramids, known as Structure II, or the Great Pyramid, was constructed by building onto an existing pyramid. It contains four separate tombs.

△ **CAPITAL OF A FOREST KINGDOM**
The Maya city of Tikal remained hidden in the lush Guatemalan rainforest until clearance of the vegetation began in the late 19th century.

THE LOST WORLD COMPLEX

A complex of monuments and temples in Tikal, the *Mundo Perdido* ("Lost World") site has the Great Pyramid at its center, around which are four distinct plazas. The site is enclosed by various structures, including, to the east, a row of eight buildings that form the boundary of the adjoining Plaza of the Seven Temples.

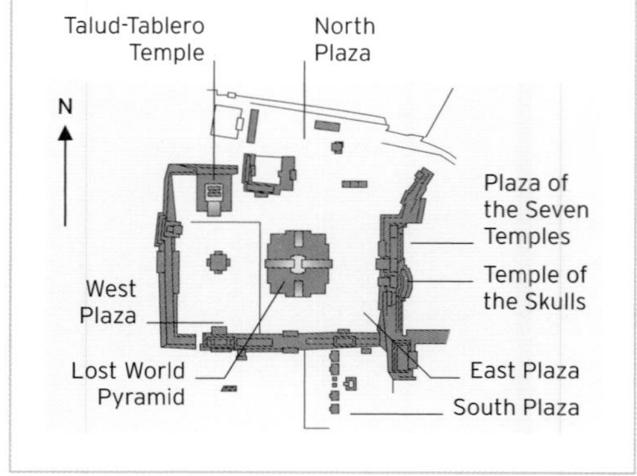

Tikal

A city of towering temples, monuments, and palaces and the center of a powerful Maya kingdom

Today, the abandoned city of Tikal—once the capital of a powerful Maya kingdom—stands in the Tikal National Park, a UNESCO World Heritage Site in the Petén province of Guatemala. While many residential areas, irrigation stations, and administrative outposts are still engulfed by the tropical forest, the center of the city, which contains the principal monuments and public buildings, has been cleared.

Lost and found

The central zone has public squares, or plazas, built on terraces connected by roadways and accessed by ramps. The squares are surrounded by temples and palaces, along with other structures such as pyramids, ceremonial platforms, and courts for the popular Maya ball games. The principal squares of the town are the Great Plaza, the Twin Pyramid complex, and the so-called "Lost World" complex (see panel, left). The Maya monuments there are among the finest yet discovered. Many of them are decorated with hieroglyphic inscriptions and paintings. These give an insight into the development of architectural and artistic styles at the site from its early settlement in the 4th century BCE to the time of Tikal's regional dominance in the 9th century CE.

▷ **EPICENTER OF TIKAL**
Tikal's Great Plaza consisted of a ball court and two temples, known as Temple 1 and Temple 2. Temple 1 (on the right in this image) was considered by the Maya to be a portal to the underworld.

Nazca Lines

Monumental geometric designs and zoomorphic figures inscribed in the surface of the Peruvian desert

The coastal plain of the Rio Grande de Nazca in southern Peru is adorned by huge geoglyphs (designs and motifs created on the ground). They are the legacy of the Nazca civilization, which flourished between 500 BCE and 500 CE. Across a vast area of desert, the Nazca Lines were scratched out of the gravel surface over about 1,000 years, forming a unique collection of designs and images on such a large scale that they cannot be properly appreciated from ground level.

Scraping through

There are mainly two types of geoglyphs on the site. The first are the straight or curved lines: straight lines stretch for several miles across the desert, sometimes forming criss-cross patterns. They range in width from 20 in (50 cm) to 16 ft (5 m) and may have held some intrinsic ritual significance or connected sacred sites. Curved lines form geometric designs, such as curves, spirals, and wavy patterns.

The second group consists of stylized representations of natural forms. The few depictions of humans and inanimate objects are generally inscribed on hillsides around the plain.

△ **THE NAZCA SPIDER**
This spider, when seen from above, shows the distinctive artistic style of the Nazca people and the astonishing accuracy with which it was rendered.

CREATING THE NAZCA LINES

The lines were made possible by the contrasting colors of the dark, iron-oxide-coated stones on the desert floor and the much lighter earth beneath them. Simply scraping these pebbles from the surface was enough to create effectively permanent marks. A selection of the designs is shown here.

LABYRINTH
236 ft (72 m) long

HANDS
170 ft (52 m) long

HERON
236 ft (72 m) long

MONKEY
360 ft (110 m) long

OWL-MAN
131 ft (40 m) long

◁ **HUMMINGBIRD TAIL**
The tail feathers of the hummingbird figure are formed by a series of long parallel lines scratched into the desert surface.

The **figurative designs** generally consist of **a single continuous line** that does not cross itself

Monte Albán

The capital of the Zapotec civilization, carved out of a mountain overlooking the Valley of Oaxaca

One of the first cities to be established in Mesoamerica, Monte Albán was continually inhabited over a period of about 1,400 years from about 500 BCE. Occupying a commanding position on a mountain ridge in southern Mexico, the city became the center of a state covering a large part of the Oaxaca valley.

City on a mountain

Most strikingly, the site comprises a number of levels and terraces cut into the top and flanks of the mountain, forming the bases for the various zones of the city. At the top is the civic center, consisting of a Main Plaza with various ceremonial structures; to the east and west of this are temples and palaces set on platforms; and at the north and south ends of the plaza are larger platforms accessed by magnificent stairways. As the city grew in importance and expanded, terraces were built on the hillsides around it and fortifications were added. The city was gradually abandoned from about the 9th century CE and only extensively excavated and restored in the 20th century.

△ **ZAPOTEC RELIEF**
A carved relief representation of the bird god Pico Ancho, this figure shows the stylized designs and patterns characteristic of Zapotec art.

The complex includes a **custom-built court** for the **ritual ball game**

▽ **PLATFORMS AND STAIRWAYS**
Grand staircases are a feature of the city center, linking the different levels at either end of the main esplanade and leading to the platforms of the monuments and ceremonial buildings.

Teotihuacán

An ancient metropolis of vast proportions that thrived for centuries and wielded immense power

The magnificent Mesoamerican city of Teotihuacán in central Mexico was built between the 1st and 7th centuries CE. At its peak, in around 600 CE, it supported a population of more than 150,000 and had tremendous economic, cultural, and religious importance in the region.

City of the gods

Little is known about the language and identity of Teotihuacán's founders. The Nahuatl-speaking Aztecs discovered the ruins of the city centuries after its decline and, believing that only gods could have created its colossal pyramids, named it Teotihuacán, or "city of the gods." Built as a model of the Mesoamerican cosmos, almost everything in it had ritual or religious significance.

The city's ceremonial center includes three vast monuments: the pyramids of the Sun and the Moon and the Temple of Quetzalcóatl, the mythical plumed serpent. Of these, the largest is the Pyramid of the Sun, the third largest pyramid in the world. Teotihuacán's architectural splendor is matched by its artistic wealth: breathtaking stone sculptures and reliefs, as well as vivid murals, are found all over the city. The Teotihuacános also produced textiles, pottery, and other artifacts, including striking masks.

The civilization fell suddenly and mysteriously sometime after 550 CE—the exact cause of its collapse continues to elude scholars. Teotihuacán was designated a World Heritage Site in 1987.

The city's **wealth** was derived from a rare **obsidian**, a volcanic glass mined in the region

URBAN PLANNING

Built on a grid plan, Teotihuacán is considered one of the world's most impressive examples of urban planning. The principal street, the Avenue of the Dead, is more than 2 miles (3 km) long and includes the three major monuments: the Pyramid of the Sun, the Pyramid of the Moon, and the Temple of Quetzalcóatl.

N
Pyramid of the Moon
Plaza of the Pyramid of the Moon
Pyramid of the Sun
San Juan River
Temple of Quetzalcóatl
Avenue of the Dead

◁ **TEOTIHUACÁN MASK**
A spectacular and emblematic piece of Teotihuacán art, this stone funerary mask was created c.200–550 CE, with further decoration added several centuries later. It is inlaid with turquoise, amazonite, coral, obsidian, and shell.

△ **THE SUN PYRAMID**
An aerial view of Teotihuacán shows the city's massive, five-tiered Pyramid of the Sun, which is 210 ft (64 m) high, with a foundation of 728 ft by 738 ft (222 m by 225 m). There are 248 steps to the top of the pyramid.

◁ **HEAVEN AND HELL**
This detail from a mural at the Tepantitla compound, Teotihuacán, c.200 CE, contrasts visions of paradise and vitality, represented by butterflies, with nightmarish scenes, including a game using a human as a ball (bottom right) and a row of sacrificial victims (top right).

Chichén Itzá

The last of the great Maya city-states, whose monuments include masterpieces of Mesoamerican architecture

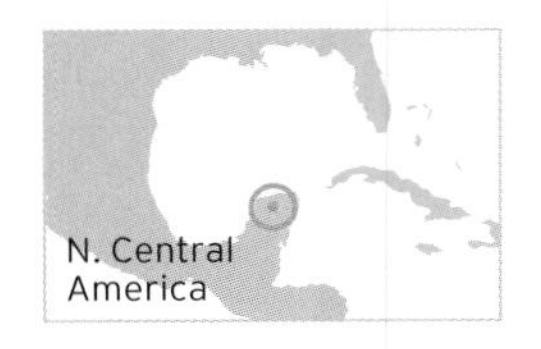

The largest Maya city in southeastern Mexico, Chichén Itzá rose to prominence in the 9th century and fell into decline in the early 13th century. This remarkable World Heritage Site of around 4 square miles (10 square km) includes pyramids, temples, cenotes (sink holes), a ball court, and an astronomical observatory to chart the movements of Venus—indicating the civilization's rich spiritual and artistic life. Warfare, torture, bloodletting, and human sacrifice were highly significant for the Maya, according to the murals, stelae, and vessels upon which these sacred practices are depicted.

Central structure

The El Castillo pyramid towers above the city's main plaza. It has a combined total of 364 steps on its four sides; the step on the top platform brings the total to 365—the number of days in a solar year. Each stairway faces one of the four cardinal directions of the compass.

▽ **CHAC-MOOL**
At the entrance to the Temple of the Warriors is a statue of the rain god Chac-mool. He holds a receptacle that is thought to be for the hearts of human sacrifices.

SOLAR ALIGNMENT

The Maya worshipped the Sun and laid out the El Castillo pyramid to align with solstices. In addition, at the spring and fall equinoxes, the Sun casts a series of shadows against the northwest balustrade, creating an optical illusion of their feathered serpent deity crawling down the pyramid.

temple platform
stairway with 91 steps
midsummer sunrise
midsummer sunset
midwinter sunrise
N
midwinter sunset

Copán

A major Maya city that provides a greater understanding of the complex civilization

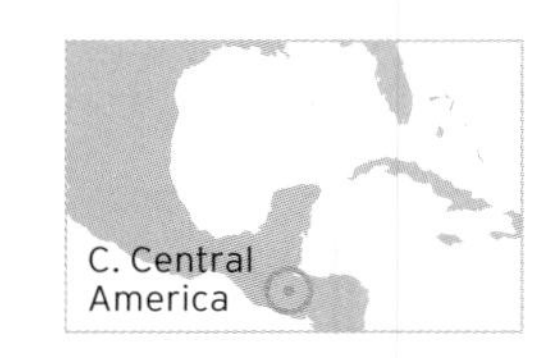

Developed between the 5th and 9th centuries CE, the Maya city of Copán, Central America, is considered to be one of the civilization's major cities. The World Heritage Site is of particular interest to scholars for the expression of Maya cosmology evident in the city's design and architecture and its various sculptures and carvings.

Astronomical inspiration

Many of the sculptures and carvings were constructed during the reign of Copán's most famous ruler, King 18 Rabbit, to reflect the patterns of the cosmos and the movement of the stars and planets, all of which helped the Maya to plan and regulate their lives. A major building in this respect is the magnificent Temple 22, conceived as a gateway to the underworld. It has numerous fascinating astronomical references, including a two-headed serpent representing the Milky Way, which symbolized the cosmic axis for the Maya. Copán's Temple of the Hieroglyphic Stairway is so called because it is beautifully carved with symbols, or glyphs, that relay the city's history. It was commissioned by King Smoke Shell and dedicated in 749 CE. Copán was abandoned in around 1200 CE. Restoration started in the 1930s and 1940s. In the mid-1970s, considerable advances were made in understanding the city's history by deciphering various hieroglyphs throughout the site.

◁ **ELABORATE CARVING**
Stela N is located in front of Temple 11, south of the Court of the Hieroglyphic Stairway. The figure on it wears a headdress, and glyphs are carved on its side.

EL CASTILLO
The pyramid rises to a height of 79 ft (24 m). At the top is a carving of Kukulcán, the plumed serpent—an important Mesoamerican deity.

Yaxchilán

Nestled deep in the rainforest, this impressive Maya site is famous for its structures, monuments, and spectacular relief carvings

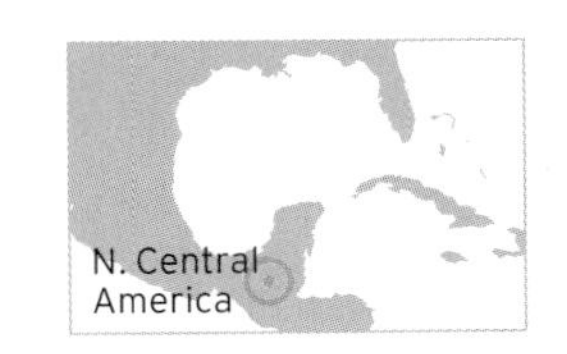

Bordered on three sides by the mighty River Usumacinta, deep in the rainforest of Mexico's Chiapas state—in the heart of what was the Maya's most densely populated region—the city of Yaxchilán commanded a strong defensive position. Many of the city's major temples were built in the hilltops above the river and may have served as astronomical observatories.

Maya art

The site flourished from around 681 CE under Lord Shield Jaguar II. It is notable for its sculptures—in particular, the spectacular carved stone lintels above the doorways, commissioned by various of the city's rulers, which often depict the bloodletting ceremonies, sacrifices, and torture that held sacred significance for the Maya. Several of the lintels are among the masterpieces of Maya art.

◁ **STELA 35**
This stela, c.600–900 CE, depicts Lady Eveningstar, a wife of Shield Jaguar the Great. She may have ruled Yaxchilán briefly, a decade after her husband's death.

△ **THE KING PALACE**
Structure 33 at Yaxchilán, built in the mid-8th century CE, is a magnificent example of Classic Maya architecture. Its roof-comb wall includes friezes, niches, and sculptural elements.

Architectural styles

PRE-COLUMBIAN AMERICA

Pre-Columbian refers to all of the indigenous civilizations that occupied the Americas prior to the arrival of Europeans in the 15th century. The most significant in terms of the architecture they produced were the Mesoamerican cultures.

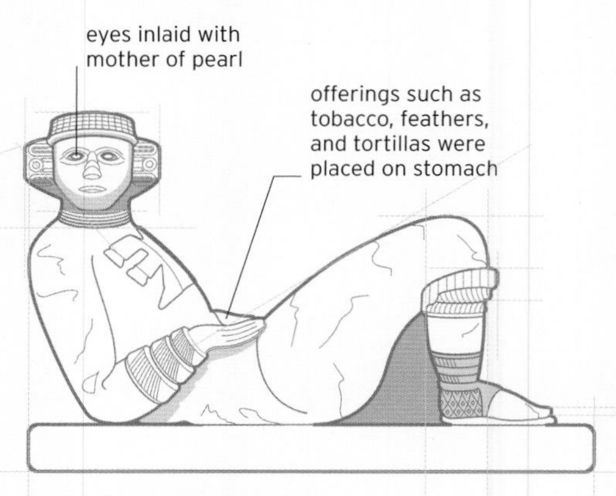

△ **CHAC-MOOL STATUE**
This statue, common in Mesoamerican cultures, depicts a reclining man holding a bowl on his torso. It is thought to symbolize a fallen warrior carrying an offering to the gods.

Mesoamerica is the historic area extending from central Mexico in the north to Costa Rica in the south. It was home to a number of major cultures, such as the Olmec, Zapotec, Aztec, and Maya, who developed their own forms of mathematics, writing, astronomy, religion—and architecture.

The stepped pyramid is the most distinctive building of Mesoamerican architecture. Unlike Egyptian pyramids, Mesoamerican pyramids did not contain burials but acted as temples, raised up toward the stars. They were a key part of the way these cultures conceived architecture as acting as the interface between the human world and the afterlife. This was also manifested in the layout of cities; temples looked out across great plazas, along with palaces and ball courts for the ritual games common to many Mesoamerican cultures. Residences and ceremonial structures were often built in separate areas of the city to create a distinction between the everyday and religious functions of the city. In some cultures, certain buildings were also aligned with the positions of stars and other objects in the sky at particular times of year.

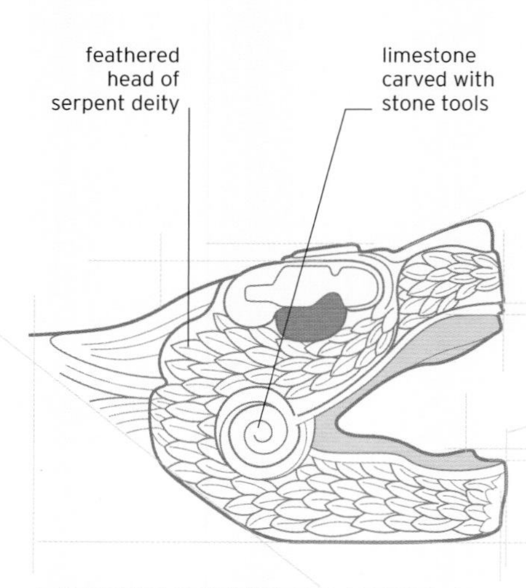

△ **SERPENT STATUES AND MOTIFS**
The serpent was an important symbol in Maya culture and is a frequent motif in sculpture. At El Castillo, shadows cast at the equinoxes make the serpent appear to undulate down the staircase.

52 panels on each side of pyramid represent number of years in a Maya sacred cycle called a Calendar Round

stone serpent head

serpent shadow cast here at spring and fall equinoxes

▲ **EL CASTILLO, CHICHÉN ITZÁ**
Chichén Itzá was a large city located in what is now Yucatán, Mexico, built by the Maya from around the 7th century CE. The centerpiece of the city complex is the limestone El Castillo stepped pyramid, topped by a temple and accessed from four external stairways.

SYMBOLIC STRUCTURES

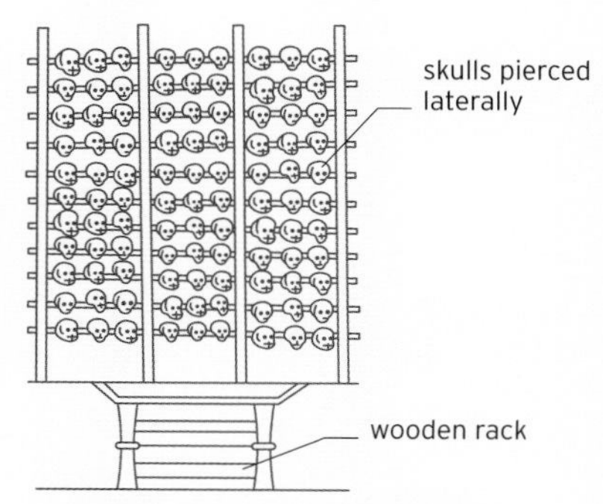

△ **TZOMPANTLI SKULL DISPLAY**
Found across several Mesoamerican civilizations, the *tzompantli* is a rack containing rows of human skulls. The skulls typically belonged to vanquished foes or victims of human sacrifice.

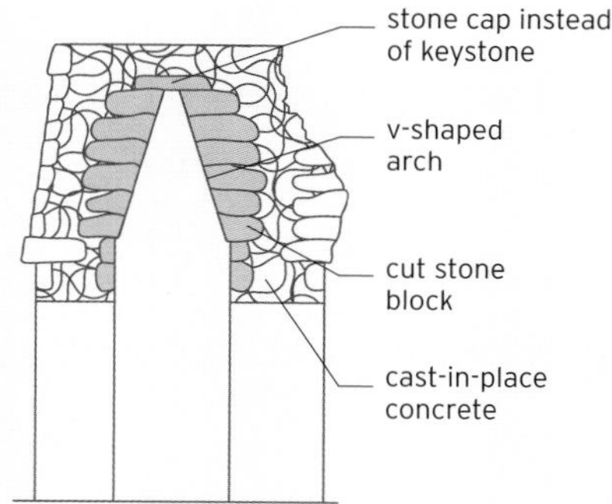

△ **MAYA ARCH**
Maya craftsmen used a kind of corbeled arch formed from adjacent overhanging tiers of masonry. They combined stone blocks with concrete that was cast within the stone structure.

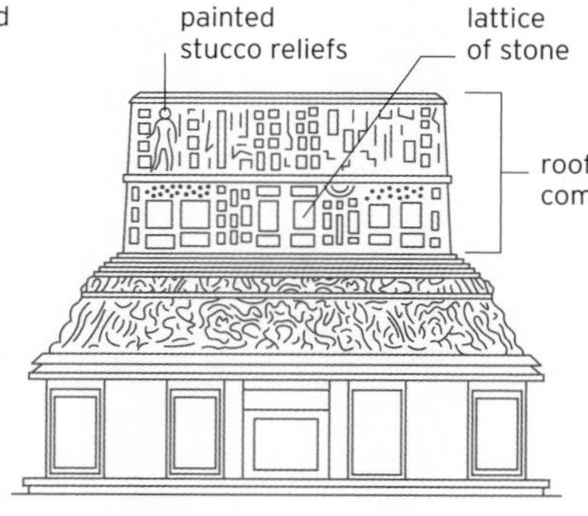

△ **ROOF COMB**
Many Mesoamerican pyramids were topped with a structure called a roof comb that served to increase the pyramid's height and to display decoration and iconography.

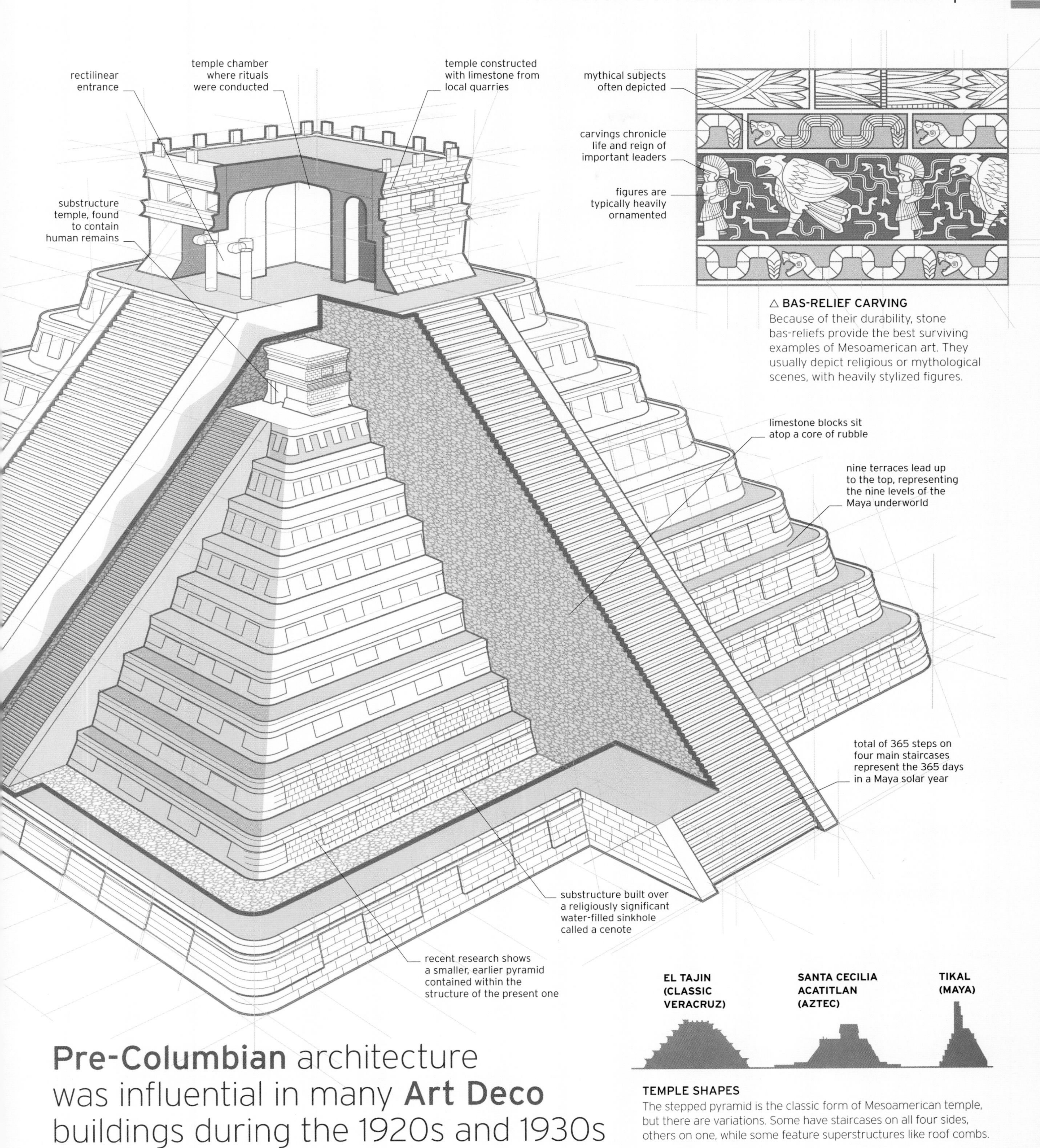

△ **BAS-RELIEF CARVING**
Because of their durability, stone bas-reliefs provide the best surviving examples of Mesoamerican art. They usually depict religious or mythological scenes, with heavily stylized figures.

TEMPLE SHAPES
The stepped pyramid is the classic form of Mesoamerican temple, but there are variations. Some have staircases on all four sides, others on one, while some feature superstructures like roof combs.

Pre-Columbian architecture was influential in many **Art Deco** buildings during the 1920s and 1930s

Temple of the Inscriptions

A funerary monument for K'inich Janaab' Pakal, ruler of the Maya city-state of Palenque

Situated in the present-day Mexican state of Chiapas, Palenque was one of the most important of the Maya cities. It became established as a trading center during the Classic Period of the Maya civilization (c.250–900 CE), when it was known by its Maya name of Lakamha, and reached its heyday during the rule of K'inich Janaab' Pakal, or Pacal the Great, who reigned from 615 CE until his death in 683 CE.

Reclaimed from the forest

Only a fraction of the city has been recovered from the forest, but what has been discovered includes some of the finest examples of late Classic Maya architecture, such as Pakal's royal palace and the funerary pyramid built to house his tomb. At the top of the nine-tiered pyramid, which is approached by a staircase of 69 steps, is the Temple of the Inscriptions, so called because of the extensive hieroglyphic texts carved on the tablets on its inner walls and the sculpted panels on the piers separating the five entrances.

The Palenque dynasty collapsed in about 800 CE, and the city was overtaken by forest; it has only gradually been rediscovered since the late 18th century.

△ **PAKAL'S BURIAL MASK**
This jade death mask was found among the treasures in Pakal's burial chamber; its eyes are made from seashells and obsidian.

THE BURIAL CHAMBER

Within the temple, one of the slabs of the floor can be lifted to gain access to the inner staircase leading down to Pakal's tomb. The sarcophagus containing his remains and funerary goods is topped with a huge stone slab, carved with a depiction of the ruler entering the underworld.

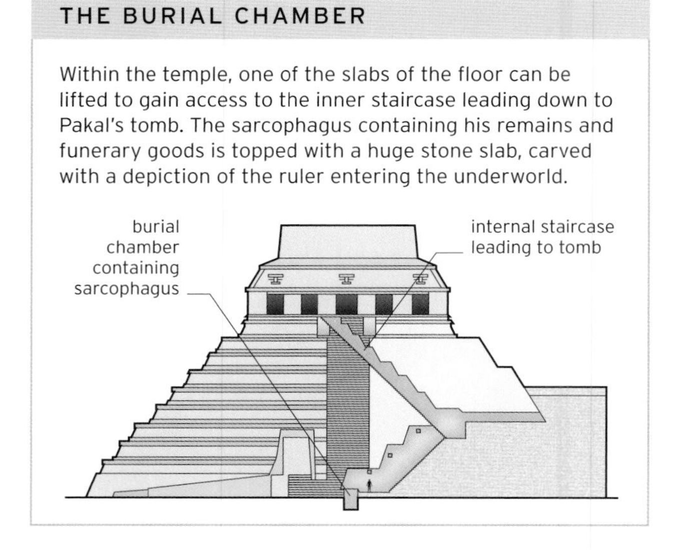

This is the only known Maya **funerary pyramid** built during its occupant's life

▽ **IMAGINING THE FACADE**
Like the other structures in Palenque, the pyramid was originally adorned with paint and stucco decorations. Of the original roof comb—a typical feature of Maya structures—only stubs remain.

Sacsayhuamán

The largest fortress-temple complex built by the Incas, remarkable for its precisely constructed dry-stone walls

The citadel of Sacsayhuamán occupies an imposing position on a hill to the north of the city of Cuzco, Peru, the capital of the Inca Empire. It was originally established by the Killke people in about 1100, but when the Incas took over the region in the 13th century, they expanded and fortified the complex.

Regarded as both a fortress and a complex of temples, the three terraced levels of Sacsayhuamán are protected by stone walls up to 59 ft (18 m) high. These defensive walls at the foot of the citadel are laid out in an unusual zigzag pattern, designed so that attackers could be fired upon from several directions.

Perfect fit

The Incas originally reinforced the citadel with simple clay and mud walls, but in the mid-15th century, they began to replace these with the massive dry-stone walls for which Sacsayhuamán has become famous. Enormous blocks of stone, weighing as much as 110 tons (100 tonnes), were quarried, then pounded and carved into polygonal shapes with such precision that they fit perfectly together without the need for mortar.

According to Inca legend, the city of Cuzco was known as the "lion city" and was laid out in the form of a puma when viewed from above; the citadel at Sacsayhuamán was intended to be its head.

▽ **INCA CITADEL**
This image shows the central area of the complex, built on terraces on the promontory above Cuzco.

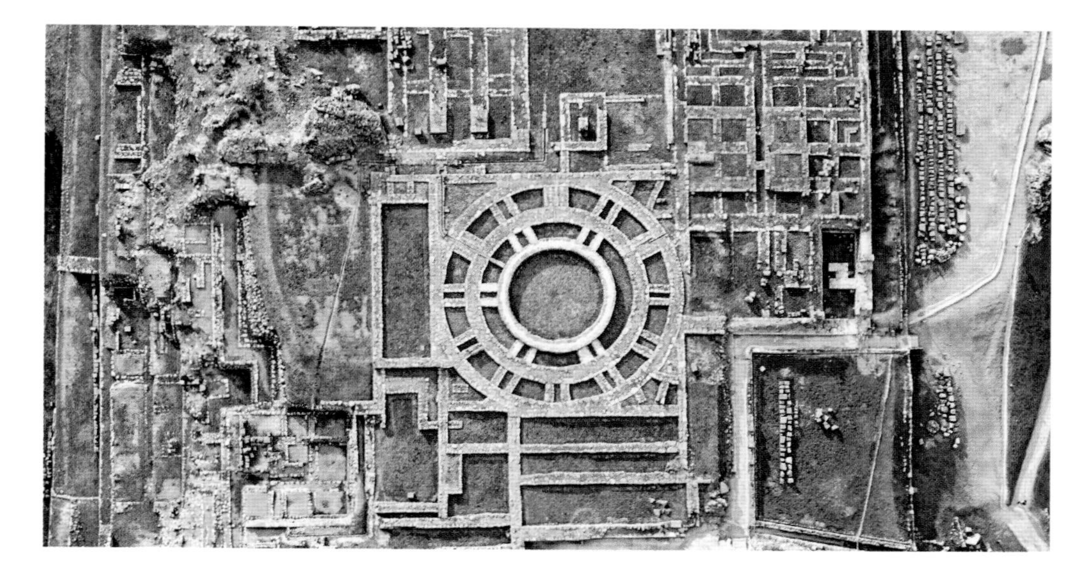

Chan Chan

The capital of the Chimú empire, with a central complex of 10 compounds separated by thick mud-brick walls

△ **VAST ADOBE CITY**
The earthen-walled city of Chan Chan covered about 8 square miles (20 square km). The compounds were well planned and enclosed by diamond-shaped sand walls.

Chan Chan (also known as Chimor) was a thriving city in the Chimú civilization of northern Peru between the 12th and 15th centuries. Situated on the coast at the mouth of the Rio Moche, it evolved from a primarily agricultural settlement through the building of canals and reservoirs that served a sprawling city arranged in distinct units. The grandest of these units are the 10 rectangular *ciudadelas* containing palace and administrative buildings made from adobe bricks of cast mud and enclosed by 33-ft- (10-m-) high walls. The walls are decorated on the outside with high-relief geometric patterns and depictions of animals and fish.

While the nature of the premolded, earthen blocks used in the city's construction renders them liable to deterioration, the dry climate has kept them essentially intact to date. Moreover, under the protection of World Heritage Site status, an ongoing program of sympathetic maintenance has preserved much of the site for posterity.

Easter Island Moai

Monumental ancestral statues created by the Rapa Nui people and unique to Easter Island in the Pacific

At some time in the first millennium, a group of Polynesians settled on the remote island of Rapa Nui, 2,300 miles (3,700 km) off the coast of Chile. Between the 11th and 17th centuries, the Rapa Nui people carved 800–1,000 colossal stone figures, each believed to embody the spirit of a prominent ancestor. Often erected on ceremonial platforms known as *ahus*, the statues (*moai*) face inland, watching over the island and its people. Most were carved from blocks of yellow-brown volcanic tuff and have an oversized head. Some also feature a red tuff *pukao*, or top-knot, and are thought to be among the later statues carved.

Walking statues

The question of how the *moai* were transported large distances from the quarry where they were carved has long been disputed. Theories about rolling tree sledges have now given way to the theory that the statues were harnessed with ropes and "walked" across the island by being tilted and pulled from side to side.

European colonists arrived on Rapa Nui in 1722. By the late 18th century, most of the *moai* had been toppled. The process of reerecting the statues began in 1978, and UNESCO declared Easter Island a World Heritage Site in 1995. In 2018, the islanders asked the British Museum in London to return a statue taken from the island in 1868.

▷ **ALL-SEEING EYES**
The only statue to have had its eyes restored is at *Ahu Ko Te Riku*. The *moai* had deep eye sockets, where pieces of white coral were placed during rituals.

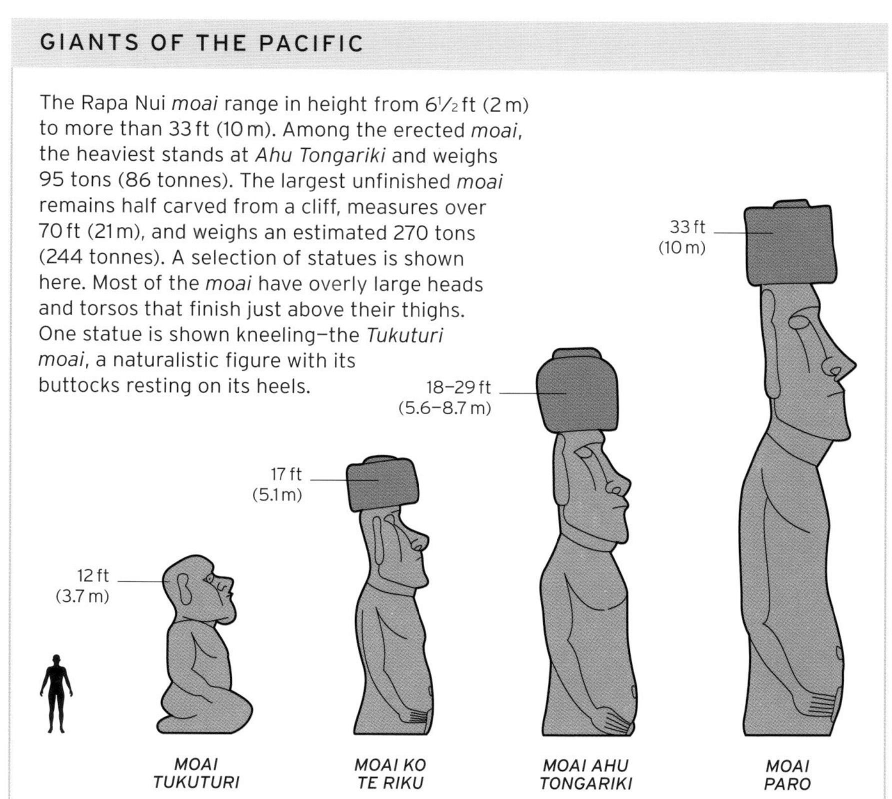

△ ***MOAI* FACTORY**
At the *Rano Raraku* quarry, nearly 900 *moai* were carved from tuff, a porous rock formed by volcanic ash. The examples seen here lack the headdresses made of red scoria stone found on some *moai*.

▽ **STONE FACED**
Ahu Tongariki is the largest ceremonial structure on Easter Island. Its 15 statues stand on a 330-ft- (100-m-) long platform, their backs to the sea, protecting the ancient village that once stood before them.

The giant stone statues at Easter Island weigh an **average** of **14 tons (13 tonnes)**

Machu Picchu

A magnificent Inca citadel and temple complex set high on a mountain ridge above the Urubamba River in Peru

W. South America

Founded by the Inca ruler Pachacuti Inca Yupanqui in the mid-15th century, Machu Picchu ("Old Hill") occupies a spectacular site in the High Andes of Peru, 50 miles (80 km) northwest of the Inca capital Cuzco. The fortified complex was primarily a sacred site dedicated to the Sun god Inti, as well as the administrative center for the region. It also enclosed a settlement of around 1,000 inhabitants (at its peak) and included accommodation for the aristocracy and a separate quarter of more modest dwellings to the south and east for the workers of the estate.

Sun temple

The buildings of Machu Picchu are made from carefully cut blocks of granite in the characteristic Inca style, fitting together so precisely that they needed no mortar and arranged in irregular patterns to withstand the effects of earthquakes. The finest of these are in the upper part of the citadel, which had religious and astronomical functions. Among them are the Temple of the Three Windows and the Torréon, or Temple of the Sun, which acted as a sort of observatory, with windows aligned with the Sun at the summer and winter solstices and with constellations important in Inca mythology.

Lost and found

Thanks to its remote location, Machu Picchu escaped the plundering of the Spanish invasion and remained largely forgotten until the American explorer Hiram Bingham came across it in his search for the legendary "lost" Inca capital in 1911. In an ongoing program of restoration, more than a third of the city has now been renovated or reconstructed.

△ **BURIAL GOODS**
Inca aristocracy were buried with valuable funerary goods, such as this golden figurine of a llama found at Machu Picchu.

▽ **MOUNTAINTOP CITADEL**
In both design and construction, the town and terraces of Machu Picchu integrate into the spectacular natural environment.

DRAINAGE ENGINEERING

Building Machu Picchu on steep slopes in an area of high rainfall demanded an astonishing level of engineering skill. The settlement had an advanced drainage system, with channels of varying profiles left in its retaining walls to allow water to escape during heavy storms and underground layers of stone chips that could act as temporary reservoirs for water. Without these features, the complex would have collapsed many years ago.

large stone slabs — gap at base of wall

OUTLET WITH DIFFERENT-SIZED STONES

foundation stones — channel cut into retaining rock wall

OUTLET WITH CHANNEL CUT IN ROCK

stone with edge cut at a slant — channel formed between foundation stones

TRAPEZOIDAL DRAIN OUTLET

◁ **THE TORRÉON** Located in the Temple of the Sun, the semicircular Torréon with its astronomically oriented trapezoidal windows was accessed through the "Serpent's Gate" in the straight wall.

The **citadel** of Machu Picchu is **7,970 ft (2,430 m)** above sea level

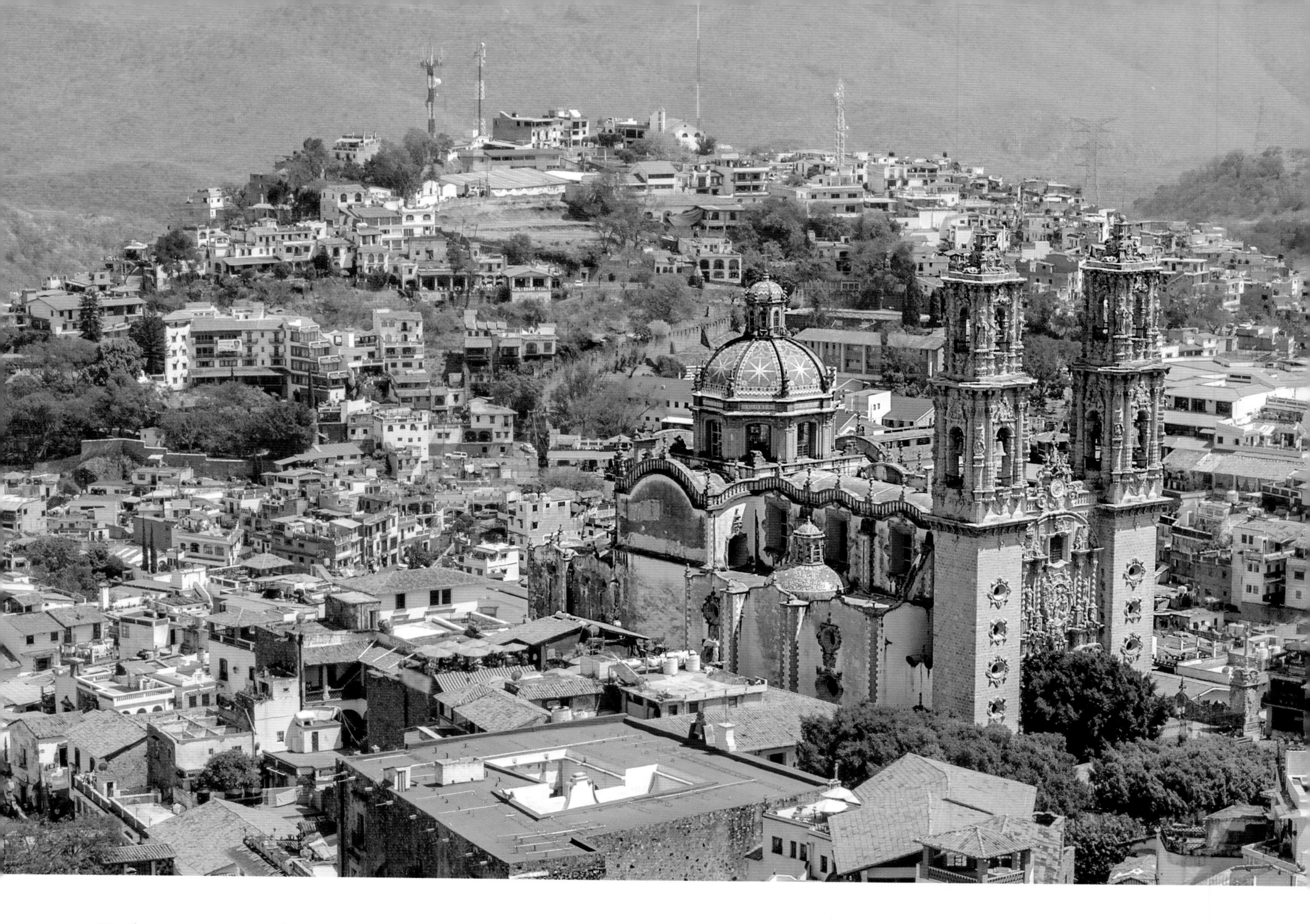

Church of Santa Prisca

A monumental twin-towered church exemplifying the colonial architecture and art of the New Spanish Baroque

The highly ornate Church of Santa Prisca in Taxco, Mexico, was built on the orders of local silver mine owner José de la Borda between 1751 and 1758 as an expression of his gratitude for his riches.

On either side of the main entrance at the west front are two lofty towers that can be seen from everywhere in the town. Their lower portions, of plain pink stone, support belfries with the elaborate sculpted decoration characteristic of the New Spanish Baroque style. The facade surrounding the entrance is also highly decorated and features a large central relief panel, as well as carvings of saints such as Santa Prisca and St. Sebastian. At the apex is a sculpture of the Assumption of Mary. At the other end of the church, there is a dome covered with brightly colored tiles.

Ornate interior

The interior of the church is no less magnificent, its decorations centering on themes of martyrdom. There are nine baroque altarpieces adorning the nave and transept, echoing the style of the main facade, and others in the chapels of the transept. On the walls are paintings by the Mexican artist Miguel Cabrera, commissioned for the church by de la Borda.

The Church of Santa Prisca was the **tallest building** in **Mexico** from **1758 until 1806**

△ **BAROQUE RELIEF**
The central panel of the facade depicting the baptism of Christ is surrounded by an oval frame featuring ornamental framework.

MARK OF NEWFOUND WEALTH
An imposing symbol of Spanish colonial prosperity, the ornate and colorful Church of Santa Prisca towers over the busy Mexican mining town of Taxco.

NARROW NAVE

Taxco is situated on a hilly landscape and had little level ground suitable for large-scale construction. To fit the site in a small ravine, the design of the Church of Santa Prisca was modified by considerably narrowing the nave, which exaggerates the height of the towers flanking the main entrance.

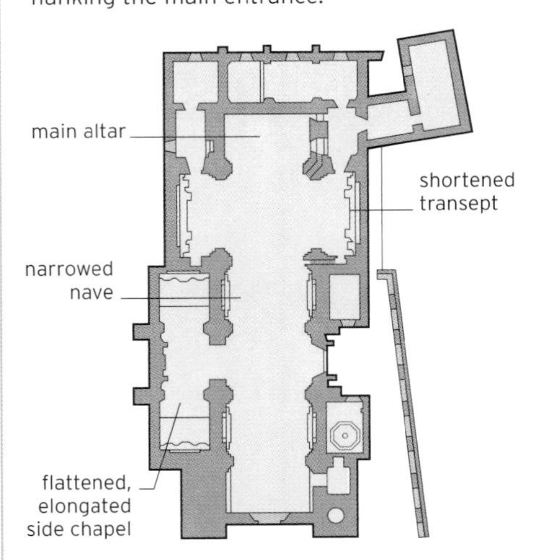

San Ignacio Miní

The red sandstone ruins of a Jesuit mission in the Argentine rainforest

Built in the mid-17th century, San Ignacio Miní was one of many Jesuit missions, known as *reducciones*, established by Spanish colonizers to introduce—and often coerce—the native South American people to a Christian way of life. It was set up in the land of the Guaraní people, now the borderland of Argentina, Brazil, and Paraguay. It was built in the contemporary Spanish Baroque style but incorporating indigenous elements, which became known as the Guaraní baroque.

Fall from grace

At the mission's peak, more than 2,000 people lived at the site, but the Jesuits eventually fell out of favor, and in the early 19th century, the mostly indigenous inhabitants attempted to destroy the buildings. Just the shell of the main mission building and the adjacent church survived, but San Ignacio Miní is nevertheless among the best preserved of the missions of the Guaraní region. The layout of the complex remains clear, and what is left of the walls and arches attests to the majesty of the architecture.

REDUCTIONS

In the parts of the Americas that came under Spanish rule in the 16th–18th centuries, many indigenous people were gathered into *reducciones* or "reductions," communities run by Jesuit missions. Following the plan of a typical Spanish village, there was often a central plaza bounded by a church, cemetery, monastery, and administrative buildings, as well as accommodation for the indigenous population.

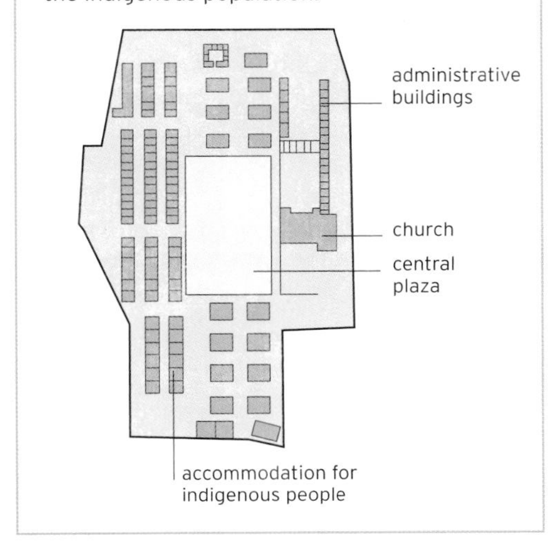

△ **REMNANT OF COLONIALISM**
The ruins at San Ignacio are magnificent even in their current state. The mix of Spanish Baroque and Guaraní architecture is still evident in the disconnected walls and arches of the church.

Citadelle Laferrière

An anticolonial fortress, now the symbol of the independent nation of Haiti

In 1791, African slaves on the West Indian island of Haiti rose in revolt against their French colonial masters. The French sent a substantial force of soldiers in 1802 to retake the island, but most died of fever. On January 1, 1804, Jean-Jacques Dessalines, leader of the revolution, proclaimed Haiti independent.

Built to impress

Fearing a renewed attack by the French, General Henri Christophe (1767–1820), in charge of northern Haiti, ordered the construction of a huge fortress. Work began in 1805 and continued until 1820. The foundation stones were laid directly into the mountain stone, held in place by a mortar of quicklime, molasses, local cows' and goats' blood, and cows' hooves cooked to a glue. Cisterns and storehouses were also built to supply enough water and food for 5,000 defenders for 1 year.

▽ **STRATEGICALLY LOCATED**
Citadelle Laferrière stands on top of a 3,000-ft- (900-m-) high mountain and covers more than 110,000 square ft (10,000 square m). The fortress's angular walls are 13 ft (4 m) thick and tower 130 ft (40 m) above the ground.

△ **STILL STACKED**
The Haitians amassed an arsenal that included about 160 cannons to defend the fortress. Piles of cannonballs still lie in stacks at the base of its walls.

About **20,000 workers** toiled for **15 years** to build this massive fortress

△ **LARGEST LOCKS**
At 110 ft (33.5 m) wide and 1,050 ft (320 m) long, the Gatún Locks at the northern Caribbean end of the canal are the largest in the canal. The locks lift ships up 87 ft (27 m) above sea level into the Gatún Lake.

Panama Canal

An engineering marvel, a shortcut between the oceans completed at an enormous human cost

△ **LATER ADDITIONS**
The 51-mile- (82-km-) long Panama Canal was expanded in 2016 and new locks were added. These can handle ships three times the size of the older locks.

The narrow isthmus of Panama that separates the Atlantic Ocean from the Pacific was long an impediment to shipping. Dangerous voyages south, around Cape Horn, deterred all but the most hardy travelers and traders. Then, French engineers, emboldened by their successful construction of the Suez Canal connecting the Mediterranean to the Red Sea, began to build a similar waterway across the Panama isthmus in 1880. However, work stopped when engineering problems and high worker mortality contributed to the bankruptcy of the construction company in 1889.

Gargantuan labors

Undaunted, the US—the country with the most to gain from such a canal—took over the project in 1904. Over the next decade, workers carved out more than 4.5 billion cubic ft (130 million cubic m) of rock and earth, creating what was then the largest man-made lake in the world—Gatún Lake—to feed the locks at either side. The new canal was an instant success, with around 350 vessels using it in 1914, its first year of operation.

THE ROUTE OF THE CANAL

Ships entering the Panama Canal from the Caribbean Sea in the north approach across Limon Bay. They then rise 85 ft (26 m) up the Gatún Locks into the inland Gatún Lake, passing through the Gaillard Cut and then descending through more locks into the Pacific Ocean next to Panama City. New sets of locks were opened at either end in 2016 to cater to larger ships.

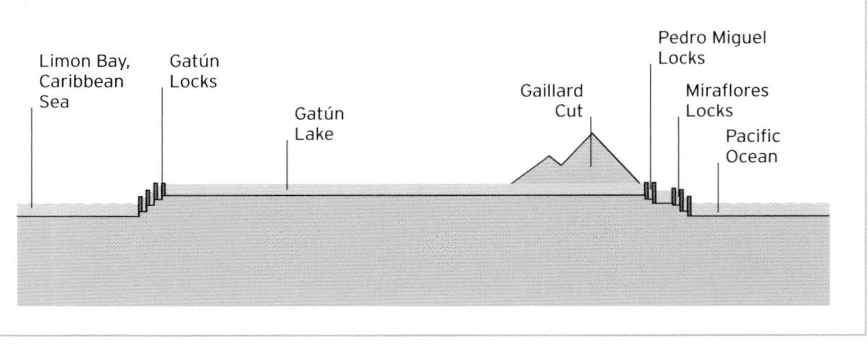

Christ the Redeemer

Rio de Janeiro's most famous landmark and the largest Art Deco style statue in the world

A colossal statue of Christ the Redeemer dominates Rio de Janeiro from the top of Mount Corcovado, its face directed at the rising Sun so that—in the words of its designer—"the statue of the divine will be the first image to emerge." The idea of building a monument on the mountain was first raised in the 1850s, but it was only in 1921 that the Catholic archdiocese proposed a statue of Christ for the summit. The foundation stone of the base was laid on April 4, 1922—the centennial of Brazil's independence from Portugal—even before the design of the statue had been agreed upon. Later in 1922, the Brazilian engineer Heitor da Silva (1873–1947) was entrusted with the design, although Brazilian artist Carlos Oswald and French sculptor Paul Landowski also influenced the final design.

An embrace for the city

The church raised most of the $250,000 required to fund the construction of Christ the Redeemer, and workers and material began making their way up the mountain by train in 1926. Built from reinforced concrete and decorated with a mosaic of triangular soapstone tiles, the statue was completed in 1931. It stands 98 ft (30 m) tall on a square base about 26 ft (8 m) high, which is covered with black granite, giving the Art Deco figure an austere footing; the arms of Christ span 92 ft (28 m) from north to south.

▷ **UNDER CHRIST'S GAZE**
The giant sculpture of Christ looks out over Rio and the Sugar Loaf Mountain, a constant reminder to the citizens of their religious duties.

Christ the Redeemer is the **fifth largest statue** of Jesus in the **world**

INSIDE THE STATUE

Within the statue are a series of reinforced floors linked by narrow staircases. The tenth floor extends into the arms of the statue, and a narrow passage reaches to the fingers. A hatch in the shoulder section gives access to the outside. A carved heart decorated with mosaic sits inside the chest of the statue. The hollow base of the statue contains the Chapel of Our Lady of Aparecida, consecrated on the statue's 75th anniversary in 2006.

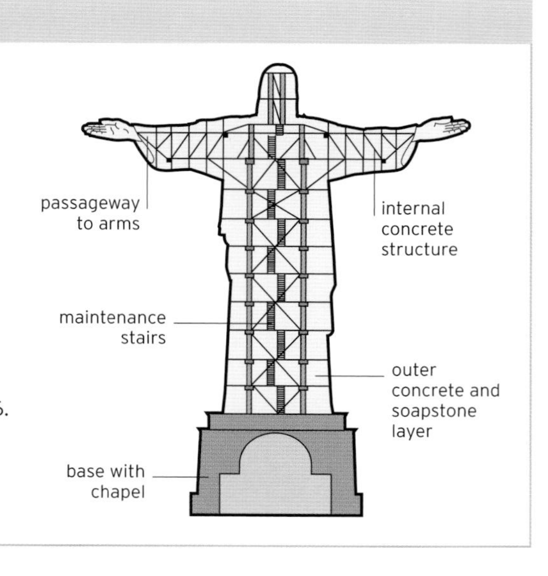

△ **CONCRETE CONSTRUCTION**
The 12-ft- (3.75-m-) high head of Christ is cast in concrete and decorated with soapstone mosaic.

Cathedral of Brasilia

A striking, crownlike Modernist cathedral with a stained-glass ceiling designed for a futuristic city

When the decision was made in 1957 to build Brasilia as a new capital city for Brazil, one of the first buildings to be started was its new cathedral. The building was designed by Brazil's most famed Modernist architect, Oscar Niemeyer (1907–2012). Its simple but stunning hyperboloid structure consists of 16 concave concrete columns, each weighing 99 tons (90 tonnes). The cornerstone was laid on September 15, 1958, and the new Metropolitan Cathedral of Our Lady of Aparecida was consecrated on October 21, 1968.

A temple of light

The approach to the cathedral is lined with 10-ft- (3-m-) tall bronze sculptures of the four evangelists: Matthew, Mark, Luke, and John. To the right is a 66-ft- (20-m-) high bell tower. Visitors to the cathedral emerge through a dark tunnel into a bright, circular hall lit by multicolored stained glass. Three sculpted angels hang from the ceiling.

ABOVE GROUND

Most of the cathedral's floor area is below ground. Only the main roof, 230 ft (70 m) wide and 138 ft (42 m) high; the ovoid roof of the baptistery; and the bell tower are visible above ground. For some observers, the roof represents two hands reaching up to heaven; others see a rendition of the crown of thorns placed on the head of Jesus at his crucifixion.

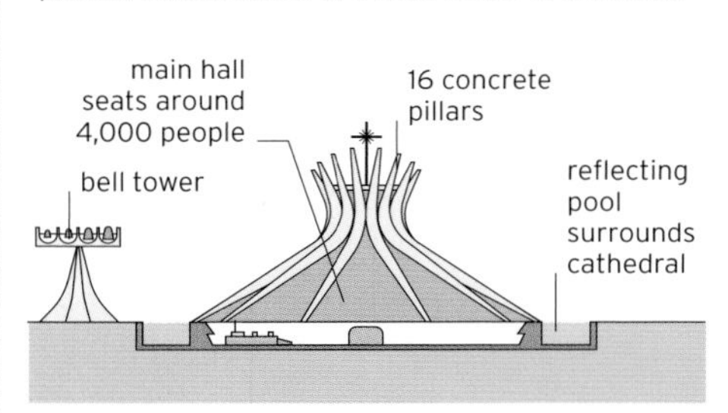

Central University City Campus of the UNAM

An ensemble of buildings that reflect the energy and aspirations of a rapidly modernizing Mexico

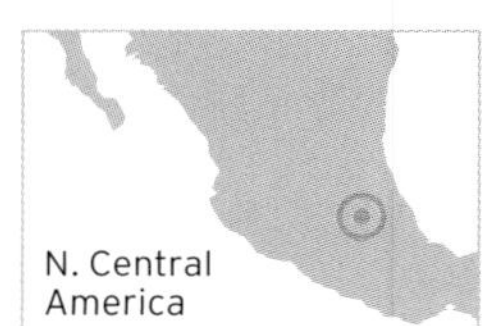

Students of the National Autonomous University of Mexico (UNAM) used to attend classes scattered around Mexico City. From 1949 to 1952, a new university was built on one campus, which became the biggest single construction project in Mexico since the Aztec empire (1325–1521). Many of the Modernist buildings incorporate the gardens and volcanic rocks on which the campus stands, but it is the Central Library that stands out for its external decoration.

Giant mosaic

The library is covered with a vast mural designed by the Mexican artist Juan O'Gorman (1905–1982) that extends over 43,000 square ft (4,000 square m). The mural contains millions of mosaics made from colored stones gathered from across Mexico. Its north wall reflects Mexico's pre-Hispanic past, the south wall its colonial history, the east wall the contemporary world, and the west wall the university and modern Mexico. In recognition of this Modernist masterpiece, UNESCO declared the entire campus a World Heritage Site in 2007.

▽ **MOSAIC MURAL**
A mural called *Historical Representation of Culture* covers the walls of the library building. It is one of the largest continuous mosaics in the world.

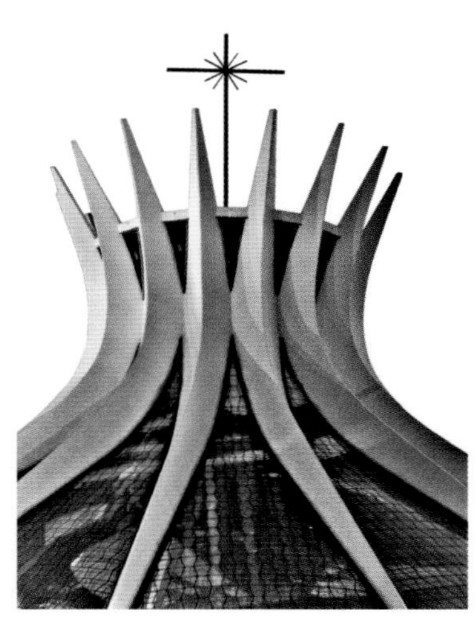

△ CROWNING GLORY
A simple metal cross stands on top of the main tower of the cathedral. The cross was blessed by Pope Paul VI, who also donated the main altar and altarpiece in 1967.

◁ ANGELS IN FLIGHT
The three angels in flight over the nave, supported by steel cables, weigh 220–660 lb (100–300 kg) each and were designed by Alfredo Ceschiatti.

Salt Cathedral of Zipaquirá

A cathedral of salt 660 ft (200 m) below ground that inspires and awes

When the Prussian naturalist Alexander von Humboldt (1769–1859) visited Colombia in 1801, he believed that the deposits of halite (rock salt) mined at Zipaquirá were larger than any of the mines found in Europe.

The mines had been exploited for centuries, during which time miners had carved sanctuaries where they prayed for protection in their dangerous labor. In the 1930s, they had excavated an underground church, and in 1950 began to turn this church into a salt cathedral, dedicated to Our Lady of Rosary, the patron saint of miners.

Structural and safety problems led to the cathedral's closure in 1992, but in response, the miners built a new cathedral 197 ft (61 m) below the old one. Inaugurated on December 16, 1995, the cathedral has three naves and 14 small chapels for each Station of the Cross, their icons and architectural details carved out of the halite rock.

△ CENTRAL FIGURE
At the center of the main vault is a marble sculpture by Colombian artist Carlos Enrique Rodriguez based on Michelangelo's *The Creation of Adam*, a fresco in the Sistine Chapel ceiling.

Up to **3,000 visitors** come to **worship** at the Salt Cathedral **every Sunday**

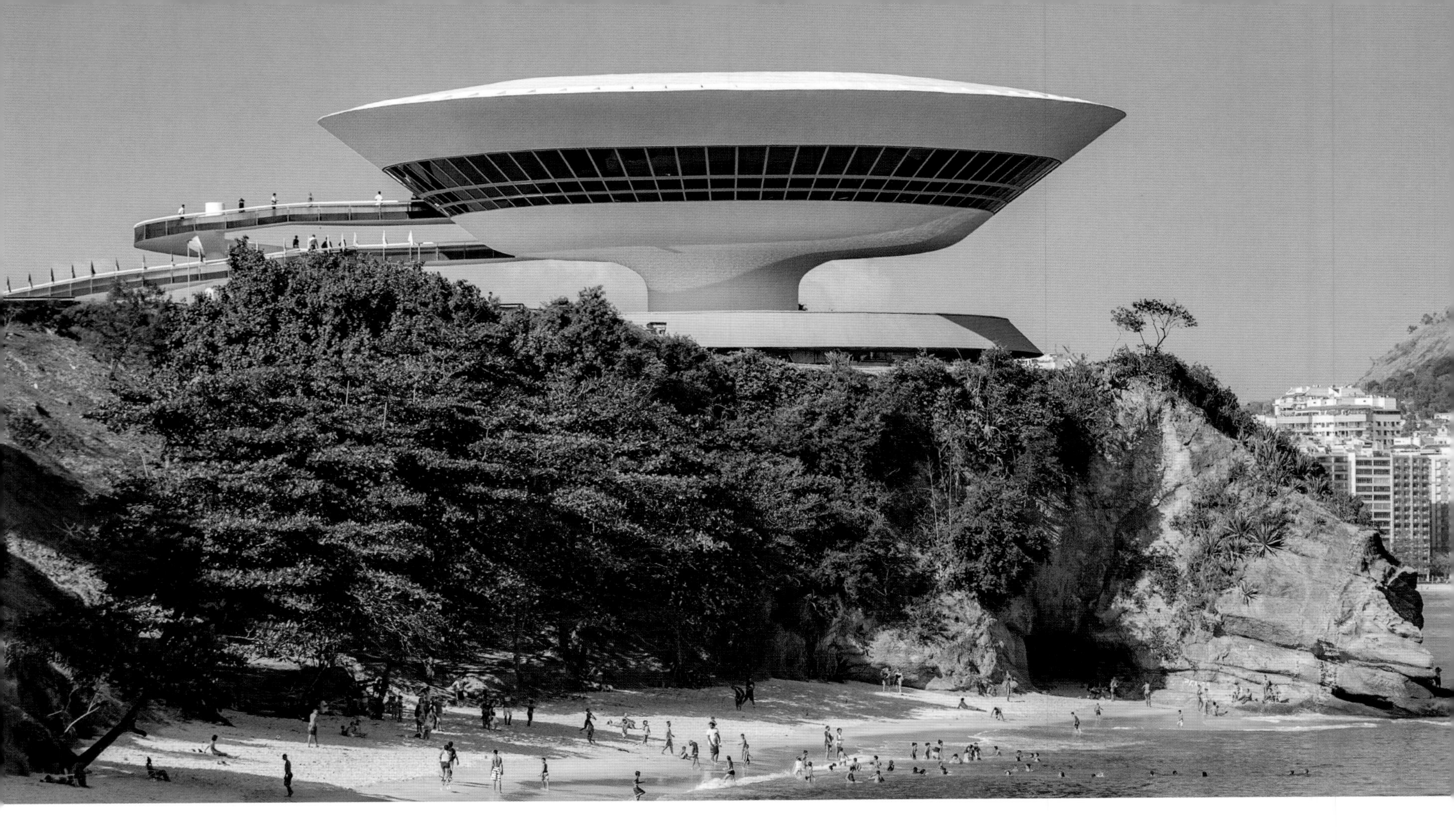

Niterói Contemporary Art Museum

Dramatically perched above a cliff, a clean-lined gallery designed by the master of Modernism

E. South America

△ **READY FOR TAKE-OFF**
The structure of the building has often been likened to a UFO poised for take-off at the edge of the waters of Guanabara Bay. The museum's location on a rocky promontory offers breathtaking views of the city of Rio de Janeiro.

Brazilian architect Oscar Niemeyer (1907–2012) is famed for his striking Modernist architecture. The Contemporary Art Museum in Niterói, outside Rio de Janeiro, is one of his simplest buildings. Completed in 1996, the museum takes the shape of a saucer, 52 ft (16 m) high with a cupola 164 ft (50 m) wide. Inside are three floors; the main, column-free Hall of Expositions has a capacity of only 60 people, making it an exclusive venue. Access to the museum from the surrounding plaza is by a 321-ft- (98-m-) long concrete ramp. The cylindrical base is surrounded by a reflecting pool covering 8,790 square ft (817 square m).

The main impact of the museum is in its location. Set on a cliff overlooking Boa Viagem beach, the museum lies in Niterói on the eastern side of Guanabara Bay, facing Rio de Janeiro on the opposite shore.

△ **MAKING AN ENTRANCE**
A boldly colored suspended walkway, recalling a red carpet, appears to be attached to the side of the saucer. It takes visitors to the Hall of Expositions.

"Niterói" comes from the local Tupi language and means **"water that hides"**

SIMPLICITY OF DESIGN

A cross-section through the museum reveals the simplicity of its design. A walkway leads up to the main building, which contains offices, the main gallery, two smaller side galleries, and a viewing gallery around the edge. Offices and other rooms are buried underground.

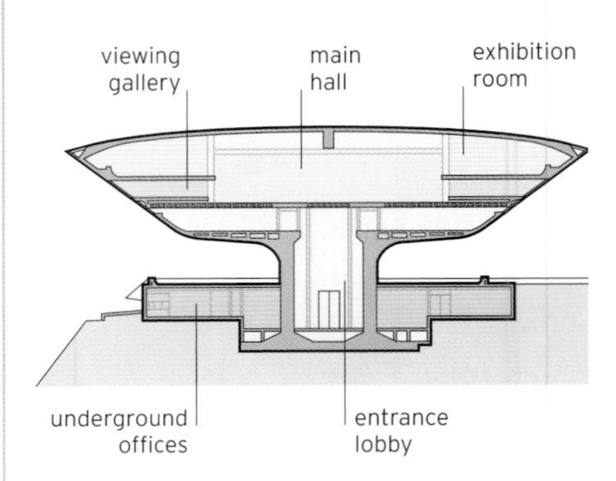

Itaipu Dam

A series of vast structures that together generate 15 percent of the electricity used in Brazil

Spanning the Paraná River between Brazil and Paraguay, the Itaipu Dam is actually a series of four different dams: there are an earthfill dam, a rockfill dam, the concrete main dam, and a concrete wing dam. With a maximum height of 643 ft (196 m), the dams together stretch for 25,981 ft (7,919 m). Behind them is a reservoir with a surface area of 520 square miles (1,350 square km) containing 7 cubic miles (29 cubic km) of water. In 2016, the dam set a world record for producing electricity of 103,098,366 megawatt hours (MWh), surpassing the Three Gorges Dam in China. Half of that electricity goes west to Paraguay, half east to Brazil, although most of the output to Paraguay is subsequently exported back to Brazil.

The sounding stone dam

The dam gets its name from an island in a river nearby. In the local Guaraní language, *itaipu* means "the sounding stone." The idea for this vast project began in the 1960s, when the neighboring countries agreed to exploit the Paraná River for hydroelectricity. Construction began in January 1971, the river was rerouted in 1978 to allow the dam to be built, and the dam opened for electricity generation on May 5, 1984.

△ **GARGANTUAN PROPORTIONS**
Around 434 million cubic ft (12.3 million cubic m) of concrete was used to build the four dams—an amount sufficient to build more than 200 football stadiums. The quantity of iron and steel used in the construction could have built 380 Eiffel Towers.

▽ **GRAND ILLUMINATION**
The illumination of the Itaipu Dam at night highlights the structural and engineering complexity of the massive dam wall.

INSIDE THE DAM

The main dam consists of large concrete segments joined together to enclose a hollow chamber. Across the four dams, water flows through 20 sets of turbines to generate electricity, while excess water from the reservoir not needed for power generation is discharged through 14 separate spillways at a rate of 2,196,572 cubic ft (62,200 cubic m) per second.

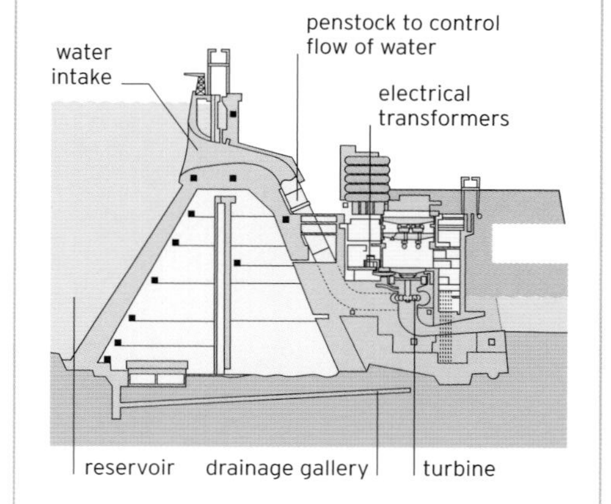

Historic line

Paris's Arc de Triomphe sits at the meeting point of 12 radiating avenues. It also lies on the Axis Historique, an imaginary line linking the courtyard of the Louvre to the Grande Arche de la Défense.

Europe

MONOLITHS TO MODERNISM
Europe

Pagan worship of natural forces inspired the first structures across Europe. These were made using ingenious techniques and knowledge of astronomy. Later, in Greece and Rome, worship of gods in human form led to new architectural wonders governed by Classical principles. A rational, mathematical approach also enabled later European architects to innovate, especially in the Romanesque and Gothic periods and later in the modern era. In the 6th century CE, Byzantines introduced Eastern elements to architecture and shifted from stone to brick—later used to great effect in the Renaissance. Materials drove creativity in later centuries: from the Baroque mastery of stucco to the Modernist and Contemporary obsession with concrete and glass.

NEOLITHIC MONUMENT BUILDERS
4000–1700 BCE

Powerful pagan beliefs underpinned many of the Neolithic buildings across Europe. Often using limited tools and without wheeled transportation, early builders shifted enormous boulders, cleaved quartzite and granite, and excavated huge volumes of earth to make tombs and religious monuments.

1 NEWGRANGE

more than 5,000 years old, Newgrange combined a long tradition of tomb building with a new idea of aligning monuments with astronomical landmarks

THE ANCIENT GREEKS
800–146 BCE

Adhering to a set of strict aesthetic values, ancient Greek builders set precedents that would be revived numerous times over the following millennia. They developed the principles of symmetry and scale and established the column, capital, and pediment as enduring architectural devices.

5 PARTHENON

THE ROMANS
509 BCE–476 CE

Although inspired by the ancient Greeks, the Romans created their own architectural models, which spread throughout their empire, influencing European cities long after their retreat. Most notable was their use of the arch and the dome as basic architectural devices.

9 THE PANTHEON

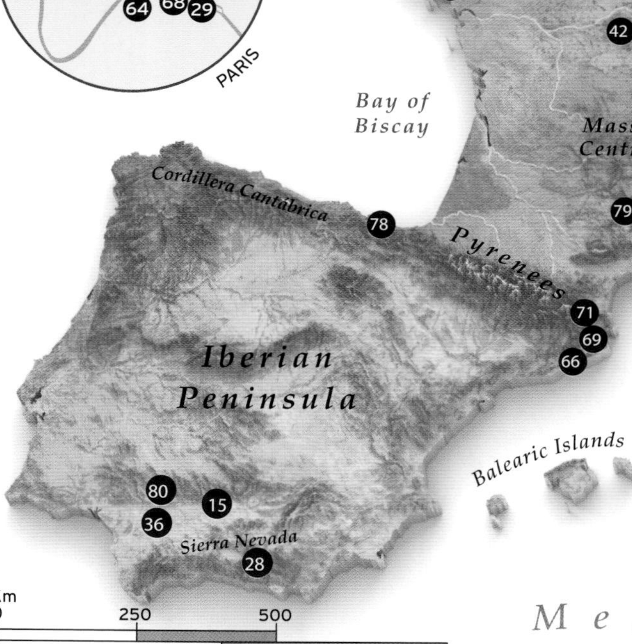

built in 537 CE, Hagia Sophia is regarded as one of the finest structures of the Byzantine era

MODERNISM
19TH–20TH CENTURIES

Industrial advances in the late 19th century made new construction materials and techniques possible, inspiring a different approach to building. Most of it was centered in rapidly expanding cities. Metal framing and concrete allowed for taller, more complex structures, and the idea of architecture as art took hold.

66 SAGRADA FAMÍLIA

THE RENAISSANCE
c.14TH–16TH CENTURIES

Architects of the Renaissance revived the rationalism of ancient Greece and Rome, revering symmetry and the laws of geometry. Among the era's innovators was Filippo Brunelleschi, who built without scaffolding, flying buttresses, or supporting arches and used herringbone brickwork to give strength to structures.

34 FLORENCE CATHEDRAL

ROMANESQUE ARCHITECTURE
c.11TH–12TH CENTURIES

As the Normans spread their political and stylistic influence, they also evolved a Romanesque building style that centered on the arch of ancient Rome. The Normans used arches on a grand scale and revolutionized architecture with their pointed, structural arch, along with a new system of ribbed vaults.

17 DURHAM CATHEDRAL

ANCIENT AND MODERN ENGINEERS

The traditions of monumental public structures and planned cities were established by the ancient Greeks and refined by the Romans as they colonized much of the continent. Later, the Industrial Revolution of Northern Europe transformed urban and rural environments with engineering feats, a process that continued in the 20th century with Modernism.

KEY SITES

1. Newgrange
2. Carnac Stones
3. Megalithic Temples of Malta
4. Stonehenge
5. Parthenon
6. Delphi
7. Pont Du Gard
8. The Colosseum
9. The Pantheon
10. Column of Marcus Aurelius
11. Diocletian's Palace
12. Basilica of San Vitale
13. Hagia Sophia
14. Mont Saint-Michel
15. Mosque-Cathedral of Córdoba
16. Aachen Cathedral
17. Durham Cathedral
18. Prague Castle
19. Westminster Abbey
20. Tower of London
21. Leaning Tower of Pisa and Duomo
22. St. Mark's Basilica
23. Basilica of Saint-Denis
24. Chartres Cathedral
25. Heddal Stave Church
26. Uppsala Cathedral
27. Palais des Papes
28. The Alhambra
29. Notre-Dame Cathedral
30. Ferapontov Monastery
31. Český Krumlov Castle
32. Bran Castle
33. Mystras
34. Florence Cathedral
35. Ponte Vecchio
36. Seville Cathedral
37. The Doge's Palace
38. Villa Capra
39. Palazzo Ducale
40. The Kremlin
41. St. Basil's Cathedral
42. Château de Chenonceau
43. Château de Chambord
44. Frederiksborg Castle
45. Wooden Churches of Kizhi
46. St. Peter's Basilica
47. Sistine Chapel
48. Blue Mosque
49. St. Paul's Cathedral
50. Palace of Versailles
51. Sanssouci
52. Blenheim Palace
53. Buckingham Palace
54. Würzburg Residence
55. Schönbrunn Palace
56. Trevi Fountain
57. Brandenburg Gate
58. The Louvre
59. Arc de Triomphe
60. Winter Palace
61. Cologne Cathedral
62. Hungarian Parliament Building
63. Palace of Westminster
64. Eiffel Tower
65. Tower Bridge
66. Sagrada Família
67. Neuschwanstein Castle
68. Musée d'Orsay
69. Casa Milà
70. Sacré-Cœur
71. Parc Güell
72. The Atomium
73. Stoclet Palace
74. Einstein Tower
75. Centre Georges Pompidou
76. Church of Hallgrimur
77. Grande Arche de la Défense
78. Guggenheim Museum Bilbao
79. Millau Viaduct
80. Metropol Parasol

Newgrange

A circular tomb that lights up on the morning of the winter solstice

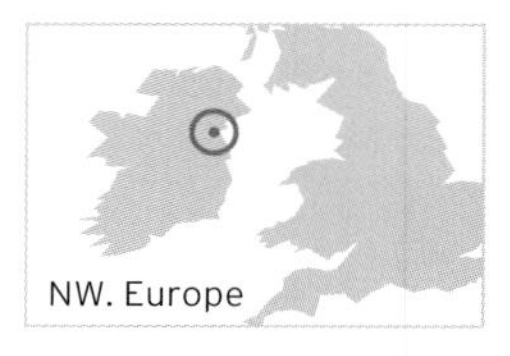

Older than the pyramids of Egypt or Stonehenge (see p.95) in England, the grand passage tomb of Newgrange in the east of Ireland was probably built in around 3200 BCE. The tomb consists of a large circular mound measuring 262 ft (80 m) in diameter, built up with alternating layers of earth and stones to a height of 36 ft (11 m). The mound has a retaining wall at the front, mostly of white quartz cobblestones, and is ringed by elaborately carved kerbstones and an outer stone circle. The 62-ft- (19-m-) long chambered passage, entered from the southeast, runs about one-third of the way into the center. At the end of the passage is a large inner chamber with a high, corbeled vault roof off which are three smaller chambers.

Quite what purpose this tomb served is still unclear; it most likely had a religious purpose, perhaps linked to the worship of the Sun. Irish mythology suggested it was home to the gods, and its power still impresses today.

SOLSTICE ALIGNMENT

At sunrise on the winter solstice, sunlight shines directly down through the passage onto the wall carvings in the inner chamber, notably the *triskelion*, or triple spiral, on the front wall. The illumination lasts for about 17 minutes. Today, the event attracts many visitors; lucky witnesses of the illumination are chosen by lottery.

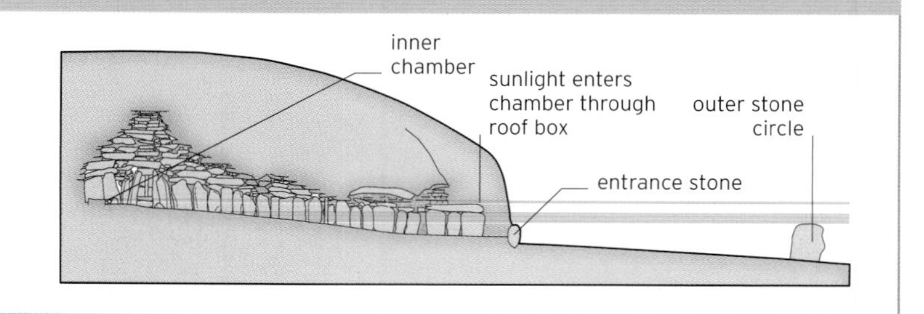

△ **MEGALITHIC TOMB**
Most of the 547 slabs of stone in the inner passage and chambers and those that form the outer kerbstones are graywacke, a type of sandstone defined by its hardness and dark color.

△ NEOLITHIC COMPLEX
Newgrange sits at the heart of the Neolithic Brú na Bóinne complex, which includes two similar passage tombs at nearby Dowth and Knowth and a further 90 monuments.

△ ENGRAVED KERBSTONE
The kerbstones that surround Newgrange are carved with a variety of curved and straight-line carvings. Their meaning is unknown, but they are thought to have some symbolic purpose.

Carnac Stones

Mysterious lines of prehistoric stones that continue to confound archaeologists

Striding across the open country around the Breton village of Carnac in western France are more than 4,000 standing stones—the largest assemblage of such stones in the world—erected by Neolithic people about 7,000 years ago, although some uncertainty surrounds their age.

The stones stretch in lines running northeast from Carnac and are arranged in three main alignments and one small alignment. The three main alignments of stones stand in converging rows or fanlike arrangements alongside prehistoric tombs (dolmens) and tumuli (mounds).

Mysterious lines

Quite why the stones were placed here is still a mystery. Some archaeologists have suggested alignments with sunsets at the solstices, or that the stones might have formed a huge observatory; others have proposed funerary uses or speculated that the stones mark a threshold between two different worlds. Local tradition says that the stones stand in straight lines because they were once a Roman legion turned to stone by the wizard Merlin.

△ RANKED BY HEIGHT
The standing stones vary in height from 2 ft (60 cm) to 13 ft (4 m), depending on their position along each alignment.

▽ ***MENÉC*** **ALIGNMENT**
The *Ménec* alignment consists of 12 converging rows of menhirs that stretch for 3,822 ft (1,165 m) and are roughly 328 ft (100 m) in width.

The **longest alignment**—the central *Kermario*—consists of **1,029 stones**

Megalithic Temples of Malta

More than 50 prehistoric temples believed to be the oldest free-standing structures in the world

S. Europe

From 6000–5000 BCE, the islands of Malta were settled by people of increasingly sophisticated culture who embarked on a thousand-year period of temple-building.

The Giants' Tower

The earliest two temples, known as *Ġgantija* (Giants' Tower), were built on the island of Gozo in about 3600 BCE. These temples were relatively simple in design, with a trilithon (a lintel supported by two uprights) entrance leading into semicircular chambers, or "apses," lined with large, upright stone slabs. Each apse contained a limestone altar, which finds of animal bones nearby suggest may have been used for sacrifices.

The megalithic temple-building period reached its peak between 3000 BCE and 2500 BCE, when temple complexes on the main island of Malta—such as those at Ħaġar Qim, Mnajdra, Skorba, and Tarxien—adopted the style of the *Ġgantija* temples, but often with a more complex ground plan of several apses.

While in the *Ġgantija* temples the two apses were approached through a small entrance chamber, in the larger temples there was a paved passageway (or in some cases a pair of passages) leading from the main entrance to the individual chambers. From the outside, the doorway into the temple from a leveled forecourt was marked by a stone archway, and the facade was made of a harder limestone than the interior. Because of this, decoration in the form of animal and plant designs sculptured in relief was generally restricted to the insides of the temples.

◁ TEMPLE ART
The temples contain numerous carved artifacts, such as this sculpted figurine, showing a fine degree of craftsmanship.

▽ TEMPLE INTERIOR
The roofs of the interconnected chambers, or apses, within the temples were supported by carved pillars and entrance doorways in the form of trilithons.

△ SACRED SITE
In this aerial image, the remains of the circle of pillars and lintels can be clearly seen around the internal horseshoe-shaped arrangement of arches and central Altar Stone.

STANDING STONES
The ring of immense pillars and lintels known as Stonehenge stands at the center of a prehistoric earthwork.

Stonehenge

A unique circle of massive standing stones with mysterious symbolic significance, and perhaps the world's most famous prehistoric monument

Situated on the chalk plateau of Salisbury Plain in the south of England, Stonehenge stands in a region rich in sites of archaeological interest. The monuments in the area had their beginnings thousands of years ago. In around 3100 BCE, the first earthworks were begun in the form of a "henge"—an area enclosed within a circular ditch and bank. In around 3000 BCE, a wooden structure was erected within the henge and used for ceremonial cremation. The stone monument began to take shape in about 2500 BCE. Eighty bluestone pillars were erected, and a huge sandstone "Altar Stone" was laid at the center of the henge. Later, the massive sarsen (dense sandstone) blocks replaced or were added to the pillars to form the present structure—a circle of 30 upright stones and lintels around a central horseshoe-shaped group of trilithons aligned with the Sun at the solstices.

Ceremonies for the dead

Scholars remain uncertain about the function of the henge. It is likely to have had a calendrical function and been used as a ceremonial and burial site, as evident from the discovery of many cremated bones at the site.

Each of the **sarsen blocks** weighed up to **33 tons (30 tonnes)**

SOLAR ALIGNMENT

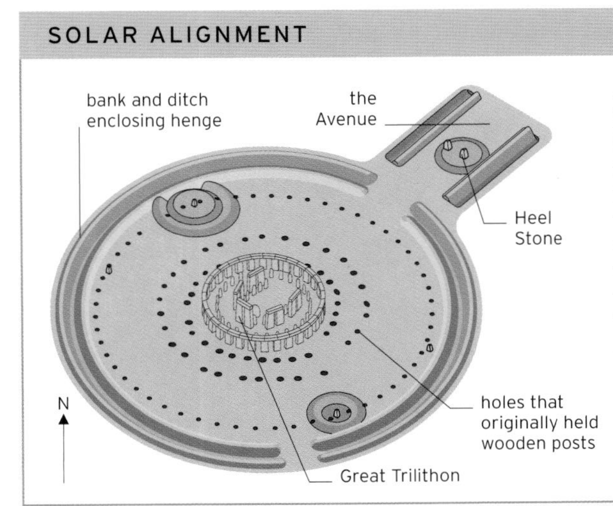

A processional approach to Stonehenge, known as the Avenue, leads from the River Avon 1.5 miles (2.5 km) away. The final stretch is aligned with the axis of the stone circle, so that when entering from the Avenue at sunrise on the summer solstice and sunset on the winter solstice, the Sun can be seen though the central Great Trilithon. The so-called Heel Stone at the main entrance is assumed to have originally been one of a pair forming a sort of gateway through which processions would have passed.

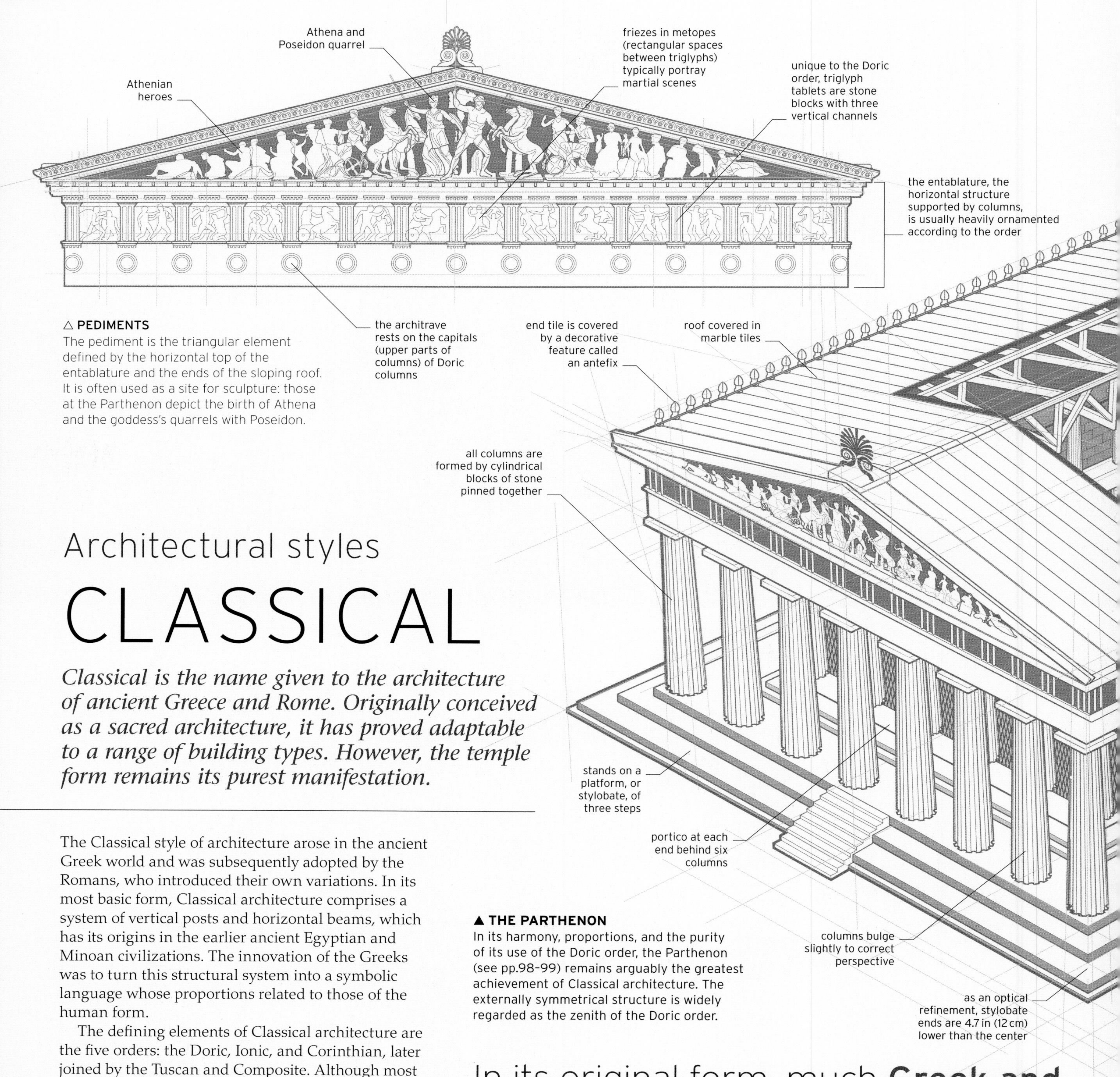

△ **PEDIMENTS**
The pediment is the triangular element defined by the horizontal top of the entablature and the ends of the sloping roof. It is often used as a site for sculpture: those at the Parthenon depict the birth of Athena and the goddess's quarrels with Poseidon.

▲ **THE PARTHENON**
In its harmony, proportions, and the purity of its use of the Doric order, the Parthenon (see pp.98–99) remains arguably the greatest achievement of Classical architecture. The externally symmetrical structure is widely regarded as the zenith of the Doric order.

Architectural styles

CLASSICAL

Classical is the name given to the architecture of ancient Greece and Rome. Originally conceived as a sacred architecture, it has proved adaptable to a range of building types. However, the temple form remains its purest manifestation.

The Classical style of architecture arose in the ancient Greek world and was subsequently adopted by the Romans, who introduced their own variations. In its most basic form, Classical architecture comprises a system of vertical posts and horizontal beams, which has its origins in the earlier ancient Egyptian and Minoan civilizations. The innovation of the Greeks was to turn this structural system into a symbolic language whose proportions related to those of the human form.

The defining elements of Classical architecture are the five orders: the Doric, Ionic, and Corinthian, later joined by the Tuscan and Composite. Although most easily identified by the particular type of column employed, the Classical order comprises the base, column, capital, and entablature, with each order having a predefined set of proportions and decorative schemes. It is this engaging combination of the structural and symbolic that lies at the heart of Classical architecture's enduring appeal.

In its original form, much **Greek and Roman** classical architecture **was not white**, as it appears now, but was painted in **bright colors**

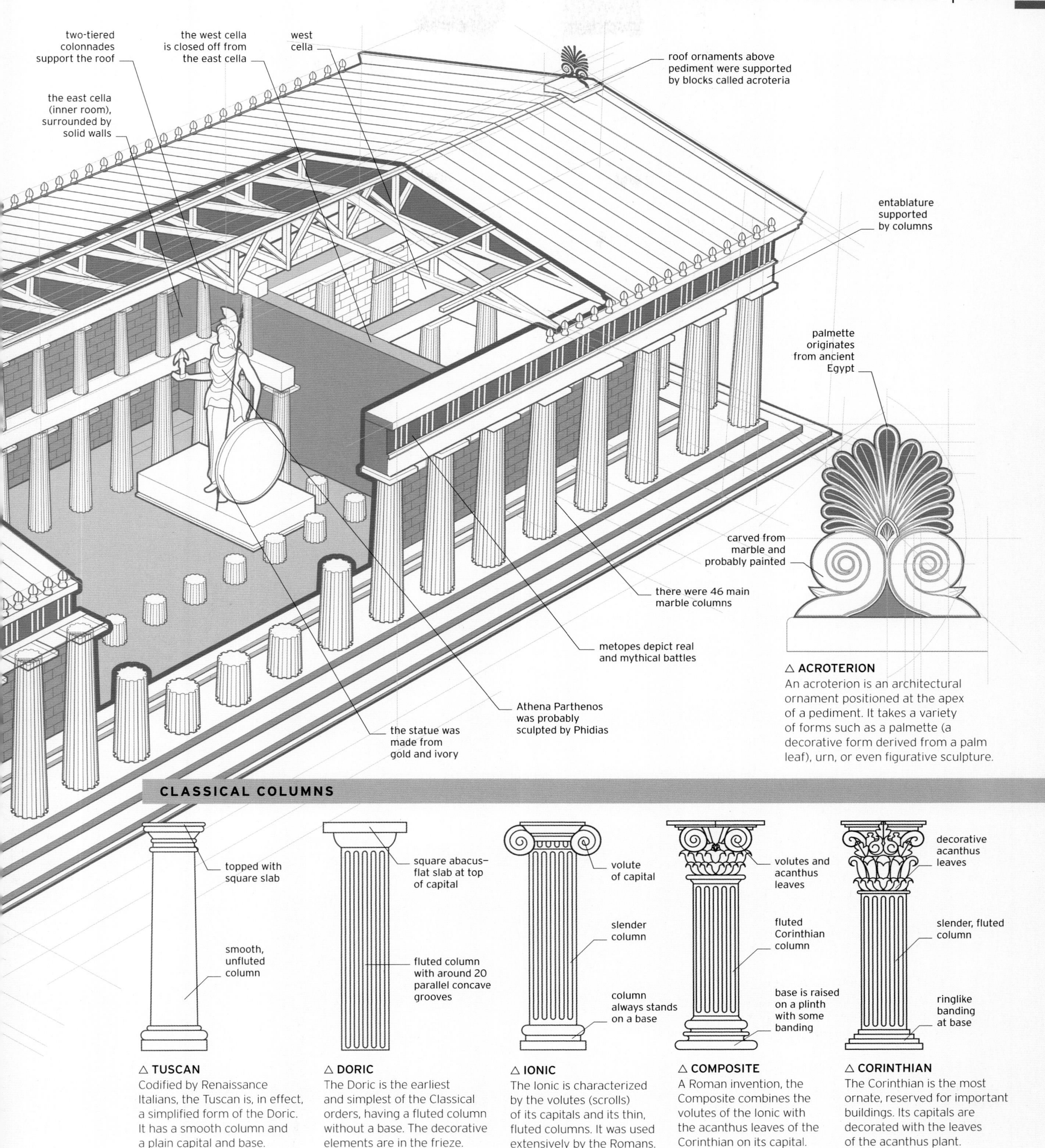

△ **ACROTERION**
An acroterion is an architectural ornament positioned at the apex of a pediment. It takes a variety of forms such as a palmette (a decorative form derived from a palm leaf), urn, or even figurative sculpture.

CLASSICAL COLUMNS

△ **TUSCAN**
Codified by Renaissance Italians, the Tuscan is, in effect, a simplified form of the Doric. It has a smooth column and a plain capital and base.

△ **DORIC**
The Doric is the earliest and simplest of the Classical orders, having a fluted column without a base. The decorative elements are in the frieze.

△ **IONIC**
The Ionic is characterized by the volutes (scrolls) of its capitals and its thin, fluted columns. It was used extensively by the Romans.

△ **COMPOSITE**
A Roman invention, the Composite combines the volutes of the Ionic with the acanthus leaves of the Corinthian on its capital.

△ **CORINTHIAN**
The Corinthian is the most ornate, reserved for important buildings. Its capitals are decorated with the leaves of the acanthus plant.

Parthenon

The temple of Athena, epitome of Classical Greek architecture, overlooking Athens from the Acropolis

The Parthenon is regarded as symbolizing the pinnacle of Classical Greek civilization, an exemplary expression of architecture of the Doric order given pride of place in the Acropolis, the citadel above the city of Athens. It was built to replace an older temple of Athena, which had been destroyed in the Persian invasion of 480 BCE. Along with other smaller temples and a grand entrance to the Acropolis, it was intended as a celebration of the Athenian victory over the invaders. The Parthenon, constructed from fine white marble from Mount Pentelicus, was the largest temple to have been built at that time, measuring 228 ft (70 m) long by 101 ft (31 m) wide.

The House of the Virgin

At the heart of the temple was an inner chamber that housed a huge statue of Athena Parthenos (Athena the Virgin) from which it derived its name, "the house of the virgin." Overseen by the sculptor Phidias, the architects Iktinos and Kallikratis developed a design incorporating elements of the new Ionic style into a basically Doric structure. This structure was then decorated with sculptures in the pediments and with friezes of relief panels above both outer and inner rows of columns.

Used as a Christian church from the 6th century and converted into a mosque after the Ottoman conquest of Athens in 1458, the Parthenon survived intact until 1687, when it was damaged during a Venetian bombardment of the city. In the early 19th century, many of the surviving sculptures were removed by Thomas Bruce, 7th Earl of Elgin, and are still, controversially, on display in the British Museum in London.

▷ **MARBLE FRIEZE**
Above the rows of columns were friezes of high-relief sculptures depicting gods and characters from Greek mythology.

The **lost statue** of **Athena Parthenos** by Phidias was made of **gold and ivory**

△ **THE TEMPLE OF ATHENA PARTHENOS**
The Acropolis is perched on a rocky hill 490 ft (150 m) above sea level, with around 20 buildings remaining today. The largest of these, the Parthenon, equaled eight tennis courts in area.

◁ **UNUSUAL COLUMNS**
Sculpted female figures known as caryatids act as columns in the Erechtheion Temple to the north of the Parthenon.

OPTICAL ILLUSION

Due to a quirk of perspective, if built to a strictly rectilinear plan, the Parthenon would have appeared oddly out of alignment. To overcome this, the architects designed the building on a slightly domed base with gently curving lines and inward-leaning columns, which bulge slightly at their middles, a technique known as entasis. The corner columns are also a little wider than and closer to their neighbors than the inner columns. To avoid appearing to slope away from the viewer at ground level, the friezes are imperceptibly tilted forward.

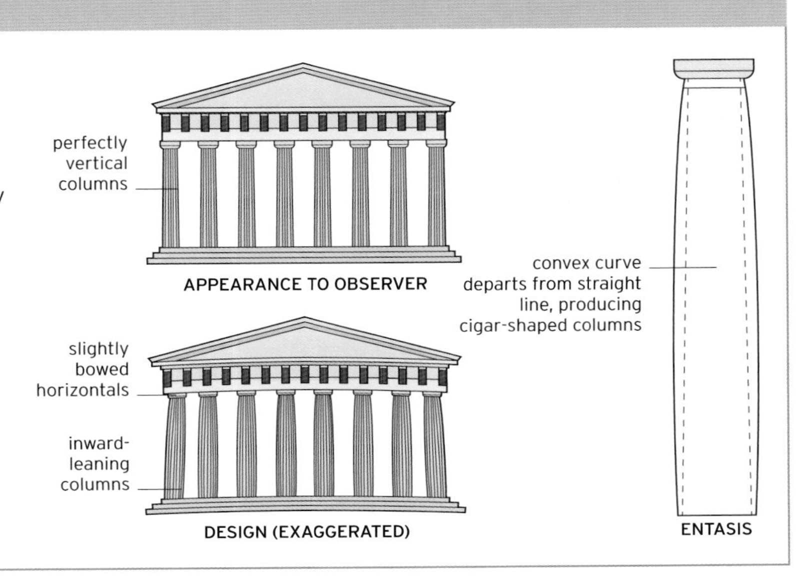

Delphi

A sanctuary dedicated to the god Apollo and considered by the Greeks to be the center of the world

Set among the magnificent peaks of Mount Parnassus, Delphi was a site of mystical importance in the ancient world. The religious sanctuary established there in the 7th century BCE was the seat of Pythia—the high priestess of the Temple of Apollo, better known as the Oracle of Delphi. Sought out for her prophecies and judgments, which drew on the spirit of Apollo, the Oracle was one of the most famous women in the ancient world.

Temple rebuilt

The Temple of Apollo was rebuilt once following a fire and again after an earthquake, and the remains there today—the temple's foundations and several Doric columns—are from the structure built in 330 BCE. Around the temple were numerous other buildings, the remains of which can still be seen. Most notable of these are a theater with a capacity of 5,000 spectators; a stadium to host the Pythian games; the treasuries; and several lesser temples, including the Tholos, a temple dedicated to the goddess Athena Pronoia.

Delphi **hosted the Pythian Games**, second to the Olympic Games in importance

△ **THE THOLOS AT DELPHI**
Delphi has been extensively excavated in modern times, and some structures have been partially restored, including the Tholos, which originally consisted of a circle of 20 Doric columns around a core of 10 Corinthian columns.

RESILIENT DESIGN

The Pont du Gard's strength results from the use of meticulously cut stones requiring a minimum of mortar. It also stems from the three-tier design: each successive level is slightly narrower than the one below; the main piers are aligned vertically; and each arch is individually constructed to compensate for possible subsidence.

lined water conduit
smaller top tier of arches
protruding blocks to support scaffolding
21-ft- (6.5-m-) thick pier
18th-century road bridge
CROSS-SECTION

Pont du Gard

A spectacular three-tiered bridge over the Gardon River, a section of the Roman aqueduct from Uzès to Nîmes

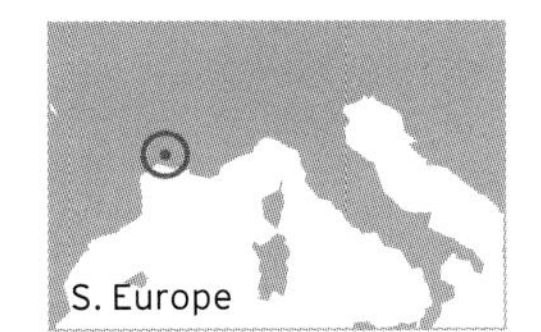

△ **BLOCKS OF STONE**
The aqueduct's stones—weighing up to 6½ tons (6 tonnes) each—were lifted into place using a human-powered winch. More than 1,000 men labored for 5 years on the construction.

At a height of almost 164 ft (50 m) above the Gardon River, the Pont du Gard is the highest elevated aqueduct built by the Romans. It is part of a 31-mile- (50-km-) long canal that carries water from a spring near the Roman city of Ucetia (present-day Uzès) to Nemasus (Nîmes), was constructed in the 1st century CE. It makes its way through very uneven terrain, sometimes in underground channels, and crosses the valley of the Gardon.

Arched crossing

The bridge crossing the gorge was built as three tiers of archways, carrying a stone-lined water conduit on its upper level. The lower level consists of six arches that are 72 ft (22 m) high. The middle tier has 11 similar arches, carrying an upper level of 47 smaller arches, 23 ft (7 m) high, of which 35 have survived. Over the bridge's span of 1,500 ft (456 m), the level of the aqueduct descends only 1 in (2.5 cm), a gradient of less than 1 in 18,000.

The Pont du Gard ceased to be used as an aqueduct after the end of the Roman Empire, but its function as a toll bridge over the river helped to keep it in reasonable repair, and in the 18th century, a road bridge was added to one side of the lower level. The bridge was renovated in 1885, and the roadway closed to traffic in 2000.

◁ **WATER FOR THE PEOPLE**
Elaborately carved drinking fountains in Nîmes spouted water supplied by the aqueduct.

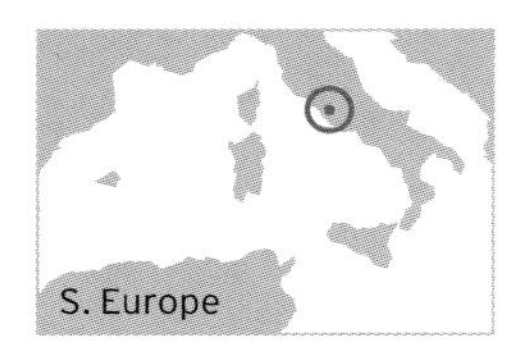

The Colosseum

The largest ever Roman amphitheater, which has become an iconic symbol of the city of Rome

The Colosseum was built in the 1st century CE as part of a regeneration of the city after the death of the disgraced Emperor Nero. It was commissioned by his successor Vespasian in 72 CE and was opened in 80 CE. A place of entertainment, it provided a huge space for combat between gladiators and animals, reconstructions of battles and mythical episodes, and public executions.

Building on a grand scale

The Colosseum was constructed mainly from local limestone with some internal brickwork. It reached a height of 159 ft (48.5 m) and covered an elliptical area that was 617 ft (188 m) long and 512 ft (156 m) wide. The exterior of the structure features Doric, Ionic, and Corinthian arcades, whose arches would have contained statues. The floor of the arena covered the hypogeum—a maze of tunnels and chambers to house the participants in the games held above.

The amphitheater fell into disrepair after 404 CE, when the games were abolished, and over time suffered from damage by earthquakes, neglect, and looting. Despite occasional attempts at restoration, it was not until the 19th century that its architectural importance was fully recognized and the Colosseum preserved for posterity.

The Colosseum was designed to hold **more than 50,000 spectators**

△ **MAGNIFICENT RELIC**
Although only half of the outer walls remain, the surviving entrance passages and hypogeum are enough to reveal the grandeur of its scale and the skill that went into its construction.

TIERED SEATING

The accommodation for spectators surrounding the arena was arranged in tiers, with separate levels for the various social classes. Ringside seats at the lower level were reserved for the emperor and senators; behind them was a zone for knights; then, the middle classes, separated into two tiers; and above that a terrace of wooden benches and standing room in aisles for ordinary citizens (plebians) and common women.

citizens and common women
middle classes
knights
senators

▷ **LIMESTONE ARCHES**
The exterior walls of the Colosseum comprise three stories of arcades decorated with half-columns and a plain fourth story with small, rectangular windows.

The Pantheon

The most perfectly preserved monument of ancient Rome, with the world's largest unsupported masonry dome

The original Pantheon, a temple to all the Roman gods, was commissioned in 27 BCE by the consul Marcus Agrippa—a great statesman and military leader, as well as a prolific builder. It was replaced by the current structure in the reign of the Emperor Hadrian some 150 years later, and, despite a number of later changes, the building completed in around 125 CE is essentially the one that still stands in central Rome today. Since 608, it has officially been a Catholic church.

Concrete construction

The design of the Pantheon is unlike that of any other ancient Roman building. Visitors pass through a massive portico with 16 granite columns and colossal bronze doors to enter an awe-inspiring circular space under a coffered dome. The internal light comes from the oculus, a hole at the apex of the dome that also admits rain in wet weather.

Made of concrete and spanning 143 ft (43.3 m), with no supporting framework, the dome is an extraordinary feat of engineering. Although no one knows exactly how it was built, one of its secrets is that the material becomes thinner and lighter toward the top—a technique that makes the dome on the outside of the building much flatter than on the inside. The weight of the dome is supported by brick-faced concrete walls 20 ft (6 m) thick. The circular interior is richly decorated with marble and includes the tombs of the artist Raphael and of two Italian monarchs.

△ **POWERFUL PORTICO**
The Pantheon's impressive columned portico blocks the view of the dome from outside the building. Each column is 41 ft (12.5 m) high and 15 ft (4.5 m) in circumference.

△ **THE OCULUS**
Natural light floods into the Pantheon through the oculus at the crown of the huge dome. Immediately below the oculus are rings of recessed panels, or coffers.

GEOMETRIC PERFECTION

The interior of the Pantheon is a cylinder capped by a roof that is half a sphere. The height of the cylinder is exactly equal to the radius of the dome, so a complete sphere could be inscribed inside the building. The coffers in the ceiling are arranged in five rows of 28, diminishing in size from bottom to top.

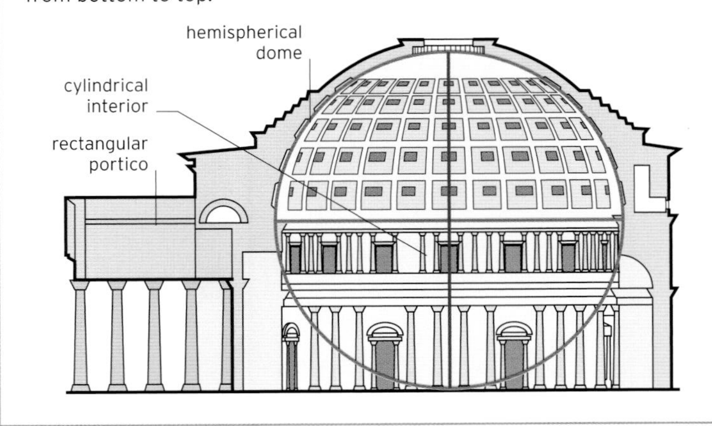

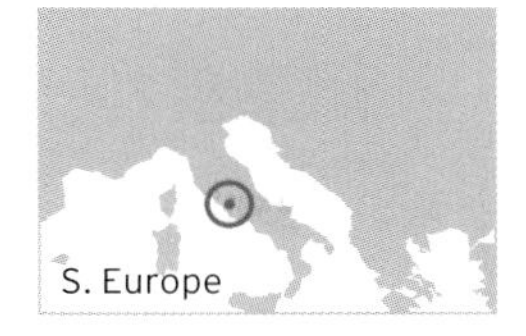

Column of Marcus Aurelius

A monumental celebration of the victories of Roman armies, illustrating ancient warfare in vivid detail

Standing in Piazza Colonna in the center of Rome, the towering Column of Marcus Aurelius was erected to honor the Roman emperor's military campaigns against "barbarian" peoples in central Europe from 172 to 175 CE. It was probably built between 176 and 193.

A tale of victory

Made of 28 blocks of Italian marble, the shaft of Marcus's column is 87 ft (26.5 m) tall. The column is hollow, with a 200-step spiral staircase rising inside. It was undoubtedly inspired by another famous monument in Rome, Trajan's Column, completed in 113 CE. Like the earlier column, it is covered in relief carvings, which unfold in a spiral from the foot of the column to the top. In the reliefs on the lower half, Marcus Aurelius is shown leading his troops in action against the Marcomanni, while the upper part illustrates his victories over the Sarmatians. One of the most famous scenes represents an incident in 172 CE when Roman troops, besieged by their enemies and dying of thirst, were rescued by a sudden rainstorm. The reliefs are of vital historical interest, providing detailed information on Roman military equipment and campaigning techniques, such as the building of pontoon bridges. Originally, statues of Emperor Marcus Aurelius and his wife Faustina stood on top of the column, but these disappeared in medieval times and were replaced by a statue of St. Paul in 1589.

◁ **VICTORY COLUMN**
Now topped by a Christian statue, the column was built to celebrate the victories of Roman armies on their imperial frontiers. The battle scenes in its reliefs are precise and graphic.

Diocletian's Palace

An extensive Roman palace and garrison complex built by the Emperor Diocletian for his retirement

Much of the central old town of present-day Split, on the Croatian coast, occupies the walled area of streets and buildings known today as Diocletian's Palace. The residence itself, however, is only a part of a fortified complex that Diocletian had built in anticipation of his retirement as Roman emperor in 305 CE.

Structure of the citadel

Protected on three sides by high walls and watchtowers, and on the south side by the sea, the fortress is divided in two by a street between the western Iron Gate and the eastern Silver Gate. To the north is the garrison and a residential area, and in the north wall is the magnificent Golden Gate, which leads onto the main north-south street, down to the central square, the Peristylum. This forms the entrance to the southern half of the complex, containing public buildings and the emperor's luxurious residence overlooking the sea. After the Romans abandoned the fortress, the buildings and spaces in the walls were taken over as private dwellings and shops, and today they are an integral part of the city.

△ **ADAPTED FORTRESS**
The main body of the Cathedral of St. Domnius is formed from the Roman Imperial mausoleum, to which a bell tower was added in the 12th century.

THE PLAN OF THE BASILICA

The blend of Roman and Byzantine elements gives the basilica a complex structure: the overall octagonal floor plan contains an inner octagon of arches supporting the tiered tower, and the narthex (entrance lobby) is unusually set tangentially to the perimeter rather than aligned with the main axis of the church.

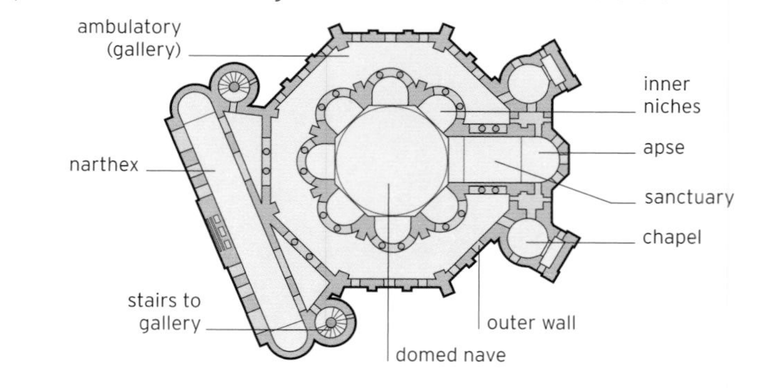

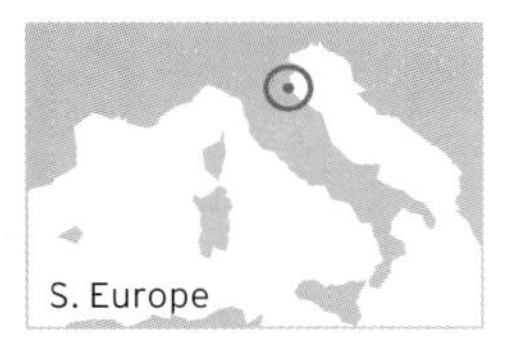

Basilica of San Vitale

A high point of Byzantine architecture in Italy, its interior is lavishly decorated with mosaics

The Italian city of Ravenna is home to a wealth of early Christian architecture, dating from the 5th and 6th centuries under successive Roman, Ostrogoth, and Byzantine rule. Among the eight buildings given UNESCO World Heritage status in Ravenna is the Basilica of San Vitale, an imposing octagonal structure built between 526 and 547 CE.

Renowned mosaics

Although not strictly a basilica in architectural form (the term refers to a rectangular building with a central nave and aisles), the church was granted the title in recognition of its ecclesiastical status. The architecture is a hybrid of Roman elements, such as the dome, and Byzantine influences—in particular, the octagonal floor plan and the use of bricks with a thick rendering of plaster. The rather austere exterior of the church, however, is misleading. Inside, the spacious polygonal nave is filled by natural light and is sumptuously decorated throughout with brilliant mosaics. These depict the saints and the life of Christ and commemorate the accession of the Byzantine emperor Justinian I and his wife Theodora as rulers of Ravenna.

△ **BYZANTINE MOSAICS**
The Basilica of San Vitale is decorated with uniquely well-preserved mosaics in the Hellenistic-Roman style. It also houses the most important collection of Byzantine art of the period.

▷ **PANEL OF JUSTINIAN**
On one wall of the apse is a mosaic panel depicting the Emperor Justinian with his court officials and clergy; opposite is a panel showing his wife, Empress Theodora.

▽ **OCTAGONAL EXTERIOR**
The typically Byzantine octagonal nave supports an upper tier with a dome. Flying buttresses are used to transmit the lateral, outward forces to the ground.

علي
محمد
الله

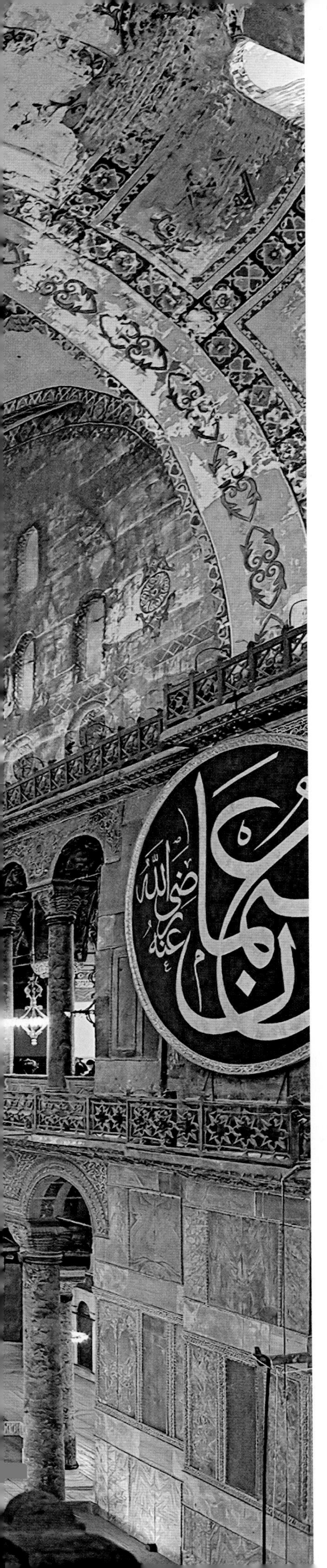

Hagia Sophia

The great Byzantine basilica of Constantinople, converted into the city's principal mosque after the Ottoman Conquest

The basilica of Hagia Sophia ("Holy Wisdom") was commissioned by Justinian the Great, Eastern Roman emperor from 527 to 565 CE, to replace the church destroyed by rioting in Constantinople (present-day Istanbul). He imagined a building worthy of being the principal church of the Eastern Empire, and his architects accordingly presented him with a revolutionary plan—a square structure topped by an enormous dome. The building was completed in 537 CE, but its untested design, combined with poor construction methods, caused the dome to collapse just 20 years later. It was rebuilt with a slightly less shallow dome and more robust reinforcements, and it has survived in this form to the present day.

Church to mosque

The interior of the basilica was richly decorated in the Byzantine style with polished marble slabs and mosaics of Jesus and the saints on walls, ceiling, and floor. Hagia Sophia served as the central basilica of the Orthodox Church until the Ottoman conquest of Constantinople in 1453. Luckily, the Muslim invaders appreciated its architectural splendor and instead of destroying it set about converting it into a mosque, adding four minarets and installing a *mihrab* (a niche indicating the direction of Mecca) and a *minbar*, or pulpit. With the collapse of the Ottoman Empire and as part of Kemal Atatürk's secularization of Turkey, Hagia Sophia closed as a mosque in 1931 and reopened as a museum in 1935.

The **dome** of Hagia Sophia was the **biggest in the world** until surpassed by the **Duomo in Florence** in the 15th century

△ **CROWNING THE CITYSCAPE**
Adorned with its four minarets, the Byzantine basilica is a distinctive landmark that dominates the skyline of the historic peninsula of Istanbul.

◁ **SPACE AND LIGHT**
Hagia Sophia's square floor plan, high walls, and massive dome give the interior a sense of openness and light.

SUPPORTING THE DOME

In order to create the spacious nave of Hagia Sophia, the architects developed an innovative design, placing a circular dome on top of a square building. The weight of the dome is borne by four piers, which are disguised by pendentives—tapering triangular pillars forming elegant curved arches beneath the dome.

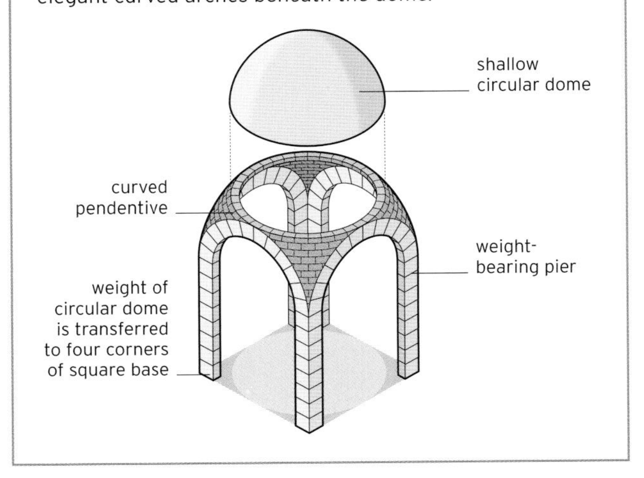

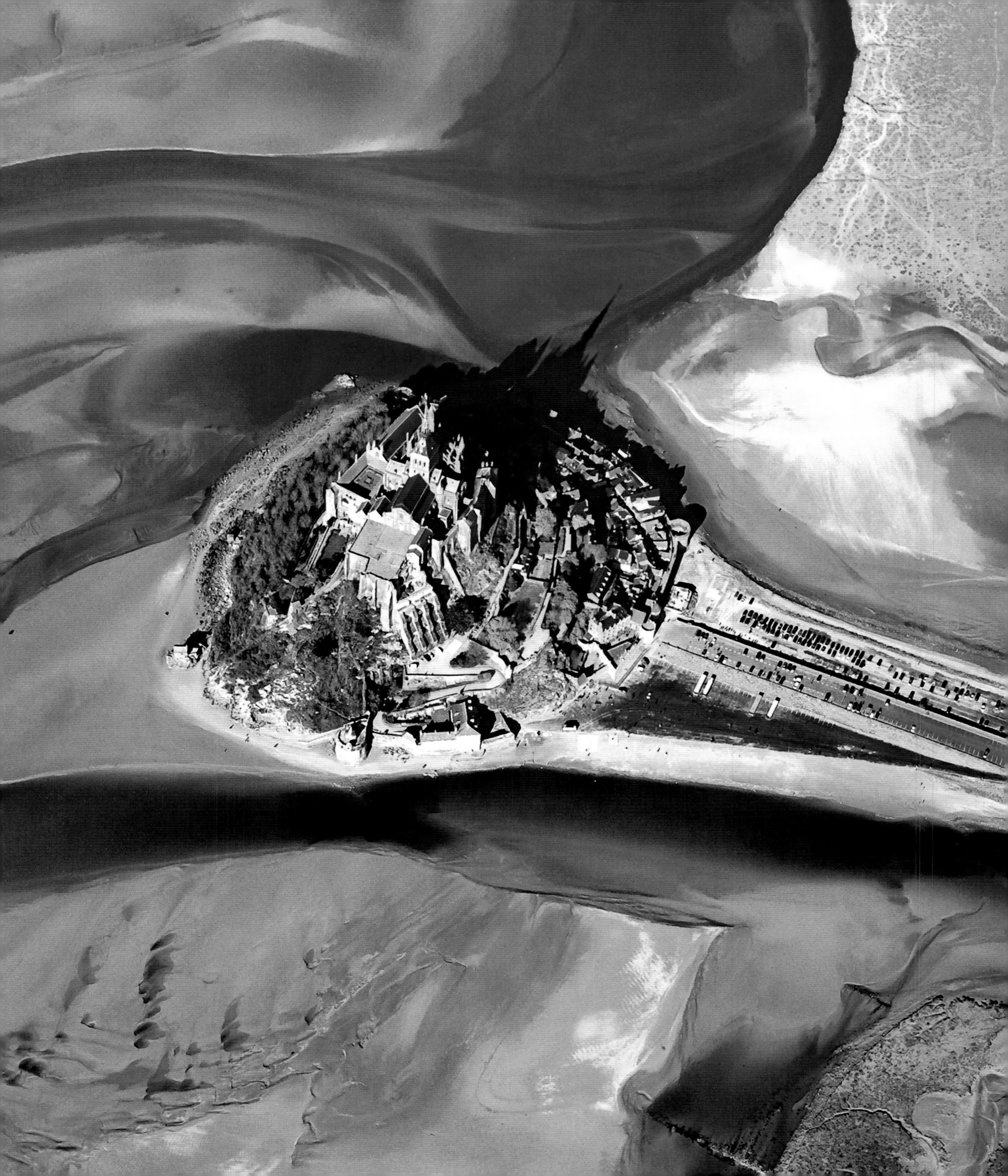

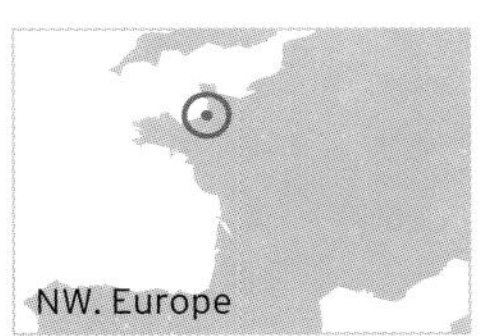

Mont Saint-Michel

A fortified medieval abbey that blends different styles in a spectacular setting

The small granite outcrop of Mont Saint-Michel lies off the coast of Normandy and fully becomes an island only at very high tides. The buildings and fortifications perched on top of the mount have a remarkable history that stretches back over 1,300 years.

Mont Saint-Michel's first church was built in 709 CE. Pilgrims soon followed, risking their lives at low tide to reach the island. In 966, a Benedictine abbey was built and, as it grew, earlier structures were incorporated into a labyrinth of foundations, crypts, and staircases. In the 13th century, a three-story masterpiece of Norman Gothic architecture, known as La Merveille ("the Wonder"), was added, and a small township began to develop on the slope below the abbey.

Fort, prison, and monument

The island was fortified in 1256 and besieged several times during the Hundred Years' War (1337–1453) and French Wars of Religion (1562–1598). The abbey's fortunes waned after the Reformation, and it was made into a prison in the 19th century. The prison closed in 1863, and the mount was classified as an historic monument soon after. The creation of a causeway in 1874 made the journey to the island less treacherous. Monks returned to Mont Saint-Michel in 1966, and it became a UNESCO World Heritage Site in 1979.

LA MERVEILLE

This view of the island shows the three stories of La Merveille, which reach a height of 115 ft (35 m). The levels reflected medieval ideas of hierarchy. Poor pilgrims would have eaten in the cellars, knights and important guests in the Knights' and Host's rooms, and—being spiritually nearest to heaven—the monks would have eaten in the refectory and relaxed in the cloisters.

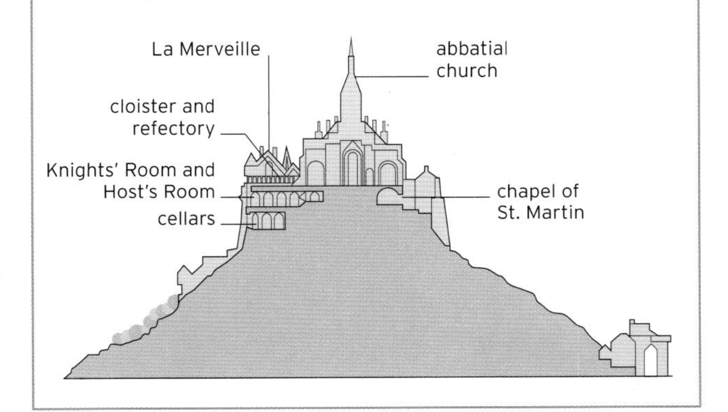

Mont Saint-Michel was the **only place** in **Normandy not to fall** to the English in the **Hundred Years' War**

◁ **CITADEL AT SEA**
Silt deposition and land reclamation mean that the island, which was once 4 miles (7 km) out at sea, is now just 1.2 miles (2 km) from the shore and might eventually become part of the mainland.

▷ **TOWN AND ABBEY**
The abbey-church's tower and spire reach 558 ft (170 m) above the bay. Large ramparts protect the town at high tide.

Mosque-Cathedral of Córdoba

A Renaissance cathedral inside an enormous mosque in a unique combination of Moorish and Christian architecture

SW. Europe

The Umayyad dynasty, originating in Mecca, established a caliphate in Damascus in the 7th century CE and invaded the Iberian peninsula in 711. The dynasty was deposed in the East in 750 but continued to flourish in Spain. In 929, a new caliphate was founded in Córdoba, elevating the Umayyad capital to one of the greatest cultural centers in Europe.

Work on Córdoba's mosque was begun in 786, on the orders of Abd al-Rahman I, the leader of the Umayyad dynasty. It was built on the site of Córdoba's Roman temple and later Visigoth church, and many of its materials were cannibalized from these older structures.

Columns and arches

The most striking feature of the mosque is the forest of marble columns and arches that extends along its main hall. Originally, there were more than 1,200 of these, but only around 850 survive. The horseshoe arches, with alternating bands of red brick and white stone, are particularly distinctive. The mosque's most opulent feature is the stunning *mihrab* (prayer niche), which was added by the Umayyad caliph al-Hakam II (r.961–976) in 962.

Part of the mosque was demolished in the 16th century to make way for the cathedral. At the same time, the minaret was transformed into a belfry, the Alminar Tower. The building is now fully protected, having been added to UNESCO's World Heritage Site list in 1984.

◁ **DECORATIVE PATTERN**
The walls and ceiling of the room-sized *mihrab* in the Córdoba mosque are decorated with plantlike patterns.

◁ **PRAYER HALL**
The great prayer hall has row upon row of horseshoe arches. The columns are of unequal length, as many were reused from earlier buildings.

△ **DOME OVER THE *MIHRAB***
Above the imposing *mihrab* is a spectacular dome with mosaic decoration in the form of an eight-pointed star.

FLOOR PLAN OF THE BUILDING

The former mosque measures 590 by 425 ft (180 by 130 m). At the entrance is the "Court of the Oranges" leading to a columned prayer hall, which ends with the *mihrab*. Since 1236, the complex has served as a Christian cathedral with the addition of a central high altar and numerous chapels along its sides.

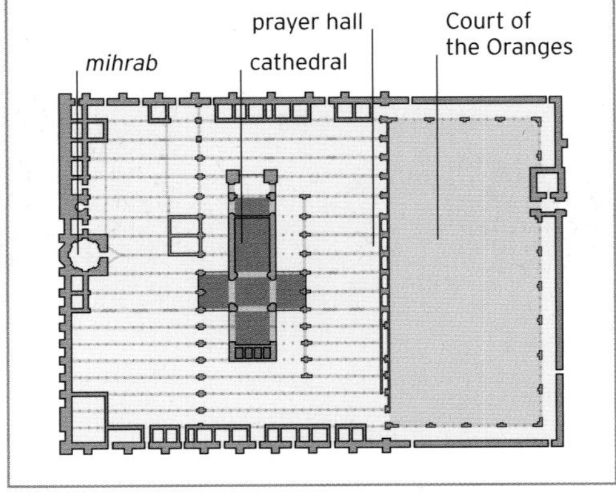

Aachen Cathedral

One of Europe's oldest cathedrals, the burial place of Charlemagne, and the coronation church of German kings for 600 years

On Christmas day 800 CE, Charlemagne (Charles the Great), King of the Franks and ruler of much of western and central Europe, was crowned Holy Roman Emperor. At Aachen, Charlemagne built a new church that drew on the architecture of both the Roman and Byzantine empires. Consecrated in 805, the octagonal Palatine Chapel, at the heart of Aachen Cathedral, was modeled on the 6th-century Byzantine-style church of San Vitale (see p.107), although its barrel and groin vaults and Corinthian columns were classically Roman.

Heavenly architecture

The chapel was to be the spiritual heart of Charlemagne's empire, its ornate mosaic, marble, and fresco decoration symbolizing the connection between Heaven and Earth. Charlemagne was buried in the chapel in 814 and canonized in 1165. As numbers of pilgrims increased, the cathedral was gradually extended.

A MILLENNIUM OF GROWTH

Aachen Cathedral grew over several centuries from the Palatine Chapel and 16-sided ambulatory built for Charlemagne, to the last major works, including the addition of the West Tower, in the 19th century. A large fire in 1656 and bomb damage in World War II meant that parts of the cathedral have been reconstructed.

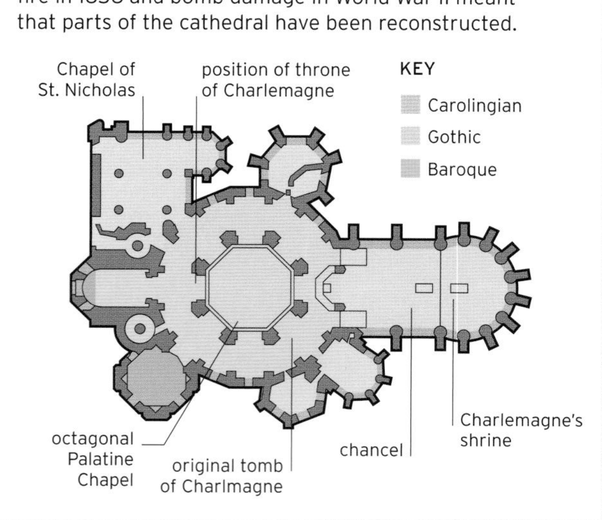

△ **CHANGING STYLE**
The original Carolingian domed chapel, its roof replaced after a fire in 1656, is nestled between the cathedral's Gothic tower (left), added around 1350, and the 14th-century Gothic chancel.

△ **EMBELLISHED DOORKNOCKER**
This is one of two lion's head doorknockers from Aachen's Carolingian "Wolf Doors." The doors were modeled on Roman temple doors and were cast around 800 CE for the Palatine Chapel.

◁ **EMPEROR'S SHRINE**
A gold-and-silver casket containing Charlemagne's remains, commissioned by the Holy Roman Emperor Frederick II in 1215, is enclosed in glass at the center of Aachen's Gothic chancel, built in 1355–1414.

30 kings and **12 queens** were **crowned** at Aachen between **936 and 1531**

Durham Cathedral

One of the finest surviving examples of Gothic architecture in Britain

Durham Cathedral was begun in 1093, less than 30 years after the Norman Conquest, and completed around 1133. It was built to house the shrine of St. Cuthbert (who spread Christianity in the north of England in the 7th century) and the tomb of the scholar and theologian the Venerable Bede (673–735 CE). The cathedral is the most complete example of Norman architecture in Britain, having retained its original Norman nave, choir, and transepts.

Innovative design

The cathedral is characterized by its solidity, sense of order and proportion, and semicircular arches and vaults built in the Romanesque style. However, the nave also features the world's first structural pointed arch—a feature that came to define Gothic architecture in the mid-12th century. The innovative use of stone ribs to form pointed arches—which provided additional support and superior weight distribution—allowed the cathedral's masons to build what is the largest surviving stone-vaulted ceiling from the period. A working church and popular attraction, the cathedral—along with Durham Castle—became a UNESCO World Heritage Site in 1986.

△ **STURDY SUPPORT**
Solid Romanesque columns support the choir's circular arches and lead to the Chapel of Nine Altars and the rose window. The chapel was built in 1242–1290 to accommodate increasing numbers of pilgrims to St. Cuthbert's shrine.

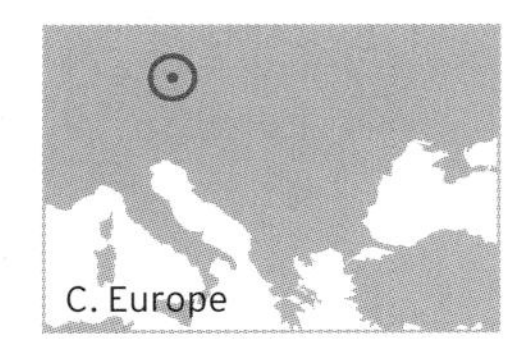

Prague Castle

The largest ancient castle in the world and a thousand-year-old seat of the Bohemian kings and later of the Czech State

Perched high above the Vltava River, Prague Castle was developed over several centuries, and its long and varied history is reflected in the mixture of architectural styles and variety of religious, royal, and military buildings found within the complex.

Changing times

The castle complex dates from the late 9th century, when Prince Bořivoj—the first Christian ruler of Bohemia—chose the site as the seat of his power, erecting a fortified wooden castle protected by a moat and ramparts. Churches dedicated to the Virgin Mary, and later St. Vitus and St. George, were among the earliest buildings erected at the site, although their original structures were subsequently destroyed or rebuilt. St. Vitus was replaced with a larger Romanesque church in the 11th century, when the castle became the seat of the Bishop of Prague. It was rebuilt again as a grand Gothic cathedral in the reign of the Holy Roman Emperor Charles IV (1316–1378), at which time Prague was the seat of power of a vast empire that stretched from modern-day southern Denmark to northern Italy and from Belgium to Krakow in Poland. The wooden castle was replaced in the 12th century by the Romanesque Old Royal Palace, which was itself rebuilt under Charles IV, who covered some of its roofs with gold-plated metal sheets.

The Renaissance left its mark in the late 15th century in Vladislav II's (r.1471–1516) Hall in the Old Royal Palace and in the Golden Lane that Rudolf II (r.1576–1612) built as a home for alchemists and servants. Rudolf also built the Spanish Hall to house his vast art collection, sadly looted in the Battle of Prague (1648).

◁ **CATHEDRAL CARVING**
This detail from the sculpted portal at the west end of St. Vitus Cathedral shows scenes from the life of Christ.

The Czech writer **Franz Kafka** lived on **Golden Lane** from **1916 to 1917**

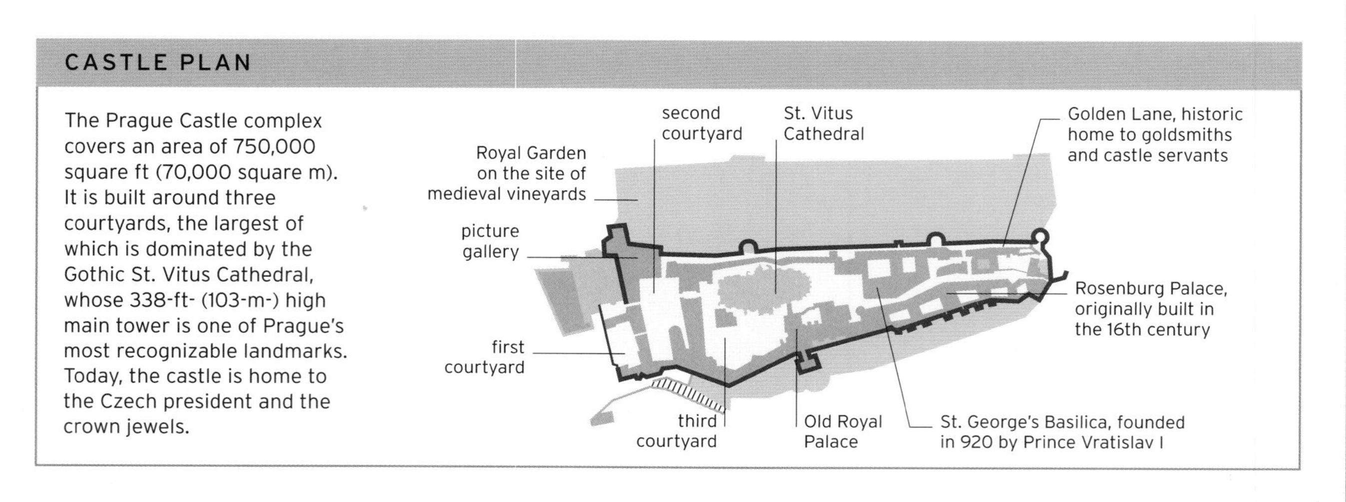

The Prague Castle complex covers an area of 750,000 square ft (70,000 square m). It is built around three courtyards, the largest of which is dominated by the Gothic St. Vitus Cathedral, whose 338-ft- (103-m-) high main tower is one of Prague's most recognizable landmarks. Today, the castle is home to the Czech president and the crown jewels.

SEAT OF POWER
Spectacularly illuminated at night, St. Vitus Cathedral rises above (from left to right) the Royal Palace, Vladislav Hall and All Saints Chapel, and the Rosenberg Palace. In the foreground are the Vltava River and Charles Bridge.

Westminster Abbey

A Gothic masterpiece, 900-year-old coronation church of the British monarchy, and mausoleum of Britain's kings and national heroes

▽ **EMERGING STYLE**
The facade of Westminster's north transept includes typical Gothic pointed arches and carved portals. The rose window reflects French influence, although the narrow, pointed "lancet" windows also signal the emergence of an English style.

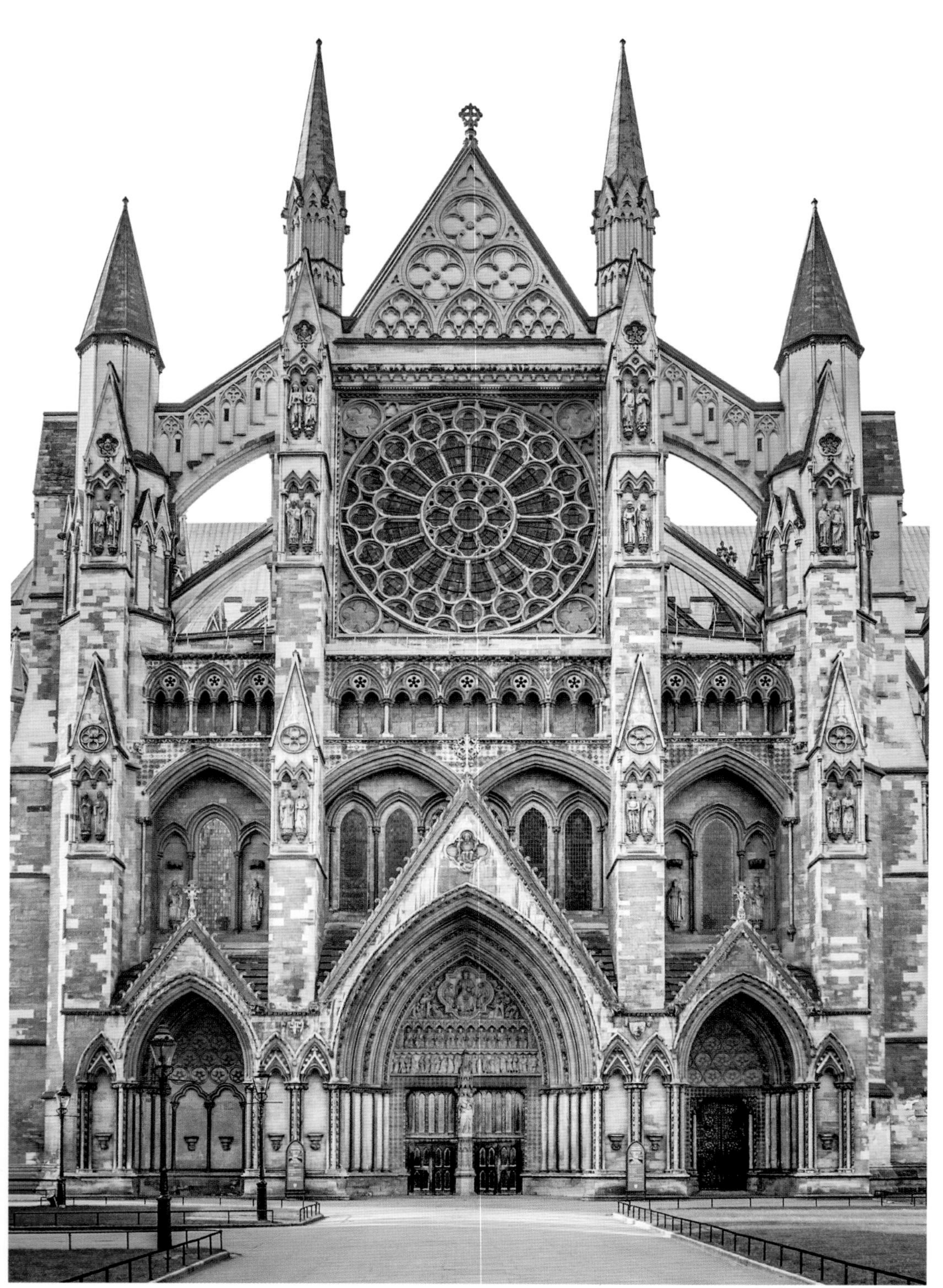

Founded by King Edward the Confessor in around 1042, Westminster Abbey has witnessed the coronation of all but two of England's monarchs (Edward V and Edward VIII) since William the Conqueror (r.1066–1087), as well as 16 royal weddings. In 1245, Edward's Norman Romanesque church—the first of its kind in England—was demolished to make way for King Henry III's new Gothic cathedral. Over the centuries, more chapels, including Henry VII's spectacular Lady Chapel, were added, and the abbey's massive west towers, designed by Nicholas Hawksmoor, were completed in 1745.

National treasures

Spreading over an area of more than 105,000 square ft (9,750 square m), the abbey is architecturally exquisite and also home to an extraordinary collection of memorial sculptures and medieval wall paintings. Around 3,300 people are buried in the abbey, and many more memorialized. British monarchs from Edward the Confessor to George II lie alongside some of Britain's most notable personages, including Charles Darwin and Isaac Newton, and those—like Jane Austen, William Shakespeare, and Charles Dickens—commemorated in Poet's Corner.

The abbey is a **"Royal Peculiar,"** subject **only** to the **Sovereign**

WESTMINSTER'S TRIFORIUM

Many Gothic cathedrals had a passageway or gallery above the side aisles, known as a triforium. Overlooking the abbey floor 52 ft (16 m) below, the triforium provided excellent views of Elizabeth II's coronation in 1953 but was long used as a storage space. In 2018, Westminster's eastern triforium was opened to the public as a gallery for the abbey's treasures.

triforium
cloister walk
nave

Tower of London

William the Conqueror's Norman fortress, royal palace, notorious prison, and treasury of the British monarchs

△ **COMPLEX CASTLE**
Inside the curtain walls, the White Tower dominates the complex. Behind it is the Waterloo Block, home to the Crown Jewels. Traitor's Gate (bottom) leads up from the Thames to the Bloody Tower just behind.

In 1066, William the Conqueror took the English crown at the Battle of Hastings. As part of the ongoing conquest of England, he built a series of castles from which his Norman followers could dominate the country. The most impressive was the fortress built on the banks of the River Thames in London. At the heart of this fortress was a solid, square keep, known from the 13th century as the White Tower after it was whitewashed on the orders of King Henry III. Constructed in stone brought from Caen, France, between c.1078 and 1100, the massive tower measured 118 by 107 ft (36 by 33 m). It had walls that were 15 ft (4.6 m) thick at the base, and it rose to a height of 90 ft (27.5 m). The Tower of London was three times as tall as its neighbors and gave its name to the whole castle complex. Dominating the landscape—it was the first structure seen when arriving in London by boat—it served as a potent reminder of the city's defeat to the Norman invaders.

Strengthened fortifications

In the 13th century, Henry III and Edward I expanded William's fortress with the addition of concentric rings of "curtain walls" (defensive walls) and towers, a moat, and the Traitor's Gate. Among the additions was the infamous Bloody Tower, where Henry VI was murdered in 1471 and where Edward IV's two sons disappeared, perhaps killed, in 1483. Later monarchs added more buildings to the Tower. In 1826, the Constable of the Tower, the Duke of Wellington, began modernizing the site, draining the moat and closing the Royal Menagerie, created by King John in the 13th century and once home to lions, bears, and elephants. Witness to battles and beheadings, the Tower has served as a prison, a royal residence, home to the Royal Mint, and an armory, and has been the repository of Britain's priceless Crown Jewels since 1600.

CONCENTRIC CASTLES

The White Tower was the earliest of the large rectangular stone keeps in England. Thereafter, many wooden castles erected in the years of the Conquest were then rebuilt in stone. Some were fortified in the 12th and 13th centuries by the addition of encircling walls. The Tower of London was one of the largest and strongest of these "concentric" castles in England.

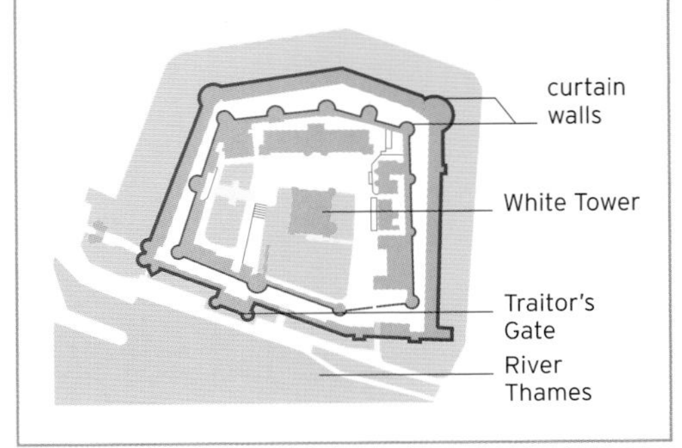

Leaning Tower of Pisa and Duomo

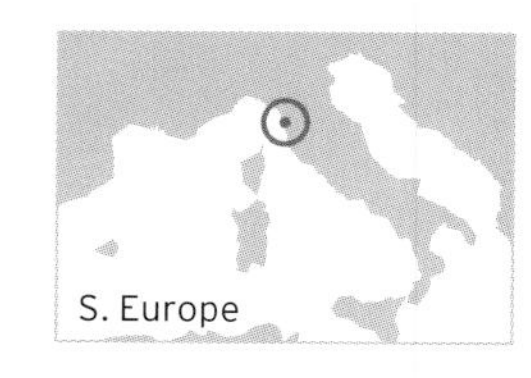

An iconic bell tower sharing Pisa's "Square of Miracles" with other marble-clad marvels of medieval architecture

Known for its unintended tilt, the Leaning Tower of Pisa is one of Italy's foremost attractions. It was planned in the 12th century as a free-standing bell tower, or campanile, for the Duomo—a magnificent cathedral built in the Romanesque style some 200 years earlier.

The crowning glory

The Duomo itself stands in a large walled piazza, with expanses of lawn between the paved areas. The piazza also contains the Baptistery, and on its northern side the Camposanto Monumentale ("monumental cemetery"). All of these buildings are of extraordinary architectural merit in their own right, but because of its tilt, the Campanile erected behind the cathedral has eclipsed their reputation.

Construction of the tower, designed to be 185 ft (56 m) tall, started in 1173. Its unstable foundations caused it to list to the south after only three floors had been built, and only after many delays was the building completed in the 14th century.

△ **RELIEF PANELS**
The pulpit of the Duomo features sculpted marble panels depicting scenes from the life of Christ, such as this Slaughter of the Innocents.

There are more than **290 steps** from the **base** of the Campanile to the **top** of the bell chamber

◁ **THE CAMPANILE**
The upper floors of the Campanile were built with one side taller than the other to compensate for the tilt.

△ **UNIQUE CLUSTER**
The Baptistery (left), Duomo (center), and Campanile (right) form an ensemble in the Piazza del Duomo regarded as the prime expression of the architectural style known as Pisan Romanesque.

CORRECTING THE TILT

By the late 20th century, the Campanile, tilting at an angle of 5.5 degrees, was in danger of toppling, despite earlier attempts at restoration. Remedial work began in 1992. First, the tower was stabilized with cables and soil was excavated from beneath its raised side to reduce the lean to less than 4 degrees, enough to stabilize the structure, without removing the tilt for which it had become so renowned.

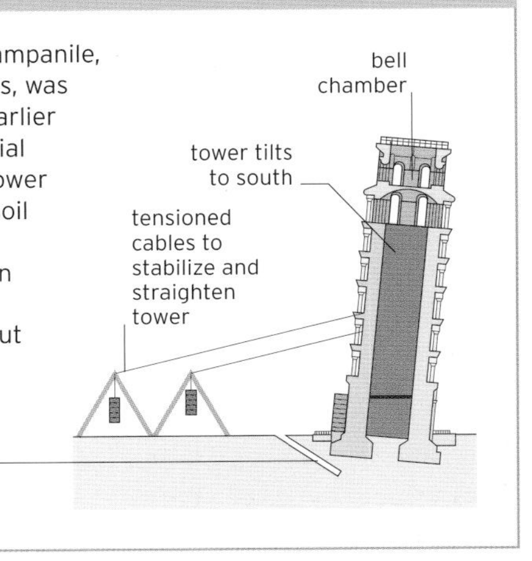

St. Mark's Basilica

An exotic church in Venice glittering with golden mosaics and filled with cultural treasures

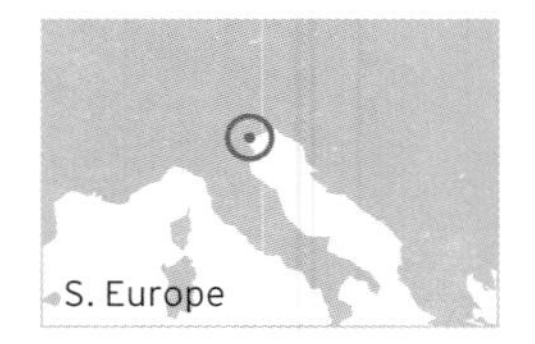

In 832 CE, two Venetian merchants presented the Doge of Venice with the purported body of St. Mark, which they had stolen from a tomb in Alexandria. The doge had a basilica built to house this sacred relic. The current basilica is the third church erected on the site. Constructed between 1063 and 1094, it served for centuries as the doge's private chapel and a setting for Venetian state ceremonies until belatedly assigned the role of Cathedral of Venice in 1807.

Byzantine influence

The basilica's five domes and the rounded arches of its facade have a distinctly eastern look, reflecting medieval Venice's close relations with the Byzantine empire. The interior glows with gold glass mosaics—mostly dating from the 12th to 14th centuries—that cover the walls and domed ceilings with vivid scenes from the Bible. The undulating, tessellated marble floor, with its bold geometric patterns and images of animals and birds, contributes to the overwhelming decorative effect.

The basilica is a treasure trove of unique cultural artifacts, some older than the building itself. Part of the glittering Pala d'Oro, an altarpiece composed of pieces of enamel set in gold leaf and almost 2,000 precious stones, dates from the 10th century. The four bronze horses of St. Mark, known as the Quadriga, which stand above the main portal on the basilica's facade, were made at the time of the ancient Roman empire, probably around 200 CE. The Quadriga was brought to Venice as loot from the sack of Constantinople (present-day Istanbul) by the Venetian-financed Fourth Crusade in 1204. The ancient Roman porphyry statue of four linked figures known as the *Portrait of the Tetrarchs*, which stands outside the basilica, is another looted artifact from Constantinople.

Retaining character

There have been later additions, such as the 15th-century Gothic pinnacles and 19th-century mosaics that have altered the facade. However, despite such changes and the impact of modern tourism, the basilica has preserved its atmosphere of age-old sanctity.

The **mosaics** inside St. Mark's cover a **total area** of **43,000 square ft (4,000 square m)**

◁ **MOSAIC OF THE CRUCIFIXION**
The mosaics in the basilica illustrate scenes from the Old and New Testaments of the Bible. This mosaic is a traditional representation of Jesus's Crucifixion.

▷ **FINE FACADE**
The basilica's lavish facade is topped by a statue of St. Mark, with the winged lion—the emblem of the saint and of Venice—below him. The four horses are replicas of originals kept in the basilica's museum.

△ **GLOWING INTERIOR**
The mosaics that line the domes and upper walls of the basilica reflect light to create a magical golden glow. The basilica is sometimes known as the Church of Gold.

GROWING DOMES

The basilica's basic structure, made largely of brick with marble facing, has changed little through almost a thousand years. It is clearly modeled on the churches of the Byzantine empire. In the 13th century, the basilica's domes were raised externally by adding wooden frames roofed with lead. The shape of the original domes can still be seen inside the building.

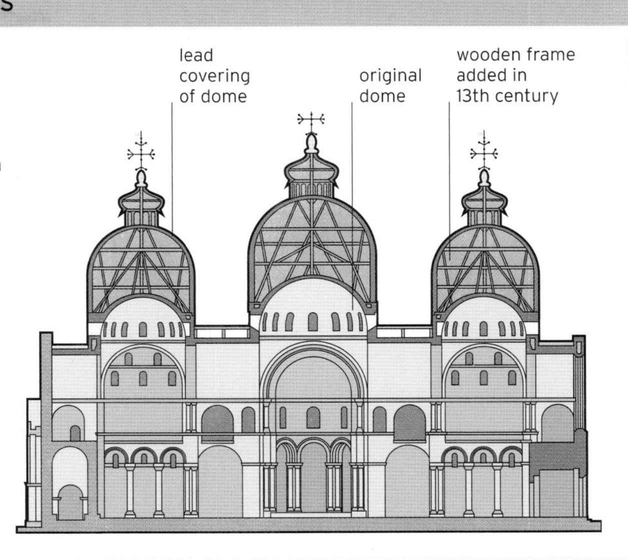

△ **THE BIRTH OF GOTHIC ART**
The basilica's choir, featuring tall, slim columns supporting a high rib vault with space for three stories of stained-glass windows, set the template for Gothic cathedrals throughout Europe.

Basilica of Saint-Denis

The 13th-century birthplace of Gothic architecture and burial place of France's kings

According to Christian tradition, St. Denis—the patron saint of France—was a 3rd-century bishop of Paris responsible for bringing Christianity to Gaul. His burial site, just outside Paris, became a focus of Christian pilgrimage, and the abbey-church built there in the 7th century grew into one of the richest in Europe. In 1135, Suger, the Abbot of Saint-Denis, decided to rebuild the church. He created a dazzling basilica in a new style, the Gothic, that would dominate European church architecture for the next three centuries.

A vision of light and space

At the church's west end, Suger built a 112-ft- (34-m-) wide facade with three doors, twin towers (the north tower was dismantled in 1846), and a rose window. At the east end, the chancel, built in 1140–1144, followed his vision for a spacious church, flooded with light. For the first time, elements such as pointed arches, rib vaults, chapels radiating around an apse, and flying buttresses that allowed for the introduction of large, upper-level (clerestory) windows were brought together in a unified Gothic style. Moreover, Suger's architects replaced the usual heavy dividing walls with slim columns so that, as he himself put it, "the whole church would shine with the wonderful and uninterrupted light of most luminous windows." The light spilled through stained-glass windows to reflect on a golden altar and great jeweled cross, both later destroyed.

◁ **SACRED SCULPTURE**
This bas-relief of the Crucifixion is from the central portal at the west end, which features seven more roundels based on Christ's life.

Chartres Cathedral

The best-preserved example of Gothic architecture in France, renowned for its Chartres blue stained glass

French Gothic architecture reached its most coherent expression in Chartres Cathedral. Many medieval cathedrals took decades or even centuries to build, resulting in a patchwork of architectural styles, but not Chartres. When the town's old Romanesque cathedral burned down in 1194, the local community pulled together to complete a new cathedral in just 26 years. Consequently, the new cathedral was wholly Gothic: cruciform in shape with two towers on the west facade, a curved apse and radiating chapels, and filled with height and light. Its vault rises 112 ft (34 m) above the nave floor; its walls are largely composed of stained glass.

Original features

Few medieval buildings have remained as untouched as Chartres Cathedral. Much of its 32,000 square ft (3,000 square m) of stained glass dates from the early 13th century, as do the narrative sculptures around the cathedral's portals and a tiled labyrinth that leads pilgrims around the floor of the nave.

▷ **ASYMMETRICAL FACADE**
The symmetry of Chartres' west end is broken by its mismatched towers; the original Gothic north tower (left) fell in 1506 and was replaced with a tower in the late Gothic Flamboyant style.

◁ **ROYAL PORTAL**
The carved figures on the jambs of Chartres' west, or royal, portal date from the mid-12th century and are thought to depict royal ancestors of Christ from the Old Testament.

FLYING BUTTRESSES

Typical of Gothic cathedrals, the flying buttresses at Chartres Cathedral work by transferring the weight of the roof, or vault, via a series of half arches that "fly" from the upper part of a wall to a buttress some distance away. The innovation allowed medieval cathedral builders to build higher than before using thinner columns and walls with large apertures for stained-glass windows, dramatically opening up the interior space and flooding it with light.

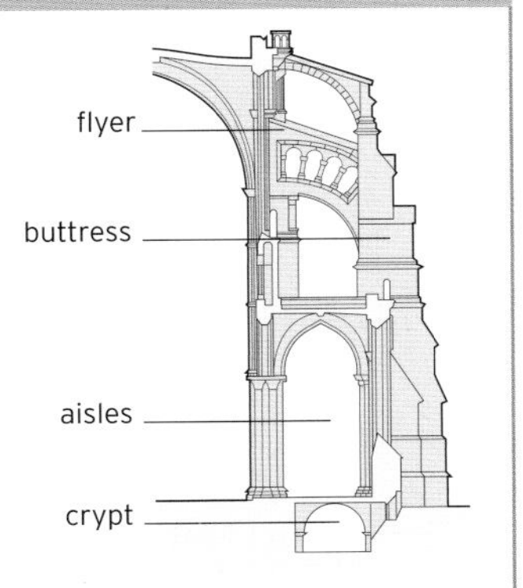

Architectural styles

GOTHIC

Gothic architecture is synonymous with the Middle Ages. Originating in France in the 12th century, it quickly spread all over Europe.

Gothic emerged from the preceding Romanesque style. While Romanesque buildings were overwhelmingly solid, with massive columns and thick walls, Gothic structures were light, open, and often elaborately decorated. This difference stemmed from several structural innovations. The most significant was the pointed arch, which was stronger and lighter than the round arch of the Romanesque, enabling masons to build higher. It also allowed the weight of the building to be carried through thin ribs and piers rather than walls, allowing for vast expanses of stained glass, which flooded the interiors with color and light.

Building higher meant that structures needed to be supported from the sides. The solution, which preserved the light interior, was the flying buttress. This was essentially the replacement of a heavy supporting wall with an open arch or series of arches.

Gothic architecture evolved considerably, with local variations, until it was superseded by Renaissance architecture (see pp.146–147) in the early 15th through the early 17th centuries. It remained closely intertwined with Christian theology, being the principal means through which everyday life connected with the divine.

gargoyles represent evil and danger

gargoyles act as a warning to those who do not follow the church's teaching

△ **GARGOYLES**
Gargoyles are grotesque or wild creatures protruding from the side of a Gothic building, with an internal spout to channel rainwater away.

NOTRE-DAME **299 ft (91 m)**

COLOGNE **515 ft (157 m)**

SALISBURY CATHEDRAL **403 ft (123 m)**

CATHEDRAL SPIRE SHAPES
A spire is the tapering structure built on top of a tower. They are an important characteristic of Gothic architecture—announcing the presence of a cathedral (or other building) from a distance.

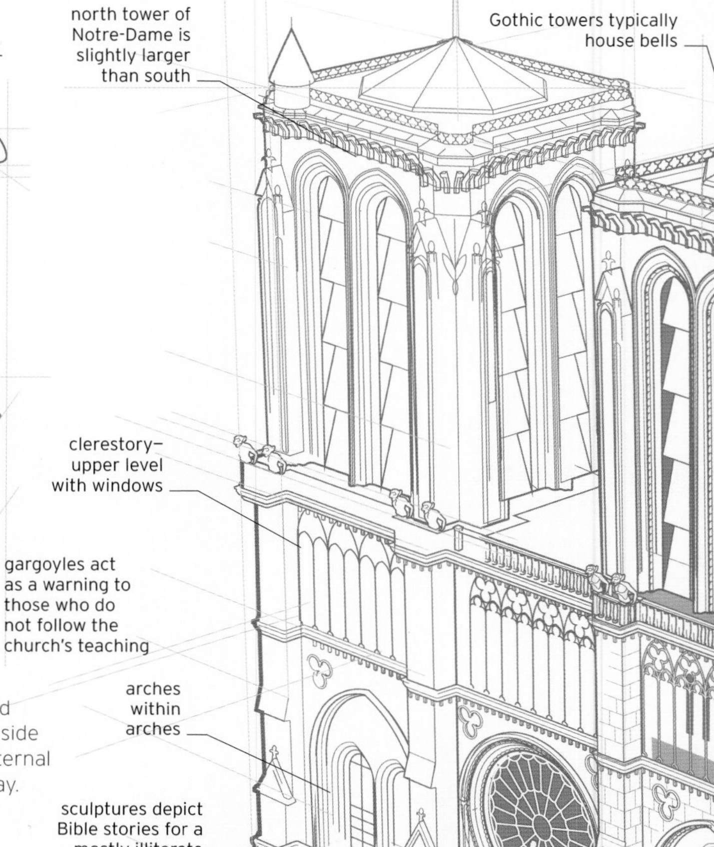

Lincoln Cathedral in the UK was the first building to surpass the Great Pyramid of Giza in height

THE PARTS OF A CATHEDRAL

baptistery, used for baptisms

transept is under spire

nave and transept meet at crossing

tombs encircle high altar

nave—long central section

chancel houses altar

transept arm

FLOOR PLAN
The floor plan of Gothic cathedrals and churches is usually in the shape of a cross. It proceeds from nave to crossing, with transept arms either side, to the chancel.

peak concludes with a sharp point

shape of arch is defined by circles with two centers

circles have radii equal to arch span

△▷ **ARCHED DOORWAYS**
Also known as portals, main doorways in Gothic buildings are often emphasized by concentric arches.

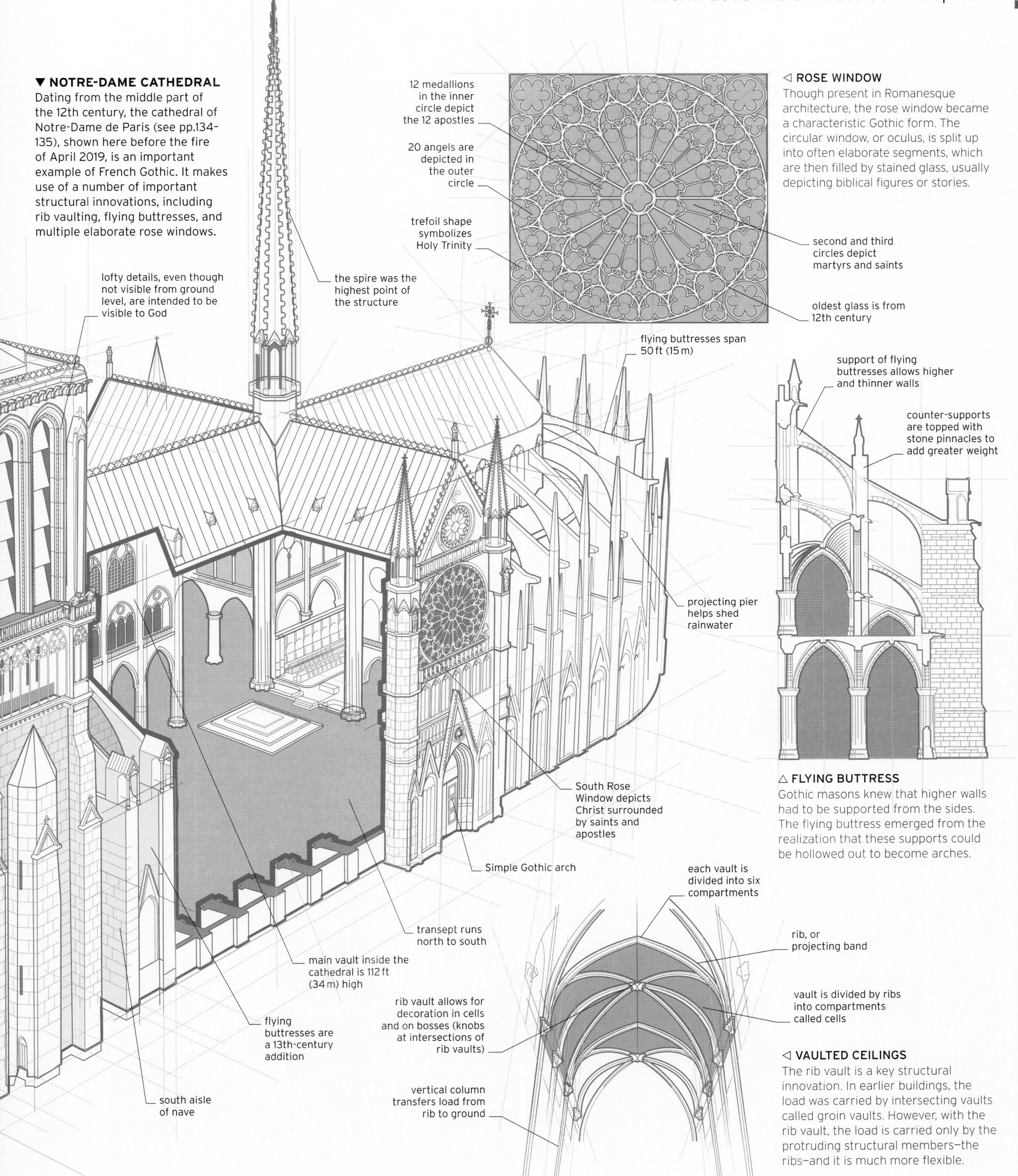

▼ NOTRE-DAME CATHEDRAL

Dating from the middle part of the 12th century, the cathedral of Notre-Dame de Paris (see pp.134–135), shown here before the fire of April 2019, is an important example of French Gothic. It makes use of a number of important structural innovations, including rib vaulting, flying buttresses, and multiple elaborate rose windows.

◁ ROSE WINDOW

Though present in Romanesque architecture, the rose window became a characteristic Gothic form. The circular window, or oculus, is split up into often elaborate segments, which are then filled by stained glass, usually depicting biblical figures or stories.

△ FLYING BUTTRESS

Gothic masons knew that higher walls had to be supported from the sides. The flying buttress emerged from the realization that these supports could be hollowed out to become arches.

◁ VAULTED CEILINGS

The rib vault is a key structural innovation. In earlier buildings, the load was carried by intersecting vaults called groin vaults. However, with the rib vault, the load is carried only by the protruding structural members—the ribs—and it is much more flexible.

△ **THE "BISHOP'S CHAIR"**
This ornately carved 13th-century chair from the chancel, ornamented with animal heads, shows the Viking tale of Sigurd the Dragon-slayer reworked as a Christian allegory.

Heddal Stave Church

The largest of Norway's elaborately carved stave churches and a masterwork of medieval wooden construction

Rising nearly 82 ft (25 m) high, Heddal Stave Church is the largest remaining stave church in Norway. Around 1,500 of these huge structures—which are held up on strong wooden posts supporting heavy horizontal beams—were built in Norway in the Middle Ages, but only 28 survive to this day.

A history in timber

Archaeologists have been able to date the church by analyzing its timbers. The wood was harvested before 1196, and the church is thought to date from around the reign of King Sverre Sigurdsson (1177–1202), a period of extensive church building. Over the centuries, the church was altered, but in 1939, it was restored and returned close to its medieval state.

Heddal's four carved portals have been preserved; the details of animal motifs, foliage, and masks on them demonstrate the height of medieval craftsmanship. While some of the church's treasures have been moved into museums, the elaborate 17th-century altarpiece showing the Crucifixion is still used in the church.

FLOOR PLAN AND CONSTRUCTION

The sophistication of the exterior of the Church masks the simplicity of its floor plan, which follows the tradition of other Christian churches except that stone columns are replaced by load-bearing wooden posts, or staves, sunk into holes in the ground. Above the floor, a system of boxes and triangles supports the roof and the vertically fixed wall planks.

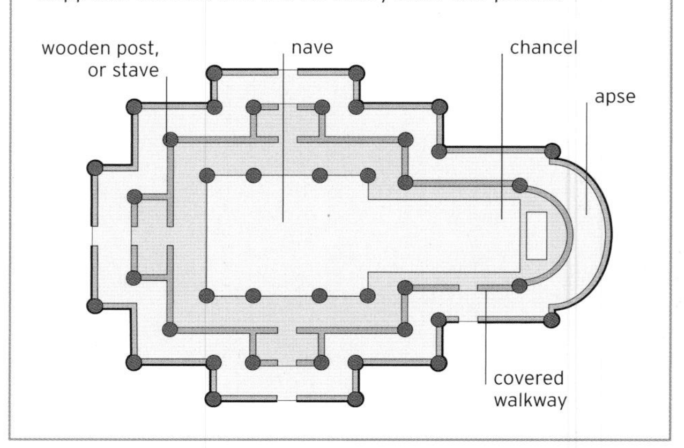

▽ **MASTERWORK OF WOOD**
Constructed around a timber framework, the church has stood for more than 800 years. Distinctive wooden shingles add to its striking appearance.

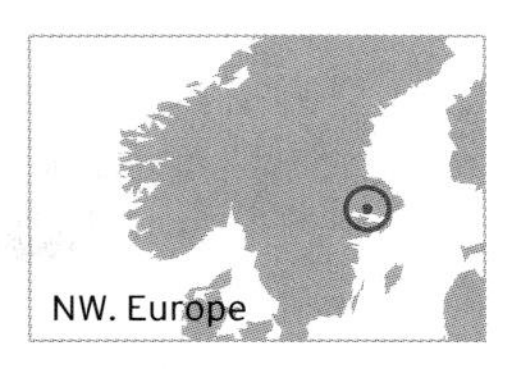

Uppsala Cathedral

The Nordic countries' largest and tallest cathedral and the seat of the primate of Sweden

Uppsala Cathedral was begun in around 1270, after an earlier cathedral located a few miles away was damaged by fire. Adverse weather, outbreaks of plague, and shortage of funds conspired to slow construction, and the building was still incomplete when it was finally consecrated in 1435. The towers at its western end were added in 1470–1489, redesigned in the 17th century, and seriously damaged by fire in 1702.

Later addition

In 1885, the Swedish architect Helgo Zettervall, a proponent of the Neo-Gothic style, made major—and often criticized—changes to the cathedral; he added the tall spires that make the cathedral as high (at 390 ft/118.7 m) as it is long. For centuries, until it lost the role to Stockholm's Strokyraas in 1719, Uppsala held the coronations of the kings of Sweden. Following the Reformation, several of Sweden's kings and queens were buried here, along with several bishops and notable scientists, including the pioneering botanist Carl Linnaeus.

△ **NEO-GOTHIC RENOVATION**
The Coronation Vault, 89 ft (27 m) high, was decorated in the 1880s in Neo-Gothic style. The renovations also uncovered some original medieval frescoes.

▷ **BRICK-BUILT CATHEDRAL**
Uppsala was built largely in brick in the Baltic Gothic style seen across northwest and central Europe, where stone for construction was scarce.

BALTIC GOTHIC

The cathedral was designed by French master builders, including Étienne de Bonneuil, and its Latin-cross floor plan is typical of other Gothic cathedrals from the 12th century. However, due to lack of locally available stone, the cathedral was constructed in red brickwork; only its chancel pillars and some details were built from the more usual limestone blocks. This distinctive style is known as Baltic or Brick Gothic.

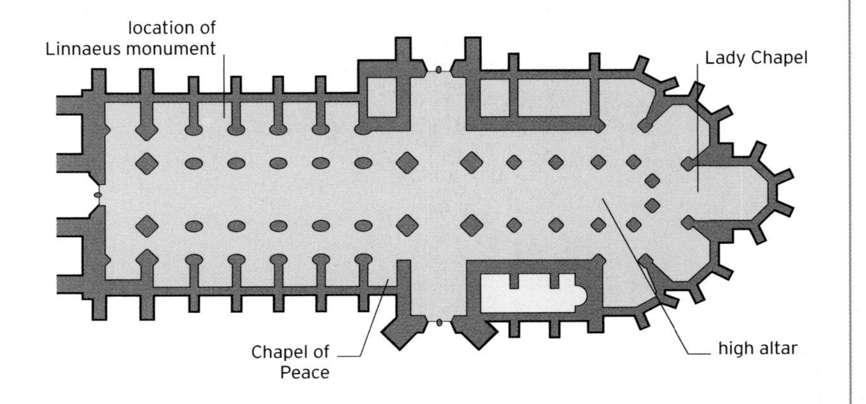

△ **MEDIEVAL GRANDEUR**
Covering an area of 161,500 square ft (15,000 square m) and ringed by 10-ft- (3-m-) thick walls and numerous towers, the Palais des Papes dominates the town of Avignon.

Palais des Papes

A grand monument to papal power in the 14th century and the largest Gothic palace in the world

In 1309, Pope Clement V abandoned Rome and moved the papal capital to Avignon on the banks of the River Rhône, where it remained until 1377. The vast papal palace was constructed by two popes in less than 20 years, from 1335 to 1352. Pope Benedict XII's Palais Vieux ("Old Palace"), completed in 1342, was built around a central cloister and housed the consistory (papal court), treasury, two chapels, and a cavernous Great Tinel—a reception and banqueting hall, where the cardinals came together to elect a new pope. The whole palace was flanked by several enormous towers. In 1342, Pope Clement VI began enlarging the palace with the addition of the Palais Neuf ("New Palace"), including the Grande Chapelle ("Great Chapel"), Audience Hall, and more towers. The palace deteriorated after the papacy returned to Rome, was sacked by French revolutionary forces in 1789, and was further damaged by military occupation afterward. Restoration began in 1906, and the palace was classified as a World Heritage Site in 1995. The surviving paintings and frescoes reveal the beauty of the original decoration.

PALACE PLAN

This plan shows the main construction phases of the palace, identifying the extents of Benedict XII's Palais Vieux and Clement VI's Palais Neuf. The palace is characterized by the number, thickness, and height of its towers and by its crenellated walls, which made it an impenetrable fortress.

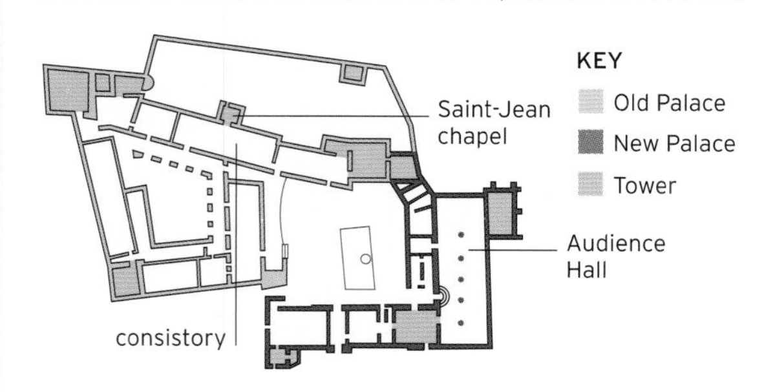

More than 1,500 church and lay **officials worked at the palace** in the early 14th century

△ PRICELESS FRESCOES
As major patrons of the arts, the popes filled the palace with paintings, including these frescoes from the Chapel of St. Martial.

▷ INSIDE VIEW
The Tour de la Campane and gilded statue of the Virgin Mary can be seen from a walkway around the cloister in the Old Palace.

The Alhambra

A spectacular citadel and palace rising above the city of Granada, the finest surviving example of Moorish architecture in Spain

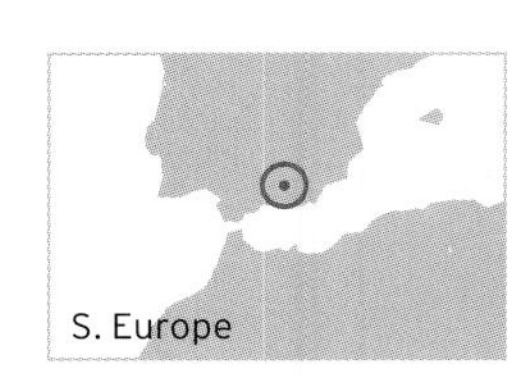

Perched strategically on a hill, with a commanding view of the surrounding region, the Alhambra was originally built in the 9th century as a military stronghold. "Alhambra" means "the red," referring to its distinctive brickwork, but it became far more than just a fortress. Under the Nasrids (1232–1492), the last Muslim dynasty to rule over Iberia, parts of it were transformed into a sumptuous palace.

The main improvements were carried out in the 14th century by Yusuf I and Muhammad V. The materials used for decoration were modest enough—mostly ceramic tiles, stucco, and timber—but the quality of the craftsmanship was outstanding. Every available surface was festooned with vegetal and geometric patterns or calligraphic inscriptions. More impressive still was the exquisite honeycomb vaulting in the large reception halls and the intricate stalactite decoration of the arches.

After the Moors

After the Reconquest of Iberia in the 15th century, the Catholic kings lived in the Alhambra for a time, adding a palace of their own, but by the 18th century, it was abandoned, falling into disrepair. Among other things, it became a gypsy encampment, a military hospital, and a prison. It was rediscovered by the Romantics in the 19th century, who were enchanted by its decaying beauty. The importance of the site was eventually recognized, and it was declared a National Monument in 1870. It has since been accepted as a World Heritage Site (1984).

▽ **COURT OF THE LIONS**
Located in the heart of the harem section of the palace, this fountain with its 12 marble lions dates from the reign of Muhammad V (1338-1391).

PHASES OF CONSTRUCTION

The Alhambra retains elements from every phase of its history. The Alcazaba is the remains of the original fortress; the Hall of the Ambassadors houses the throne room and reception space where the sultan received visiting dignitaries; and the Church of Santa María de la Alhambra was converted from a mosque after the Reconquest.

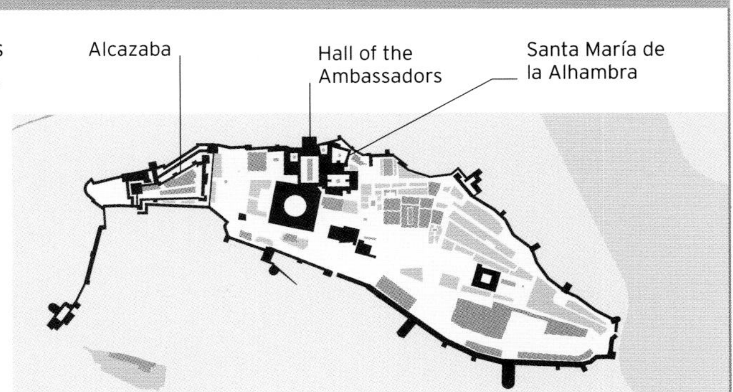

Many of the Alhambra's **interior arches** have **no structural function**; they were designed solely for **ornamental purposes**

◁ **MOORISH DECORATION**
Seen here is a part of the decoration in the Relaxation Hall of the Royal Baths, situated near the Court of the Lions.

▽ **MAGNIFICENT VISTA**
The walls and towers of the Alhambra overlook the city of Granada and beyond it the Plain of Granada.

Notre-Dame Cathedral

France's famous medieval Gothic cathedral, renowned for its rose windows and grotesque statues popularized by Victor Hugo

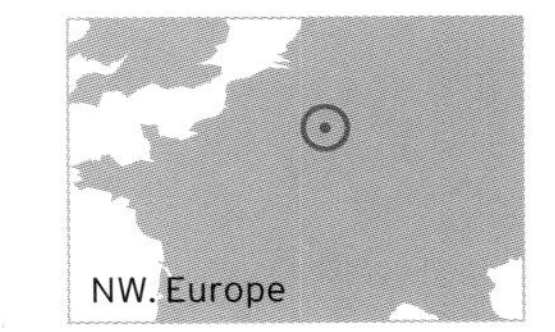

Standing on the eastern end of the Île de la Cité in Paris, Notre-Dame Cathedral was begun by Maurice de Sully, bishop of Paris, in 1163. A succession of four master builders worked to complete the choir, western facade, and nave by 1250; more chapels were added over the following century.

A Gothic masterpiece

The cathedral's two massive three-story towers, single-arch flying buttresses, and three large rose windows make it characteristically Gothic (see pp.126–127). However, it was set apart from contemporary church buildings by the naturalism and scale of its sculptures. Its three huge portals were decorated with vividly carved Biblical scenes, as well as symbols of medieval science and philosophy; these created a *liber pauperum* ("poor people's book") through which Biblical stories could be understood even by the illiterate. The cathedral's exterior was also decorated with extraordinary carved grotesques, fantastical monsters that served either to ward off evil or channel rainwater. They also provided inspiration for Victor Hugo's novel *The Hunchback of Notre-Dame* in 1831. The cathedral was desecrated during the French Revolution of 1789, but Hugo's novel revived interest in the building, and in 1845, French architect Eugène Viollet-le-Duc began restoring it. In April 2019, the spire and most of the roof were destroyed by fire, although the cathedral's main stone structure remained standing.

The **largest** of the cathedral's **10 bells** is known as **Emmanuel** and weighs **14 tons (13 tonnes)**

◁ **SPIRE AND SAINTS**
Twelve copper statues of the Apostles surround the 19th-century, lead-covered oak spire, which was destroyed in the 2019 fire.

△ **HOLY ISLAND**
This view of Notre-Dame and the Île de la Cité from the east shows the large flying buttresses around the apse and the south transept's rose window.

GOTHIC FACADE

Notre-Dame's Gothic west facade spans 134 ft (41 m). It incorporates three portals—the largest is the Portal of the Last Judgment—and a 31-ft- (9.6-m-) wide rose window. The verticality of the high towers is balanced by horizontal galleries to create an elegant but powerful elevation.

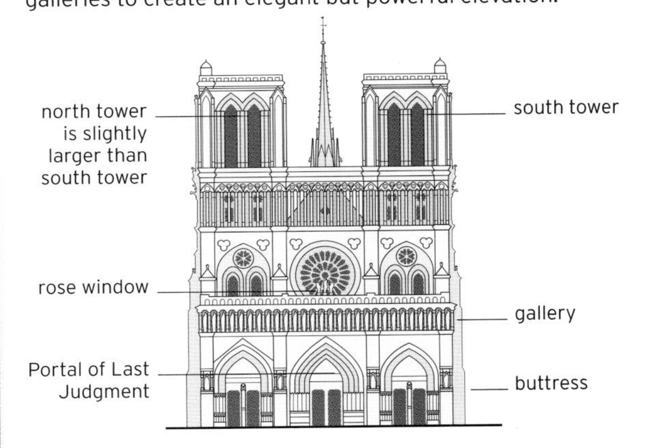

Ferapontov Monastery

Home to the works of a Russian fresco master, the most complete example of a Russian monastic complex from the 15th–17th centuries

Founded by St. Ferapont in 1398, in the Vologda region north of Moscow, Ferapontov became one of Russia's most powerful monasteries under the patronage of the descendants of Ivan III (1440–1505), the first Grand Prince of All Rus. In 1490, Abbot Ioasaf, spiritual advisor to the Grand Prince, rebuilt the original wooden Church of the Nativity of the Virgin in brick and commissioned the renowned artist Dionisy to paint its interior. Covering every wall and ceiling of the church, Dionisy's frescoes remain largely undamaged and richly colored, presenting depictions of an array of miracles, archangels, saints, church fathers, and figures of Christ Pantocrator ("Ruler of All"). The Church of the Annunciation, a refectory, the treasury, and household buildings were added in the early 16th century, and—having survived the ravages of the Time of Troubles (1598–1613)—the monastery grew to include the Church of St Martinian, the gateway church, and a bell tower. With its elegant white facades, tented towers, and onion domes, Ferapontov Monastery provides a beautiful ensemble of the features that dominated Russian architecture from the 15th–17th centuries.

RUSSIAN ROOFS

Onion domes (from the 13th century onward) and polygonal, steeply pitched tent towers (from the 16th century) were common features of Russian church architecture until the late 17th century, and—as shown here—Ferapontov included examples of each. Constructed around a wooden frame, they were invaluable where stone for construction was scarce and useful in preventing the build-up of snow on church roofs.

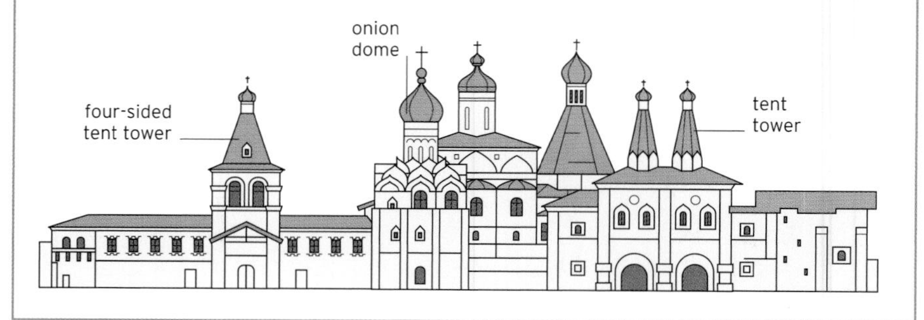

△ **BEAUTY IN SIMPLICITY**
This image reveals the beautifully pure and simple exteriors of Ferapontov's buildings: from left to right, the Church of St. Martinian, the Church of the Nativity of the Virgin, the Bell Tower, and the Church of the Annunciation.

△ **VIBRANT INTERIOR**
The fresco of Christ Pantocrator ("Ruler of All") inside the dome of the Church of the Nativity of the Virgin at Ferapontov Monastery was painted around 1502 by Russian artist Dionisy.

Český Krumlov Castle

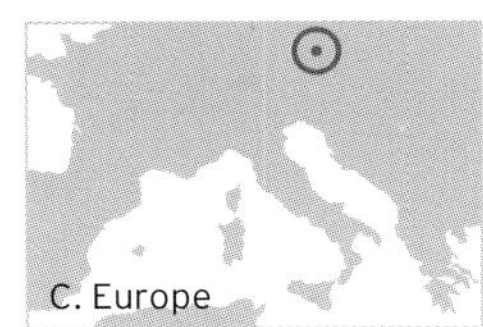

A castle complex built in the 14th–19th centuries on a rocky promontory shaped by the Vltava River

Český Krumlov Castle was founded in the mid-13th century by the powerful Witgonen family, then lords of Krumlov. It passed through the hands of three more noble Czech families—the Rosenbergs (1302–1602), the Eggenbergs (1602–1719), and the Shwarzenbergs (1719–1947)—before being handed to the Czech state in 1947. Declared a national monument in 1989, it became a UNESCO World Heritage Site in 1992.

Homage to history

The castle grew to incorporate over 40 buildings—including palaces, stables, a salt-house, a brewery, and a dairy—built around five courtyards and a large formal garden. Its buildings reflect a range of architectural styles, from the Gothic to the Renaissance and Baroque. Among them is the Gothic *Hrádek* ("Little Castle") with its beautiful Renaissance facade, tower, and frescoes, and a Baroque theater from 1766 that is the most complete example of its kind in the world. Krumlov's massive ramparts are a reminder of the castle's original—and never seriously tested—defensive function, while the theater and the bears that inhabit the moat are a curious reminder of its history as an important social and cultural hub in Southern Bohemia.

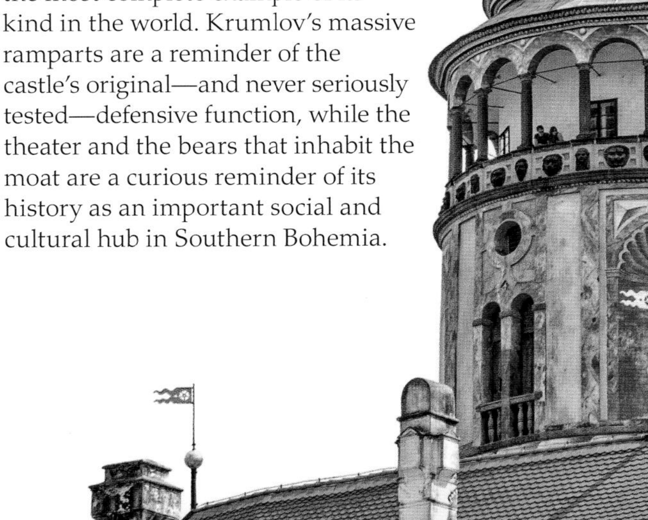

▽ **TOWERING ABOVE**
The brightly painted, part-Gothic, part-Renaissance six-story Castle Tower has a belfry and a galleried balcony leading up to the clockworks, which enclose the clock bells.

CLIFFTOP CASTLE
Bran Castle sits on a cliff at a height of 2,500 ft (762 m). It is arranged over four floors with narrow, winding stairways leading through 60 rooms, some of which are connected by secret passages.

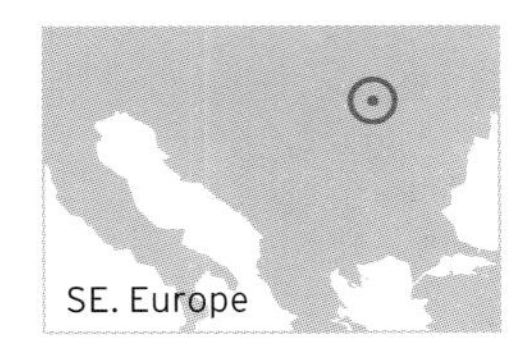

Bran Castle

Popularly known as Dracula's castle, a Gothic fortress and palace in the Carpathian mountains

For more than 600 years, Bran Castle has dominated a mountain pass on the border between the historic regions of Transylvania and Wallachia. Now in Romania, in the 14th century, it lay within the domains of the kings of Hungary. German settlers from Saxony built the castle, with royal permission, between 1377 and 1388. The tall, thick outer walls proved a necessary defensive feature; the fortress, on the exposed edge of Christian Europe, was vulnerable to the armies of the Muslim Ottoman Empire that rampaged northward from Turkey and the Balkans.

Dracula's castle?

The castle's association with Vlad III the Impaler, the notoriously cruel 15th-century ruler of Wallachia said to have inspired the blood-sucking fictional character Dracula, is dismissed by most historians as an invention of the modern tourist industry. Vlad the Impaler earned his nickname by impaling his enemies on pointed stakes. Dracula's creator, the Irish novelist Bram Stoker, never visited the castle. It is possible, however, that Vlad may have been imprisoned briefly in the castle's vaults in 1462.

After World War I, when the castle passed from Hungary to Romania, it became the favorite palace of the Romanian Queen Marie, who was attracted to its romantic solitude. The building now exhibits the queen's valuable collection of furniture and art. Although the castle's connection with Dracula may be fictitious, its haunting atmosphere and spectacular setting guarantee its place as one of Romania's most successful tourist destinations.

RESHAPING THE CASTLE

The original 14th-century fortress was built from wood and stone and had a skewed rectangular footprint; its walls were studded with fire holes, now converted into windows. The 16th century saw the building of new towers that transformed the building plan and the addition of glass windows and roof tiles.

Prince Mircea Chapel

inner courtyard

GROUND FLOOR

guard room

Mystras

Perfectly preserved ruins of a medieval city on towering heights above the Spartan plain

In 1249, William of Villehardouin, a French crusader knight, built a castle at Mystras on Mount Taygetos in the Peloponnese, southern Greece. His fortress was absorbed into the Greek Byzantine Empire, and a town spread across the slopes. By the 14th century, Mystras was a flourishing center of culture and learning, its richly endowed monasteries and churches embellished with beautiful frescoes.

Eventual decline

From 1460, when it was taken over by the Ottoman Turks, the city went into decline. As a consequence, it changed little through the centuries until 1832, when the last of its shrinking population had abandoned the lofty mountain site. Mystras has survived as an almost perfect ensemble of late Byzantine architecture, both religious and secular. It has been a UNESCO World Heritage Site since 1989.

◁ **PERIBLEPTOS MONASTERY**
Built into the side of a cliff, in the mid-14th century, the castlelike Peribleptos Monastery is one of many religious buildings in Mystras.

▽ **MYSTRAS FROM ABOVE**
Villehardouin's castle stands on the hilltop above Mystras, with the palace of the Byzantine rulers of the city—known as the Despots of Morea—below it.

◁ **FLORENTINE MASTERPIECE**
Florence Cathedral presents an imposing structure both through its size and its beauty. Four million bricks were used to construct the famous dome.

Florence Cathedral

A masterpiece of Renaissance architecture with a soaring dome that dominates the skyline of the city of Florence

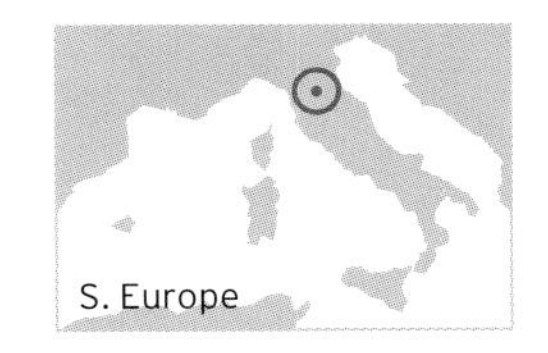

In 1296, the citizens of Florence, then a wealthy and independent city-state in Tuscany, announced a plan to build a cathedral, "an edifice ... so magnificent in its height and quality that it shall surpass anything of its kind produced ... by the Greeks and the Romans." The chosen architect, Arnolfo di Cambio (c.1240–1310), designed a building on an impressive scale, with massive pillars along the nave supporting an arched stone roof. At the east end, a vast octagonal space was to be capped by a dome. Unfortunately, although the ancient Romans had built large domes, their engineering knowledge had been lost with the passage of time.

Reviving Classical knowledge

Florence Cathedral remained unfinished for over a century, until the task was handed to architect Filippo Brunelleschi (1377–1446). True to the spirit of the Renaissance—the rediscovery of the learning of the ancient world—Brunelleschi traveled to study the ruins of ancient Rome. He returned to Florence primed to begin work on the dome in 1420. The result, declared complete in 1436, was a triumph of engineering, as well as a thing of beauty. The dome was subsequently topped by a stone lantern, designed by Brunelleschi but completed after his death by his follower Michelozzo.

The interior of the cathedral is austere, but the exterior is decorated with bands of colored marble, matching the nearby Giotto Campanile and Florence Baptistery. The church's extremely elaborate facade is a 19th-century addition. The Cattedrale di Santa Maria del Fiore, as it is formally known, is at the center of a UNESCO World Heritage Site and is one of the world's most visited tourist attractions.

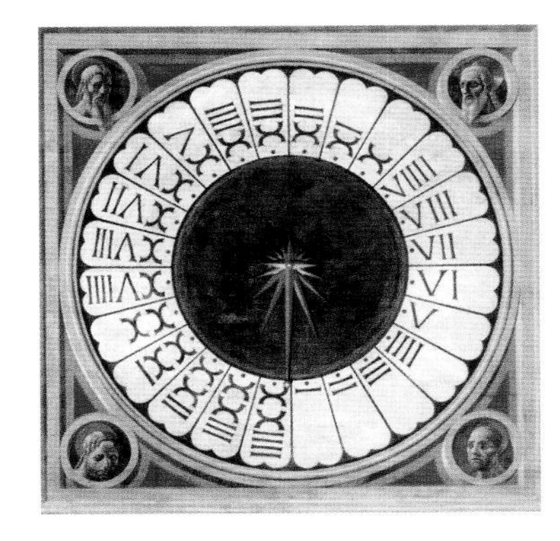

△ **CATHEDRAL CLOCK**
A one-handed clock inside the cathedral measured time from sunset to sunset. The clock face was painted by Paolo Uccello (1397–1475).

▽ **BAPTISTERY CEILING**
The octagonal Baptistery standing next to the cathedral has a richly decorated interior. The mosaics on the ceiling date from the 13th century.

The **cathedral dome** spans **138 ft (42 m)** and rises to **374 ft (114 m)** above the ground

BRUNELLESCHI'S DOME

The dome of Florence Cathedral is constructed on an octagonal base. From this rises a lightweight skeleton of stone ribs, over which an inner and an outer skin are laid, chiefly of brick. The lantern, plugging a 20-ft (6-m) hole at the top of the dome, helps to hold it together, as well as adding to the height.

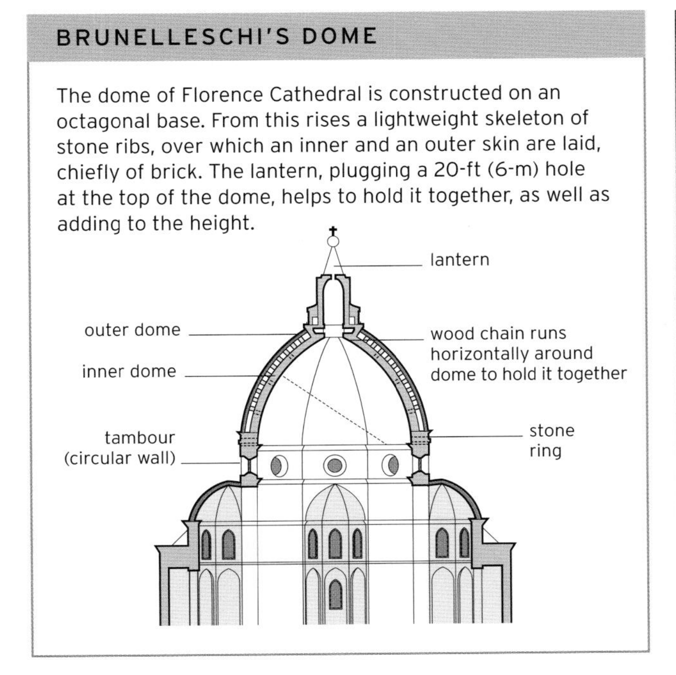

△ **ARCHED CROSSING**
The three-arched Ponte Vecchio connects the center of Florence with the district of Oltrano to its south; Oltrano means "across the Arno."

Ponte Vecchio

A medieval survivor spanning the Arno River as a monument to commerce

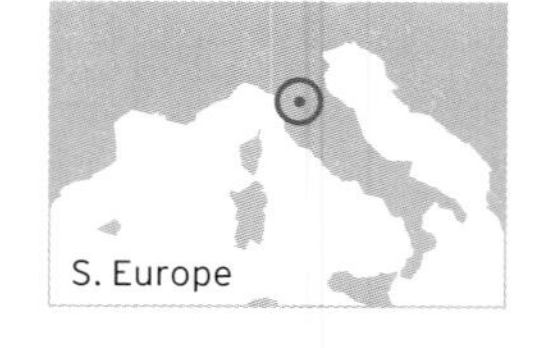

When the Ponte Vecchio—the shop-lined bridge that crosses the Arno River at its narrowest point in Florence, Italy—was built in 1345, it would not have been that special. At that time, most urban bridges were covered with shops and stalls, as well as spaces for tables where merchants could show their wares. What makes the Ponte Vecchio special is that the shops have survived to the present day. Initially, butchers plied their trade on the bridge; today, it is jewelers, art dealers, and tourist souvenir sellers. Interestingly, the concept of bankruptcy started on the bridge, as a money-changer's table (*banco*) would be broken (*rotto*) by soldiers if he could not pay his debts.

Third time lucky

The Ponte Vecchio is the third bridge to have stood on this site. The first was built by the Romans, with stone piers and a wooden superstructure. After that bridge was destroyed in a flood in 1117, it was rebuilt, but again swept away in 1333. The current bridge has a main arch span of 98 ft (30 m), with two side arches each 89 ft (27 m) in width. The arches rise 11½ ft–14½ ft (3.5–4.4 m) above the river. Running along the top of the bridge is Vasari's Corridor, an enclosed, private corridor built by Giorgio Vasari, court architect to the Medici ruling family, in 1565. The corridor connects the Palazzo Vecchio, Florence's town hall, with the Palazzo Pitti, the Medici family palace, on the south bank.

△ **INTERESTING ADDITIONS**
Shop owners have made alterations to the bridge's superstructure by adding windows and external shutters, giving it a colorful and chaotic look.

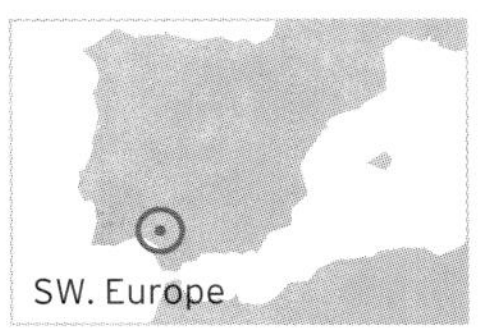

Seville Cathedral

A vast Christian edifice built to supplant one of the Moorish treasures of Islamic Spain

In 1172, the Almohad caliph of southern Spain ordered the building of the Great Mosque in Seville. After the city was reconquered by the Christian Castilians in 1248, it was converted into a cathedral that served until 1401, when the city leaders decided to build a new place of worship on the same site. This new cathedral, dedicated to St. Mary, was completed in 1528.

Built in Gothic style, it was covered with ornament and carving and filled with sculpture, paintings, tombs, and memorials. Two parts of the original mosque were retained: the bell tower (La Giralda) of the cathedral was built on the minaret of the Great Mosque, and the courtyard entrance was inherited from the Moors.

▽ **MASSIVE PROPORTIONS**
At 443 ft (135 m) long, 328 ft (100 m) wide, and 138 ft (42 m) high, Seville Cathedral was the biggest cathedral in Christendom and remains the largest to this day.

LA GIRALDA

The original Islamic minaret (left) was built in 1198 to resemble the minaret of the main mosque (see p.228) in Marrakech, Morocco. After the Christian reconquest of Seville, it was converted into a bell tower for the cathedral and then extended higher in 1568.

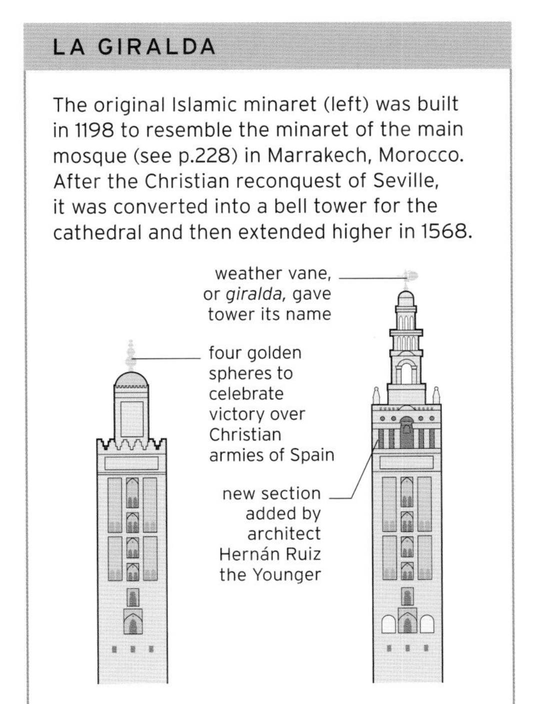

The Doge's Palace

A lavish building that embodies the glory and cruelty of the historic Venetian republic

For more than a thousand years, from 726 to 1797, the maritime city-state of Venice was a republic ruled by an elected doge, or duke. The luxurious palace built for the doge and his administration in the 14th and 15th centuries displays the extraordinary wealth and influence the city had attained by that point in its history through the skill of its sailors and the rapacity of its merchants. Much of Venice's trade was with the Muslim world, and this is reflected in the unique style of the palace's facades, which give a distinctively Islamic twist to European Gothic architecture.

Center of law and government

The palace fulfilled a range of functions, serving as living quarters for the doge, government offices, a venue for receiving foreign ambassadors, a parliament, a courthouse, and a prison. The vast rooms of the state apartments are embellished with scores of paintings by Renaissance artists, mostly trumpeting the glories and triumphs of the Venetian republic.

The palace also holds plentiful evidence of the darker aspects of the Venetian state, which earned a fearsome reputation for ruthlessness and secrecy. Political prisoners, tried in secret, were incarcerated in cells under the roof of the palace, known as the *Piombi* (the "leads"), or in the gloomy basement rooms of the *Pozzi* (the "wells"). A new prison was built alongside the palace in 1600, connected to it by the elegant Bridge of Sighs, named for the prisoners' presumed expressions of woe. After the end of the era of the doges, the palace still housed administrative offices until 1923, when it became a museum and found its modern vocation as a magnet for tourists.

△ **LAVISH CEILINGS**
The palace's state rooms have ornate gilded ceilings embellished with paintings by Venetian Renaissance masters such as Veronese and Tintoretto.

The palace contains one of the **world's largest** oil paintings, **Tintoretto's *Paradise***

◁ **ANGEL STATUE**
This stone angel is one of the figures placed at the corners of the facade of the Doge's Palace. The startlingly original building created a distinctive Venetian Gothic style of architecture.

△ **FLOATING PALACE**
The palace seems to float on its own reflection. Built of brick with marble facing, it is a light structure suited to stand on foundations in island mud.

VENETIAN GOTHIC STYLE

The facades of the palace have open arcades on the lower two floors, with typically Venetian Gothic pointed arches. The delicately carved stonework of the capitals and quatrefoils above the columns gives a magical effect of lightness. The upper wall is faced with pink and white marble in a bold decorative pattern.

quatrefoil design

carved capital

slender column

Gothic pointed arch

Architectural styles

RENAISSANCE

Originating in Italy, the Renaissance saw the rebirth of Classical learning in Europe. In architecture, this manifested through a decisive break with medieval traditions and the embrace of the architectural styles of the ancient Roman world.

The word Renaissance derives from the Italian *la rinascita*, meaning "rebirth." Coming to the fore in the 15th century, the rebirth in question was of Classical models of learning, philosophy, literature, art, and architecture (see pp.96–97). Central to this was the emergence of Humanist thought, which prioritized human agency over received wisdom and ecclesiastical doctrine. Architecturally, this was manifested through a revival of interest in the architecture of the ancient world and the attempt to reposition architecture as an intellectual, liberal art as opposed to a craft.

The architect Leon Battista Alberti was the foremost theorist of the early Renaissance, with his influential treatise *De re aedificatoria* (1454) arguing for Classical architecture's fundamental basis on geometry deriving from nature. Subsequent treatises further codified Renaissance architectural principles, allowing for their dissemination and application across Europe.

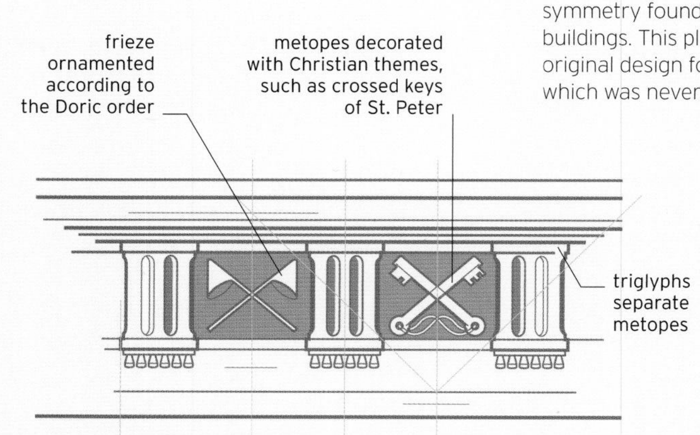

△ **FRIEZE**
Some Renaissance architects took the opportunity to tailor Classical building codes and systems to particular building types and patrons—for example, in the ornamentation of a frieze.

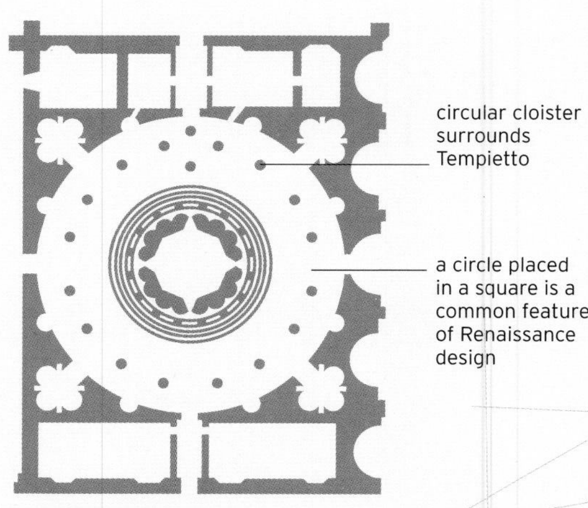

△ **SYMMETRY**
Renaissance architects were deeply concerned with incorporating the symmetry found in nature in their buildings. This plan shows Bramante's original design for the Tempietto, which was never fully realized.

The **only surviving book on architecture** from the ancient world was *De Architectura* by the Roman architect **Vitruvius**

CLASSICAL ARCHITECTURE REBORN

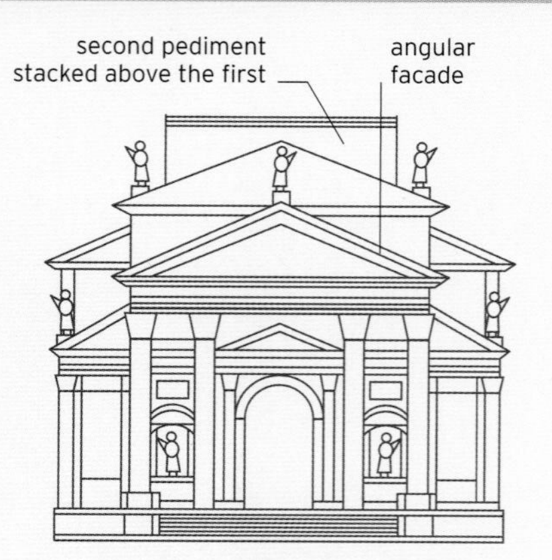

△ **SUPERIMPOSED PEDIMENTS**
A superimposed pediment is one pediment placed above or behind another. It allows ceilings of two different heights but is also used as an expressive device.

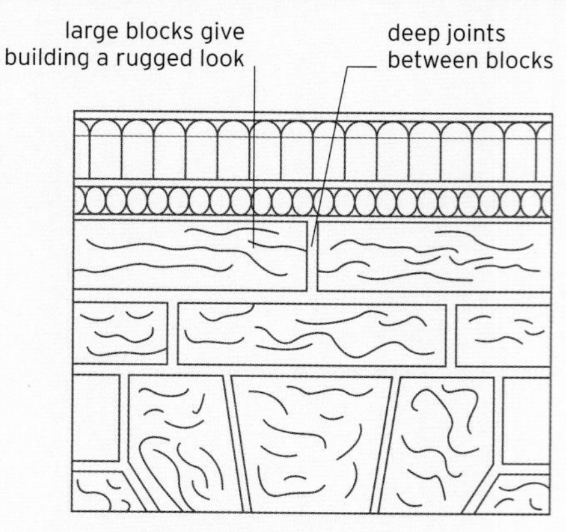

△ **RUSTICATED STONEWORK**
Rustication is the accentuation of the joints between adjacent stone blocks to emphasize their weight and solidity. It is frequently used in lower stories.

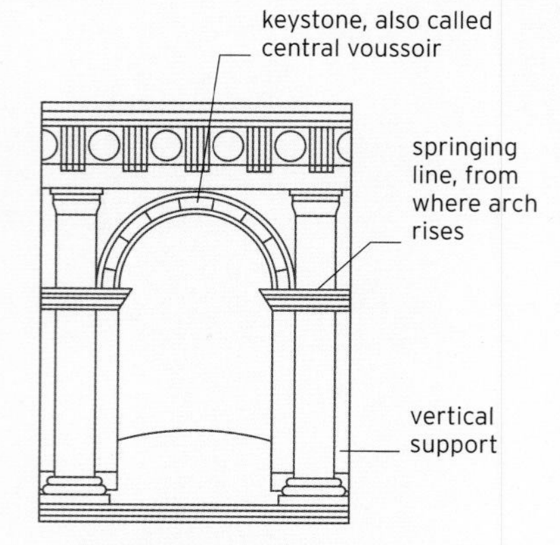

△ **ROUNDED ARCH**
Gothic architecture was founded on the pointed arch. Renaissance architecture, in contrast, reverted to the rounded arch of the Romans, which it used in new ways.

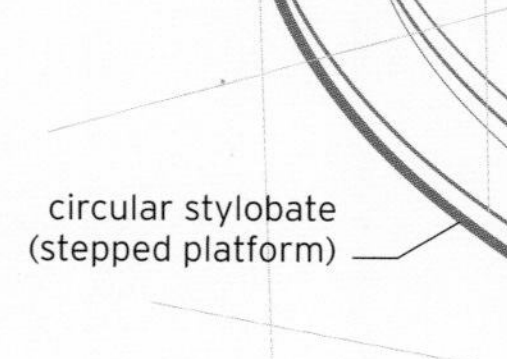

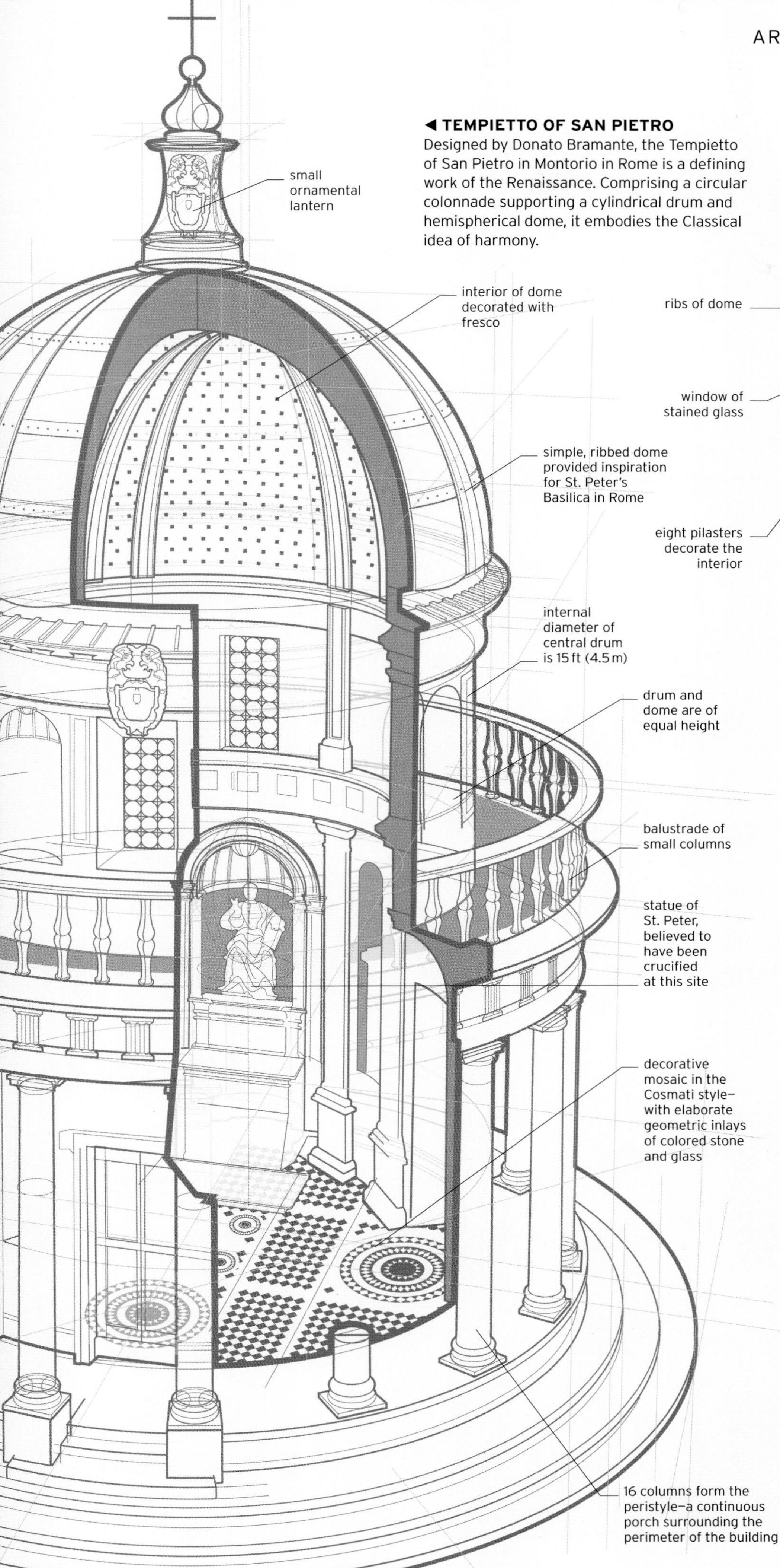

◀ TEMPIETTO OF SAN PIETRO
Designed by Donato Bramante, the Tempietto of San Pietro in Montorio in Rome is a defining work of the Renaissance. Comprising a circular colonnade supporting a cylindrical drum and hemispherical dome, it embodies the Classical idea of harmony.

ribs of dome
window of stained glass
eight pilasters decorate the interior

△ DOMED CEILING
The underside of a dome was typically articulated, or segmented, in one of two ways: architecturally with sunken panels of diminishing size called coffers, or, as seen in the Tempietto, figuratively via mural painting in fresco (painting on wet plaster).

interior entablature echoes the exterior
pilasters are ornamental, only giving the appearance of support
statue of St. Peter, holding the keys of heaven in one hand and the gospel in the other

▷ PILASTERS
Taking the form of a column flattened against a wall, the pilaster is a key element in Renaissance architecture. Without any structural purpose, it is used as a decorative element in articulating the extent of a wall.

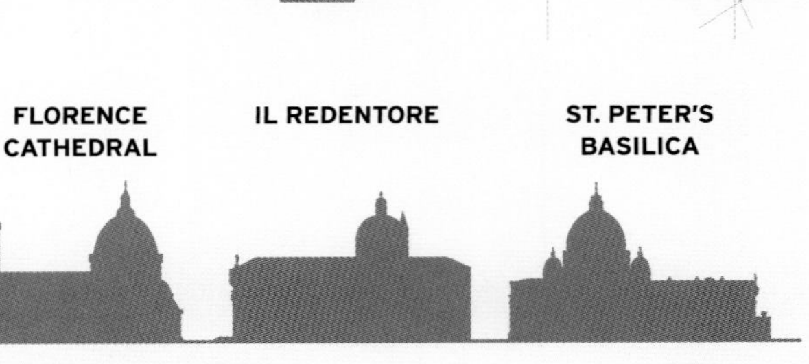

RENAISSANCE DOMES
While Gothic architecture made use of the spire to top a cathedral or church, Renaissance architects favored the dome for its symmetry, geometrical purity, and structural stability.

Villa Capra

A Palladian villa that is among the most influential buildings in the history of Western architecture

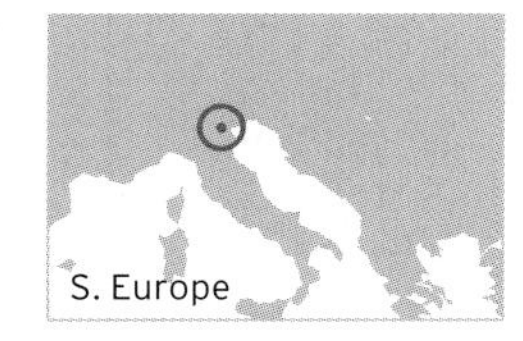

Standing on a hill outside Vicenza in northern Italy, the Villa Capra, also known as La Rotonda, was designed by Venetian architect Andrea Palladio (1508–1580). It was built as a country house for a high-ranking clergyman, prelate Paolo Almerico, who had returned to Vicenza after a career in the Vatican Court, and its vibrant interior was decorated with spectacular frescoes celebrating Christian virtue.

Harmony and serenity

Begun in 1567, Villa Capra reflects Palladio's profound study of the buildings of ancient Rome. It is an almost perfectly symmetrical structure, with a domed cylindrical central hall surrounded by identical columned porticos, modeled on Roman temples such as the Pantheon (see pp.104–105), on all four sides. As a result, the villa appears balanced and sophisticated from whichever angle it is viewed, and its careful placement in the landscape gives a harmonious view of the "natural theater" surrounding it. Completed after Palladio's death in 1580 by his pupil Vincenzo Scamozzi, the building has since been an inspiration to architects across the Western world, who have been fascinated by the perfection of its formal beauty.

PRINCIPLES OF PERFECTION

Palladio's design is based on strict mathematical proportions intended to create a mystic harmony between the elements of the building. The basic floor plan is a circle within a square–forms that symbolized the perfection of the Universe to Renaissance thinkers. The building is precisely oriented so that the north-to-south and east-to-west axes pass through the corners of the square.

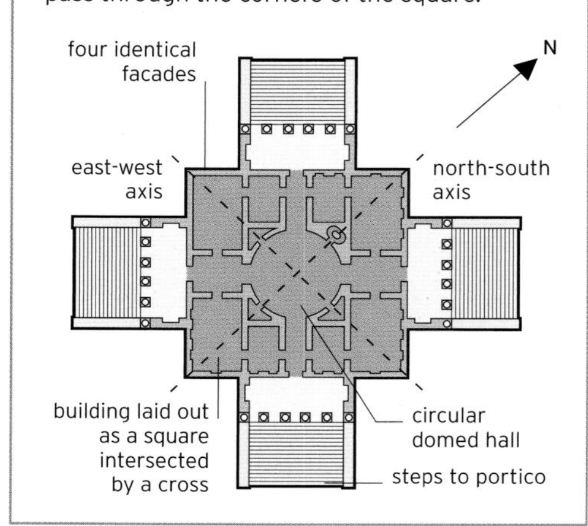

▽ **CLASSICAL STYLE**
Palladio's inspired use of Classical proportion and natural space made him one of the most successful architects of his time. He built more than 20 villas in the Veneto region.

THE CITY PALACE
Called a "city in the shape of a palace" by Castiliogne, the palace was already huge when Laurana took over construction in 1464. The Dalmatian architect added the facade with the twin turrets, the courtyard, and other parts of the complex.

PLAN OF THE PALACE

The Ducal Palace was built around a courtyard, with the main living quarters on the first floor, or *piano nobile*, shown here. The facade overlooking the valley had a balcony between twin towers. The Duke's study, the *studiolo*, was a tiny room only 12 ft (3.6 m) wide, but exquisitely decorated with inlaid wooden panels.

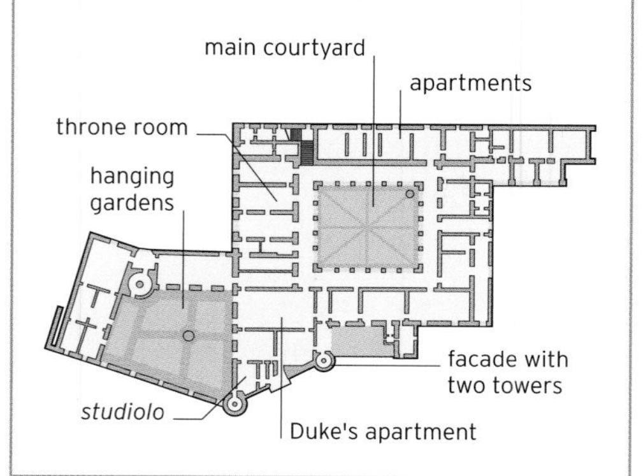

Palazzo Ducale

An architectural masterpiece designed as a setting for the life of a cultured Renaissance court

Dominating the walled town of Urbino in the Marche region of Italy, Palazzo Ducale ("Ducal Palace") was created to satisfy the social and cultural ambitions of Federico da Montefeltro (1422–1482), a minor aristocrat who had won fame and fortune as a mercenary general. Duke Federico aspired to make his court a renowned center of humanist learning and civilized manners.

Courtly life

From 1465, Federico employed architect Luciano Laurana to rebuild and extend the existing castle at Urbino. Laurana created two slender towers, with a loggia (covered area) between them, providing fine views over the surrounding hills. Inside the building, a monumental stairway led from an arcaded courtyard to the upper stories. Federico's devotion to Classical learning was evident in a magnificent library, an exquisite personal study (the *studiolo*), and a "temple of the muses" dedicated to the deities of ancient Rome.

Succeeding Laurana after 1472, engineer Francesco di Giorgio Martini installed an innovative plumbing system that showed regular baths were a part of the Duke's ideal of civilized living. The wit and elegance of the courtly life at Urbino were later celebrated by Baldassare Castiglione in his *Book of the Courtier* (1528). The palace now houses an outstanding collection of Renaissance art.

△ **MAIN COURTYARD**
The arcaded courtyard, with its Classical proportions, is among the oldest parts of the palace. Built in pale stone and brick, it is topped with an inscription that celebrates the glory of Federico.

CITY WITHIN A CITY
Behind the red walls of the Kremlin can be seen the gold-domed towers of its cathedrals, as well as the vast Grand Kremlin Palace. The Grand Palace, completed in 1849, includes the 17th-century Terem Palace, nine churches, and more than 700 rooms.

△ **TYPICALLY RUSSIAN**
The Grand Kremlin Palace is adorned with Russian-style window surrounds, which have double arches and drop ornaments and are decorated with the Russian eagle.

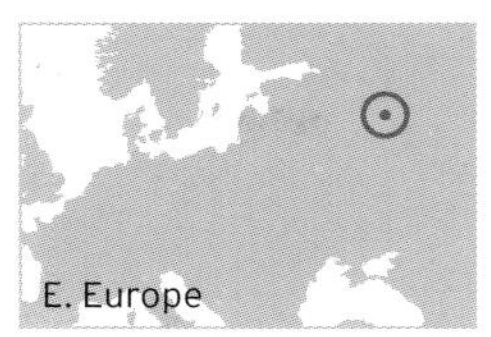

The Kremlin

The symbolic heart of Russian power for over 600 years

The Moscow Kremlin is a fortified triangle on the banks of the River Moskva that served as the seat of the Grand Dukes of Moscow, the residence of the Russian czars, and the home of the Soviet leaders, and is now the official residence of the President of the Russian Federation. Its crenellated, red brick walls, 1.5 miles (2.5 km) long and punctuated by 20 towers, were built at the end of the 15th century at the request of Ivan III (the Great), Grand Prince of Moscow. They enclose an area of nearly 3,200,000 square ft (300,000 square m), within which are an eclectic assemblage of buildings, including three cathedrals and four palaces.

Fusion of styles

The Kremlin's architecture ranges from the simple white, Italianate-Byzantine beauty of the 15th-century Cathedral of the Assumption built for the Grand Prince, to the 17th-century Moscow Classical-style Senate building built in the reign of Catherine the Great, and the Modernist glass and concrete Soviet-era State Kremlin Palace.

△ **POWERFUL PORTAL**
The *Spasskaya* ("savior") Tower, once the Kremlin's main entrance, was believed to have the power to protect the Kremlin from invaders.

A "kremlin" is a **fortified complex** within a town; **five Russian kremlins**, including the one in Moscow, are **UNESCO World Heritage Sites**

◁ **COLORFUL CATHEDRAL**
Frescoes of Biblical scenes and saints' lives cover the interior of the Kremlin's Dormition Cathedral, the coronation church of Russian monarchs from 1547 to 1896.

EAGLES AND STARS

After the 1917 Russian Revolution, the Soviet leaders wanted to destroy all imperial symbols. In 1935, they removed the metalwork imperial eagles from four of the Kremlin's towers and replaced them with red five-pointed stars. The original stars soon dulled and were replaced with 1-ton (0.9-tonne) glass stars measuring up to 12 ft (3.75 m) across that light up and revolve.

DOUBLE-HEADED EAGLE

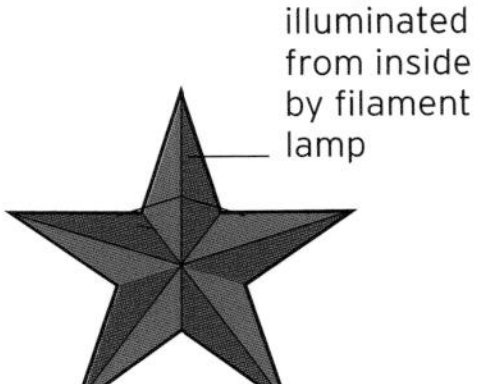

KREMLIN STAR

St. Basil's Cathedral

Ivan the Terrible's unparalleled, colorful 16th-century Orthodox church named after a Muscovite shoemaker and saint

In 1554, the first Czar of Russia, Ivan Vasilyevich—known as Ivan the Terrible—ordered work to begin on the construction of a cathedral to honor his conquest of the independent khanates of Kazan and Astrakhan. The cathedral was completed in 1561 and was officially known as the Cathedral of the Intercession of the Virgin on the Moat, because it was built on the moat that separated the Kremlin (see pp.150–151) from the town. It is more popularly known as St. Basil's, after the "Holy Fool," Basil the Blessed, who was buried in the church after his death in 1557. Legend says the Czar blinded the cathedral's architect, Postnik Yakovlev, to prevent him from building a more beautiful church.

Religious and secular uses

Set on white stone foundations, St. Basil's was built in red brick around an internal wooden frame; where stone was used, it was painted to look like brick. Its onion domes were originally hemispherical and gilded; the cathedral's bright colors were added from the 17th century on. Later, its interior was also covered with intricate murals and icons. St. Basil's survived many fires and escaped the destruction planned by Napoleon's army in 1812 and Stalin's regime in the 20th century. A museum since 1929, it now forms part of the Red Square and Kremlin World Heritage Site. Religious services are held once a year on the Day of the Intercession in October.

St. Basil's **symbolized** the Bible's **Heavenly City**, depicted as a **turreted citadel** in old manuscripts

THE CATHEDRAL'S CHAPELS

The cathedral complex is composed of nine chapels. The Central Chapel of the Intercession of the Virgin is the largest, rising to an internal height of 151 ft (46 m). Four large chapels are aligned to the points of the compass, and four smaller chapels are raised on platforms. The chapel dedicated to St. Basil was added in 1588.

St. Basil the Blessed Church

N

Central Chapel of the Intercession of the Virgin

△ ONION DOMES
An aerial view shows the chapels' onion domes, the tented tower of the belfry (top right), and the green-roofed galleries that connect the buildings.

TOWERING CATHEDRAL
The Central Chapel of the Intercession of the Virgin towers over St. Basil's other chapels, whose heights indicate their saints' importance in the heavenly hierarchy.

Château de Chenonceau

The loveliest of the châteaux on the Loire and, after Versailles, the most visited of all French castles

Standing slightly apart from the main body of the château, a 15th-century circular tower is the only surviving remnant of a fortified manor house, which in 1513 passed to Thomas Bohier, a royal chamberlain. His wife, Katherine Briçonnet, rebuilt the manor into the château seen today.

The Ladies' Castle

Chenonceau is known as the Ladies' Castle because women played a major role in its construction and preservation. Diane de Poitiers (1500–1566), the mistress of Henri II, had the greatest impact. In the 1550s, she commissioned architect Philibert de l'Orme to build the fine, arched bridge and long gallery spanning the Cher River; she also oversaw the creation of the magnificent gardens. Queen Catherine de' Medici, Henri's widow, later forced Diane to exchange Chenonceau for another château, and under her the upper floor of the gallery was added by architect Jean Bullant in 1576–1578. Mary, Queen of Scots, stayed at the château briefly, but the most tragic inhabitant was Louise de Lorraine. After the assassination of her husband, Henri III, she became a recluse here, surrounding herself with somber tapestries decorated with skulls, crosses, and burial shovels. The château is currently owned by the Meniers, a famous family of chocolatiers.

CHAMBERS OF THE RIVAL LADIES

The two great rivals favored the same living area in the château. Diane de Poitiers' chamber is on the ground floor and Catherine de' Medici's bedroom is directly above it. The latter is currently furnished with early Flemish tapestries and an ornate four-poster bed.

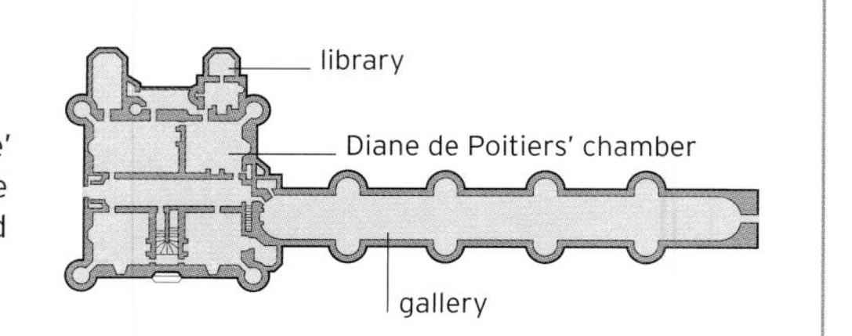

◁ **GRAND GALLERY**
Chenonceau's stunning gallery is 197 ft (60 m) long. With its elegant chalk-and-slate tiled floor, it served as a ballroom in its heyday.

▽ **RIVAL GARDENS**
Diane de Poitiers' garden, shown here, was planted on the right side of the castle, while Catherine de' Medici created hers on the left.

RENAISSANCE SPLENDOR
The overall design of Chambord displays an elegant simplicity and symmetry. In contrast, its profusion of chimneys and turrets is remarkably varied.

Château de Chambord

Built as a hunting lodge, the largest of the Loire châteaux and a masterpiece in the Renaissance style

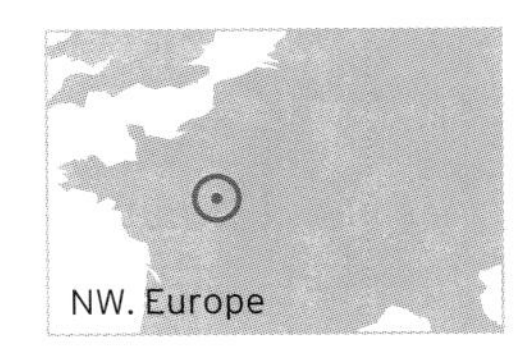

Chambord was the brainchild of the young king François I (1494–1547). In 1519, he demolished a fortress on the site and replaced it with a monumental hunting lodge. Size was everything, as he wanted to impress visiting guests and dignitaries. The work took many years and cost a fortune; when it was completed, Chambord contained 440 rooms, 365 fireplaces, and 83 staircases. In addition, the fence around its well-stocked deer park was the same length (20 miles/33 km) as the *périphérique*—the ring road around Paris.

Impressive skyline

Chambord's most spectacular feature is its double-spiral staircase—a masterpiece of design and engineering. Its roof is also remarkable, with an astonishing array of chimneys, cupolas, and gilded lanterns (roof turrets). François, it seems, wanted to rival the skyline of Constantinople. In spite of the care he lavished on the place, François only stayed for around 7 weeks. Chambord was never meant to be a home, and its high ceilings and open loggia made it cold. There were later inhabitants, but for long periods, it lay empty. Chambord was added to UNESCO's World Heritage Site list in the 1980s.

◁ **SPIRAL STAIRCASE**
Chambord's staircase is set in a huge, lanternlike case in the central axis of the château. It has no parallel among other buildings of the time.

DOUBLE HELIX

The ingenious staircase enables two groups of people to use the stairs without ever meeting, although they can see each other through gaps in the structure. Rumors persist that the designer was Leonardo da Vinci, who was a guest of François I around the time of construction.

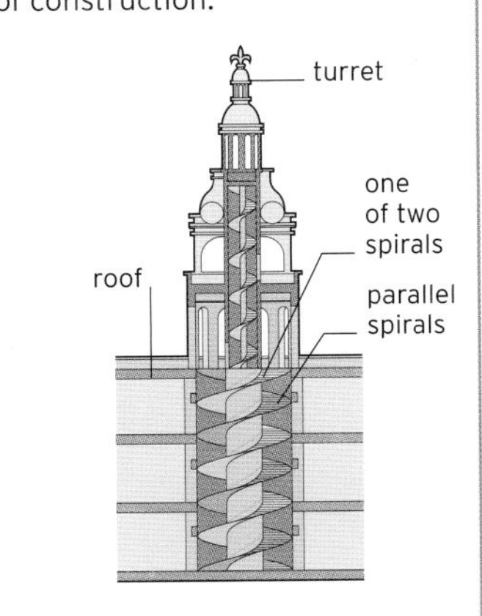

Wooden Churches of Kizhi

A pair of multidomed churches and a bell tower constructed entirely of wood in traditional Russian style

On a spit of land on the island of Kizhi, in Lake Onega, northwest Russia, is the site of two remarkable 18th-century churches, the Church of the Transfiguration and the Church of the Intercession, together with a 19th-century bell tower. The three buildings are constructed entirely of wood: the walls are built from logs of Scots pine, the roofs are made of spruce planks and shingles, and the domes are covered with shingles of aspen—weathering has made them appear silvery. It is the distinctive use of multiple onion domes that makes the Kizhi churches unique. Unlike other traditional Russian wooden churches with a simple pyramidal roof, the octagonal Church of the Transfiguration features 22 cupolas with decorative shingles.

BUILDING WITH WOOD

The churches were constructed using traditional Russian carpentry techniques—the horizontal logs of the walls were fitted together with interlocking or dovetail joints. The only nails used in the construction were for securing the wooden roof tiles that were layered together to form the onion domes.

central post

layered shingles

strut

scarf joint joins two pieces of wood

▽ TRADITIONAL TIMBER CONSTRUCTION
The Church of the Transfiguration (center), along with the Church of the Intercession (left) and the Belfry (right) behind it, are masterpieces of Russian joinery. They were built using only basic tools, such as axes and chisels.

Frederiksborg Castle

The Renaissance-style residence of King Christian IV of Denmark-Norway, now the Danish Museum of National History

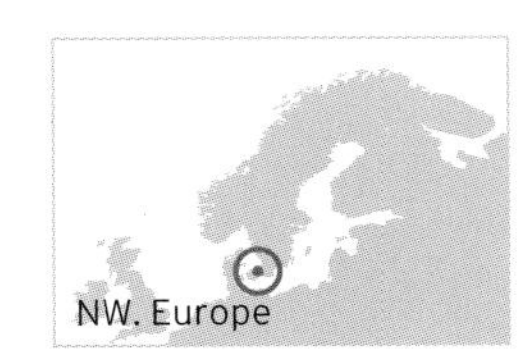

△ **THE KNIGHTS' HALL**
The spacious central hall, like the rest of the interior of the castle, is rich in elaborate architectural features. It is ornately decorated in the Baroque style and contains a wealth of tapestries and artwork.

Built in the early 17th century, the imposing Frederiksborg Castle stands on one of the three small islands in the Castle Lake at Hillerød, north of Copenhagen. The main building, the King's Wing, is symmetrical, with a quadrangle at its center. However, the complex as a whole, which includes the Chapel Wing, Princess's Wing, and Terrace Wing, is not symmetrical.

Castle to museum

The buildings are primarily made of red brick, with sandstone decoration. They have steep roofs and gables in the Dutch Renaissance style but also feature several towers and spires, and a wealth of sculpted decorations celebrating the reign of King Christian IV (r.1588–1648). The castle was almost completely destroyed by a fire in 1859 but was reconstructed as nearly as possible to the original design, inside and out. Once completed, the castle was opened to the public as a museum in 1882.

BAROQUE GARDEN

Frederiksborg Castle stands in extensive grounds, originally a park for hunting deer. Frederik IV (r.1699–1730) commissioned court gardener Johan Cornelius Krieger to create a formal garden to complement the Baroque style of the castle. This featured a fountain and cascades running through terraces of geometric formal flowerbeds.

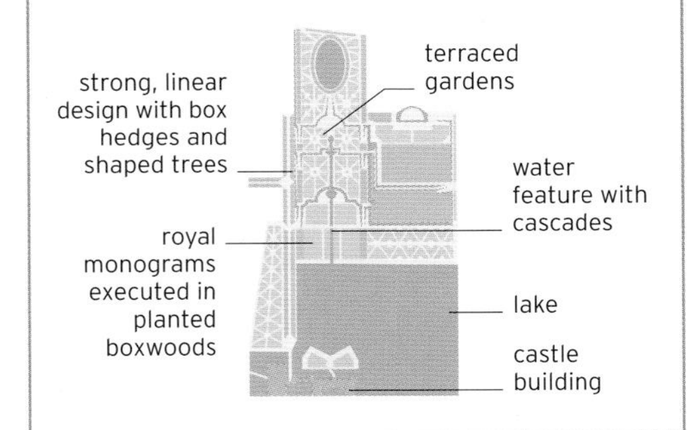

△ **THE KING'S WING**
Overlooking the lake and gardens, the facade of the main castle building combines the seriousness of the Dutch Renaissance style with Baroque ornamentation and flourishes.

St. Peter's Basilica

The most celebrated Roman Catholic place of worship, awe-inspiring in its scale and lavish decoration

St. Peter's Basilica is built over the presumed site of the tomb of the apostle Peter. Located in Vatican City, an independent city-state in Rome ruled by the Pope, it attracts millions of pilgrims and tourists every year. The interior of the building, enriched with marble, gilt, and mosaics, impresses above all with its size: its nave is 610 ft (186 m) long, and its dome rises 446 ft (136 m) above the floor.

Making the second basilica

The first basilica on the site was built in around 330 CE. In 1506, Pope Julius II made the radical decision to replace this decaying edifice with a new structure, albeit one made using stones plundered from ancient Roman ruins. His favorite architect, Donato Bramante (1444–1514), set to work on a church of epic size.

Both Julius II and Bramante were long dead by the time the project was completed. Many famous Renaissance figures worked on St. Peter's, including Michelangelo (1475–1564), who designed the dome. The long nave and the imposing facade, with its array of columns and statues, were completed by 1615. Work continued on the embellishment of the interior under Baroque sculptor and architect Gianlorenzo Bernini (1598–1680), who was responsible for the huge bronze *baldacchino* (canopy) over the tomb of St. Peter.

△ **WORK OF MANY HANDS**
Statues line the facade of St. Peter's with the mighty dome rising behind. Many of Italy's most famous artists worked on the creation of the basilica.

ST. PETER'S PIAZZA

Located directly in front of the basilica, St. Peter's Piazza (or St. Peter's Square) provides a gloriously theatrical setting. Designed by Bernini and built between 1656 and 1667, the piazza is surrounded by two sweeping lines of Tuscan columns, each line four rows deep. The columns, topped by Baroque statues, gather the faithful toward the facade of the basilica. The arrangement allows the maximum number of people to see the Pope speaking from a balcony above the main door.

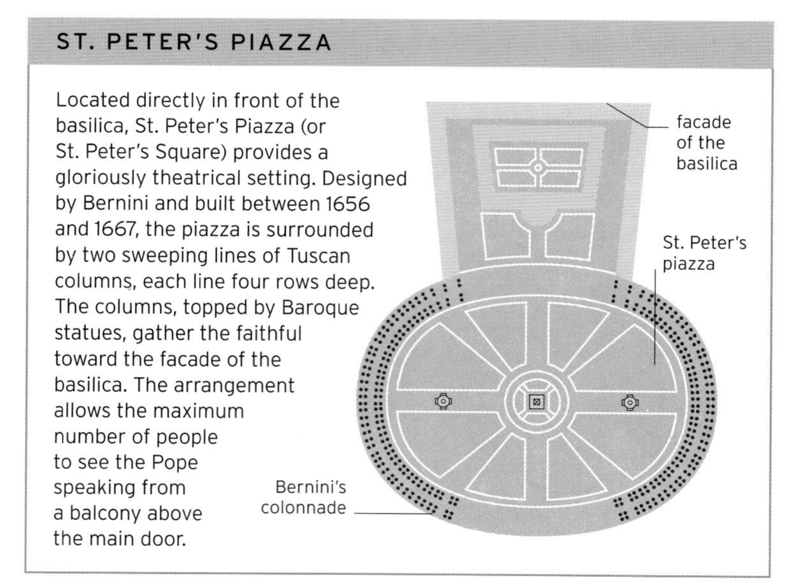

△ **BIRD'S-EYE VIEW**
This view shows the splendid setting for the basilica designed by Bernini. The ancient Egyptian obelisk was erected in the center of the piazza in 1586.

EXTRAVAGANT DECOR
The interior of St. Peter's is a treasure trove of Renaissance and Baroque art. On the left is one of the piers of Bernini's bronze *baldacchino*.

Sistine Chapel

Vatican chapel embellished by some of the greatest masterpieces of Renaissance art

Forming part of the Papal palace at the Vatican in Rome, the Sistine Chapel was built on the orders of Pope Sixtus IV in 1475. It is the room in which the conclave of cardinals meets to elect a new Pope, but its fame rests even more upon the quality of the art it contains. Sixtus commissioned some of the leading painters of his time, including Pietro Perugino, Sandro Botticelli, and Domenico Ghirlandaio, to cover the chapel walls with frescoes. In 1508, Sixtus's nephew Pope Julius II tasked the Florentine artist Michelangelo Buonarroti with painting the chapel ceiling. It took Michelangelo 4 years to complete this astonishing work, which covers an area of more than 5,400 square ft (500 square m). Between 1536 and 1541, as a much older man, he returned to paint the wall fresco of the Last Judgment, often considered his masterpiece.

THE CEILING AND ALTAR WALL

Michelangelo's ceiling of the Sistine Chapel illustrates scenes from the biblical Book of Genesis, from the creation of the world through the fall of Adam and Eve to the story of Noah. The altar wall fresco depicts Christ's second coming and God's final judgment on humanity.

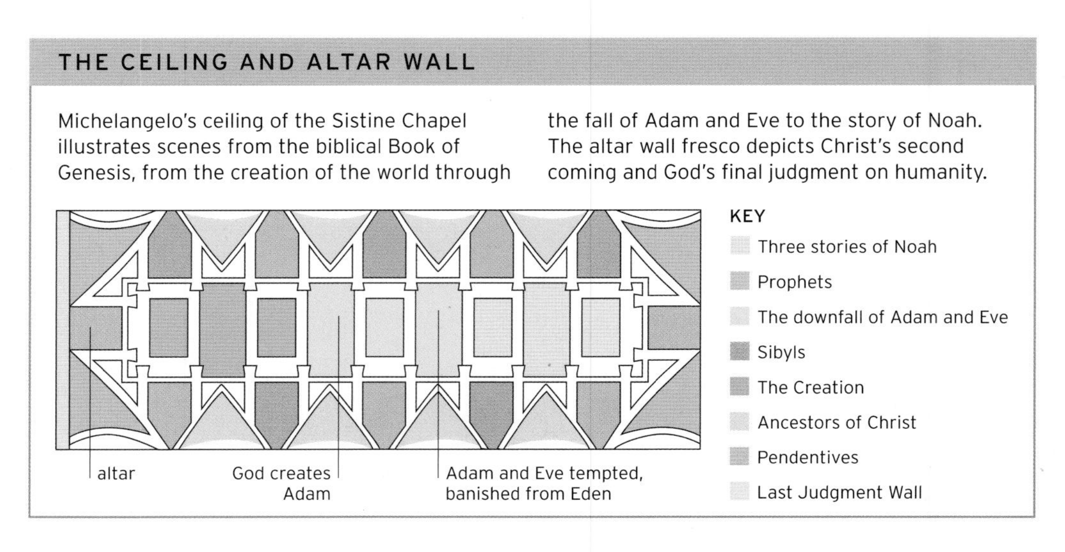

The **painting** on the Sistine ceiling contains **more than 300 figures**

▽ **CEILING FRESCOES**
Michelangelo's Sistine Chapel ceiling is brightly colored and characterized by massive, muscular figures.

Blue Mosque

An imposing Ottoman place of worship especially famed for its lavish blue-tiled interior

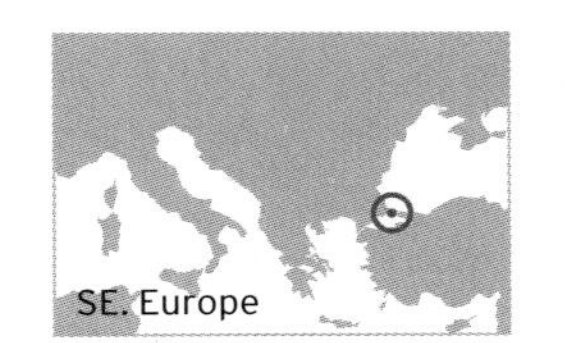

△ **OTTOMAN SPLENDOR**
The Blue Mosque was the last major mosque of the Classical period of Ottoman architecture. Here, its six minarets thrust into the sky of Old Istanbul.

Dominating the skyline of the old quarter of Istanbul, Turkey, the Sultan Ahmet Mosque, also known as the Blue Mosque, was the late flowering of an empire in decline. When the 19-year-old Ottoman sultan for whom the mosque is named ordered its construction in 1609, his empire was reeling from defeats in the wars against Austria and Iran. Creating a vast Muslim building to rival the nearby Christian-built Hagia Sophia was an act of defiance, ignoring the shrunken state of the Ottoman empire's finances.

Blue-tiled interior

Sultan Ahmet's architect, Sedefkar Mehmed Aga (c.1540–1617), designed the mosque complex, incorporating a religious school (*madrasa*) and a charitable hospice, as well as a prayer hall and courtyard. Built on a grand scale, it is topped by a central dome surrounded by four half-domes and six slender minarets. The austere gray-stone exterior contrasts with the brilliantly decorative interior of the prayer hall, lit by 200 stained-glass windows—a gift from the Ottomans' Christian ally, Venice—and lined with 20,000 blue tiles from the imperial ceramic production center at İznik. The mosque was completed shortly before Sultan Ahmet's death, at the age of 27, in 1617.

▽ **TILED INTERIOR**
The brightly colored İznik tiles lining the mosque are shown to advantage by light from the many windows in the domed ceiling.

St. Paul's Cathedral

Christopher Wren's Baroque masterpiece, whose dome has defined London's skyline for more than 300 years

The cathedral of St. Paul's, the seat of the Bishop of London, stands on top of Ludgate Hill, the highest point in the City of London. It is the fifth cathedral to St. Paul to stand on this site. Built by the English architect and polymath Christopher Wren, the cathedral was consecrated for use on December 2, 1697.

Classical features

Wren's design for St. Paul's needed to satisfy the requirements of the Anglican church and the demands of his royal patron, Charles II. Built in the English Baroque style, with a landmark dome (as on St. Peter's Basilica in Rome; see pp.158–159) and a Classical portico on the west front, the cathedral stands on eight massive piers to support its weight.

The cathedral continues to play a key role in British national life in the modern era. It became a symbol of resistance in World War II, and it was the location of the funeral of Winston Churchill in 1965 and the wedding of Prince Charles to Diana Spencer in 1981.

THE DOME

The dome of St. Paul's is among the highest in the world. It is double-shelled, with a brick cone between the two domes that supports the timbers of the outer, lead-covered dome and the weight of the ornate stone lantern that stands on top.

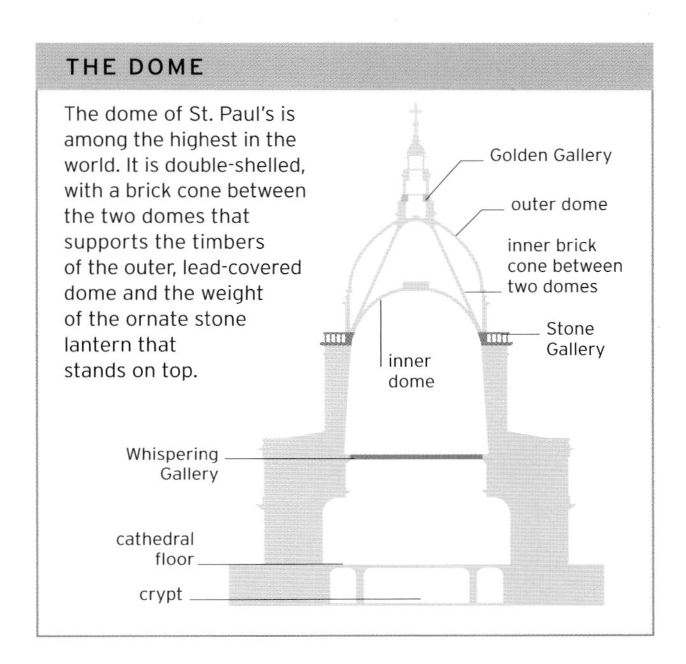

At **364 ft (111 m) high**, St. Paul's was the **tallest building** in London from **1710 to 1967**

▷ **INNER DOME**
The dome, painted by James Thornhill (1675/6-1734), features eight scenes from the life of St. Paul with a central oculus, or eye-shaped opening, inspired by the ancient Pantheon (see pp.104-105) in Rome.

◁ **STANDING ALONE**
Planning laws in the City of London limit the height of new buildings to protect the view of St. Paul's Cathedral from prominent places around the city.

Architectural styles

BAROQUE AND ROCOCO

Characterized by the effects of illusion and drama, the Baroque emerged in the 17th century as a manifestation of the spirit of the Counter-Reformation. Originating in Italy, it spread through Europe, followed by its variant form, Rococo.

The origins of Baroque are closely bound up with the doctrines of the Counter-Reformation. Emerging in response to the Protestant Reformation of the 16th century, the Counter-Reformation sought to reaffirm the principles of the Catholic faith, which the Protestants had challenged. Art and architecture became vital tools in this process, with a number of church commissions used as opportunities to express the Catholic Church's power. Architects such as Bernini, Borromini, and Guarini deployed bold forms, sweeping curves, and dramatic effects of light and shade to create buildings of amazing power and drama.

As Baroque spread over Europe, local variations emerged. Although it had originated in the context of ecclesiastical architecture, Baroque was often used in palace architecture, where it functioned more broadly as an architecture of power. And it was in this context that it developed into Rococo in the 18th century.

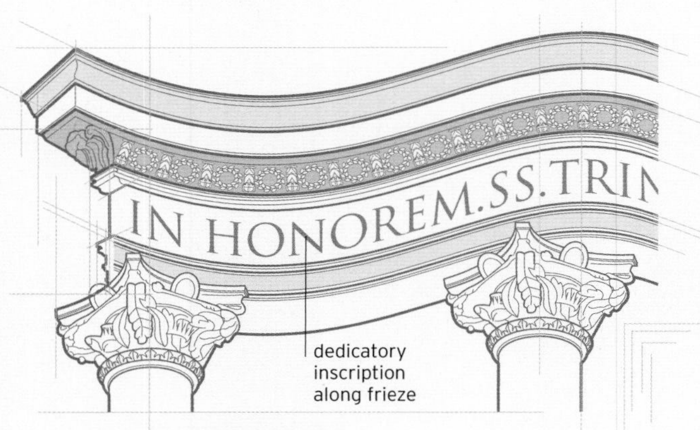

△ **CURVED ENTABLATURE**
While still depending on the Classical language for its architectural vocabulary, Baroque exaggerates its forms and proportions. One of the most recognizable examples is the way a straight entablature might be transformed into a concave or convex curving surface.

The word **Baroque** derives from the Portuguese for a **misshapen pearl**

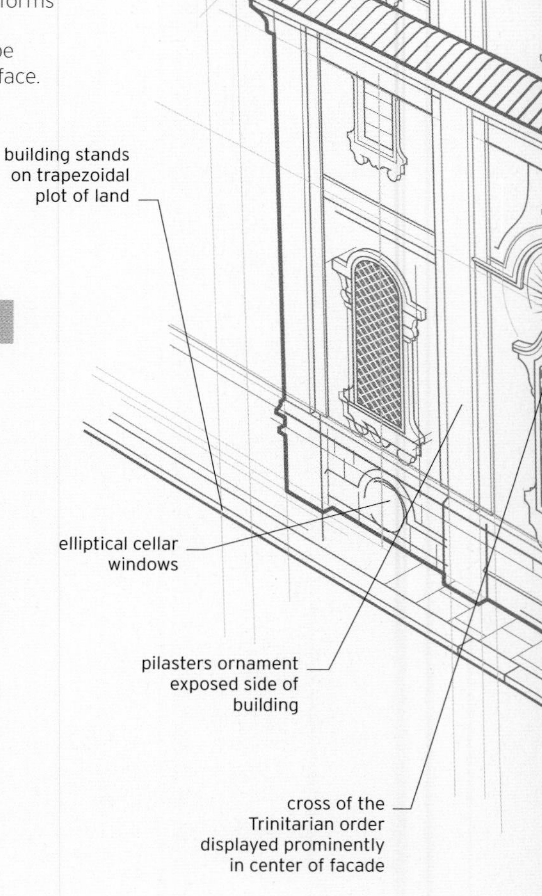

STATUES, DECORATION, AND STAIRCASES

△ **CHERUBS**
A cherub or *putto* is a small winged, usually quite chubby, boy, frequently found in scenes of the divine in Baroque art and architecture.

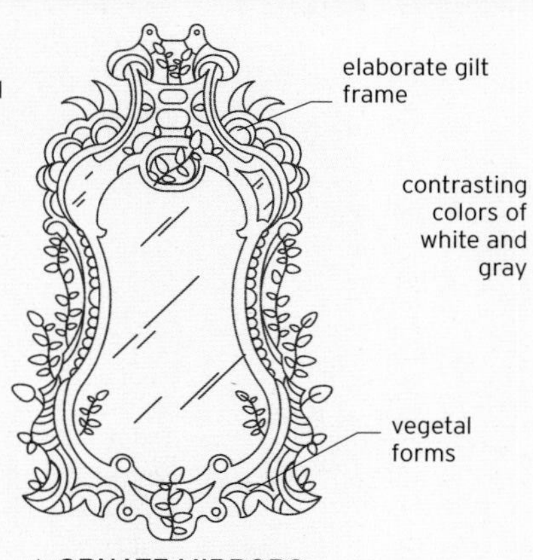

△ **ORNATE MIRRORS**
Rococo is an interior architecture style in which the forms of Classical architecture are subsumed into an almost abstract decorative language.

△ **TIERED EXTERNAL STAIRS**
The tiered staircase is a key example of the layering, bold ornamentation, and intersecting volumes that characterize Baroque.

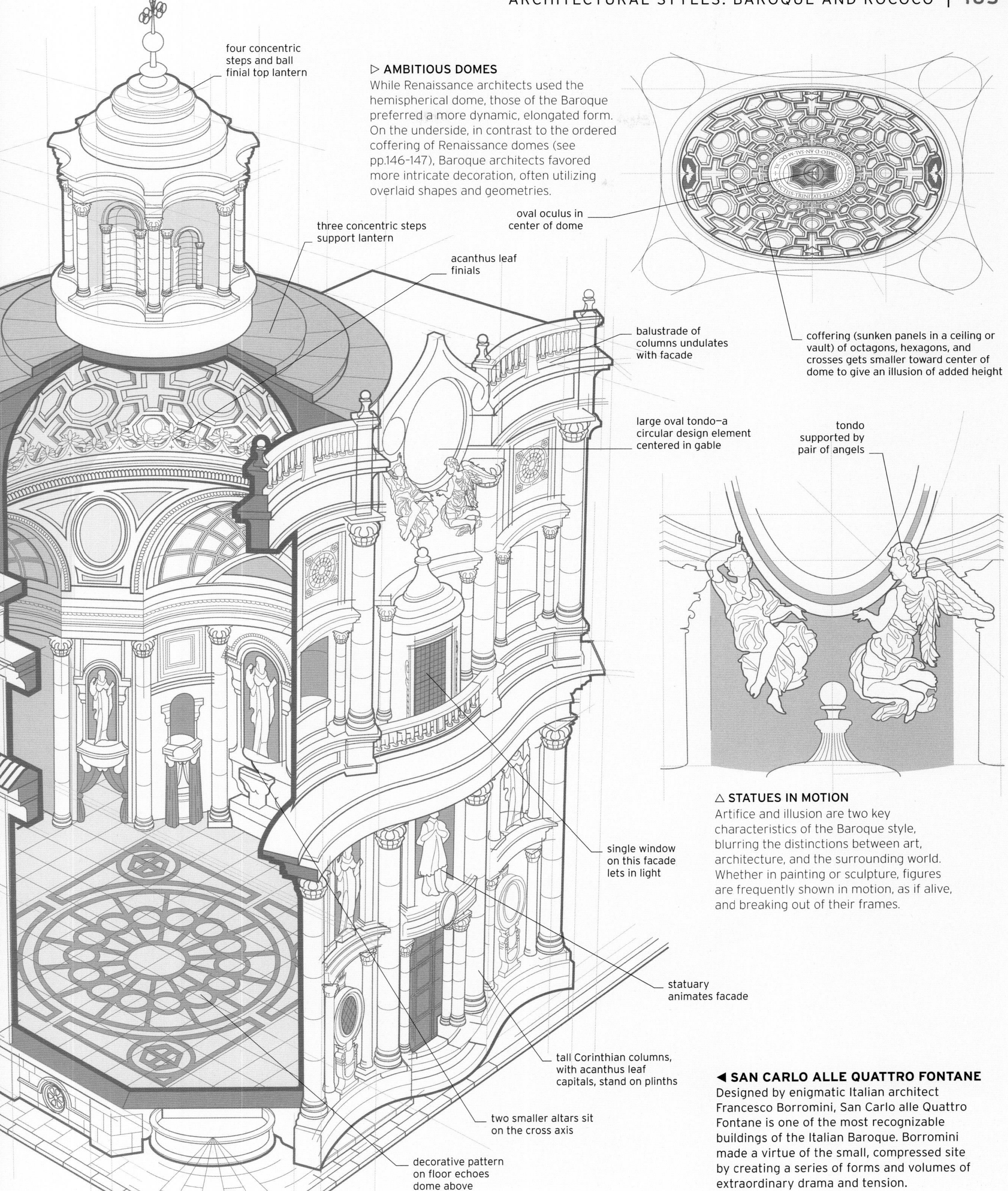

▷ **AMBITIOUS DOMES**

While Renaissance architects used the hemispherical dome, those of the Baroque preferred a more dynamic, elongated form. On the underside, in contrast to the ordered coffering of Renaissance domes (see pp.146–147), Baroque architects favored more intricate decoration, often utilizing overlaid shapes and geometries.

△ **STATUES IN MOTION**

Artifice and illusion are two key characteristics of the Baroque style, blurring the distinctions between art, architecture, and the surrounding world. Whether in painting or sculpture, figures are frequently shown in motion, as if alive, and breaking out of their frames.

◀ **SAN CARLO ALLE QUATTRO FONTANE**

Designed by enigmatic Italian architect Francesco Borromini, San Carlo alle Quattro Fontane is one of the most recognizable buildings of the Italian Baroque. Borromini made a virtue of the small, compressed site by creating a series of forms and volumes of extraordinary drama and tension.

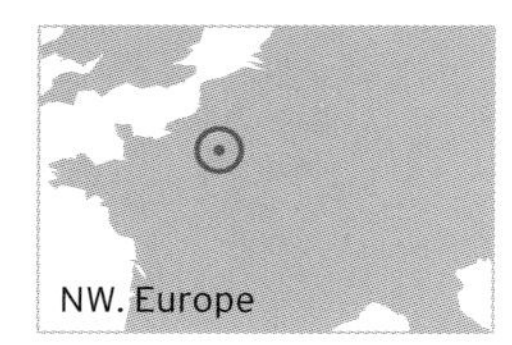

Palace of Versailles

A palace built to showcase the wealth and power of France's Sun King, Louis XIV

In the 1630s, King Louis XIII transformed a hunting lodge near the village of Versailles, outside Paris, into a modest château. His successor, Louis XIV, had ambitions on an altogether larger scale. In 1661, he ordered architect Louis le Vau to turn the building into a palace fit for an absolute monarch ruling by divine right, while the decoration of the interior was supervised by the artist Charles le Brun.

Courtly living

Louis XIV made Versailles the hub of his kingdom—the center of government and a place of court ritual where the French nobility were turned into slaves of etiquette. A new architect, Jules Hardouin-Mansart, added the splendid Hall of Mirrors, completed in 1684, as the palace continued to grow, its maze of corridors and staircases linking more than 700 rooms. Built in the grounds, the Grand Trianon—a small palace clad in pink marble—was a place of escape where the king could relax with his intimates. The Palace of Versailles served as a royal residence until the French Revolution of 1789 overthrew the monarchy. Today, it has a new vocation as one of the world's most visited tourist sites.

△ STATUE PARK
The gardens have more than 250 statues of mythological figures, many incorporated into spectacular fountains. Such statuary, with its Classical allusions, was an essential element of Baroque garden design.

△ MARBLE COURT
The original château at Versailles, built for Louis XIII, still survives at the heart of Louis XIV's far grander palace. It surrounds a courtyard paved with black-and-white marble.

△ WATER FEATURES
The palace has views across the formal gardens to the Grand Canal, where Louis XIV and his courtiers held elegant boating parties. Water pumped from the River Seine fed the fountains.

◁ HALL OF MIRRORS
Completed in 1684, the Hall of Mirrors is most famous as the site where the Treaty of Versailles was signed at the end of World War I. The hall contains 357 mirrors in total.

THE GARDENS OF VERSAILLES

Celebrated landscape architect André le Nôtre originally laid out the extensive gardens for the palace at Versailles, with fountains, an orangery, graveled walks, and an ornamental canal. His work is considered the supreme masterpiece of the formal French garden style. Various changes were made to the grounds over time. King Louis XVI gave his queen Marie Antoinette her own château on the grounds, the charming Petit Trianon. She also had a fake hamlet constructed in which to act the part of a shepherdess.

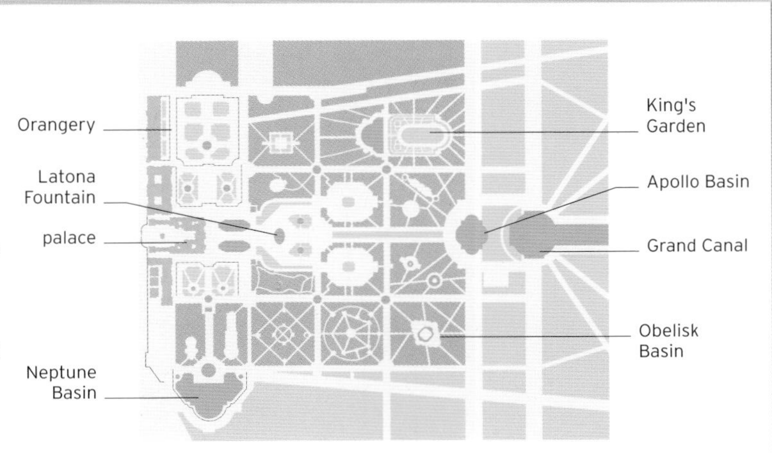

In Louis XIV's day, the palace was crowded with up to **7,000 courtiers and servants**

Sanssouci

A pleasure palace built for Prussia's "enlightened despot," Frederick the Great

King Frederick II of Prussia (r.1740–1786), also known as Frederick the Great, had the small palace of Sanssouci created as a retreat from the burdens of government and warfare—the French phrase *sans souci* means "without worry." Built between 1745 and 1747 in Potsdam on the outskirts of Berlin, it was designed by court architect Georg Wenzeslaus von Knobelsdorff. The king also played a direct role in the conception and execution of the project. He eventually fell out with Knobelsdorff, and the architect was fired before the building was completed.

Lighter spirit

Sanssouci was designed in the lightweight Rococo style (see pp.164–165), which was the height of fashion in Europe in the mid-18th century. Derived from Baroque but decidedly less ponderous and pompous, Rococo typically exploited mythological themes to create a world of delightful fantasy. Hedonistic and self-consciously frivolous, it was wholly appropriate to a residence meant for pleasure rather than pompous public display.

Parkland views

The main part of the palace, consisting of 10 principal rooms, is on a single floor. Set upon a hill above a terraced vineyard, the palace is surrounded by parkland dotted with playful, ornamental temples and pavilions. Originally Rococo, the interior was later redecorated in the Neo-Classical style.

Frederick regularly spent his summers at Sanssouci through his lifetime and requested to be buried there on his death. This wish was ignored until 1991, when his remains were brought from Potsdam Garrison Church and interred on the vineyard terrace.

A **dog lover**, Frederick asked to be **buried at Sanssouci** alongside his **favorite greyhounds**

ROCOCO FACADE
The main facade of the palace is amusingly enlivened with carved figures of bacchants, the unruly followers of the ancient Roman wine god Bacchus.

Blenheim Palace

An historic country house designed as a tribute to British war hero the Duke of Marlborough

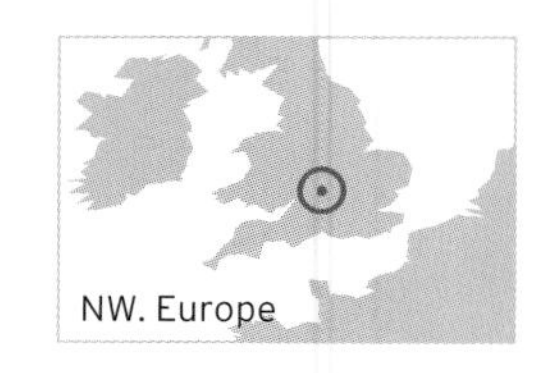

In 1704, the Duke of Marlborough, who commanded the British army at the Battle of Blenheim, was awarded a grant of royal land in Oxfordshire. He engaged John Vanbrugh, a playwright with little architectural experience, to design a palace. Vanbrugh, along with Nicholas Hawksmoor, had already begun work on an extravagant house in Yorkshire, and Blenheim was to be its equal in grandeur. More a memorial than a home, Blenheim was built in the English Baroque style. It was set in gardens that were later transformed by the famed landscape architect Lancelot "Capability" Brown.

▽ **LAVISH BAROQUE**
Vanbrugh designed Blenheim on a grand scale, inspired by the example of the Baroque palaces that were being built in Louis XIV's France.

THE *PIANO NOBILE*

The palace has the state rooms on the first floor, the *piano nobile*. It is decorated with many paintings celebrating Marlborough's battlefield successes. The chapel houses the sarcophagus in which he and his wife Sarah are interred.

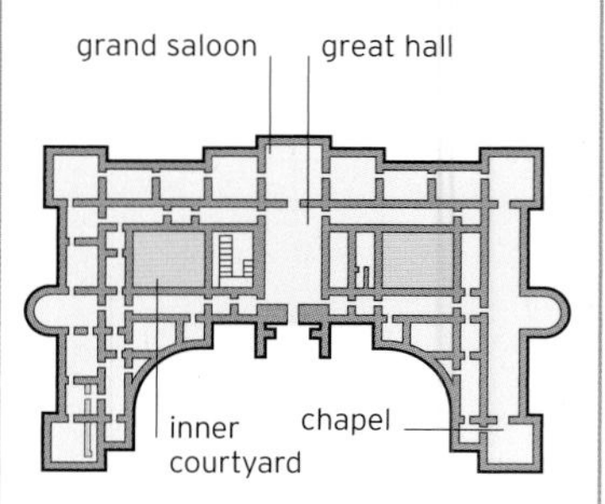

Buckingham Palace

The principal London residence of the British monarchy since the time of Queen Victoria

King George III bought Buckingham House, a mansion alongside St. James's Park, in 1761 as a convenient London home for his family. In 1826, his successor, George IV, ordered architect John Nash to expand the building into a royal palace. The first monarch to occupy the new palace was Queen Victoria, after her accession in 1837.

One palace, many styles

The building has undergone many changes over time. Marble Arch, which now stands alongside Hyde Park, was originally a grand entrance to the palace courtyard. The eastern facade, with its famous balcony where the royal family appears on occasions such as weddings, did not take its current form until the early 20th century. The interior revels in a mix of styles, from Nash-designed rooms in the French Neo-Classical style to the Belle Epoque cream-and-gold decor favored by Edward VII. The state rooms are open to tourists, but the palace still serves as a royal residence, an administrative building, and a venue for official ceremonies and banquets.

◁ **HUB OF THE MONARCHY**
An aerial view shows Buckingham Palace facing east toward the 1911 Victoria Memorial and the Mall. The parkland to the rear and side of the building is used for royal garden parties in summer.

Würzburg Residence

An extravagant showcase of the exuberant arts and crafts of the Rococo period

Now a city in Bavaria, southern Germany, in the 18th century, Würzburg was a small independent state ruled by prince-bishops. In 1720, Prince-Bishop Johann Philipp Franz von Schönborn ordered his architect Balthasar Neumann to begin construction of a lavish new palace. Aided by other prominent architects from Germany and France, Neumann created a building that was a fusion of European Baroque styles, costing 1.5 million guilders (when a guilder was a worker's weekly wage). The exterior was completed by 1744, but work continued for another quarter century on the interior—a masterpiece of ornamental carving, stucco, glass, and gilt dominated by the vast, colorful ceiling and wall paintings of Venetian Rococo artist Giovanni Battista Tiepolo.

Rebuilding history

In 1945, much of the building was destroyed by Allied bombing. The sections that survived intact fortunately included the Imperial Hall, the main staircase and vestibule with their irreplaceable Tiepolo frescoes, and the White Hall with its prodigious stucco work. Inspired efforts by modern craftsmen have since recreated other parts of the original interior. The residence was declared a UNESCO World Heritage Site in 1981.

◁ **FOUNTAIN FIGURE** This statue of Walther von der Vogelweide, a renowned medieval German love poet, forms part of the Frankonia fountain in the residence's forecourt.

THE GARDENS

The gardens of the Würzburg Residence had to be designed to fit within the shape of the town's fortifications, so they follow the outline of angular bastions. The formal Court Garden in Baroque style next to the house extends into meadow and woods in the English style.

pointed bastions
East Garden
nursery grounds
palace buildings
East Court Garden
Rosenbach Park
Frankonia fountain
South Garden
Landscape Garden

△ **SURVIVING DESTRUCTION**
The magnificent Imperial Hall, or *Kaisersaal*, is decorated with paintings by Giovanni Battista Tiepolo and his son Giovanni Domenico. It survived the bombing of Würzburg in 1945.

◁ **COURT GARDENS**
An aerial view shows the Würzburg Residence and the formal East Garden, created by Bavarian garden designer Johann Procop Mayer.

◁ **PALACE AND MONUMENT**
The vast honey-colored Schönbrunn Palace is overlooked by the Neo-Classical Gloriette monument on the crest of a hill in the park behind it.

Schönbrunn Palace

The sumptuous summer residence of the Habsburg rulers of the Austrian Empire

In the 1740s, work began on transforming a hunting lodge on the outskirts of Vienna into a summer palace for the Habsburg Austrian Empress Maria Theresa. The Habsburgs had the habit of building on a monumental scale, and Schönbrunn was no exception. Developed from an original design by architect Johann Fischer von Erlach, the vast Baroque palace eventually boasted a total of 1,441 finely decorated rooms, from vast reception and banqueting halls to the residential suites of members of the imperial family. The building sits in extensive gardens and landscaped parkland. The grounds include a menagerie, a palm house, an orangerie, a folly imitating Roman ruins, and the much-admired Gloriette—a Neo-Classical monument on the crest of the hill behind the palace, built to the design of architect Johann Hetzendorf von Hohenberg in 1775.

A rich history

Schönbrunn Palace is a site full of historical resonance. As a child prodigy, Wolfgang Amadeus Mozart performed there for Empress Maria Theresa at the age of 6. In the second half of the 19th century, it became the favorite palace of the much-loved Empress Elizabeth, who was tragically assassinated by an anarchist in 1898, and finally it was the site of the abdication of Austria's last emperor, Karl I, in 1918. Restored to their full imperial splendor after damage in World War II, the palace and gardens are now one of Vienna's major tourist attractions. The gardens extend from the palace and reflect the Baroque ideal that architecture and nature should be intertwined. With their highly ordered beds, the gardens were as much a symbol of Habsburg power as the house itself.

Trevi Fountain

An exuberant Baroque fountain that is one of the most popular tourist sites in Rome

Built of white travertine and marble, the Trevi Fountain is a monumental fountain in the heart of Rome. Its creator was architect Niccolo Salvi (1697–1751), selected for the job after a competition staged by Pope Clement XII in 1732. Salvi produced a flamboyant theatrical design with a basin that is 65 ft (20 m) wide, embellished by huge statues and a high backdrop formed by adding a decorative facade to the palace behind. The sparklingly clear water for the fountain is brought from a spring outside Rome by the Aqua Vergine, a Renaissance update of an original ancient Roman underground aqueduct. Salvi did not live to see the fountain finished, as he died in 1751. His successor, Giuseppe Pannini (1691–1765), completed the project in 1762.

The theme of the fountain's group of statues is "the taming of the waters." The massive figure of Oceanus—one of the ancient immortal Titans, lord of the seas—rides a shell chariot drawn by sea horses and guided by tritons blowing conches. Standing in niches to each side of Oceanus are allegorical statues representing Abundance and Health. The figures of Oceanus and the tritons are by Italian sculptor Pietro Bracci (1700–1773).

The fountain in popular imagination

The romantic potential of the fountain has been exploited in several famous films, most notably William Wyler's *Roman Holiday* (1953) and Federico Fellini's *La Dolce Vita* (1960). Throwing a coin into the basin, to guarantee luck or a return to Rome, is an established tourist custom.

Coins thrown into the fountain **each year** have added up to over **$1 million**

△ **ALLEGORICAL FIGURES**
A triumphal arch on the rear wall frames the statue of Oceanus. To his left is the figure of Abundance, holding the horn of plenty, and to his right is Health, holding a broad cup from which a snake drinks.

◁ **MIGHTY OCEANUS**
The central figure of Oceanus is 19 ft (5.8 m) high and holds a staff of authority in his right hand. Other sculptural details on the fountain include depictions of more than 30 species of plants.

△ **MONUMENTAL GATEWAY**
The Brandenburg is Berlin's only surviving historical city gate. Made from sandstone, the Neo-Classical structure is 85 ft (26 m) high, 215 ft (65.5 m) long, and 36 ft (11 m) deep.

Brandenburg Gate

An imposing, monumental structure that provides a ceremonial focus for the city of Berlin

C. Europe

In 1788, Frederick the Great of Prussia ordered his architect Carl Gotthard Langhans (1732–1808) to build a monumental gateway in Berlin's city wall. In line with 18th-century German taste for ancient Greek culture, the design imitated the propylaeum (monumental gateway) at the Acropolis in Athens (see pp.98–99). Completed in 1791, the structure has six massive Doric columns on each side. Prussian sculptor Johann Gottfried Schadow (1764–1850) created the Quadriga—a four-horsed chariot with the goddess of peace—placed atop the gate in 1793.

The Brandenburg Gate has had a stormy history. The Quadriga was stolen by the French Emperor Napoleon in 1806 but returned after France's defeat in 1814. The gate was badly damaged in World War II and then became a central feature of the wall dividing Berlin from 1961 to 1989. Restored to its former glory, it now symbolizes German reunification.

▽ **THE QUADRIGA**
The Quadriga was originally built with a statue of Eirene, the Roman goddess of peace, driving the chariot. She was replaced by Victoria, goddess of victory, after Prussia defeated France in 1814.

The Louvre

Now a museum housing the world's most famous painting, but also, in its time, a fortress, a palace, and a storehouse of stolen treasure

Located on the banks of the River Seine in Paris, the Louvre attracts over 8 million visitors a year, who come to marvel at its fabulous collection of art and antiquities. It began its existence as a fortress built in the 1190s by Philippe-Auguste, King of France, from 1180 to 1223, and became a royal residence in the 14th century, before François I (r.1515–1547) decided to rebuild it as a magnificent Renaissance palace. François invited Leonardo da Vinci to come to France, and the artist brought the *Mona Lisa* with him.

François' palace was designed by Pierre Lescot (c.1515–1578), with superb bas-reliefs by Jean Goujon (c.1510–1568). Later monarchs left their own mark on the place, the finest addition perhaps being Claude Perrault's elegant, Classical colonnade, commissioned by Louis XIV and built in 1667–1673. His extremely original design was chosen from a competition held in 1664.

A museum of art

French rulers kept their private art collections in the Louvre, but it was only in 1793 that its doors were opened to the public. It was renamed the Musée Napoléon in 1803, when Bonaparte used it to display the artworks he had looted during his campaigns. Most of these were later returned, but the collection and the visiting crowds continue to multiply, so the Louvre itself has been expanded in recent times.

The Louvre Pyramid

In the 1980s, President Mitterrand initiated his "Grand Louvre" project, an ambitious scheme to renovate the museum. Its most spectacular component was the steel and glass Pyramid in the Cour Napoléon. Designed by the Chinese American architect I. M. Pei (1917–2019), it forms the main entrance to the museum.

△ **ORNATE FACADE**
A detail in the Cour Napoléon shows graceful caryatids (female figures) supporting allegorical figures representing the history of poetry.

The *Mona Lisa* has her own mailbox at the Louvre to receive the love letters sent to her

▽ **LATER ADDITIONS**
Originally built as a fortress, the Louvre was enlarged by a succession of French rulers over the centuries. The striking glass pyramid was added to the main courtyard in 1989.

Arc de Triomphe

A monumental arch, a symbol of French patriotism, located in the heart of Paris

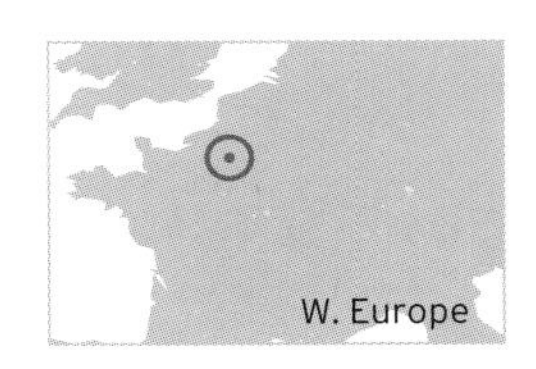

△ FOCAL POINT
The current prominence of the arch owes much to the work of Baron Haussmann, who redesigned the *Place de l'Etoile* completely as part of his radical transformation of Paris in 1853–1870.

Built between 1806 and 1836, the Arc de Triomphe is an imposing legacy of the Napoleonic era, commissioned by Bonaparte himself to celebrate his victory at the Battle of Austerlitz (1805). He never lived to see it, although his cortège passed through it when his body was returned to France in 1840.

National significance

Designed by Jean-François Chalgrin, the arch is 164 ft (50 m) high, affording spectacular views of the city from its summit. The design is loosely based on a Roman triumphal arch, and the monument is dedicated to France's military achievements. On the exterior are beautifully sculpted panels and a frieze commemorating famous victories, while the names of distinguished officers are listed on the inner walls. The Tomb of the Unknown Soldier lies underneath the arch, with an eternal flame that is rekindled every day. The Arc de Triomphe has become the nation's focal point during times of celebration, as well as mourning. It is also the traditional starting point for the annual Bastille Day parade.

PLACE DE L'ETOILE

The arch is situated prominently on a roundabout where 12 avenues converge. The road layout is symmetrical, forming the shape of a star and giving rise to the name *Place de l'Etoile* ("place of the star"), but the name changed to Place Charles de Gaulle in 1970.

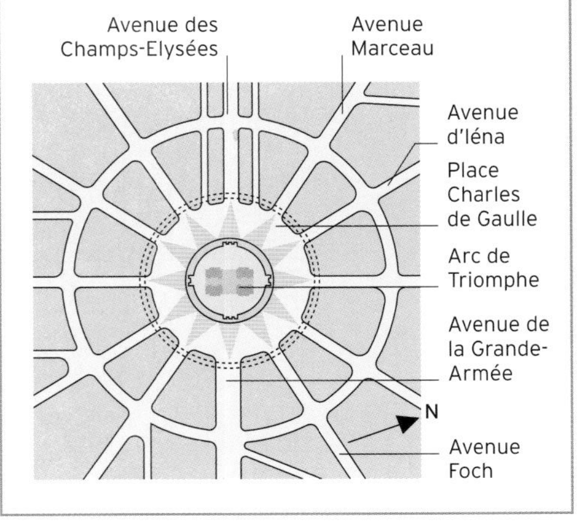

△ INNER WALLS
The names of 660 military commanders—most dating back to the Napoleonic era—are inscribed on the internal walls of the arch.

Winter Palace

A magnificent Baroque palace that was once the St. Petersburg residence of the Russian czars

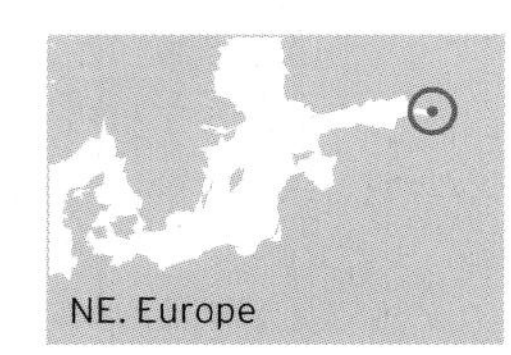

Built on the banks of the River Neva in the heart of St. Petersburg, the Winter Palace was designed by Italian architect Bartolomeo Rastrelli for Empress Elizabeth, who ruled Russia from 1741 to 1762. By the time the building was completed, the formidable Empress Catherine the Great (r.1762–1796) had taken the throne—she is the ruler with whom the palace is most associated.

Rastrelli created a vast three-story building with elongated facades. He decorated the interior in lavish Rococo style, although much of this work was destroyed by a disastrous fire in 1837. The palace was extended in the later years of Catherine's reign by the addition of the Hermitage wings.

△ IMPERIAL EAGLE
The double-headed eagle was adopted as a symbol of Russian imperial power. This gilded version appears on one of the gates of the Winter Palace.

From seat of power to museum

The imperial family continued using the palace until 1880, when the explosion of a bomb in the dining room showed it was too easy for would-be political assassins to penetrate. During the Russian Revolution of 1917, it was briefly the seat of a provisional government, which was driven out by the Bolsheviks in the storming of the palace. The Winter Palace is now the site of the Hermitage Museum, containing one of the world's foremost collections of painting and sculpture.

▽ ROCOCO STAIRS
The Jordan Staircase is one of the few elements of Rastrelli's original Rococo interior of the Winter Palace that has survived intact to the present day.

The Winter Palace contains **1,500 rooms and halls; 1,945 windows; and 117 staircases**

THE CATHEDRAL CRANE

When work on the cathedral halted in the 15th century, the builders left a crane on top of the unfinished southern steeple. A fine example of medieval engineering, the crane stayed on its lofty perch for 400 years. It was not dismantled until work on completing the steeple was finally resumed in 1868.

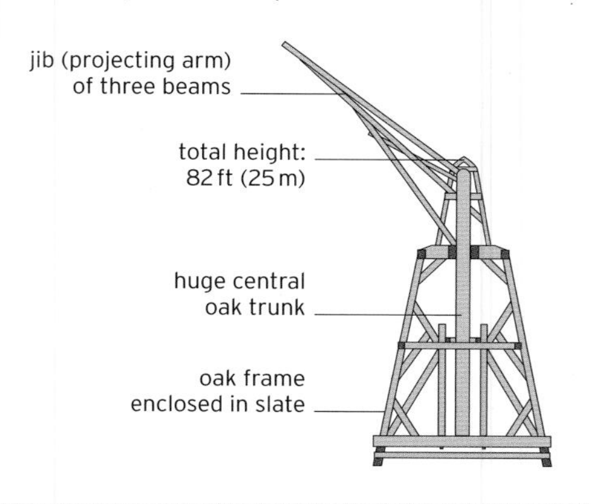

Cologne Cathedral

A medieval Gothic masterpiece that miraculously survived wartime bombing

△ **CATHEDRAL PORTALS**
The three massive portals on the cathedral's west facade, copiously decorated with carvings and statues, are flanked by the towers of the two lofty spires.

In 1164, the archbishopric of the German city of Cologne acquired the purported remains of the Three Kings who, according to the Bible, had brought gifts to the infant Jesus. Enclosed in a golden reliquary made by German goldsmiths, these sacred relics attracted thousands of pilgrims to Cologne from all over Europe. In 1248, work began on building a vast new cathedral to house the reliquary and receive the influx of visiting worshippers.

Interrupted construction

Designed in the elegant High Gothic style then fashionable in Europe, the cathedral was a supremely ambitious building. By the end of the 15th century, it was still not completed, and both money and inspiration had run out. Work on the cathedral did not resume until 1842. The twin spires that are the building's crowning glory were finally completed in 1880: at 515 ft (157 m) high, they were then the tallest man-made structures in the world.

During World War II, Cologne was heavily bombed, and the cathedral suffered multiple hits by high-explosive and incendiary devices. Astonishingly, the building survived essentially intact, and restoration has since returned it to its original glory.

◁ **COLORFUL GLASS**
The cathedral is renowned for its stained-glass windows, some dating as far back as the 14th century. This image represents St. Apollinaris of Ravenna.

◁ **IMPRESSIVE FACADE**
The Parliament Building revels in an eclectic mix of architectural styles, its Gothic spires and pointed arches capped by a Renaissance-style dome. The facade overlooking the Danube River is 880 ft (268 m) long.

▽ **CENTRAL DOME**
The interior of the building's central dome is lavishly decorated. Beneath the dome lies a 16-sided hall that houses an exhibition of royal regalia, including the 12th-century Crown of St. Stephen once used for the coronation of Hungarian monarchs.

Hungarian Parliament Building

An architectural splendor on the bank of the Danube in Budapest that reflects the golden age of the Hungarian nation

Erected between 1884 and 1904, the Hungarian Parliament Building is a dazzling expression of the confidence and prosperity of the city of Budapest at the time. Recognized as equal partners in the dual monarchy of Austria–Hungary, the Hungarians were experiencing a national revival. The architect chosen for the project was a Hungarian, Imre Steindl (1839–1902), and almost every piece of material used in the building's construction was sourced in Hungary. It was embellished with 88 statues of historic Hungarian rulers.

Grand scale

The Gothic Revival style of the long facade is an homage to London's Palace of Westminster (see opposite), but the scale of the building is much larger. Its construction and decoration required 40 million bricks, about half a million precious stones, and some 88 lb (40 kg) of gold. One of the building's two parliament halls is still in use as a National Assembly. The other, the richly decorated Chamber of Peers, is open to tourists, who can also view the Hungarian crown jewels, now displayed in the Dome Hall. Since 1987, the Parliament Building has been part of a UNESCO World Heritage Site that also includes the Buda Castle quarter on the opposite bank of the Danube River and the 19th-century Chain Bridge that crosses the river.

The **dome** of the Parliament Building is **315 ft (96 m) tall**

Palace of Westminster

A world-famous landmark in London and the seat of the United Kingdom's parliament

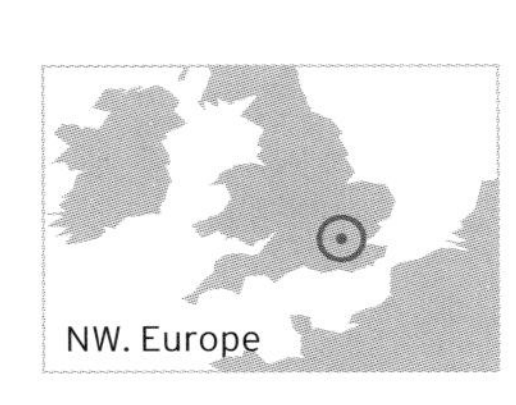

On October 16, 1834, the old Palace of Westminster was destroyed by a fire. Built as a royal residence in medieval times, it had been used as a meeting place for parliaments since 1265 and as a permanent location for the House of Lords and the House of Commons since the 16th century. The only substantial section of the old palace to survive the fire was Westminster Hall, where law courts were located. A competition to design a new building for the site was won by British architect Charles Barry (1795–1860). With the inspired assistance of young designer Augustus Pugin (1812–1852), Barry created the striking edifice in the Gothic Revival style that still stands today.

Palace of a thousand rooms

Situated on the north bank of the River Thames, the building is most famous for its clock tower. Pugin's masterpiece, completed in 1859, the tower has clock dials 23 ft (7 m) in diameter and contains five bells, the largest known as Big Ben. The Victoria Tower, at the other end of the building, stands taller at 323 ft (98.5 m) and houses the parliamentary archives. The interior of the palace, which has over a thousand rooms, is a riot of Victorian painting, sculpture, frescoes, mosaics, carved wood, and stained glass.

PLAN OF THE PALACE

The debating chambers of the UK parliament's two houses, the Lords and Commons, occupy only a small part of the palace. The Sovereign's Entrance, used by the Queen at the state opening of parliament, is at the base of the Victoria Tower. Westminster Hall serves for ceremonial occasions.

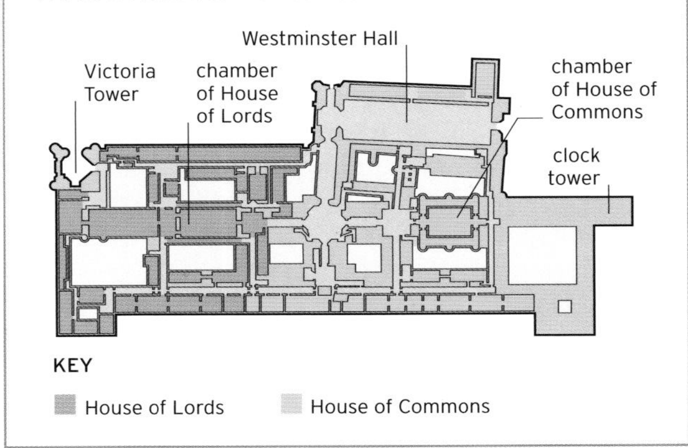

▷ **SYMBOL OF LONDON**
Designed by Pugin, the palace's Gothic clock tower has become an icon of its city. The tower is built of limestone-clad brick with a cast-iron spire.

Eiffel Tower

Once the world's tallest building, a flamboyant feat of 19th-century engineering that became a Parisian icon

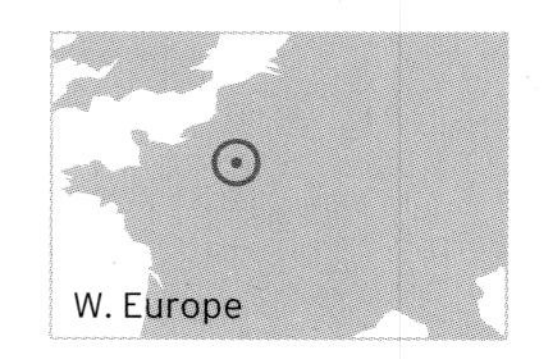

The Eiffel Tower was created as a spectacular entrance for the Paris Exhibition of 1889, held to celebrate the centennial of the French Revolution. The tower was designed by employees of the company run by Gustave Eiffel (1832–1923), an engineer who had made a name for himself as a builder of bridges, viaducts, and railway stations. It was selected out of more than 100 competing designs for the exhibition entrance because it was thought to best symbolize a century of scientific and technological progress.

Factory construction

Rising to a height of 984 ft (300 m), the Eiffel Tower stood almost twice as tall as any previous human structure. It was constructed of 18,000 factory-made wrought-iron pieces, brought to the site along the River Seine for assembly and joined together with 2.5 million rivets. Its wrought-iron latticework marked a radical departure from traditional monumental buildings in brick or stone. Parisian aesthetes pretended to hate it, protesting against the "useless and monstrous" tower that would ruin "the hitherto untouched beauty of Paris." However, the public loved it from the start, almost 2 million people flocking to buy tickets to mount the tower during the 6 months of the exhibition.

New purposes

Gustave Eiffel had a 20-year contract to capitalize on the tower commercially, after which it was supposed to be dismantled. However, the placing of a radio transmitter on the top in 1904 gave it a practical use that saved it from destruction. The tower was surpassed as the world's tallest structure by the Chrysler Building (see p.42) in New York in 1930, but its prestige as the symbol of the city of Paris has continued to grow. In the 21st century, it is one of the world's most visited tourist attractions.

Four **restaurants** on the **first floor** of the tower served hungry visitors during the **Paris Exhibition**

VIEWING LEVELS

The Eiffel Tower is about as tall as a modern 90-story skyscraper—the tallest structure in Paris—but appears less imposing due to its delicate lattice structure and because there are no buildings nearby with which to compare its height. Visitors to the tower can ascend to three viewing levels; on a clear day, the town of Chartres—48 miles (77 km) to the southwest—is visible from the highest level, as one ascends on elevators that move through a curved arc at the lower levels.

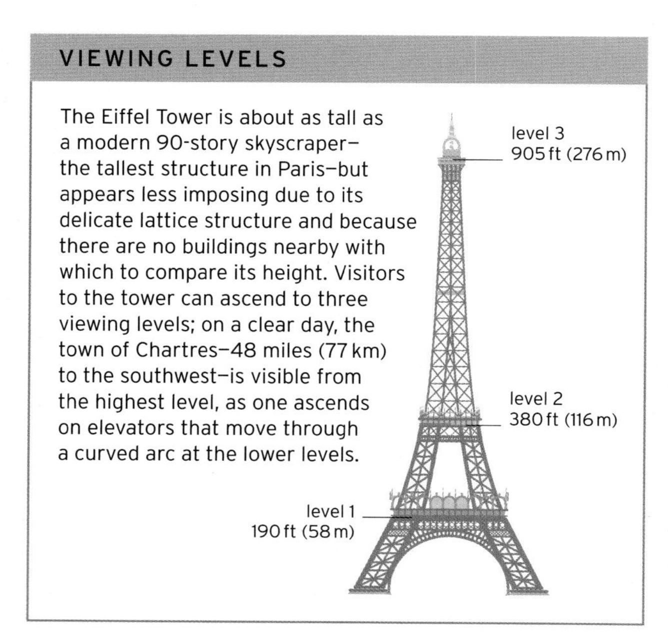

△ **IN THE MAKING**
A few small steam cranes; 12 temporary wooden scaffolds, each 98 ft (30 m) high; and four larger scaffolds, each 130 ft (40 m) high, were used to assemble the first level of the tower.

INTRICATE LATTICEWORK
The decorative arches at the base of the Eiffel Tower made a stylish entrance to the Paris Exhibition of 1889. The building proved wrought-iron could be beautiful.

Tower Bridge

A Neo-Gothic bridge, masterfully engineered to allow the passage of shipping

Traffic on London's roads grew in the late 1800s, outstripping the capacity of the capital's bridges and highlighting the need for a crossing downstream of the congested London Bridge. However, a new bridge at such a location would need to allow the passage of merchant ships on their way up to the Pool of London—the wharves on either side of the river. In 1884, after much discussion, an ingenious design by engineer John Wolfe Barry (1836–1918) and architect Horace Jones (1819–1887) was approved, and its specifications were set out in an act of Parliament the following year. Its roadway is divided into two leaves known as bascules (from the French term for "balance scale"), which are lifted up to allow shipping to pass.

A grand opening

The bridge was built in the Neo-Gothic style fashionable at the time in order to harmonize with the nearby Tower of London. Construction began in 1886 and took 8 years. Two massive piers were sunk into the riverbed and filled with more than 78,400 tons (71,120 tonnes) of concrete to support the bridge's two towers and walkways, which were made of steel and clad in Cornish granite and Portland stone. The bridge was finally opened by Edward, Prince of Wales (later Edward VII), on June 30, 1894, and has been open to traffic—and raised for ships—ever since.

△ **THE ENGINE ROOM**
A workman oils the engines beneath Tower Bridge that pump steam into the hydraulic accumulators in order to raise or lower the bascules of the bridge.

▷ **THE BRIDGE AT SUNSET**
Tower Bridge is lit each evening to show off its many fine architectural features, particularly the brickwork of the two towers and the two upper walkways.

OPENING THE BRIDGE

When the bridge is opened, the weight of its leaves, or bascules, is counterbalanced by weights beneath each of the bridge's two towers. The power to raise the bascules originally came from two 360-horsepower steam engines that pressurized water in accumulators; the water was released to drive the lifting engines when needed. Opening and closing the bridge could take as little as 5 minutes. A new electrohydraulic drive system was installed in 1974.

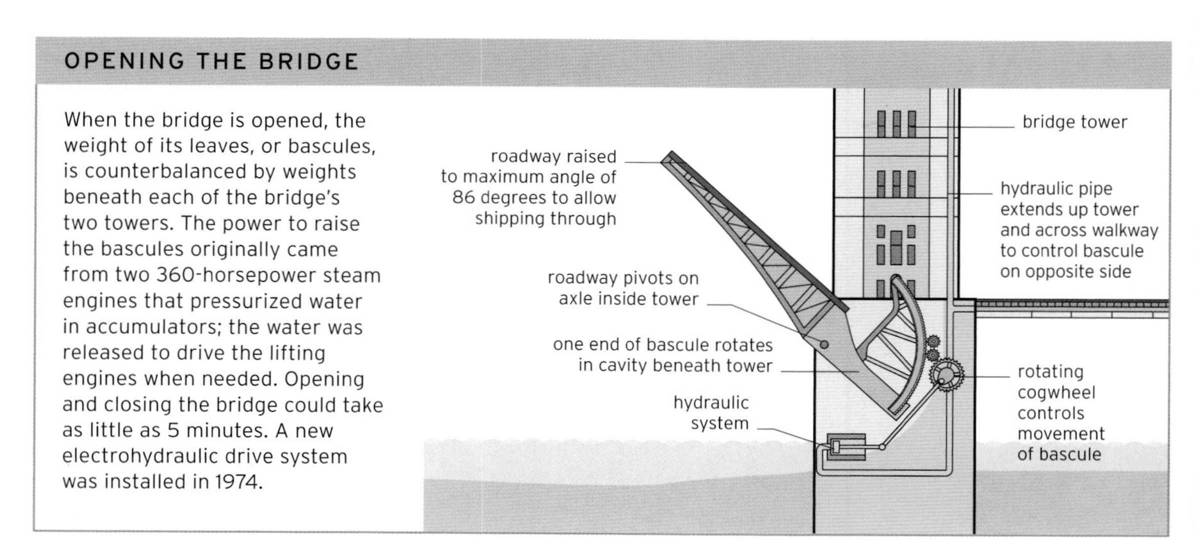

Sagrada Família

Antoni Gaudí's unfinished, experimental masterwork, a symbol of his love for God and his native Catalonia

The towering spires and immense portals of Barcelona's Sagrada Família (or Expiatory Temple of the Holy Family) skillfully combine elements of Gothic church architecture with the fluid, organic, and personal Art Nouveau style of Antoni Gaudí (1852–1926). Work on the church began in 1882 to a traditional cruciform Neo-Gothic design by Francisco de Paula del Villar but, following disagreements about the project's direction, Gaudí took over as chief architect the following year.

An expression of belief

Gaudí saw the church's construction as an act of faith and devoted his life to the work, even living in a hut on the site. Villar's design soon gave way to Gaudí's extraordinary imagination. Every inch of the structure reflects his love of light, color, and natural forms and is rich with Christian symbolism. By the time Gaudí died in 1926, killed in a trolley accident outside the church, work was well advanced on the vast Nativity Transept. Adorned with stalactitelike gables and towering cone-shaped spires topped with twisting finials of colored broken tiles, it was unlike anything seen in a church before. In Gaudí's highly experimental central nave, treelike columns hold up a carved canopy of sunflower shapes to create an inner forest. Mass has been held daily since the Sagrada Família was consecrated by Pope Benedict XVI in 2010, and millions visit the site—recognized as a UNESCO World Heritage Site in 2005—each year. Their donations help fund the $28 million needed each year for ongoing construction work. It is hoped that structural work will finally finish in 2026 and decorative work in 2032.

When **completed**, the Sagrada Família will be the **tallest church building** in the **world**

SPIRES AND SYMBOLISM

Each of the church's 18 spires will represent an important Biblical figure. The tallest tower, almost 590 ft (180 m) tall, represents Christ and is flanked by four spires representing the Gospel writers. Mary's spire has a central place in the nave, while the 12 spires for the apostles top the Nativity, Passion, and Glory facades.

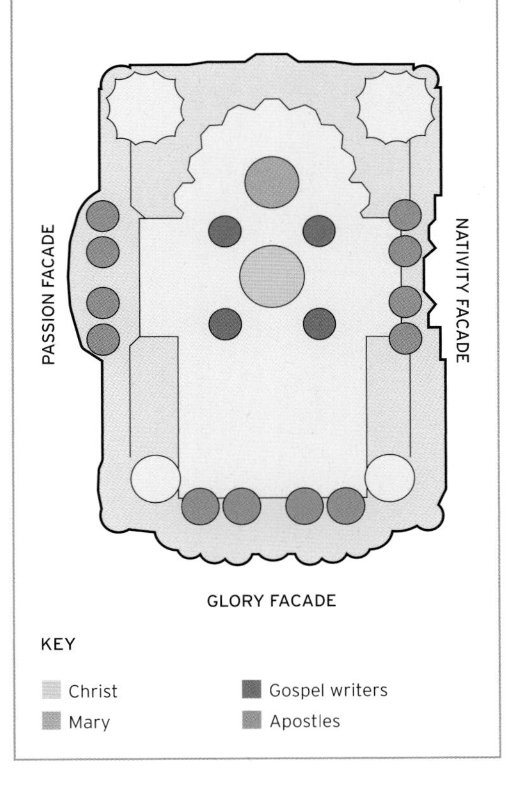

▷ **COLOR AND LIGHT**
Stained-glass windows and painted roof bosses flood Gaudí's extraordinary 148-ft- (45-m-) high nave, among the highest in the world, with color and light.

◁ **WORK IN PROGRESS**
Work on the church has continued nonstop since 1882 and will finally finish in 2032, after 150 years. The vast building site is Spain's most visited tourist attraction.

△ **SYMBOLS AND SCULPTURE**
The distinctive "dripping wax"-style carving of the Nativity Facade combines exotic natural forms with traditional sculpted figures symbolizing Christ's birth.

Neuschwanstein Castle

A fairy-tale castle from a Wagnerian fantasy created by King Ludwig of Bavaria at a spectacular clifftop location

Neuschwanstein is famously known as the model for the Sleeping Beauty Castle in Disneyland, California. It was an apt choice because the German castle, situated on a rocky outcrop in southern Bavaria, is a romantic wonderland in its own right. It is a theatrical tour de force that looks like a medieval castle, although it was built in the 19th century. It has a Minstrels' Hall, where musicians never performed, and a Throne Room, where no throne was ever installed.

Medieval fantasy

Neuschwanstein was the creation of Ludwig II, King of Bavaria (r.1864–1886). He had a difficult childhood and developed a fixation with medieval history. In particular, he became fascinated by Lohengrin, a figure from Arthurian literature. Ludwig decided to create a Grail Castle, where the Holy Grail was reputedly kept in Arthurian lore, and hired Christian Jank, a scene-painter who had helped to stage the dramatist and composer Richard Wagner's opera *Lohengrin*, to design his ambitious project.

Sadly, Ludwig did not live to see his dream completed. He was deposed by the Bavarian authorities and died under mysterious circumstances. During World War II, Neuschwanstein was used as a storehouse for Nazi loot. Since then, it has become a hugely popular tourist destination.

◁ **CONSERVATION EFFORTS**
Ludwig's Castle is constructed in a dramatic setting. However, its sheer rock walls and foundations are closely monitored and secured periodically. The limestone facades are also prone to weathering.

A MODERN BUILD

Despite its appearance, the castle is a thoroughly 19th-century building that sits on concrete foundations and is made of limestone-clad brick. The structure is 427 ft (130 m) long, and its highest tower reaches 213 ft (65 m) into the sky. More than 200 rooms over six floors were planned, but only 14 were completed, including the Byzantine-style Throne Room, which features a gilded cupola.

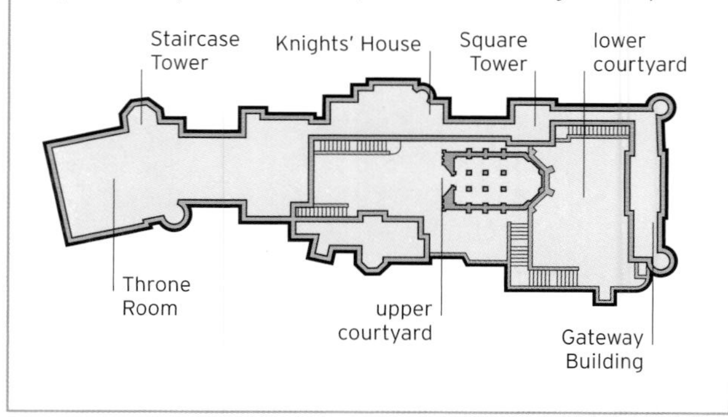

Musée d'Orsay

An architectural transformation of a former railway station into a museum known for its Impressionist paintings

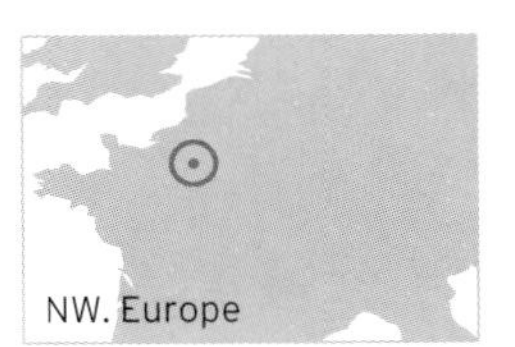

△ **CAST-IRON ARCHES**
Laloux's magnificent great hall is now the main artery of the museum. Fittingly, it provides a wonderful setting for a wide range of artworks that were created during the heyday of the old Gare d'Orsay.

Situated on the Left Bank of the River Seine, the Gare d'Orsay was built in 1900 to cater to the huge influx of visitors expected at the Paris Exposition in the same year. It was the world's first all-electric rail terminal; because there was no steam circulating within, the architect, Victor Laloux, was able to house the station in a huge enclosed vault of iron and glass. Sadly, the station had a limited lifespan, as its platforms were short, and by 1939, it was only running suburban services. The threat of closure loomed large. There were plans to demolish the building and erect a hotel in its place. However, building permission was denied due to a public outcry, and the station was placed on the Inventory of Historical Monuments in 1973.

A new lease of life

While the station lay empty, it was used as a car lot, a film set, an auction house, and a theater. By the late 1970s, work on its transformation to a museum was underway. An architectural firm, ACT Architecture, transformed the building, while the Italian architect and designer Gae Aulenti created the new interior. The museum was laid out on three levels. The great hall contains the central nave and the terraces on the mezzanine, which open out into further galleries at the side. The top level was installed above the lobby. The various levels were connected by glass walkways behind the main clock. Key features, such as the iconic clocks, were retained, while others were revamped. The decorative rosettes in the vault, for example, were turned into air-conditioning vents, while the station's former buffet has become a bookstore. The museum, which opened to the public in December 1986, focuses on artworks from 1848 to 1914, one of the most fertile periods in French culture. Uniquely, it encompasses all branches of the arts—painting, sculpture, photography, architecture, and the decorative arts.

△ **TELLING TIME**
The massive, Belle Epoque clocks from the old station have become trademark features of the museum. This clock presides over the main gallery, while the glass on the facade provides spectacular views of Paris.

△ **INTERNAL COURTYARDS**
In Gaudí's ingenious design, the entire complex of the Casa Milà was arranged around two large courtyards. The irregularly shaped shaft channeled extra light and ventilation into all of the apartments.

Casa Milà

A stunning example of Modernist architecture, groundbreaking biomorphic design, and Gaudí's last private commission

Located in the heart of Barcelona, the Casa Milà is one of the greatest works of Catalan architect Antoni Gaudí. It is one of the seven properties designed by him that UNESCO have placed on their World Heritage List owing to their "outstanding creative contribution to the architectural heritage of modern times."

The Casa Milà was commissioned in 1906 by Pere Milà, a wealthy textile industrialist. He and his wife admired another of Gaudí's buildings—the Casa Batlló—and wanted something similar. Unfortunately for them, Gaudí was a visionary genius, not given to repeating his experiments. He altered the design during the construction process, creating something very different, and they were unhappy with the result. The reaction of the general public was equally negative, and the building was dubbed *La Pedrera*, or the "stone quarry."

◁ **UNDULATING FACADE**
The surfaces of Gaudí's buildings convey a sense of movement, undulating like waves. The facade is like a curtain, however, carrying no weight and concealing the solid structure behind.

Organic inspiration

The Casa Milà is essentially an apartment block on the corner of two streets. The Milà family occupied the entire first floor, while the upper floors were split into smaller apartments and rented out. Gaudí did not use the colorful tiles and mosaics that characterize most of his work. Instead, he opted for an extraordinary biomorphic design: the stonework facades undulate like waves; the sculptural wrought-ironwork resembles plant forms; and the ceiling of the laundry area in the attic calls to mind the ribcage of a huge, whalelike creature.

Behind his flights of fancy, however, Gaudí's ideas served a practical purpose. The strange figures on the roof terrace actually conceal air vents and stairwells, while in the basement is Barcelona's first garage.

Sacré-Cœur

A famous Parisian landmark built as a symbol of unity after national disasters and which offers outstanding views over the city

NW. Europe

Sacré-Cœur is one of the most popular tourist sites in Paris. Located in the heart of Montmartre, it offers commanding views of the French capital—even more so for those who climb the 237 steps to the top of the dome.

A beacon over the city

The church, which was created in the wake of France's humiliating defeat in the Franco–Prussian War (1870–1871), was seen as a symbol of atonement and hope. The project was sanctioned by Parliament in 1873, and a competition was held to decide on the design. The winner was Paul Abadie, well known as a restorer of ancient buildings. His design has been likened, rather unkindly, to a baby's bottle or a confection of whipped cream. Abadie's masterstroke, however, was to use Château-Landon stone, which hardens and whitens with age. As a result, the basilica gleams above the French capital. Sacré-Cœur was eventually consecrated in 1919. The church is dedicated to the Sacred Heart of Jesus (also commemorated by a Catholic feast day), as depicted in the huge mosaic by the French artist Luc-Olivier Merson, which adorns the apse behind the altar.

△ **INTERIOR OF DOME**
The dome above the chief mosaic has a clerestory and two encircling balconies. Visitors can reach this point by climbing a daunting spiral staircase.

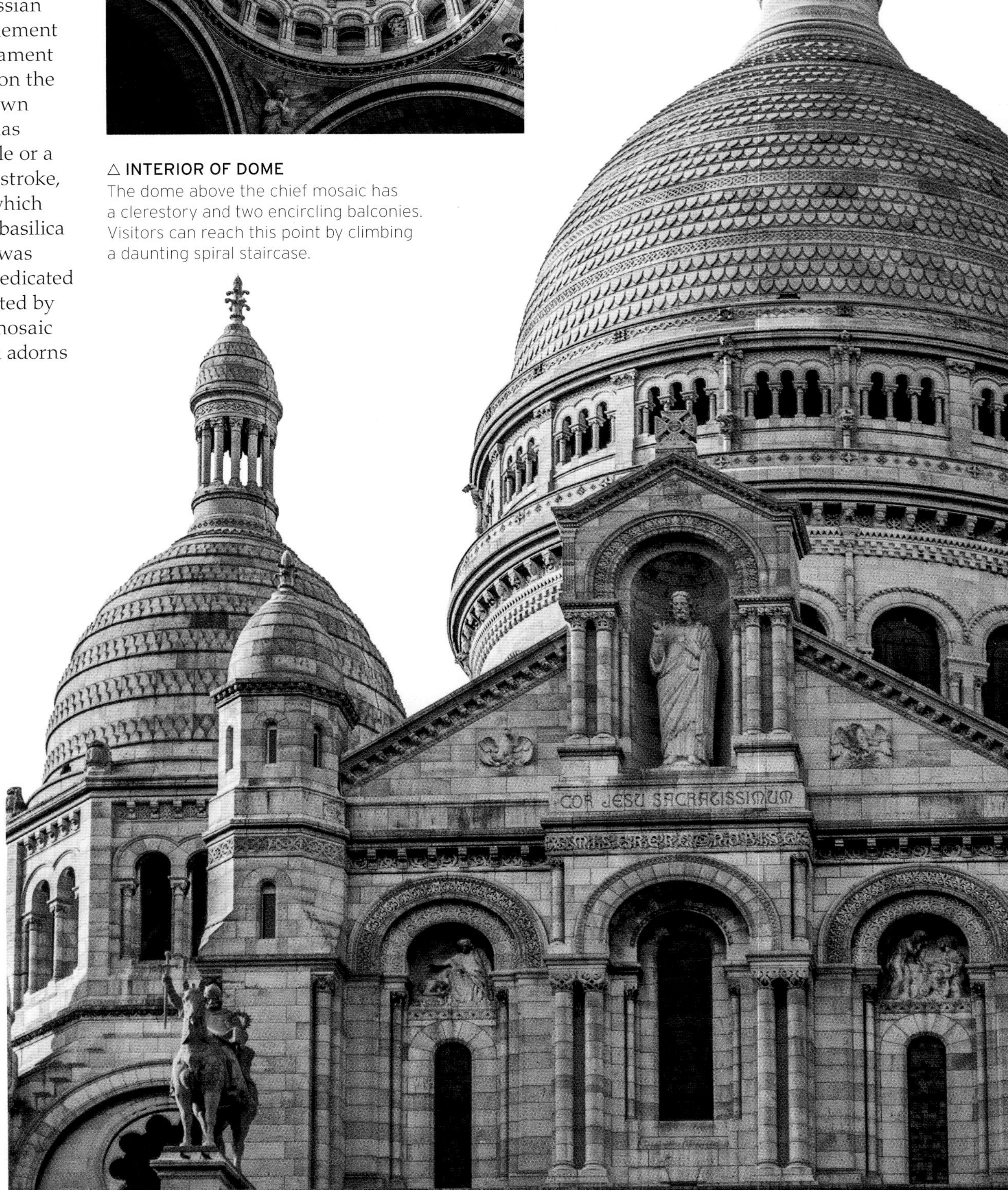

▷ **PARISIAN BASILICA**
This view of Sacré-Cœur clearly shows two of its famous beehive domes. The bronze statue above the porch, depicting St. Louis holding his sword aloft, is by French sculptor Hippolyte Lefèbvre.

SOMBER STRUCTURE

After the brilliant white exterior, the interior of Sacré-Cœur seems dark and still. The mix of Romanesque and Byzantine styles reflects Abadie's role on the Commission for Historical Monuments. The huge mosaic in the apse behind the altar depicts Christ in all his majesty revealing his golden heart.

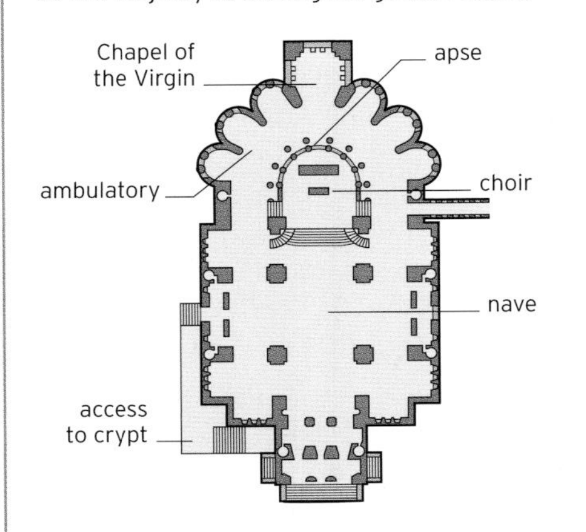

◁ **ENTERING THE PARK**
This aerial image taken over the park's entrance shows the Warden's House (bottom right) and steps leading to the Hypostyle Hall and the terrace above.

△ **HALL OF COLUMNS**
The Hypostyle Hall was designed to be a market area. Its roof is supported by 86 columns and adorned with a tile-shard mosaic made by Gaudí's assistant.

A giant **mosaic-covered salamander** called **El drac** greets visitors to the park

▷ **MOSAICS AND ORGANIC SHAPES**
Colorful mosaics decorate the organic forms and mushroom-shaped pinnacle on the roof of the Warden's House at the park's entrance. The mosaics combine tile-shard decoration with geometric patterns.

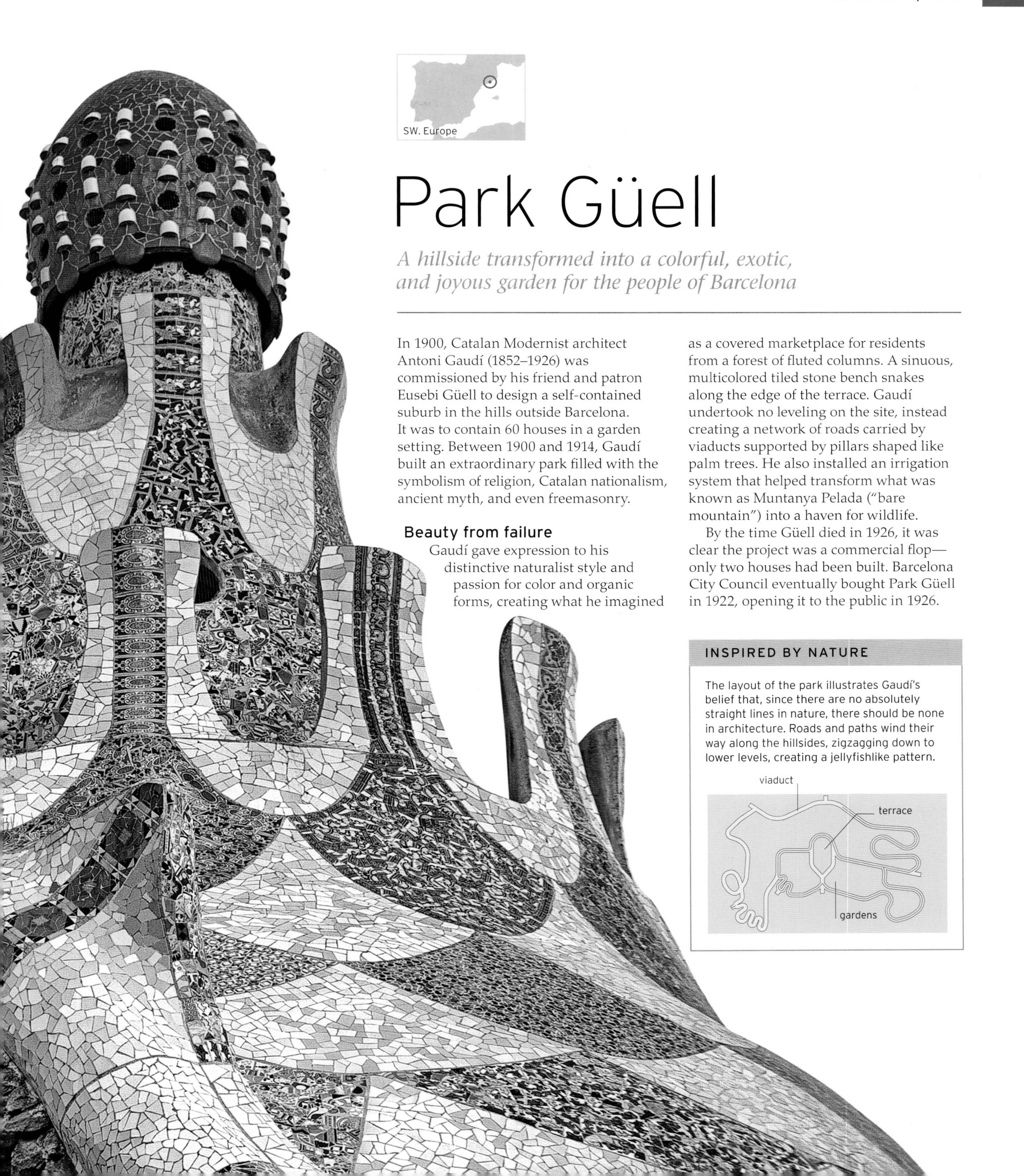

Park Güell

A hillside transformed into a colorful, exotic, and joyous garden for the people of Barcelona

In 1900, Catalan Modernist architect Antoni Gaudí (1852–1926) was commissioned by his friend and patron Eusebi Güell to design a self-contained suburb in the hills outside Barcelona. It was to contain 60 houses in a garden setting. Between 1900 and 1914, Gaudí built an extraordinary park filled with the symbolism of religion, Catalan nationalism, ancient myth, and even freemasonry.

Beauty from failure

Gaudí gave expression to his distinctive naturalist style and passion for color and organic forms, creating what he imagined as a covered marketplace for residents from a forest of fluted columns. A sinuous, multicolored tiled stone bench snakes along the edge of the terrace. Gaudí undertook no leveling on the site, instead creating a network of roads carried by viaducts supported by pillars shaped like palm trees. He also installed an irrigation system that helped transform what was known as Muntanya Pelada ("bare mountain") into a haven for wildlife.

By the time Güell died in 1926, it was clear the project was a commercial flop—only two houses had been built. Barcelona City Council eventually bought Park Güell in 1922, opening it to the public in 1926.

INSPIRED BY NATURE

The layout of the park illustrates Gaudí's belief that, since there are no absolutely straight lines in nature, there should be none in architecture. Roads and paths wind their way along the hillsides, zigzagging down to lower levels, creating a jellyfishlike pattern.

MATERIALS AND DETAILS

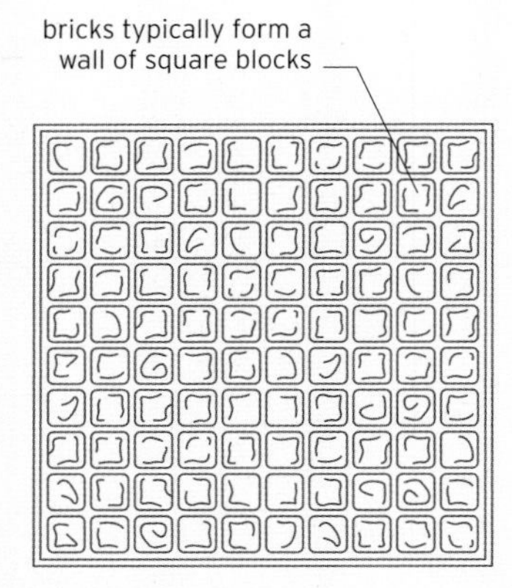

△ **GLASS BRICKS**
Glass bricks are a common feature in Modernist buildings. They allow for a literal wall of glass, while their semiopaqueness ensures some privacy for the building's interior.

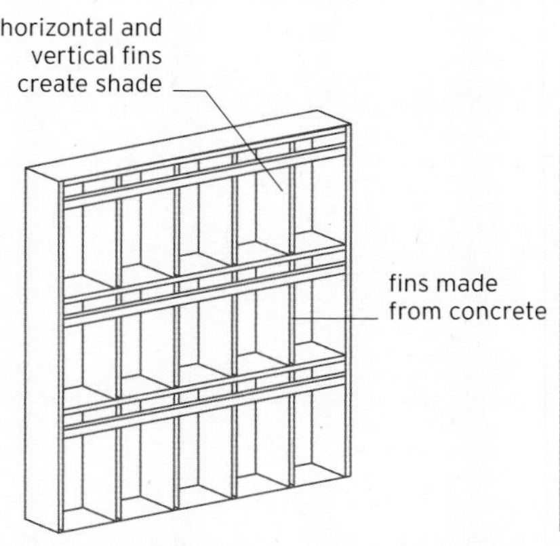

△ **BRISE-SOLEIL**
Meaning "sun blocker," a brise-soleil is a structure attached to the face of a building, intended to block direct sunlight—important when a facade contains large expanses of glass.

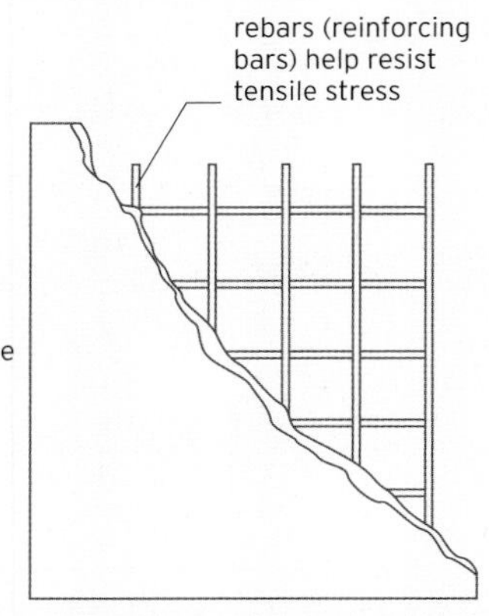

△ **REINFORCED CONCRETE**
A composite material of concrete and steel reinforcement bars, reinforced concrete remains strong under both compression and tension, making it a vital modern building material.

▼ **VILLA SAVOYE**
Villa Savoye embodies Le Corbusier's Five Points. Supporting walls are replaced by pilotis (point 1). This allows for complete freedom in the internal plan (2), a correspondingly free facade (3), ribbon windows to maximize natural light (4), and a roof garden (5).

Architectural styles

MODERNISM

Modernism embraced the use of new building materials and technologies to forge an architecture that reflected the new spirit of the 20th century. Rejecting decoration in favor of abstraction, Modernism transformed cities across the world.

By the beginning of the 20th century, a number of architects began to argue that architecture as it was currently practiced had reached the end of the road. While the industrial revolution had transformed almost every aspect of everyday life, architecture had continued to look to increasingly anachronistic historic styles. Modernism emerged as a concerted attempt to forge an architecture that would not only reflect industry's transformations but also harness them toward positive and progressive ends.

One of the most influential architects attempting to articulate this new direction was the Swiss-French Charles-Édouard Jeanneret, better known as Le Corbusier. He famously outlined Five Points of a New Architecture, which would help show the way forward structurally and aesthetically. In effect, Le Corbusier advocated that form should follow function, that the appearance and layout of a building should be determined by its intended use. Ornament and decoration were replaced with a fresh, clean, and supposedly rational aesthetic.

Modernist principles proved influential all over the world, notably in the rebuilding effort after World War II. While Modernism's legacy has been profound, it has also been highly contested, especially when it was aligned with particular political ideals.

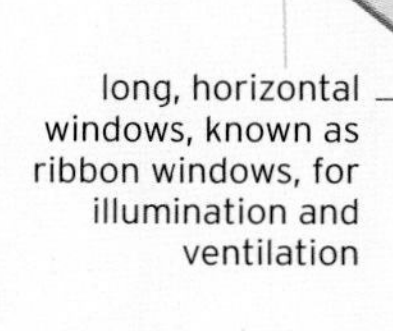

◁ **CURVED GLASS**
Many Modernist architects placed huge importance on the amount of natural light entering buildings, with large curved or ribbon windows a common feature across a range of building types.

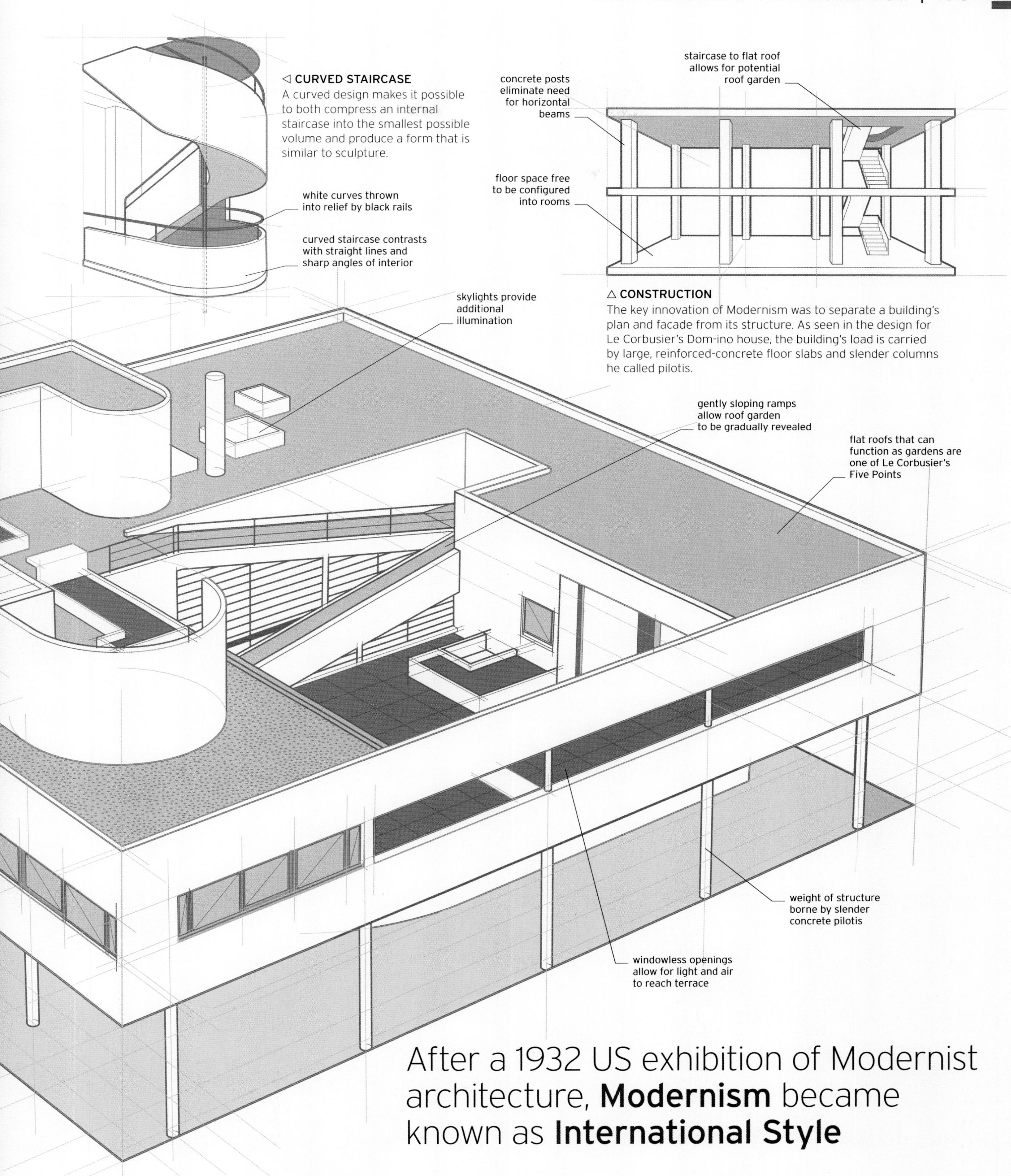

◁ **CURVED STAIRCASE**
A curved design makes it possible to both compress an internal staircase into the smallest possible volume and produce a form that is similar to sculpture.

△ **CONSTRUCTION**
The key innovation of Modernism was to separate a building's plan and facade from its structure. As seen in the design for Le Corbusier's Dom-ino house, the building's load is carried by large, reinforced-concrete floor slabs and slender columns he called pilotis.

After a 1932 US exhibition of Modernist architecture, **Modernism** became known as **International Style**

The Atomium

A monument to the Atomic Age, its nine shining spheres arranged to represent the atoms in a single crystal of iron

Built as the main pavilion for Expo 58, a world's fair held in Brussels in 1958, the Atomium was designed to stand only for the duration of the event. However, its startlingly futuristic design and popularity with the public ensured its survival and its status as one of the most visited attractions in the city. For an exhibit planned to last for only 6 months, the Atomium remains an engineering wonder more than 60 years after it was built.

Emblem of progress

The Atomium is 335 ft (102 m) tall and stands on the Heysel Plateau to the north of Brussels. It was designed by engineer André Waterkeyn (1917–2005) as a symbol of postwar faith in technological innovation and a vision of a peaceful future guided by scientific knowledge. The structure follows the Atomic Style—popular in the 1950s—that recalls the molecular models of high-school chemistry classes. It takes the shape of nine iron atoms arranged inside a unit cell of an iron crystal magnified 165 billion times. The structure consists of nine spheres, each 60 ft (18 m) in diameter, connected by tubes that are each 10 ft (3 m) wide. These enclose stairs, a central elevator, and escalators that are among the longest in Europe. Only six of the nine spheres are accessible to the public; each has two main floors, containing exhibition halls and other public spaces, and a lower service floor. The top sphere includes a restaurant with panoramic views of Brussels.

STRENGTHENING THE ATOMIUM

When originally conceived, the Atomium was to be free-standing, resting on its lowest sphere. Wind tunnel tests suggested, however, that it would have toppled over in winds of 50 mph (80 kph). Hence, three pairs of support columns were added to stabilize the structure by supporting the three main lower spheres. Emergency stairs were also added as a precaution.

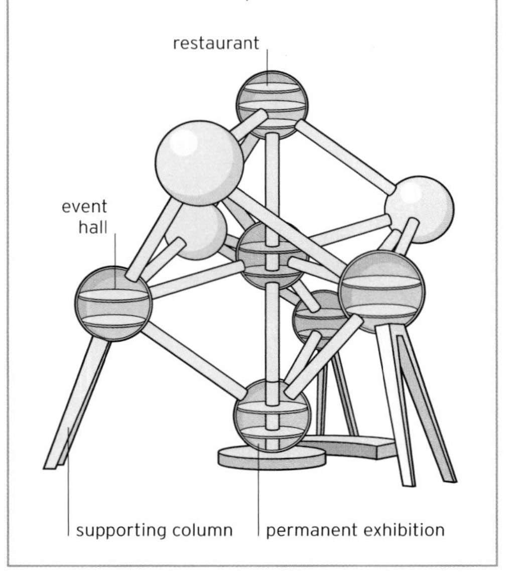

◁ **NEW SKIN**
The spheres of the giant iron atom were originally coated with aluminum, which began to corrode. The coating was replaced with stainless steel in 2001 to ensure that it was corrosion-resistant and matched modern architectural standards.

△ **ACCESSIBLE ATOMS**
Six of the nine spheres in use are connected by stairs, escalators, and an elevator that runs up the vertical central tube.

Stoclet Palace

A Modernist Viennese mansion, a "total work of art" set in the heart of Brussels

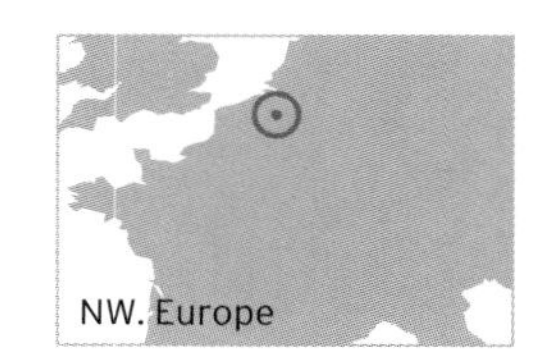

Adolphe Stoclet (1871–1949) was a wealthy Belgian industrialist and director of the Société Générale de Belgique, one of the largest investment companies in Belgium. While in Vienna to oversee the construction of a railway, he met one of the leading lights of the Vienna Secession art movement, the architect Josef Hoffmann (1870–1956). Stoclet shared Hoffman's avant-garde taste and asked him to design a house for him in Brussels.

Attention to detail

The resulting house is Hoffmann's masterpiece, a remarkable synthesis of the arts (*Gesamtkunstwerk*) and one of the most luxurious private houses of the 20th century. Asymmetrical in outline, its stark, rectangular exterior hides a refined, almost theatrical interior, with geometric furniture and rich decoration. Hoffmann designed every aspect of the building, right down to its door handles and wall paintings. Although a UNESCO World Heritage Site, the house is still occupied by the Stoclet family and thus not open to visitors.

△ **SOFTENING THE LINES**
The straight-lined exterior of the palace is softened by the windows that break through the eaves and the upright Art Nouveau balustrades that surround the balconies.

Einstein Tower

An Expressionist observatory designed to test Einstein's revolutionary theories

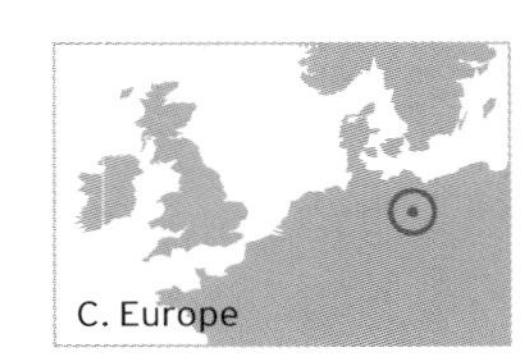

In 1911, Albert Einstein (1879–1955) published his *General Theory of Relativity*. One of the theory's predicted effects was a slight increase in the wavelength of light escaping from a strong gravitational field—a phenomenon known as gravitational redshift. German astronomers led by Erwin Finlay-Freundlich (1885–1964) decided to test the theory and commissioned a new solar observatory in Potsdam, outside Berlin, to be named after Einstein.

Dedicated to physics

The German architect Erich Mendelsohn (1887–1953) set out to create a building with a dynamic structure that would align with Einstein's radical theories. He opted to build the tower from reinforced concrete, suitable for its sculpted form. Construction difficulties caused by the complexity of the design, together with material shortages during World War I, meant that the observatory was actually built of brick and then covered with stucco. The designer achieved his vision for the building, however, with Einstein himself remarking on its "organic" form.

△ **SCULPTED APPEARANCE**
The building is curvilinear without any right angles, the soft corners and sinuous lines giving it a flowing look. It has been likened to the building in illustrations of the nursery rhyme about the old lady who lived in a shoe.

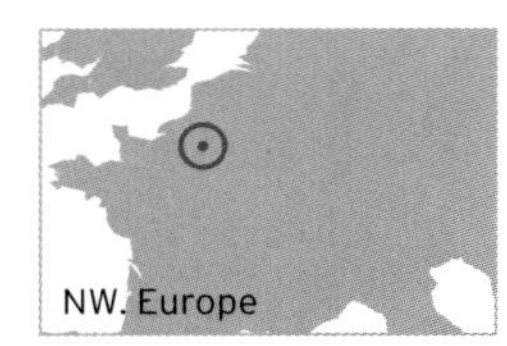

Centre Georges Pompidou

A high-tech structure with inside-out construction that startled its Parisian neighbors

Most French presidents like to leave their own architectural mark on Paris, none more so than Georges Pompidou, Gaullist prime minister from 1962 to 1968 and president from 1969 until 1974. It is ironic that this conservative statesman is associated with one of the most radical, futuristic buildings of the age. Designed by Briton Richard Rogers and Italian Renzo Piano, the building established Rogers's signature mark of exposing a building's services, such as water and air conditioning, in pipes on the outside. Rudely, some critics dubbed this style Bowellism, and the building was unpopular in many architectural circles when it opened in 1977.

An artistic success story

The building—known colloquially as the Beaubourg due to the area of Paris in which it stands—is 149 ft (45.5 m) tall. Its outer framework is covered with brightly colored pipes, ducts, and external elevators. Inside, the building houses the largest modern art museum in Europe, an experimental music center, and a library.

A SIMPLE BOX

While the Centre Georges Pompidou is highly decorated outside, its inside is relatively straightforward. Two basement levels house reception areas and ticket sales; the six floors above contain galleries, cinemas, restaurants, a library, offices, and a bookstore. With the load-bearing structures restricted to the outer framework, all the floors are open plan and can be divided up and reorganized at will for maximum flexibility.

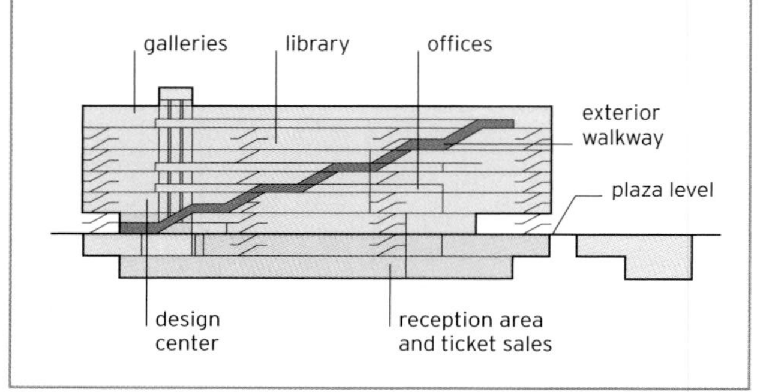

More than **16,500 tons (15,000 tonnes)** of steel were used to construct the **metal framework**

◁ **COLOR CODED**
The exterior pipes are all color coded: white for ventilation, blue for air conditioning, green for plumbing, and yellow for electricity. The exterior stairs and elevators are red.

▷ **INSIDE-OUT**
The external clutter of structural towers and supports, stairs, elevators and escalators, and service pipes obscures the outside of the building but ensures that the inside is kept clear for exhibitions and other uses.

MUSÉE,
EXPOSITIONS
Visiteurs sans billet
BOUTIQUES

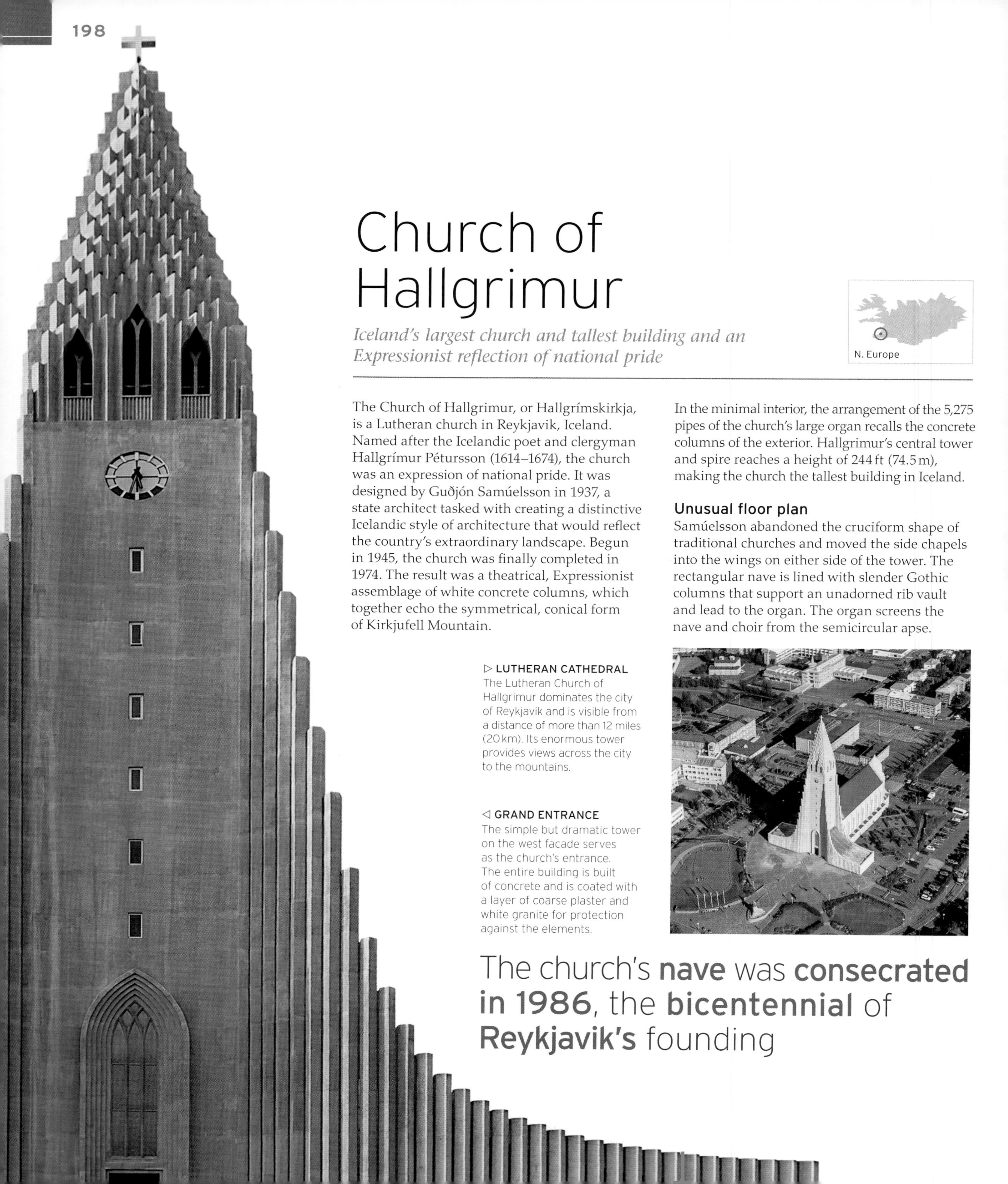

Church of Hallgrimur

Iceland's largest church and tallest building and an Expressionist reflection of national pride

The Church of Hallgrimur, or Hallgrímskirkja, is a Lutheran church in Reykjavik, Iceland. Named after the Icelandic poet and clergyman Hallgrímur Pétursson (1614–1674), the church was an expression of national pride. It was designed by Guðjón Samúelsson in 1937, a state architect tasked with creating a distinctive Icelandic style of architecture that would reflect the country's extraordinary landscape. Begun in 1945, the church was finally completed in 1974. The result was a theatrical, Expressionist assemblage of white concrete columns, which together echo the symmetrical, conical form of Kirkjufell Mountain.

In the minimal interior, the arrangement of the 5,275 pipes of the church's large organ recalls the concrete columns of the exterior. Hallgrimur's central tower and spire reaches a height of 244 ft (74.5 m), making the church the tallest building in Iceland.

Unusual floor plan

Samúelsson abandoned the cruciform shape of traditional churches and moved the side chapels into the wings on either side of the tower. The rectangular nave is lined with slender Gothic columns that support an unadorned rib vault and lead to the organ. The organ screens the nave and choir from the semicircular apse.

▷ **LUTHERAN CATHEDRAL**
The Lutheran Church of Hallgrimur dominates the city of Reykjavik and is visible from a distance of more than 12 miles (20 km). Its enormous tower provides views across the city to the mountains.

◁ **GRAND ENTRANCE**
The simple but dramatic tower on the west facade serves as the church's entrance. The entire building is built of concrete and is coated with a layer of coarse plaster and white granite for protection against the elements.

The church's **nave** was **consecrated in 1986**, the **bicentennial** of **Reykjavik's** founding

Grande Arche de la Défense

An elegant symbol of France and an important modern architectural landmark on Paris's Triumphal Way

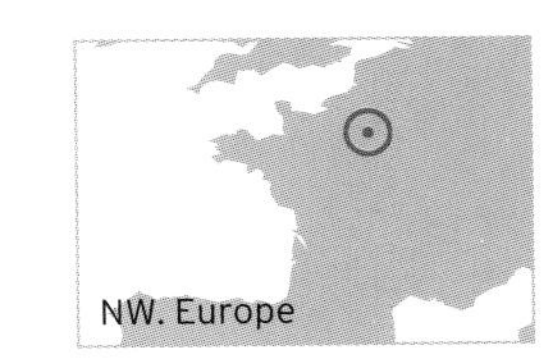

△ **A MIRROR TO THE CITY**
Seen at night, the mirrored cladding of the Grande Arche's outer walls reflects the bright lights of La Défense; its white inner surfaces provide a spectacular contrast.

In 1982, President François Mitterrand (1916–1996) launched *Grands Projets Culturels*, an architectural program designed to revitalize Paris and create a series of monuments emblematic of France at the end of the 20th century. Among the eight projects that transformed the Paris skyline over the next 20 years was the Grande Arche de la Défense. It was designed by Danish architect Otto von Spreckelsen and engineer Erik Reitzel, who had competed against 424 others to create a monument for La Défense, a business district west of Paris.

Tribute to the past

Originally known as La Grande Arche de la Fraternité ("the Arch of Fraternity"), the vast cuboid was to be a monument to humanity and friendship. Work began in 1985, with the building of a concrete frame, which was then clad in glass and marble. The arch was inaugurated on July 14, 1989, the 200th anniversary of the French Revolution. The exhibition space, restaurant, and viewing platform were closed in 2010 but reopened in 2017, with a walkway offering views over La Défense and down the Triumphal Way (see panel, right) to the Louvre.

THE TRIUMPHAL WAY

The Grande Arche de la Défense sits at the west end of the Axis Historique, or Triumphal Way, a line of important historic and modern monuments, avenues, and buildings that begins at the Louvre Museum (see p.174). Built above a metro station and expressway, the arch was placed at a slight angle to the axis to allow room for its foundations.

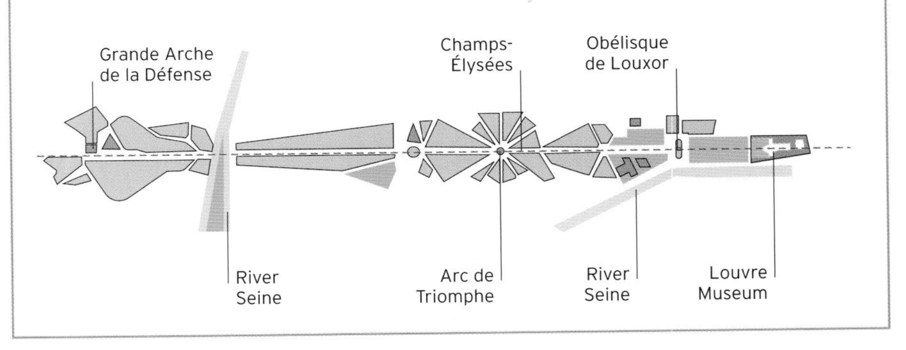

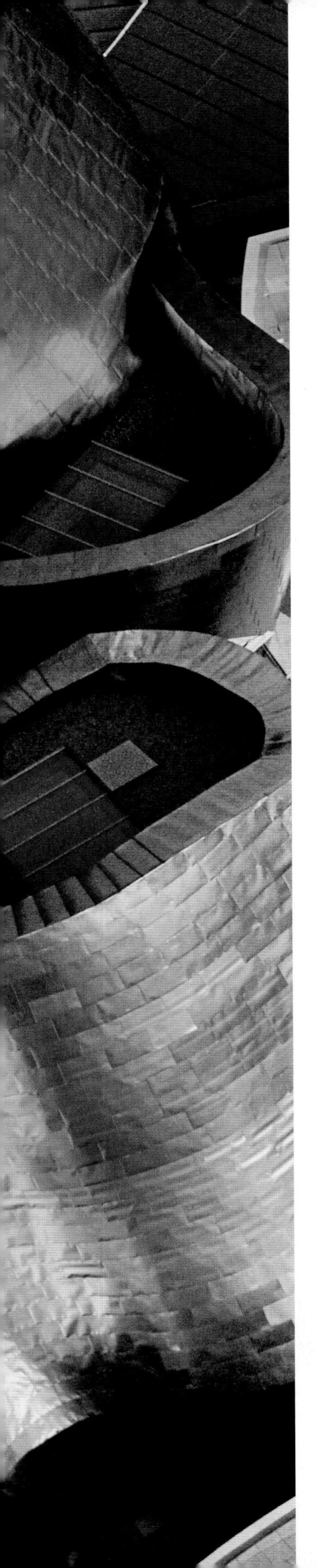

Guggenheim Museum Bilbao

An art museum that is itself also a sculpture—a building that rejuvenated a city in decline

Hailed as the most influential work of architecture in the second half of the 20th century, Frank Gehry's groundbreaking contemporary art museum was commissioned in the early 1990s to reinject life and cultural vitality into the once-thriving port of Bilbao, northern Spain. Its doors opened in 1997.

Perspectives on a city

The Canadian-born US architect made full use of the site's industrial urban context and its riverside location: from one perspective, with the city as a backdrop, it presents right angles and windows, combined with stone; from another view, with the Nervión River in the foreground, more fluid, undulating shapes harmonize with the surrounding water. In outline, the structure alludes to a ship, with curved panels suggesting sails, its titanium-clad exterior changing color according to the light and weather conditions. A metal flower adorns the highest point.

With 20 galleries on three levels, the massive structure combines titanium, glass, and limestone in an audacious experiment with volumes and perspectives. All the galleries converge on a large central atrium, which acts as the organizing heart of the museum.

Gehry's masterpiece has redefined Bilbao, bringing not only visual and cultural impact but also millions of visitors—and with them, considerable economic growth.

Fishlike forms appear around the building, such as **glass scales** that conceal stairs and elevators

◁ **THE EXTERIOR**
Magnificent curves and undulating forms in titanium, limestone, and glass define the museum's exterior, creating a modern space of immense force and originality.

△ **CIRCULATION HUB**
The monumental atrium—with its soaring shapes in steel, glass, limestone, and plaster, intersected by horizontal walkways—forms the heart of the museum.

THE ATRIUM

The magnificent atrium in the museum, a hall of curved volumes and twisting glass and steel, reaches a height of 164 ft (50 m). This vast empty space, flooded with light, acts as a central point from which the museum's various rooms, halls, and pathways can be accessed.

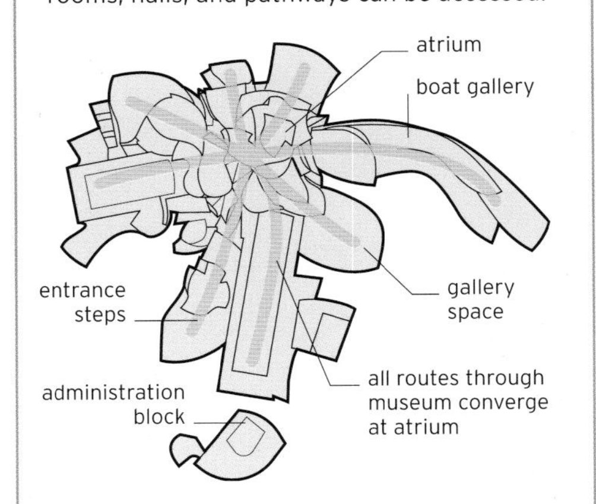

Millau Viaduct

A sublime cable-stayed bridge, built in response to the problem of modern traffic congestion

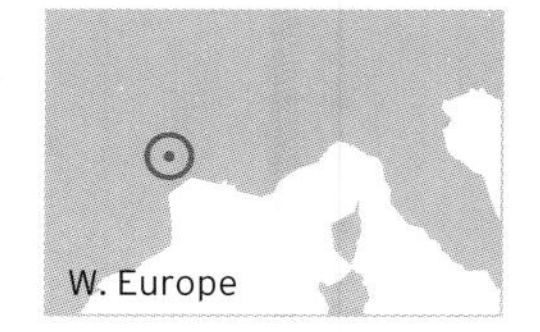

Until the opening of the Millau Viaduct in 2004, vacation traffic from Paris south to the Riviera and Spain caused regular congestion in the narrow Tarn valley near Millau in southern France. The first plans for a bridge to bypass the town were discussed in 1987, and four possible routes were surveyed. Each presented its own technical, geological, or environmental problems, but in 1991, the decision was made to build a viaduct over the River Tarn to the west of Millau.

French-British cooperation

The resulting 8,200 ft (2,500 m) bridge, designed by British architect Norman Foster and French structural engineer Michel Virlogeux, stands more than 660 ft (200 m) above the river below. The deck on which the vehicles run is a lightweight steel structure that was prefabricated and then assembled on site. It is shaped like an inverted trapezoidal box that is cambered (sloped) on each side. Built at a total cost of $442 million, the bridge was inaugurated on December 14, 2004, and opened for traffic 2 days later.

BUILDING THE DECK

Construction of the deck started on both sides of the Tarn valley, with sections rolled out from one pier to the next until they met in the middle. The deck is supported from below by piers and from above by stays attached to vertical pylons. There are seven piers and pylons, creating two outer spans of 669 ft (204 m) each and six inner spans of 1,122 ft (342 m) each. The deck itself is 91 ft (28 m) and made of metal sheets ½ in (14 mm) wide.

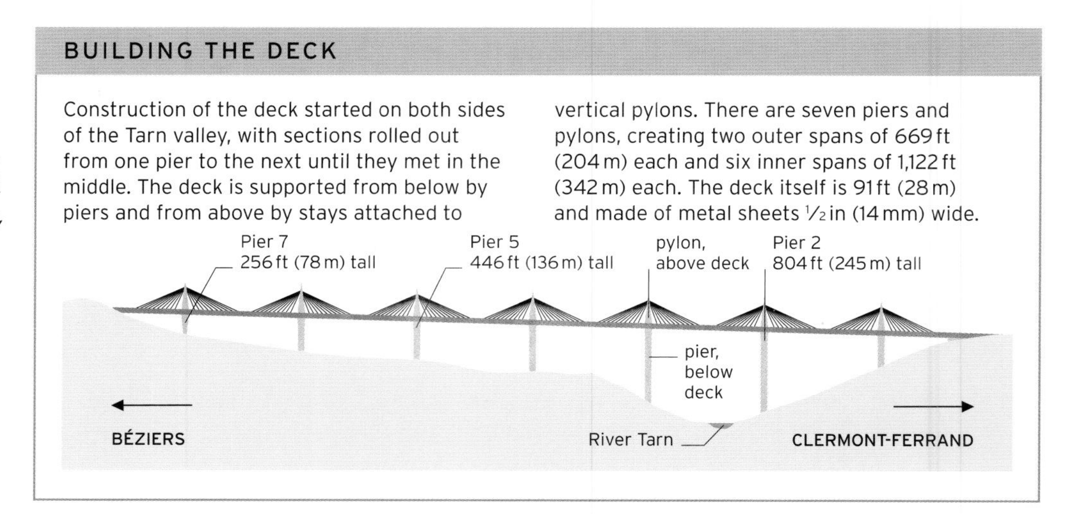

Pylon 2 of the viaduct is **75 ft (23 m)** taller than the **Eiffel Tower**

THE TALLEST IN THE WORLD
Standing at a height of 1,125 ft (343 m), the Millau Viaduct is the tallest bridge in the world, 69 ft (21 m) taller than its nearest rival—the Bosphorus Bridge in Istanbul, Turkey.

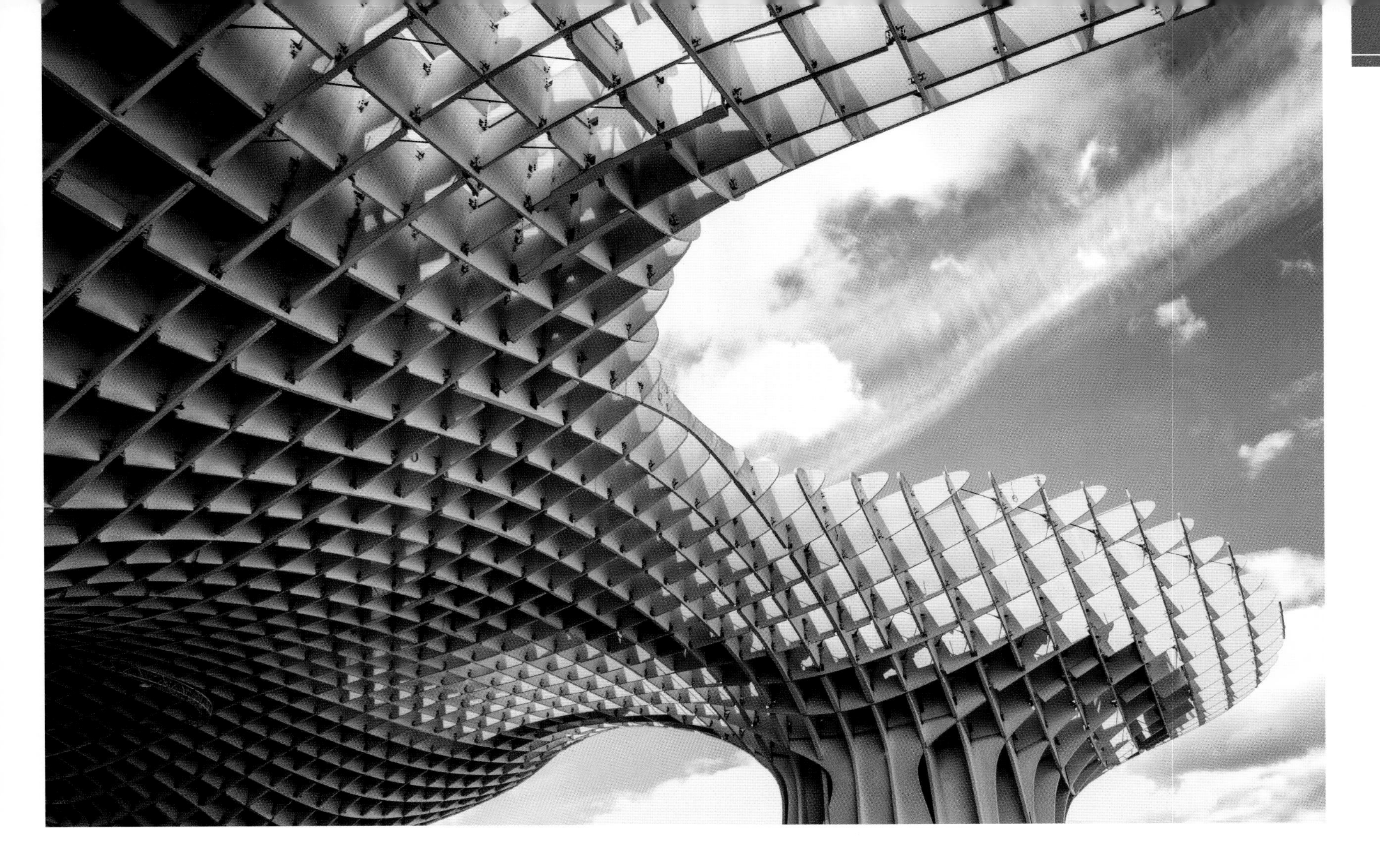

Metropol Parasol

A light and spacious wooden canopy that stands on ancient remains and houses one of Seville's most vibrant markets

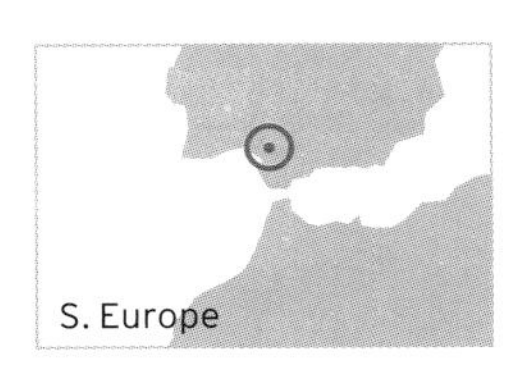

△ **GIANT MUSHROOMS**
Because of its distinctive shape, the new market is nicknamed *Las Setas de la Encarnación*—the Spanish for "mushrooms" is *las setas*.

Seville's Plaza de la Encarnación has been home to a food market since the mid-19th century. The original, dilapidated market buildings were pulled down in 1973, and the traders moved into one corner of the square. Their temporary exile lasted for nearly four decades, when a new market, over an underground car park, was planned. However, construction stopped when ancient Roman and Moorish-era ruins were discovered at the site.

Mushrooms and fig trees

A competition for a structure to cover the ruins and provide a new site for the market was won by the German architect Jürgen Mayer, who was inspired by the vaults of Seville's Cathedral of St. Mary of the See (see p.143) and by the weeping fig trees that grow in the nearby Plaza de Cristo de Burgos.

The Metropol Parasol is built on four levels. Underground Level 0 contains a museum housing the Roman and Moorish remains discovered on site. Level 1 houses the market, and its roof holds an open-air public plaza sheltering under the wooden parasols above. Levels 2 and 3 hold panoramic terraces and restaurants that offer some of the best views of the ancient city of Seville.

△ **A MODERN INTERLOPER**
One of the largest timber structures ever built, the wooden canopy of the new market is in great contrast to the older, more sedate buildings of Seville that surround it.

BIRCHWOOD STRUCTURE

The Metropol Parasol has been described as the largest wooden building in the world. It consists of six connected parasols or umbrellas shaped like giant mushrooms. Each parasol consists of a mesh of straight wooden panels and beams glued together and sculpted into curves. The graceful structure opened for trading and enjoyment in 2011.

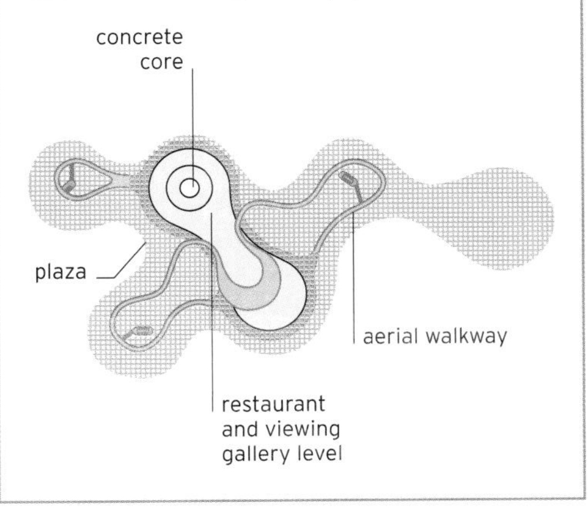

Ancient wonder and modern city
The Great Pyramids of Giza are the only survivors of the Seven Wonders of the Ancient World. They stand on a small plateau with the densely packed modern city of Cairo wrapping around its edges.

Africa

MASTERS OF BRICK AND STONE

Africa

Adapting to a wide range of climates, the builders of Africa's man-made wonders have relied on ingenious uses of materials. North of the Sahara, the Nile Valley's limestone and sandstone were the building blocks of ancient temples and pyramid tombs. The Egyptians excelled at crafting stone columns, positioned close together to support heavy loads, which enabled them to build large-scale structures. Some of the continent's best-known architectural achievements are in North Africa, which benefitted not only from easily quarried stone but also from frequent cultural and trade exchanges with other parts of the world. South of the Sahara, the masons of Ethiopia carved entire buildings from solid stone, while other African architects used the earth itself, baked into bricks under the Sun.

KEY SITES

1. The Great Sphinx
2. Pyramids of Giza
3. Karnak Temple Complex
4. Abu Simbel Temples
5. Meroë Pyramids
6. Timgad
7. Royal Mausoleum of Mauretania
8. Obelisks and Royal Tombs of Aksum
9. Leptis Magna
10. Amphitheater of El Jem
11. Great Mosque of Kairouan
12. Kano City Walls
13. Lalibela Rock Churches
14. Koutoubia Mosque
15. Great Zimbabwe
16. Royal Palaces of Abomey
17. Great Mosque of Djenné
18. Afrikaans Language Monument
19. Basilica of Our Lady of Peace

ISLAM MOVES WEST
632–681 CE

The movement of Islam toward northeast and northwest Africa sparked a wave of mosque building, with 7th-century examples in Eritrea, Ethiopia, Egypt, Somalia, Tunisia, and Algeria. These early mosques borrowed from the style of the time, with innovations such as square or rectangular rather than circular minarets.

11 GREAT MOSQUE OF KAIROUAN

CHRISTIANITY BRINGS CHURCHES
1ST CENTURY–EARLY 15TH CENTURY

As Christianity spread through the Horn of Africa, places of worship and monasteries were built to cement the Christian presence and attract converts. The stonemasons of Ethiopia applied indigenous tradition with remarkable results in carving the Church of St. George (below) and other churches from solid rock.

13 LALIBELA ROCK CHURCH

THE ROMAN OCCUPATION
146 BCE–698 CE

Wherever the Romans expanded, they introduced their model of city life in order to placate and attract their new subjects, as well as to provide a home for soldiers and expatriate citizens. Even remote outposts such as Timgad display the characteristic Roman grid of streets, with footpaths, colonnades, and public facilities.

6 TIMGAD

MONUMENTS OF ANCIENT EGYPT
3100–30 BCE

The pyramid was a trademark of Egyptian monument building, but the Egyptians also built with columns and pediments, which were later adopted by the Greeks. They worked with massive blocks of limestone and granite—using no pulleys, iron tools, or the wheel—and also leveled their foundations extremely accurately.

2 PYRAMIDS OF GIZA

Madeira
Canary Islands
Atlas Mountains
Grand Erg Oriental
Sahara
Western Desert

the Great Mosque of Kairouan was established in 670 CE and is the oldest place of Muslim worship in Africa

the pyramids of Giza are the only surviving Wonders of the Ancient World

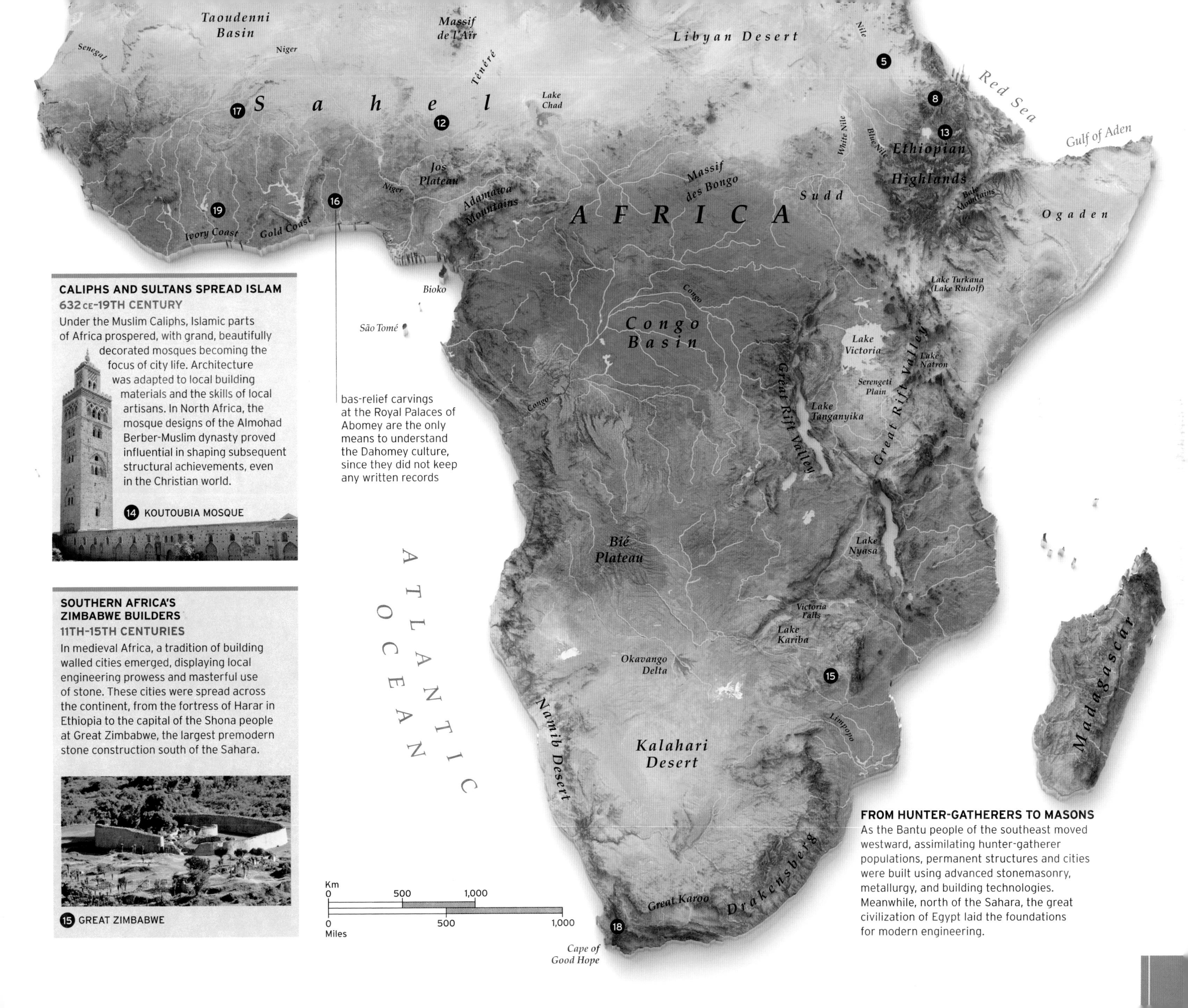

CALIPHS AND SULTANS SPREAD ISLAM

632 CE–19TH CENTURY

Under the Muslim Caliphs, Islamic parts of Africa prospered, with grand, beautifully decorated mosques becoming the focus of city life. Architecture was adapted to local building materials and the skills of local artisans. In North Africa, the mosque designs of the Almohad Berber-Muslim dynasty proved influential in shaping subsequent structural achievements, even in the Christian world.

14 KOUTOUBIA MOSQUE

SOUTHERN AFRICA'S ZIMBABWE BUILDERS

11TH–15TH CENTURIES

In medieval Africa, a tradition of building walled cities emerged, displaying local engineering prowess and masterful use of stone. These cities were spread across the continent, from the fortress of Harar in Ethiopia to the capital of the Shona people at Great Zimbabwe, the largest premodern stone construction south of the Sahara.

15 GREAT ZIMBABWE

16 bas-relief carvings at the Royal Palaces of Abomey are the only means to understand the Dahomey culture, since they did not keep any written records

FROM HUNTER-GATHERERS TO MASONS

As the Bantu people of the southeast moved westward, assimilating hunter-gatherer populations, permanent structures and cities were built using advanced stonemasonry, metallurgy, and building technologies. Meanwhile, north of the Sahara, the great civilization of Egypt laid the foundations for modern engineering.

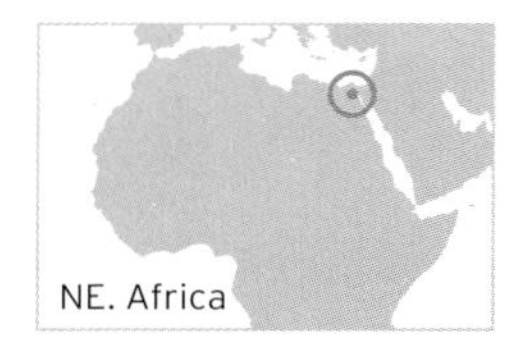

The Great Sphinx

One of the world's biggest sculptures, which stands as a tomb guardian in Giza's sands

The Great Sphinx—a massive limestone statue of the mythical creature with the body of a lion and the head of a king—stands in front of the pyramid of the Pharaoh Khafre at Giza and adjacent to his valley temple. Built during Pharoah Khafre's reign (2520–2465 BCE), approximately 240 ft (73 m) long and 66 ft (20 m) high, it is dynastic Egypt's first ever monumental statue.

The statue may originally have included the distinctive blue-and-yellow striped Nemes headdress often worn by Egyptian pharaohs, with its two flaps that fall over the shoulders. The sphinx's facial features are generally thought to resemble those of Khafre, although several commentators have suggested that the statue depicts Khufu, Khafre's father.

A symbol of omnipotence

A fusion of the lion (king of the beasts, a symbol of royalty, and a guardian figure) and the divine ruler, the Sphinx may symbolize the pharaoh as an omniscient and omnipotent ruler. The lion is also associated with the Sun: the sphinx at Giza faces the rising Sun and, when viewed from the east-southeast, its head appears flanked by the two pyramids behind it. Some scholars argue that this configuration resembles the hieroglyph for "horizon" (a Sun disk between two mountains) and is associated with the Sun god Hor-em-akhet ("Horus in the Horizon"), the name that, centuries later, the Egyptians of the New Kingdom (c.1539–1075 BCE) gave the Sphinx.

By the New Kingdom, the Great Sphinx was in disrepair. A red granite stela, known as the Dream Stela, erected by Thutmose IV (r. c.1400–1390 BCE), records a dream he had when he was a prince. In it, the sphinx appears to Thutmose, promising him the throne if he makes repairs to the statue. Thutmose did indeed make various repairs to the sphinx before he became king, including removing sand that was covering it, replacing stones, and building a protective wall. The inscription is considered a propaganda tool that justified his right to the throne.

◁ **SIGNIFICANT AXIS**
An icon of ancient Egypt, the Great Sphinx was carved out of a single piece of limestone. Built on an east-west axis, it aligns with Khafre's temple.

△ **ANCIENT AND MODERN**
In this aerial view, taken from the south of modern-day Cairo, Giza's massive, majestic pyramids tower over the densely packed high-rise buildings of the sprawling metropolis, creating a juxtaposition of Egypt's ancient and modern worlds.

A HIDDEN CHAMBER

In 2017, a vast cavity 98 ft (30 m) or more in length, thought to be a secret chamber, was detected above the Grand Gallery in Khufu's Great Pyramid. The breakthrough was made by means of muon analysis—a technique that uses sensors to detect subatomic particles called muons in order to generate three-dimensional images of volumes.

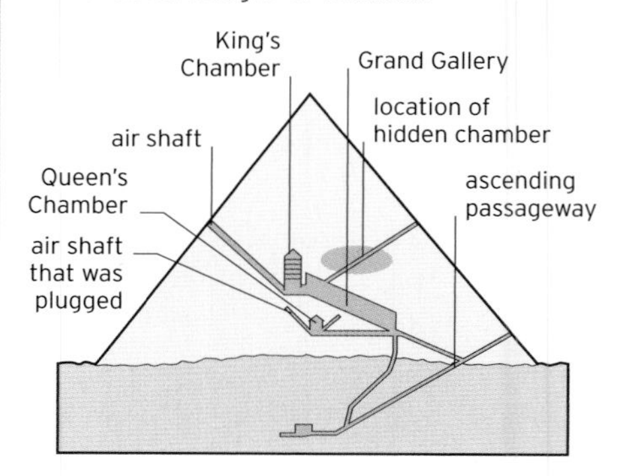

Pyramids of Giza

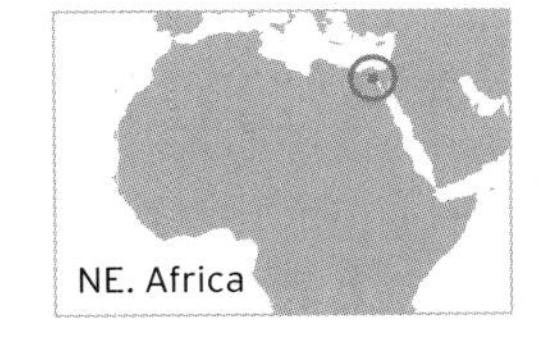

Astonishing feats of engineering and imagination that have become monuments to a remarkable ancient civilization

The pyramids at Giza on the west bank of the Nile are among the world's most familiar landmarks. They include the Great Pyramid, which was one of the Seven Wonders of the Ancient World. Tremendous feats of engineering, craftsmanship, and labor, they are thought to have been built during Khufu's reign (c.2545–2525 BCE). The massive pyramid complexes included burial chambers for the kings and queens, mortuary temples, valley temples, and connecting causeways. More than just royal burial tombs, the pyramids housed all the items that the pharaohs would need to ensure their comfort in the afterlife, in the belief that after death they would achieve immortality and divine status. Full-sized boats were buried near the pyramids as transport for the pharaoh in the afterlife.

Ramps to the sky

The oldest and largest of Giza's three main pyramids is Khufu's Great Pyramid, which rose to a height of 480 ft (146 m), with a base width of 755 ft (230 m). The colossal structure contained around 92 million cubic ft (2.3 million cubic m) of stone blocks, which were arranged with astonishing precision. The core of the pyramid is made of yellowish limestone and was originally surfaced with smooth, white Tura limestone. (This was later removed to build mosques.)

The second, smaller, pyramid at Giza is 471 ft (144 m) high and was built by Khufu's son, Khafre. The third pyramid, built by Menkaure and unfinished at the time of his death, is the smallest of the three (about 213 ft/65 m in height) but has a more complex mortuary temple. These tombs have immense religious and mythic significance and symbolize not only ancient Egypt's veneration of the dead, but also the might of the pharaoh and the centrality of the Sun god Ra. The pyramid's triangular shape, for example,is often said to derive from the belief that, following the death of a pharaoh, the Sun god would intensify his rays, creating a ramp on which the spirit of the deceased could easily ascend to the sky.

△ **KHUFU'S GRAND GALLERY**
This spectacular corbel-roofed passageway known as the Grand Gallery, inside the Great Pyramid, leads to Khufu's burial chamber, which contained his mummified corpse.

Karnak Temple Complex

A site of immense cultural significance and a principal source of knowledge about ancient Egyptian civilization

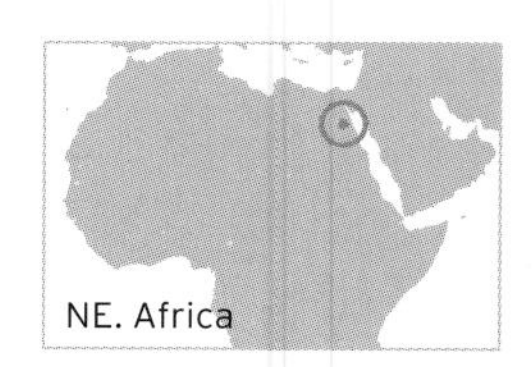

The massive temple complex on the east bank of the Nile at Karnak, near Luxor, is Egypt's most visited site after the pyramids at Giza (see pp.208–209). It includes a vast range of structures—from temples, chapels, giant obelisks, and massive gateways (pylons) to sphinxes, sacred lakes, and a hypostyle (columned) hall (see pp.212–213).

Sites of worship

The site has three existing main temple precincts dedicated to gods: those of Mont, the god of war, in the north; Mut, the Earth goddess and wife of Amun, in the south; and between these two, by far the largest and most important, the precinct of Amun, the greatest of all gods. The prominence of Karnak derives from the Temple of Amun, particularly from the New Kingdom (c.1539–1075 BCE), when it assumed preeminent religious significance as the center dedicated to his cult.

One of Karnak's most spectacular features is its enormous hypostyle hall; around 54,000 square ft (5,000 square m) in area, it includes 134 massive columns—the majority 33 ft (10 m) high, with architraves that weigh 154,000 lb (70,000 kg)—notable for their deeply incised hieroglyphs, designed to prevent erasure by subsequent rulers.

Karnak is unique among Egypt's sites for the length of time it took to build—the complex was under continual construction for some 2,000 years, beginning during the Middle Kingdom (c.1980–1630 BCE) and ending in Ptolemaic times (305–30 BCE). It was designated a World Heritage Site in 1979.

▷ **ANCIENT COLUMNS**
The columns at Karnak were built to massive proportions and deeply inscribed to ensure the pharaohs' place in history. Originally, they would have been vibrantly painted.

Some **30 pharaohs** were involved in Karnak's **construction**

KARNAK TO LUXOR: THE SACRED ROAD

The temples of Karnak and the nearby Luxor are linked by an avenue of human-headed sphinxes, almost 2 miles (3 km) long. The avenue originally served as the outbound processional route of the annual Opet festival, which reenacted the marriage of the gods Amun and Mut; the spectacular procession then returned on the Nile on barques. A series of magnificent reliefs at Karnak record the movement of the barques.

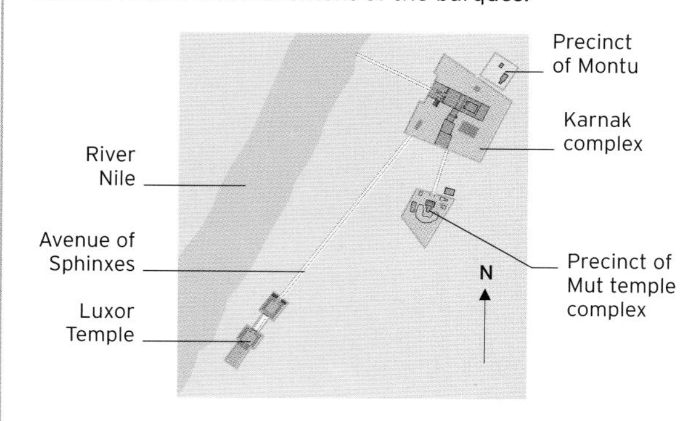

△ **GREAT HYPOSTYLE HALL**
The monumental hypostyle hall, one of Karnak's highlights, was used for religious ceremonies and would only have been accessed by priests and rulers.

◁ **STATUE OF AMUN**
The god Amun, depicted here in human form, would have been honored daily by priests at his spectacular temple at Karnak.

Architectural styles

ANCIENT EGYPT

Although the pyramid is the most recognizable building type of ancient Egyptian architecture, a range of extraordinary temples and palaces also survive, generally as a result of their massive size and robust construction from stone.

The ancient Egyptian civilization came into existence around 3000 BCE, before beginning to decline around 1000 BCE, and ending with the death of Cleopatra in 30 BCE. Its strength and longevity rested on its utilization of the River Nile for large-scale agriculture, which created the wealth and stability necessary for society and culture to flourish and to undertake construction projects often on a colossal scale.

Egyptian architecture made extensive use of stone and mud-brick, generally relying on a post-and-lintel construction system. Flat roofs constructed of huge stone slabs required closely spaced columns to carry the load. Surfaces, including columns and entrance pylons, were highly decorated with hieroglyphics and motifs deriving from Egyptian religion. Though our knowledge of ancient Egypt's architecture is informed by the survival of mostly religious buildings and tombs, hieroglyphics tell us a huge amount about Egyptian society, daily life, history, and the central importance of their religion.

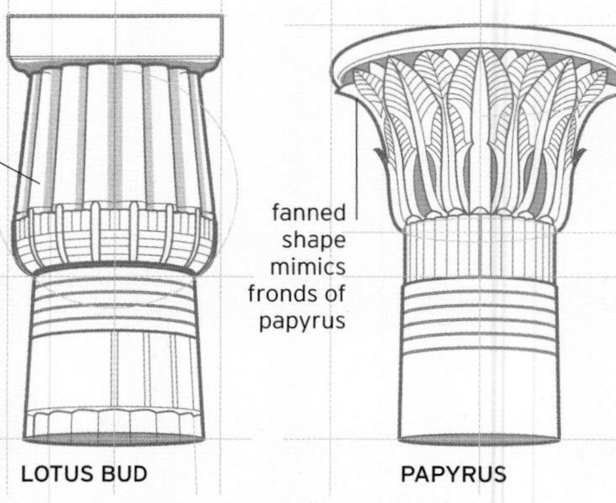

△ **LOTUS BUD AND PAPYRUS CAPITALS**
A capital is the uppermost part of a column that splays out to support the load above. In Egyptian architecture, capital forms were typically derived from plants that held significance in Egyptian culture, such as the lotus bud and papyrus.

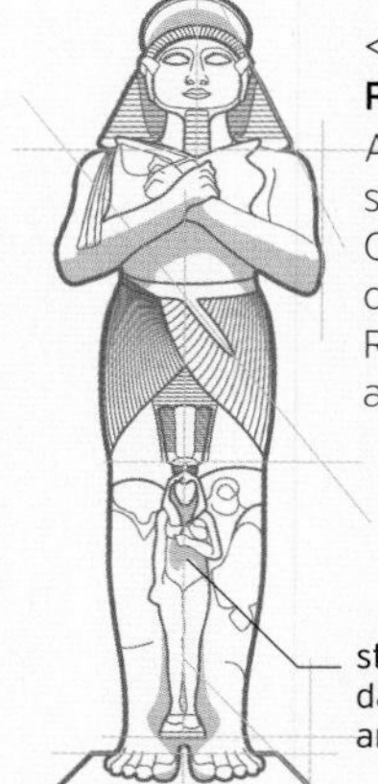

◁ **COLOSSUS OF RAMESSES II**
A colossus is a massive statue of a pharaoh or deity. One of many at Karnak, this colossus commemorates Ramesses II, often known as Ramesses the Great.

Hypostyle Hall ceiling supported by 134 internal columns

granite Colossus of Ramesses II at entrance to Hypostyle Hall

columns of reddish-brown sandstone

Kiosk of Taharqa, used for public ceremonies

vertical groove would have housed wooden flag pole

entrance pylons remain unfinished and of unequal height

Temple of Amun measures 1,200 x 360 ft (366 x 110 m)

Amun is associated with rams

statuettes of Ramesses II stand between forelegs of sphinx

▽ **SPHINXES**
A sphinx is a mythical creature, usually combining a lion's body with a human's head. It acted as a guardian at the entrance of a temple. The most famous sphinx, the Great Sphinx of Giza (see p.208), is 240 ft (73 m) long and 66 ft (20 m) high.

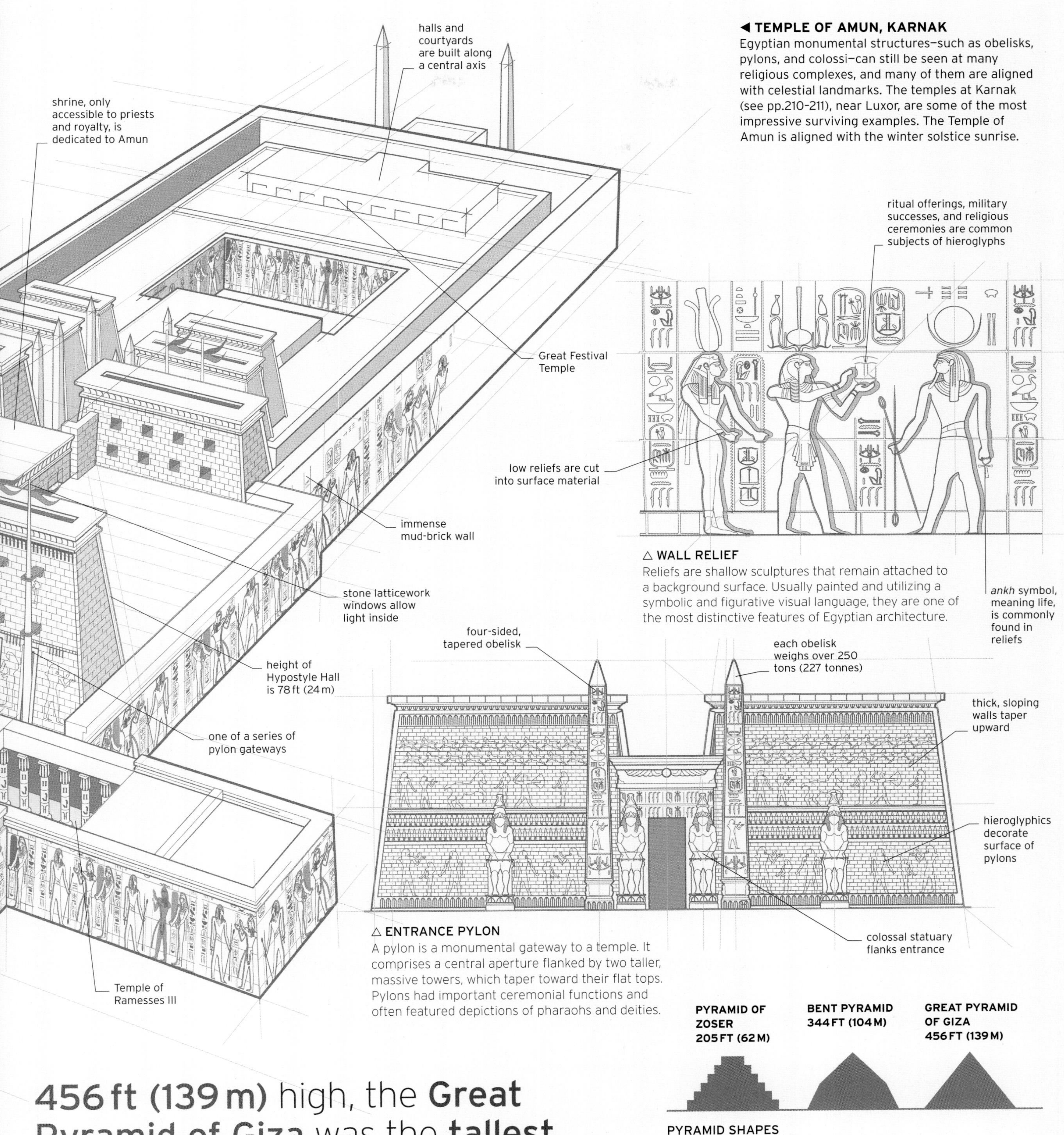

◀ TEMPLE OF AMUN, KARNAK
Egyptian monumental structures—such as obelisks, pylons, and colossi—can still be seen at many religious complexes, and many of them are aligned with celestial landmarks. The temples at Karnak (see pp.210–211), near Luxor, are some of the most impressive surviving examples. The Temple of Amun is aligned with the winter solstice sunrise.

△ WALL RELIEF
Reliefs are shallow sculptures that remain attached to a background surface. Usually painted and utilizing a symbolic and figurative visual language, they are one of the most distinctive features of Egyptian architecture.

△ ENTRANCE PYLON
A pylon is a monumental gateway to a temple. It comprises a central aperture flanked by two taller, massive towers, which taper toward their flat tops. Pylons had important ceremonial functions and often featured depictions of pharaohs and deities.

PYRAMID SHAPES
Pyramids are burial monuments for pharaohs. The Great Pyramids of Giza have smooth, sloping sides; in others, the slopes may change angle or take the form of pronounced steps.

456 ft (139 m) high, the **Great Pyramid of Giza** was the **tallest building in the world** until 1300 CE

△ COLOSSUS OF RAMESSES II
This view of the immense limestone statue of the deified Ramesses II outside the main temple shows his finely carved features and distinctive headdress.

▷ TEMPLE GUARDIANS
Four magnificent rock-hewn statues of Ramesses II, depicting the pharaoh in different guises, guard the doorway to his temple at Abu Simbel. Although three-dimensional, they were designed to be viewed head-on.

△ INSIDE THE TEMPLE
The hypostyle hall inside Ramesses II's temple features eight carved pillars of the deified pharaoh as Osiris, god of the underworld.

Abu Simbel Temples

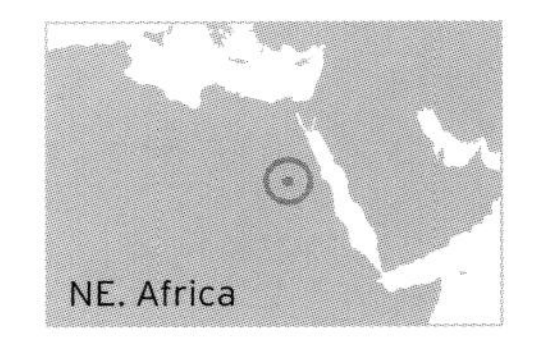

Ancient temples that are a tribute to the power and authority of ancient Egypt

The two spectacular cliff-cut temples of Abu Simbel are among the most famous of Egypt's ancient monuments. Situated on the west bank of the River Nile in the south of the country, the colossal structures were commissioned by Ramesses II, whose reign (c.1279–1213 BCE) was the longest in Egyptian imperial history.

An astute statesman, the pharaoh built the two sandstone temples as symbols of his power and the might of his kingdom. The date when construction began is the subject of debate, but the temples are thought to have taken 20 years to build. Ramesses dedicated the main temple, or the Great Temple, to himself and a smaller temple, the Small Temple, to his principal wife, Queen Nefertari, the personification of the goddess Hathor.

Symbols of power

The Great Temple is a huge edifice, 108 ft (33 m) high, with an imposing facade: two pairs of massive statues, 66 ft (20 m) high, of the enthroned Ramesses flank the entrance. The interior of the temple is triangular in layout and includes various chambers and three halls of decreasing size, which extend around 184 ft (56 m) into the cliff. The most breathtaking is the hypostyle hall, 30 ft (9 m) high, with its mighty pillars depicting Ramesses. Carved reliefs on the walls narrate events in the pharaoh's reign.

The Small Temple (39 ft/12 m high), immediately north of the Great Temple, has two statues of Ramesses and one of Nefertari on each side of its entrance. The temple interior is notable for its own hypostyle hall, with its six decorated pillars and walls that include images of Hathor and of the king and queen making offerings to various deities.

The UNESCO World Heritage Site attracted attention in the 1960s, when, under threat from flooding from the building of the Aswan High Dam, the temples were removed and reassembled further inland. The extraordinary feat of skill and engineering took around 4 years to complete.

▷ **THOTH AS A BABOON**
A statue, depicting the god Thoth as a baboon, was one of four that decorated the solar chapel, which formed part of the Great Temple.

The **relocation** of Abu Simbel involved cutting the site into around **16,000 huge blocks** and **reassembling** it piece by piece

SACRED ALIGNMENT

The architects and engineers involved in the complex relocation of Abu Simbel were careful to ensure that the entrance to the main temple was correctly aligned so that twice a year, the Sun's rays would continue to penetrate the structure's deep, innermost sanctuary and illuminate the face of the statue of the deified Ramesses II, just as they had done since the 13th century BCE.

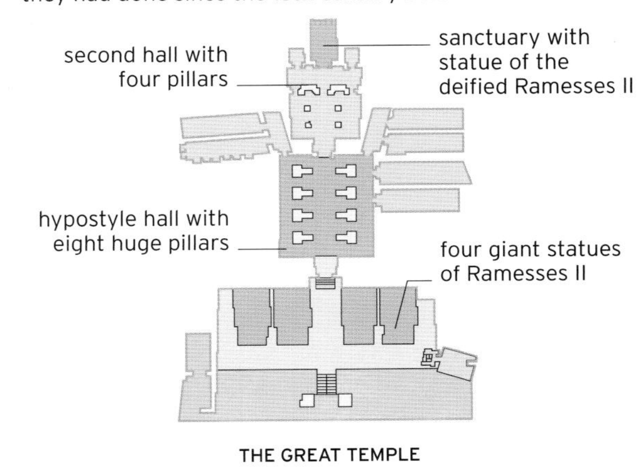

THE GREAT TEMPLE

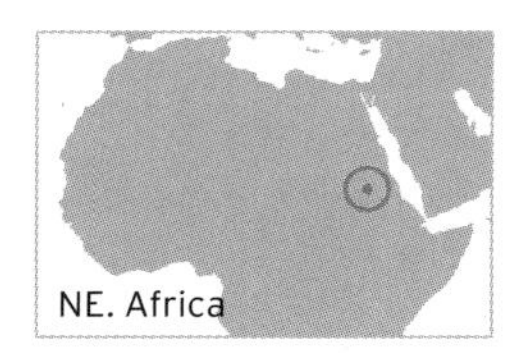

Meroë Pyramids

A collection of more than 200 pyramids built over the royal tombs of the Kingdom of Kush

From the 8th century BCE, the Kingdom of Kush ruled over Nubia, the land along the Nile south of Egypt, for almost 900 years. Meroë, in present-day Sudan, was chosen as the capital of this empire in the 3rd century BCE, and through trade—especially in ironware and ceramics—became established as a prosperous city, as well as the home of the royal family.

The Kingdom of Kush developed a distinctive civilization in parallel with that of neighboring Egypt while incorporating some elements of Egyptian culture, most conspicuously the construction of pyramids over the tombs of their rulers and prominent citizens. These can be seen today in three excavated cemetery sites around the town of Meroë, where many of the Meroite rulers are buried.

Nubian outline

The Meroë pyramids are recognizably different from the more famous ones at Giza in Egypt (see pp.208–209), with a characteristically Nubian steep-sided shape and often an imposing entrance structure. Although not as massive as the Egyptian pyramids, the sheer number of pyramids in close proximity in the Meroë burial sites makes an equally impressive sight.

The burial chambers beneath the pyramids originally housed the remains and funerary goods of the royalty of Kush, and they were decorated with paintings, relief carvings, and hieroglyphics showing Egyptian influence. Unfortunately, the tombs have been repeatedly raided over the ages, and little of the treasure they contained has survived to modern times.

△ **NUBIAN JEWELRY**
Among the goods found in the tombs were many items of jewelry, including this gold and enamel hinged bracelet decorated with an image of the goddess Hathor and geometric patterns.

DESTRUCTION AND REBUILDING
Of the more than 200 pyramids at Meroë, many were partially dismantled in Italian explorer Giuseppe Ferlini's search for treasure. However, some, such as those in the foreground of this image, have been fully reconstructed.

PYRAMID DESIGN

The pyramids at Meroë are in the typically Nubian style, smaller and steeper than those found in Egypt. They are constructed from blocks of sandstone, rising at an angle of 65–70 degrees from a narrow base to a height of between 20 ft (6 m) and 98 ft (30 m). Entrance to each pyramid is through a covered passage behind a monumental gateway, but the burial chambers are dug separately into the ground at the foot of an earth ramp.

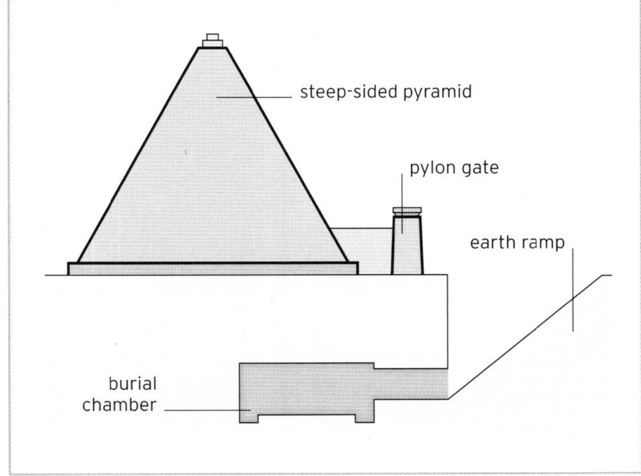

△ **ENTRANCE PASSAGEWAY**
A covered stone passageway of relatively simple rectilinear design leads from the entrance to the interior of each pyramid.

◁ **PYLON GATEWAY**
A feature of the Meroë pyramids is the monumental gateway, modeled on Egyptian temple gates known as pylons, which leads to the entrance passageway.

ROMAN URBAN PLANNING
The Forum, Theater, and other public buildings are at the heart of the city, close to the junction of the two colonnaded main thoroughfares.

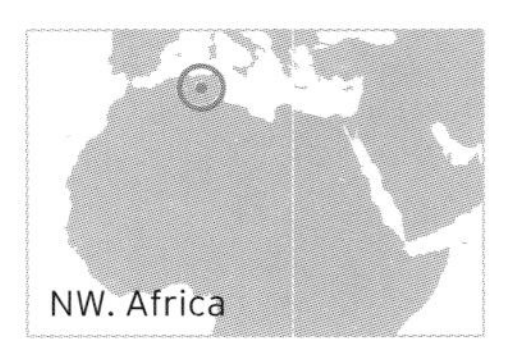

Timgad

A magnificent colonial city exemplifying the grid design used in Roman urban planning

Preserved by centuries of burial in Saharan sand, the ruins of Timgad in the Aurès mountains of present-day Algeria are sometimes referred to as the Algerian Pompeii. Although sacked by invaders in the 5th and 8th centuries, the foundations of the Roman walled city have survived, along with some of the original structures, and the site today shows the sophistication of Roman town planning at its height.

Limestone city

Timgad was founded in about 100 CE by the Roman Emperor Trajan, probably as a residence for troops defending the Roman colony in Numidia from Berber attack. Working on a site where there had been no previous settlement, the city's planners had free reign and an opportunity to design a city on a strictly regular grid. Local limestone was used for paving the streets and as the primary building material; many of the buildings were lavishly decorated with mosaics.

Originally designed to house around 15,000 people, Timgad rapidly expanded beyond the city walls as it became a trading center, as well as a garrison town, and later, in the 3rd century, an important regional center of Christianity. The Romans were forced out in the 5th century, and the town was ransacked and left uninhabited some 300 years later.

ROMAN GRID PLAN

The rectangular perimeter wall encloses a city constructed with an orthogonal grid plan of streets. The principal streets are the *Decumanus*, which runs east-west through the center of the town, and the *Cardo*, the main road running north-south from the northern entrance to where it meets the *Decumanus* at the Forum in the center of the town.

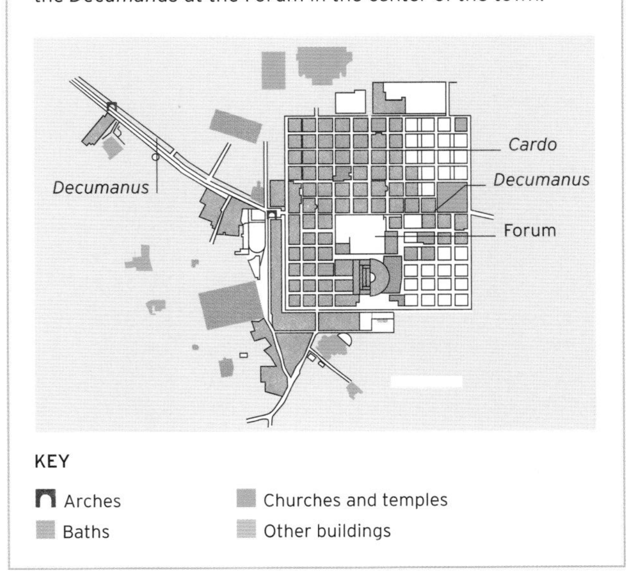

KEY

- Arches
- Baths
- Churches and temples
- Other buildings

△ TIMGAD THEATER
The theater, cut into the side of a small hill, seats up to 3,500 people and is still used to stage performances. It was largely rebuilt by French archaeologists.

△ DECORATIVE MOSAIC
Mosaics were used to decorate the walls and floors of the limestone buildings, as marble—which the Romans favored—was not available locally.

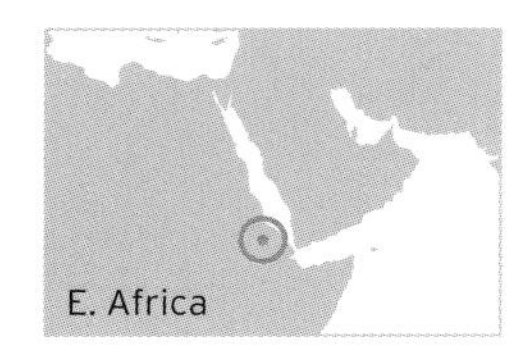

Obelisks and Royal Tombs of Aksum

Massive monolithic stelae standing among the ruins of the ancient capital of the Kingdom of Aksum

The city of Aksum, in the present-day Tigray region of Ethiopia, was the center of a powerful civilization that flourished from the 1st to the 8th centuries. Numerous monuments and buildings from all periods of its history have survived, but undoubtedly the most impressive are the giant stelae, popularly known as obelisks—although they do not have the characteristic pyramidal cap of a true obelisk—which date from the 3rd and 4th centuries. These are arranged in two sites: the Northern Stelae Park, with the nearby tombs of the 6th-century King Kaleb and his son, King Gebre Meskel; and the smaller Gudit Stelae Park.

Ruins at Aksum

The collection of hundreds of stelae in the northern park is dominated by the Obelisk of Aksum, which stands 81 ft (24.6 m) high, but even this would have been dwarfed by the 109-ft (33-m) Great Stela, which lies in pieces where it fell, apparently in the process of erection. Each stela is decoratively sculpted with door- and windowlike panels, and many are topped with carved ornaments. Excavations at Aksum have also revealed the ancient megalithic tomb of King Bazen, the Ta'akha Maryam and Dungur palaces, and a reservoir known as the "Queen of Sheba's Bath"; and among the relics now in the Aksum museum is the 4th-century Ezana Stone, with inscriptions in ancient Greek, Sabaean (an old South Arabian language), and the Classical Ethiopian language, Ge'ez.

The **fallen Great Stela** in Aksum **weighs 573 tons (520 tonnes)**

▷ **KING EZANA'S STELA**
The largest unbroken stela in Aksum, with characteristic false window and door carvings, this structure is believed to mark the burial site of the 4th-century King Ezana.

△ **FALSE DOOR**
This cross shape inscribed on a false door of the monument led to it mistakenly being named the *Tombeau de la Chrétienne* ("tomb of the Christian woman").

Royal Mausoleum of Mauretania

An imposing funerary monument for the first sovereigns of Mauretania

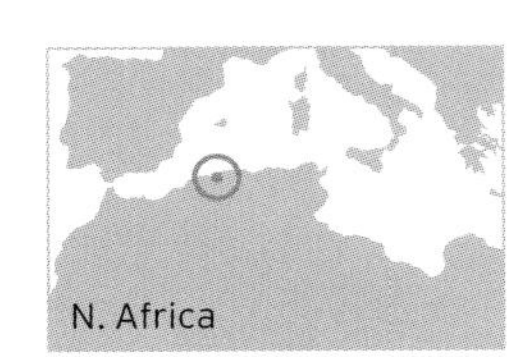

△ **CONE-SHAPED MONUMENT**
Built of stone blocks, the circular mausoleum stands on a square base and is capped with a tiered conical structure rising to a height of about 130 ft (40 m).

In Algeria, between Cherchell and Algiers, is the UNESCO World Heritage Site of Tipasa, containing numerous archaeological sites dating from Phoenician, Roman, Early Christian, and Byzantine times. Among them is the imposing Royal Mausoleum of Mauretania, built in 3 BCE on the orders of Juba, formerly King of Numidia, then the client king of the Roman province of Mauretania. Now known to be the tomb of King Juba II and his wife, Queen Cleopatra Selene II, it was thought to have been intended as a monument for Cleopatra and is known locally as *Kubr-er-Rumia*, "Tomb of the Roman Woman"—a reference to her parents, Mark Antony and the Egyptian Queen Cleopatra. The couple's vaulted burial chambers at the heart of the monument are located behind heavy stone doors accessed via an internal spiral gallery.

Damaged heritage

The mausoleum incorporated elements of Numidian, Greco-Roman, and Egyptian architecture as befitted the lineage of its occupants. The vaulted burial chambers—reminiscent of those of the Egyptian pharaohs—are housed in a windowless round building, decorated on the outside with 60 Ionic columns. Typically, Numidian buildings of this type were topped with a solid monumental cone or pyramid of stone blocks, but because of the poor state of repair, this is no longer discernible on the Royal Mausoleum. The capitals of the columns have been removed, and the remains of the royal couple are missing, presumed stolen by tomb-raiders. Although under UNESCO protection since 1982, the monument has suffered badly from looting, vandalism, and neglect.

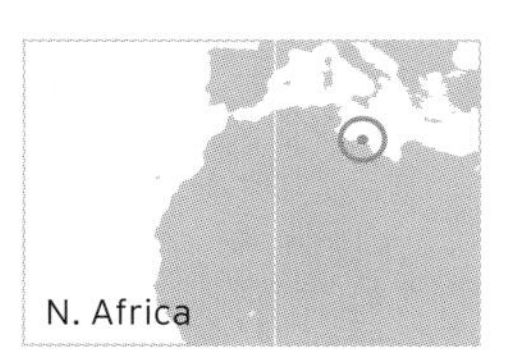

Leptis Magna

An ancient Phoenician port transformed into a major Roman city by the Emperor Septimius Severus

Situated on a natural harbor in present-day Al Khums at the mouth of the Wadi Lebdah on the coast of Libya, Leptis Magna was well placed to be a thriving Mediterranean port. It became a Roman colony in the 2nd century CE and underwent extensive expansion and regeneration during the reign of Septimius Severus (r.193–211 CE), a native of Leptis. As a result of his ambitious building program, Leptis became one of the most important—and most handsome—cities of Roman Africa. Among the improvements Severus initiated was the enlargement of the harbor and improvement of its piers and facilities, along with a substantial development of the infrastructure of the town, and its expansion to the south and west.

Past glories

Today, the ruins of Leptis are a testament to the vision Severus had for the town: the grid plan of streets, with a colonnaded thoroughfare linking the town center and the harbor; provision of public facilities—often using marble and granite—including the forum, bathhouses, and a theater; and grand monuments such as the basilica and the magnificent central archway.

Leptis fell to Arab invasion in the 7th century, and was abandoned soon after, but remained in a remarkable state of preservation as revealed by its excavation during the 20th century.

◁ **FEARSOME DECORATION**
Found in the Severan Forum, this carved ornament depicts the snake-haired Medusa, a fearsome gorgon reputed to have originated in Libya.

Severus built an **aqueduct** almost **13 miles (20 km)** long to supply the city with **water**

◁ **GRAND BASILICA**
The grandest of the many well-preserved structures at Leptis Magna is the basilica, a building for official, legal, and public functions, commissioned by Severus.

△ **NEW CITY**
The scale of Severus's development of Leptis Magna is evident in the theater and the Severan Forum above the old port.

CITY DESIGN

Under the reign of Septimius Severus, the old harbor town was substantially redeveloped, and a "New Town" was added above it to the southwest. This was designed on a typically Roman grid plan, with the main streets leading down to the old center and renovated harbor.

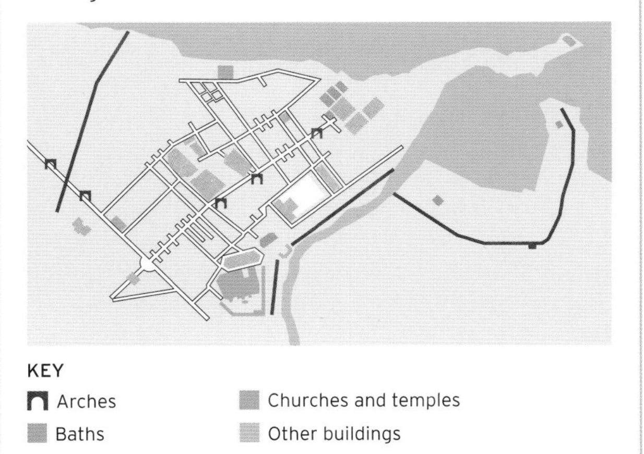

Amphitheater of El Jem

The largest amphitheater outside Italy and one of the best preserved Roman ruins in the world

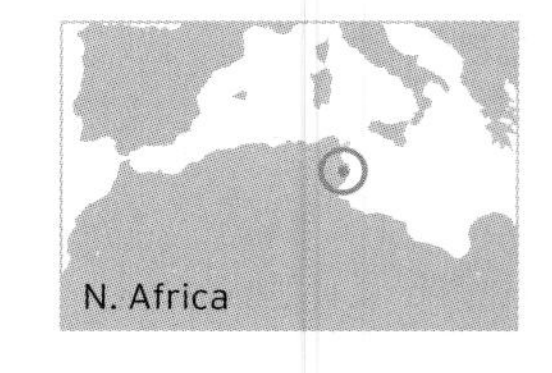

The city of Thysdrus, present-day El Jem (or El Djem) in Tunisia, was one of the most important in Roman North Africa. The ruins of this once-great trading center—and most notably of its impressive amphitheater—have been preserved by burial under desert sands.

Projecting Rome's power

The building of the amphitheater is thought to have begun in 238 CE. Capable of holding up to 35,000 people, it was intended to provide a venue for spectacular events. However, construction did not go to plan due to political turmoil in the province—the year 238 saw six rulers vying for recognition as emperor. Some archaeological evidence suggests that building work was inadequately funded and possibly not even entirely finished.

Unlike others in Roman Africa, the amphitheater's structure is free-standing rather than cut into a hillside, adding to its grandeur. Its three levels of arched galleries are constructed of sandstone blocks built onto the bedrock of the plain with no foundations.

Thysdrus was occupied by Vandals in the 5th century, and then by Arabs in the 7th century, when it served as a fortress. The amphitheater remained almost intact until the 17th century, when some stones were removed for use as building materials.

▷ ARCHED GALLERIES
The surviving three tiers of arcades feature arches and columns in the Corinthian style, echoing those of Rome's Colosseum (see pp.102–103).

EL JEM AND THE COLOSSEUM

The amphitheater at El Jem was clearly modeled on the Flavian Colosseum in Rome and is only slightly less grand in scale. Both have three levels of Corinthian or composite arcades and similar tiered seating around the oval arena. A significant but less obvious difference is the less sophisticated complex of underground passageways in the El Jem amphitheater.

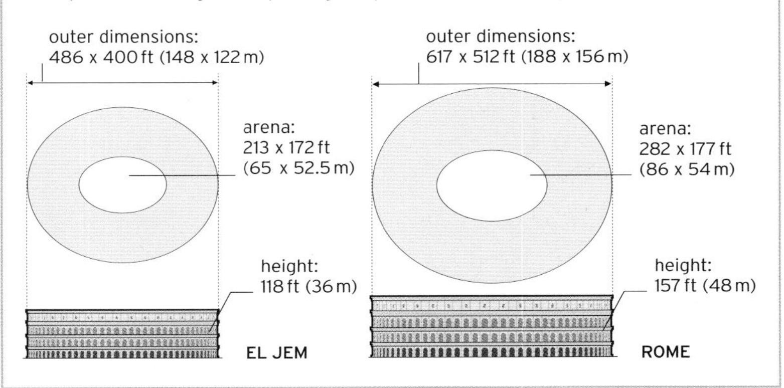

The amphitheater is the **third** to have been **built** at El Jem–ruins of an **earlier structure** can still be seen

△ **SYMBOL OF AUTHORITY**
The El Jem amphitheater, the third largest in the Roman Empire, was intended as a conspicuous symbol of Roman wealth and power in North Africa.

◁ **RED LIMESTONE**
The local limestone blocks used in the construction of the amphitheater remain well preserved thanks to the arid climate, but the building has suffered war damage and some looting.

Kano City Walls

Impressive earthen fortifications of a city in northern Nigeria that have survived the ravages of time

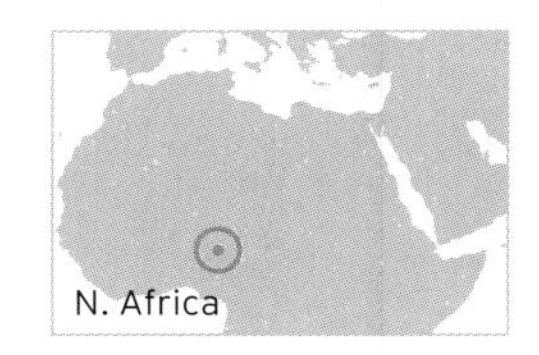

△ **MEDIEVAL STRUCTURE**
Built to enclose the medieval city of Kano, the walls are a fine example of traditional West African mud architecture. Parts of the city walls have been adapted for modern uses such as housing.

The walls of the ancient city of Kano were built primarily for defense. They were begun by Kano's medieval ruler Gijimasu in around 1095 and extended in stages until the 16th century. Made of sun-baked red clay, the walls stretched for at least 11 miles (17 km) around the city and are broadly triangular in cross-section. In places, the walls are more than 23 ft (7 m) high and 30 ft (9 m) thick at their base. They were pierced by 15 narrow gateways, with wooden gates strengthened by hammered iron strips.

Rapid expansion

The medieval city protected by Kano's walls was a bustling trading center. The Kurmi Market, which still exists alongside the walls today, lay at the southern end of trade routes across the Sahara Desert to the Mediterranean. In recent times, the growth of the city has far outstripped its origins. By the 21st century, Kano had become the second largest city in Nigeria and one of the world's most rapidly expanding urban areas.

With a population approaching 4 million by 2020, the modern city has engulfed its ancient walls in a wave of new development. Some of the gateways have been demolished to make way for modern roads, and other parts of the structure have fallen into disrepair. Yet, despite a chronic lack of funding, valiant attempts have been made to hold back this tide of destruction. Enough of the walls still remain to give an awesome impression of past glories and hold out hope for future preservation.

▷ **MAINTAINING THE WALLS**
Determined efforts are required to preserve and restore Kano's ancient walls. The mud-brick has to be resurfaced annually with fresh clay to protect against damage by heavy seasonal rainfall.

△ **PRAYER HALL**
Lines of columns divide the mosque's finely decorated, richly carpeted prayer hall into a series of parallel naves. Many of the columns were taken from earlier Roman and Byzantine buildings.

◁ **HOLY CITY**
The Great Mosque stands in the heart of Kairouan's medieval walled quarter. Kairouan is ranked as the fourth most sacred city of Islam, surpassed in holiness only by Mecca, Medina, and Jerusalem.

Great Mosque of Kairouan

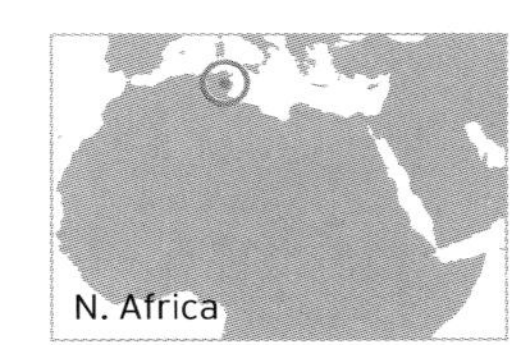

The oldest mosque in Africa, which exhibits the decorative and architectural splendor of medieval Islamic civilization

The first mosque at Kairouan, Tunisia, was founded in 670 CE by Uqba ibn Nafi, one of the Arab warriors spreading the newly born Islamic faith by conquest across northern Africa. Although the current mosque is often referred to as the Uqba Mosque, its buildings date from the mid-9th century, when Kairouan flourished under the Aghlabid dynasty (800–909). The Aghlabids made the Great Mosque not only a place of worship but also a renowned center of learning.

Intricate decoration

From the outside, the mosque seems to be an austere building, dominated by its massive, three-tiered stone minaret. Inside its extensive prayer hall, however, austerity is replaced by exuberant decoration. The space is filled with marble, granite, and porphyry columns. The wooden ceiling is painted with elaborate designs; the *minbar* (pulpit) is finely carved in Indian teak; and the *mihrab*, the focus of prayer, is an elaborate display of marble and tilework. By the 11th century, Kairouan had gone into decline, but the mosque has remained one of the most sacred places in the Islamic world.

IMPOSING STRUCTURE

The mosque covers an area of more than 116,000 square ft (10,800 square m). The towering minaret, 105 ft (32 m) high, is mounted by a staircase of 129 steps. The vast inner courtyard, surrounded by an arcade of horseshoe-shaped arches and columns, has six side entrances. The entire building has about 500 columns, most of them in the prayer hall.

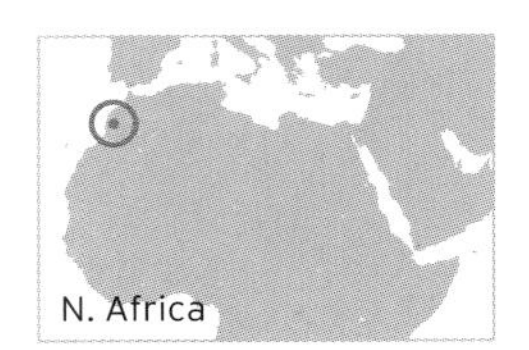

Koutoubia Mosque

A medieval mosque whose towering minaret dominates the skyline of the Moroccan city of Marrakech

The Koutoubia Mosque was built in the 12th century as an expression of the religious fervor of Berber tribesmen inspired by a radical version of Islam. Known as Almohads, these fierce but pious warriors swept down from the Atlas mountains to conquer all of northern Africa from Morocco to Egypt, as well as the Muslim areas of southern Spain.

Almohad leader Abd al-Mu'min ordered the first Koutoubia Mosque to be built as soon as he took possession of Marrakech in 1147. Unfortunately, a defect was discovered in the building—its *mihrab,* supposed to point the praying faithful toward Mecca, was slightly out of geographical alignment. So even before the first mosque was finished, work began on a second mosque alongside it, identical in every way except for its orientation. The first mosque was allowed to decay, the second being the one that stands today.

Mosque and medina

Completed during the reign of Almohad Caliph Yakub al-Mansur (r.1184–1199), the current mosque stands close to the narrow streets of the medina, Marrakech's atmospheric old town, and the famous Jemaa el-Fna square with its snake charmers and acrobats. The minaret is 230 ft (70 m) tall, made chiefly of pink-red sandstone, decorated with strips of colorful ceramic tiling and elaborately patterned window surrounds. The mosque's prayer hall contains fine examples of Moroccan craftwork, but it is not accessible to non-Muslims.

The name **Koutoubia Mosque** means **"booksellers' mosque,"** for a trade much practiced nearby

△ INSIDE THE MOSQUE
The interior walls of the mosque are painted plain white, contrasting with the colorful mats that carpet the floors. Non-Muslim tourists have never seen this sight, as the building is open only to Muslim worshippers.

◁ NIGHT BEACON
Standing tall behind the food stalls of the bustling Jemaa el-Fna square, in the heart of old Marrakech, the mosque's distinctive minaret is lit up at night. It is visible from a distance of 18 miles (29 km).

COLUMNED HALL

The sandstone walls of the mosque enclose a courtyard created around a fountain and a large prayer hall. Measuring about 58,000 square ft (5,400 square m) in area, the hall is divided into a series of parallel naves by 112 columns supporting horseshoe-shaped arches. The minaret stands at the northeast corner of the building.

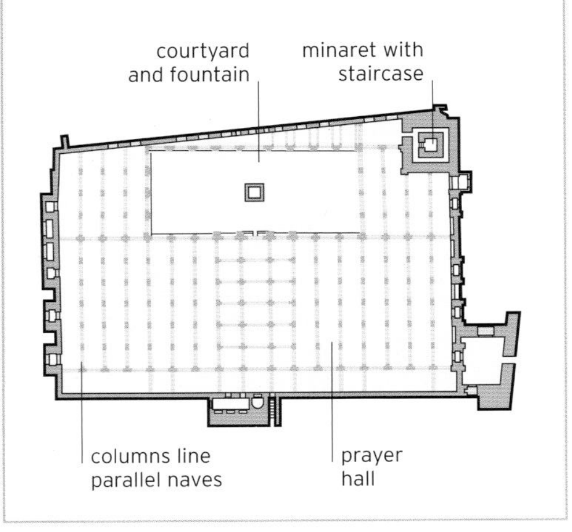

Lalibela Rock Churches

Orthodox Christian churches carved from solid rock in the mountains of Ethiopia

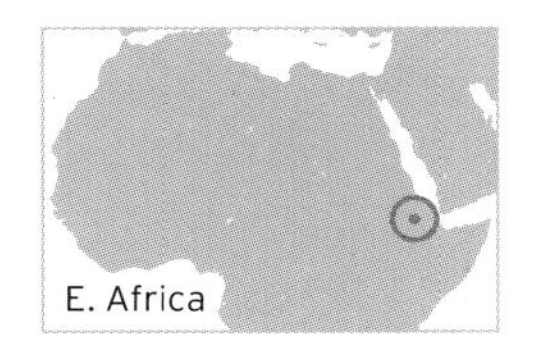

The 11 rock churches at Lalibela in the Amhara district of northern Ethiopia were probably built by King Gebre Meskel Lalibela, after whom the town is named. A member of the Zagwe dynasty, he ruled this mountainous area of Ethiopia about 900 years ago. A devout Christian of the Ethiopian Orthodox faith, he is said to have created the churches in response to the fall of Jerusalem to a Muslim army in 1187, intending the complex as a recreation of the holy city.

Exceptional technique

Each of the churches was made by chiseling away soft, reddish volcanic rock to isolate a rectangular block. The workers then cut inward, hollowing the block to form a room inside. The largest church, Bete Medhane Alem, has rows of outer columns like a Greek temple. Bete Gyorgis, the best preserved, is in the shape of a cross. Designated a UNESCO World Heritage Site in 1978, the churches are a living place of pilgrimage visited by thousands of worshippers each day.

NEW JERUSALEM

The layout of the Lalibela churches was intended as a symbolic representation of Jerusalem. The river that crosses the site is called the Jordan, after the river that flows through parts of the Holy Land. The principal groups of churches are connected by tunnels. Bete Gyorgis stands separate, allegedly added to the scheme after St. George appeared to King Lalibela in a dream.

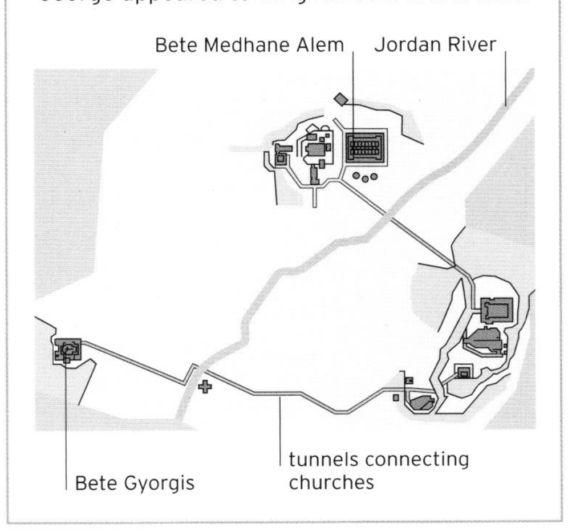

▽ STANDING TALL
Bete Gyorgis, or the Ethiopian Coptic Church of St. George, is about 40 ft (12 m) tall. Twelve windows are cut through an upper story, perhaps representing the 12 Apostles of Christ.

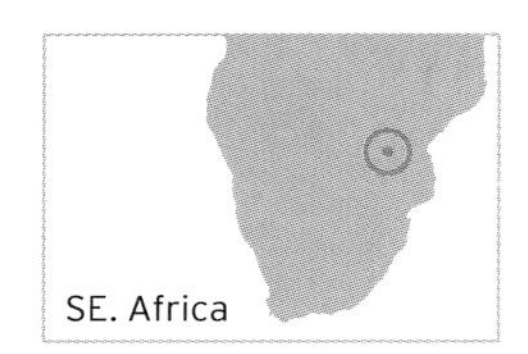

Great Zimbabwe

The ruined capital of the medieval Kingdom of Zimbabwe, once linked to King Solomon and the Queen of Sheba

Extending over 3 square miles (8 square km) near the town of Masvingo in the eastern hills of Zimbabwe, the stone ruins of Great Zimbabwe are among the oldest and largest man-made structures in sub-Saharan Africa. Great Zimbabwe was built by the Shona—a population of Bantu people—between the 11th and 15th centuries. The site is divided into three main areas: the Hill Complex, the Great Enclosure, and the Valley Ruins. The Hill Complex and Great Enclosure are characterized by extensive stonework, comprising granite bricks laid without mortar. They also feature the *daga* (earthen and mud-brick) buildings that dominate the Valley Ruins, where thousands of goldsmiths, potters, weavers, blacksmiths, and stonemasons once lived.

A center of trade

Archaeological finds from Persia and China suggest that Great Zimbabwe was a great trading center, its wealth built on the gold and copper mines that pepper the area. However, distinctive soapstone bird carvings, known as birds of heaven and thought to represent messengers from *Mwari* (God), suggest the site also had religious significance. Great Zimbabwe was abandoned in the 15th century. On encountering the site in the 19th century, European explorers thought they had found the legendary King Solomon's mines.

The word ***zimbabwe*** is a Shona word meaning **"stone houses"**

◁ **BOULDERS AND BLOCKS**
The stone walls at Great Zimbabwe were built using boulders and granite slabs from the surrounding hills, broken to be made portable and laid in layers without the use of mortar.

STURDY WALLS
This aerial view of the Great Enclosure shows the thick boundary and internal walls. The 33-ft- (10-m-) high conical tower is obscured by vegetation.

THE GREAT ENCLOSURE

The Great Enclosure is the largest structure in Great Zimbabwe. Its circular 16-ft- (5-m-) thick wall has a circumference of 820 ft (250 m) and was built from 900,000 granite blocks. Within the wall, a narrow passage leads to a solid conical tower, over 16 ft (5 m) in diameter at the base and over 30 ft (9 m) tall. It also has a series of further enclosures, community areas, and *daga*-hut living quarters.

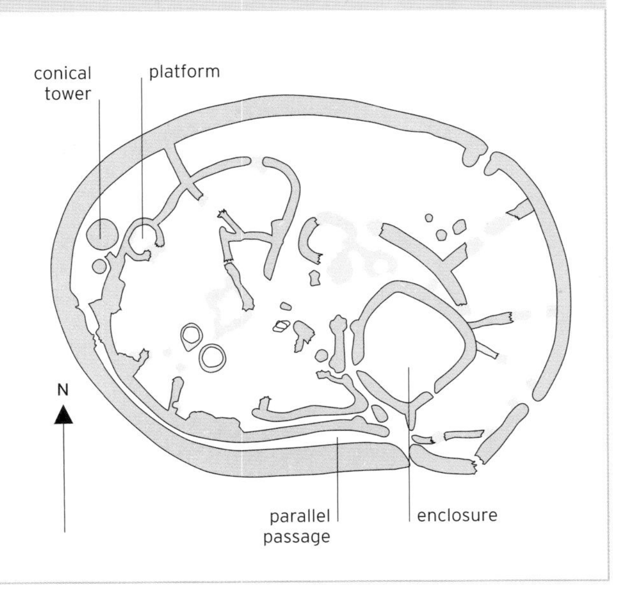

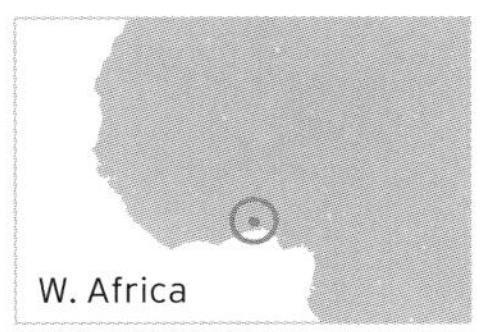

Royal Palaces of Abomey

A palace complex built between the 17th and 19th centuries by the powerful kings of the vanished West African kingdom of Dahomey

From 1625 to 1900, the Kingdom of Dahomey was one of the most powerful in West Africa, its success built largely on trading slaves with Europeans. At their capital in Abomey in modern-day Benin, the 12 successive kings of Dahomey constructed a vast complex of royal palaces, covering 5,000,000 square ft (470,000 square m).

△ **SYMBOLIC SCULPTURE**
This detail from one of the colorful bas-relief earthen panels at the Abomey palaces shows a lion, the symbol of King Glélé (r.1858–1889).

Many purposes

Built as a series of courtyards, the palaces included the royal mausoleums, religious buildings, and council and public meeting rooms. They were both the administrative center of the kingdom and its treasury and cultural heart, decorated with colored bas-relief earthen panels depicting the kingdom's battles, myths, beliefs, and customs. In 1892, the last independent king of Dahomey ordered his troops to burn the palaces but the 19th-century palaces of Ghezo and Glélé survived and have since been restored.

△ **TRADITIONAL CONSTRUCTION**
The courtyard and buildings of King Ghezo's palace were made from mud-brick, and the palace walls were decorated with bas-relief panels.

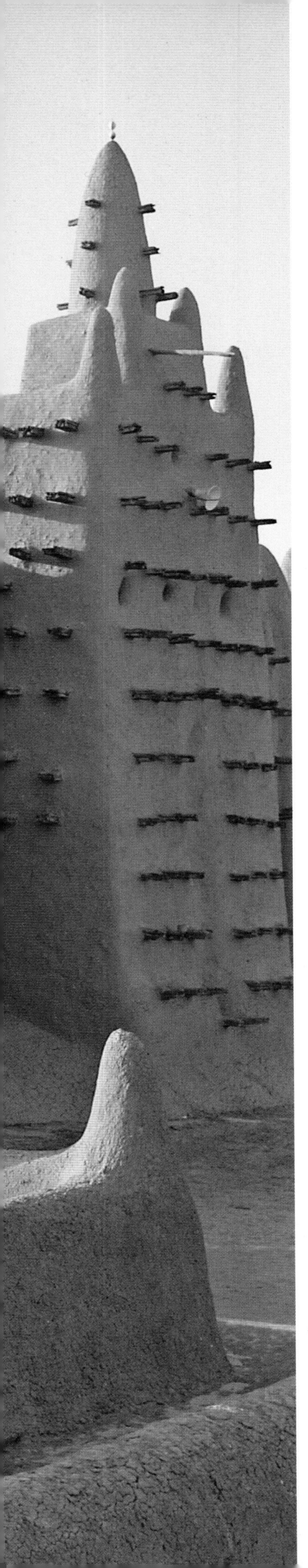

Great Mosque of Djenné

A sacred mud-brick building that towers over an ancient trading city in West Africa

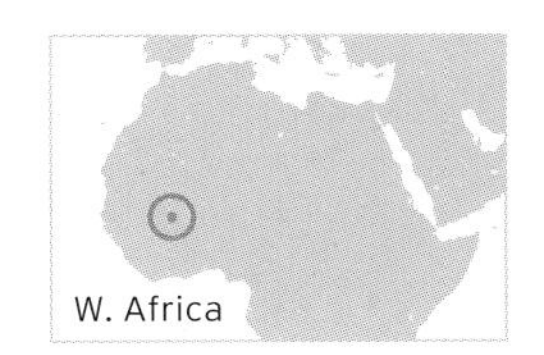

The first mud-brick Great Mosque at the Malian town of Djenné was built around 700 years ago. The current building was constructed in 1906–1907, when Mali was ruled by a French colonial administration. A local master-builder, Ismaila Traoré, was responsible for the work, employing long-established mud-building techniques.

A changing structure

The bricks of the mosque were made by mixing mud and sand with husks and straw, then drying them in sunshine. The flat earthen roof is supported by pillars rising from the sand floor. The walls are protected against the elements by a layer of plaster, composed of river silt mixed with other materials that may include dry rice husks and cow manure. The plaster has to be reapplied almost every year, an activity that has become a ceremony in which the whole community joins. The constant repairs to the structure mean that the detail of its appearance is ever changing.

THE PARTS OF THE MOSQUE

The mosque is a striking example of Sudano-Sahelian architecture. Built on a raised platform to protect against floods, it consists of a walled courtyard and a large prayer hall. Inside the prayer hall, 90 pillars in 10 rows support the mud-and-palm roof. Three large towers protrude from the front of the building.

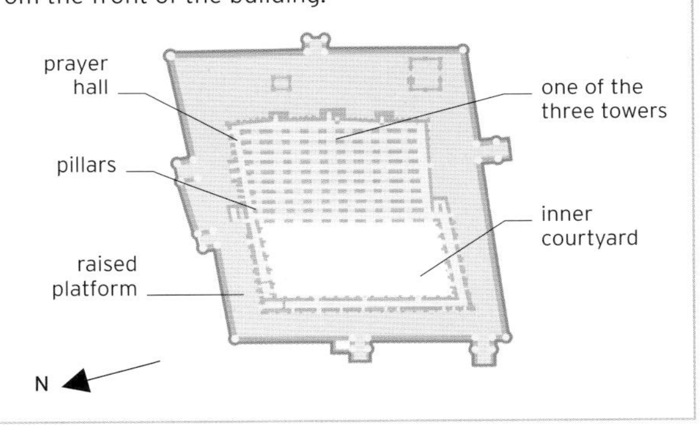

◁ **MUD FACADE**
The high mud walls and towers of the Great Mosque are patterned with protruding bundles of palm wood. An ostrich egg is placed at the apex of each tower to symbolize fertility.

▷ **IMPRESSIVE SCALE**
An aerial view shows the great extent of the mosque. With a prayer hall 165 ft (50 m) long and towers around 52 ft (16 m) high, the mosque is the world's largest mud-built structure.

Basilica of Our Lady of Peace

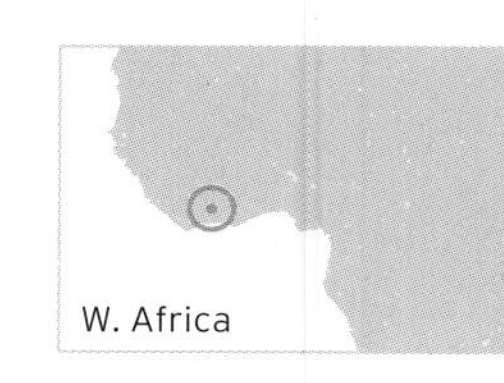

With its massive dome and towering columns, this is the world's largest church, heavily inspired by St. Peter's Basilica in Rome

The Basilica of Our Lady of Peace was constructed in Yamoussoukro, the administrative capital of Côte d'Ivoire (Ivory Coast), between 1986 and 1989. Designed by the celebrated Lebanese-born Ivorian architect Pierre Fakhoury, it was inspired by the Basilica of St. Peter (see pp.158–159) in the Vatican City. The building was commissioned by Félix Houphouët-Boigny, Côte d'Ivoire's first president (in office in 1960–1993), who is said to have personally funded the construction of the lavish building at an estimated cost of $200–$300 million.

The Catholic basilica stands around 518 ft (158 m) high and occupies an area of about 320,000 square ft (30,000 square m). It is supported by 128 Doric columns, and its vast marble and granite plaza can hold up to 300,000 people. The structure's 36 stained-glass windows were handmade in France.

At the request of Pope John Paul II, Fakhoury lowered the height of the dome to make it smaller than the dome of St. Peter's—nonetheless, the 30-ft- (9-m-) high cross on top of the dome makes the structure significantly taller than its Italian equivalent.

▷ **DOMINATING THE LANDSCAPE**
The building's immense dome, with its gilded cross, dominates the landscape, but the huge church—in a Muslim region—has always been sparsely attended.

Around **1,500 workmen** were employed on the basilica, which took just **3 years to build**

△ INSIDE THE BASILICA
Spectacular stained-glass windows and imposing columns are among the church's most distinctive features. Its pews, which can seat around 7,000 people, were made from West African Iroko wood, which turns from yellow to a rich copper color with age.

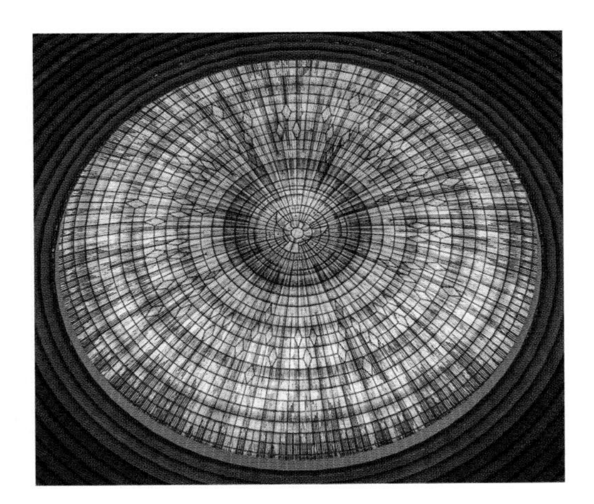

△ CENTER OF THE DOME
A dove with outstretched wings, measuring around 23 ft (7 m) in width, appears in the central position of the stained glass at the heart of the dome.

Afrikaans Language Monument

An iconic South African landmark that celebrates one of the world's youngest languages

S. Africa

The Afrikaans Language Monument was erected in 1975 to mark the 50th anniversary of Afrikaans as an official language of South Africa. Situated on Paarl Mountain, Western Cape, it was designed by architect Jan van Wijk, who sought to establish a link between the evolution of the language and the form of the monument. To this end, its tallest spire, 187 ft (57 m) high, indicates the growth of Afrikaans, while smaller, adjoining structures suggest the key influence of Africa and, by contrast, the declining influence of European languages on Afrikaans.

Cultural construction

Constructed over 2 years, the structure—also known as Taalmonument—symbolizes the convergence of nationhood, language, and landscape. It is made from Paarl granite, white sand, and cement, which has been hammered to evoke the texture of the surrounding rocks.

The monument initially attracted controversy as it was built during the apartheid era (1948–1994) and was perceived to privilege Afrikaans as South Africa's primary language, with English being considered less important and with no regard for indigenous local languages, which were not recognized in what was then "white" South Africa.

▽ BLENDING IN
The sculpture was designed to suggest a dialogue with nature. Its lines and curves replicate the peaks and dips of the landscape, and the material used in its construction helps it blend in with its rocky surroundings.

LAYOUT AND SYMBOLISM

The monument's main column represents Afrikaans as a developing language. A smaller column symbolizes South Africa. Three obelisks of varying height suggest the contribution of European languages in shaping Afrikaans, while different-sized domes indicate the role of Africa. Indonesian influences are shown by a wall on the stairway.

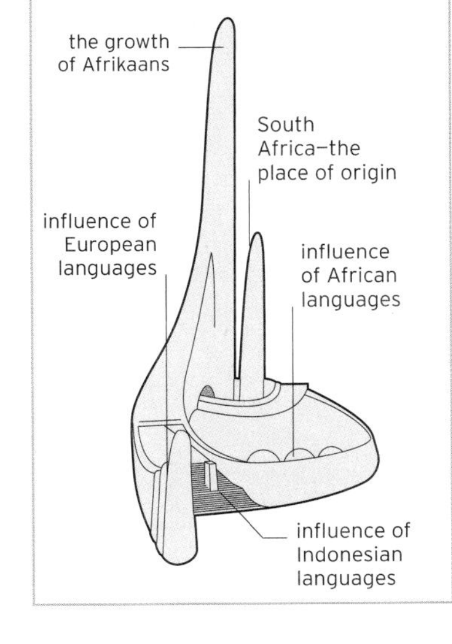

Building green
Gardens by the Bay is a futuristic park in Singapore. Its 18 steel-framed supertrees not only provide support for real plants and flowers but also harvest rainwater and incorporate air intakes, exhaust vents, and solar panels.

Asia and Australasia

KEY SITES

1. Göbekli Tepe
2. Palmyra
3. Ephesus
4. Mohenjo-Daro
5. Ziggurat of Ur
6. Petra
7. Terracotta Army
8. Ajanta Caves
9. Ellora Caves
10. The Hanging Temple
11. Persepolis
12. Dome of the Rock
13. Great Mosque of Samarra
14. Leshan Giant Buddha
15. Borobudur
16. Vardzia
17. Dambulla Cave Temples
18. Sacred City of Kandy
19. Ancient Buildings of Wudang Mountains
20. Candi Prambanan
21. Chand Baori
22. Temples of Bagan
23. Bahla Fort
24. Minaret of Jam
25. Angkor Wat
26. Monastery of Geghard
27. Palace of the Shirvanshahs
28. Himeji Castle
29. Gyeongbokgung Palace
30. Forbidden City
31. Great Wall of China
32. Hampi
33. Mehrangarh Fort
34. Palitana
35. Great Mosque of Mecca
36. Registan Square
37. Shah Mosque
38. Golden Temple of Amritsar
39. Taj Mahal
40. Toshogu Shrine
41. Potala Palace
42. Paro Taktsang
43. The Grand Palace
44. Pink Mosque
45. Royal Exhibition Building
46. Sydney Harbour Bridge
47. Temple of the Golden Pavilion
48. Lotus Temple
49. Sydney Opera House
50. National Assembly Building of Dhaka
51. Wat Rong Khun
52. Akashi Kaikyo Bridge
53. Sheikh Zayed Mosque
54. Burj Khalifa
55. Gardens by the Bay
56. Tianjin Binhai Library

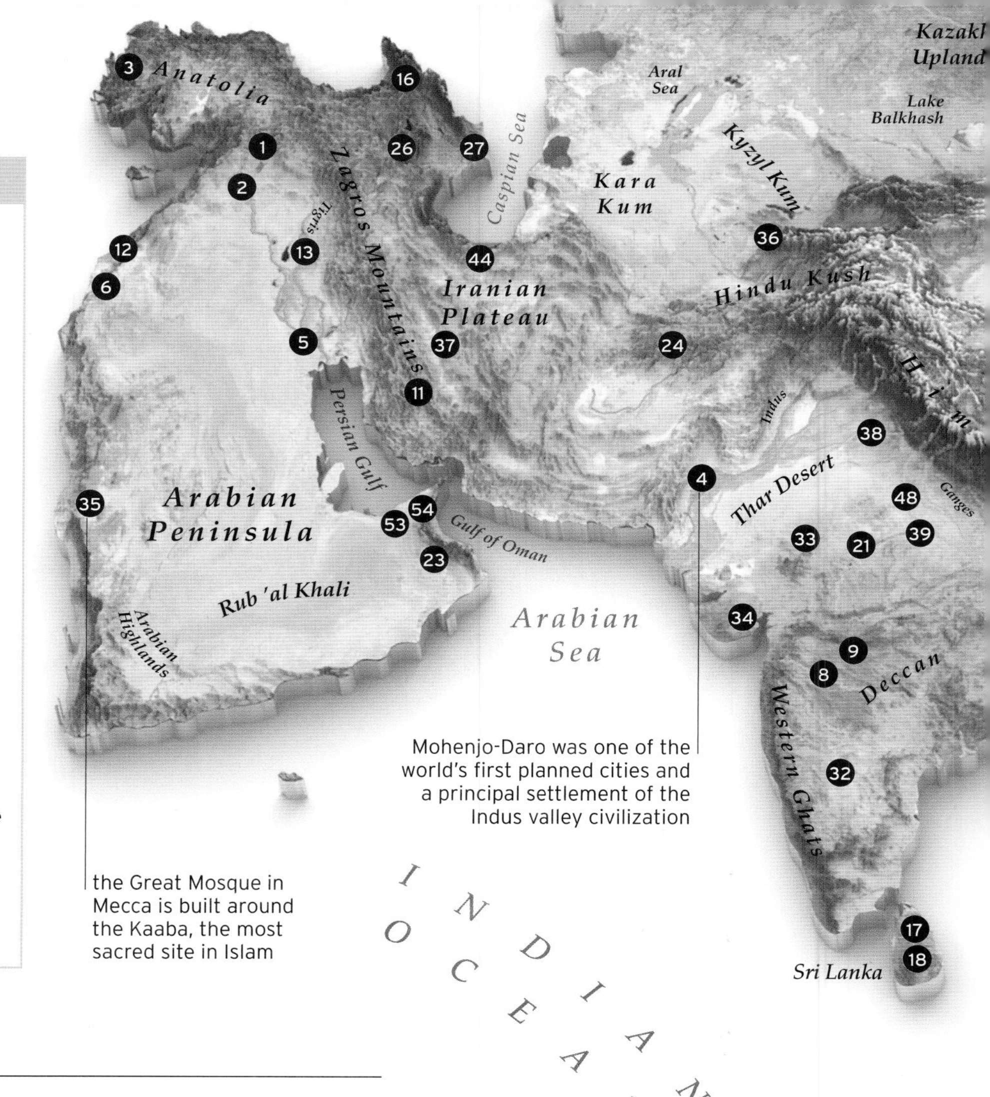

Mohenjo-Daro was one of the world's first planned cities and a principal settlement of the Indus valley civilization

the Great Mosque in Mecca is built around the Kaaba, the most sacred site in Islam

EMPIRE BUILDERS AND INNOVATORS

Asia and Australasia

As waves of political and religious empire builders moved eastward across the vast Asian continent and south to Australasia, they spread engineering techniques and stylistic influences and unleashed building ingenuity. Persian taste and knowledge migrated to the Indian subcontinent in the 6th century, followed by Islamic arts and architecture, as the Umayyad Dynasty expanded its reach to India.

In turn, India's religions, Hinduism and Buddhism, introduced new aesthetic ideals to China, Japan, and Southeast Asia, informing the architecture of temples, palaces, and shrines. In recent times, devotion to the idea of sustainability has launched a new generation of intelligent, energy-saving wonders across Asia and Australasia.

HINDUISM SPREADS THROUGH SOUTHEAST ASIA

6TH–14TH CENTURY CE

As traders from India expanded their business in Southeast Asia, they brought Hindu culture with them, notably the religious architecture of the Indian temple. The building styles merged with local traditions, evident in the step pyramids of the Khmer empire and the Indonesian temple of Prambanan.

25 ANGKOR WAT

URBAN PLANNING AND IDEALISM

The earliest forms of urban planning and advanced stone construction emerged in the Fertile Crescent and Indus Valley. In East Asia, there is also evidence of early urban planning in China, and Buddhism inspired architecture that sought to replicate heaven on Earth. Colonialism left its mark on Southeast Asia and Australasia, where ecologically sensitive building is now considered critical.

THE MING DYNASTY INSPIRES CULTURAL GROWTH

1368–1644

Political stability in China during the Ming dynasty enabled large-scale feats of construction, from the Forbidden City and the Great Wall to the Chang Mansion near Myanmar. These structures expressed the reassertion of imperial power after Mongol rule through their monumental scale and the strict visual hierarchy of classical Chinese architecture.

30 FORBIDDEN CITY

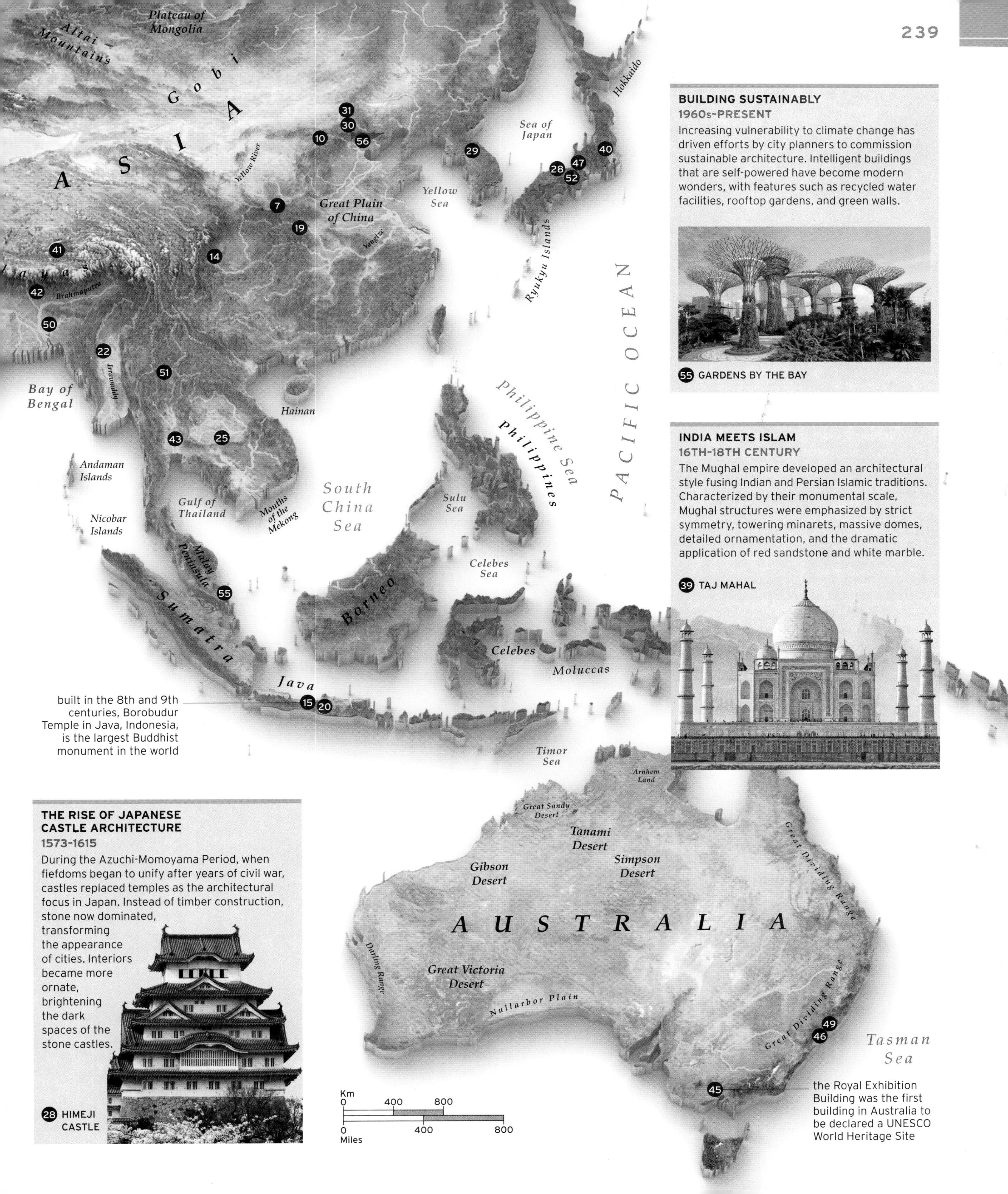

BUILDING SUSTAINABLY
1960s-PRESENT

Increasing vulnerability to climate change has driven efforts by city planners to commission sustainable architecture. Intelligent buildings that are self-powered have become modern wonders, with features such as recycled water facilities, rooftop gardens, and green walls.

55 GARDENS BY THE BAY

INDIA MEETS ISLAM
16TH-18TH CENTURY

The Mughal empire developed an architectural style fusing Indian and Persian Islamic traditions. Characterized by their monumental scale, Mughal structures were emphasized by strict symmetry, towering minarets, massive domes, detailed ornamentation, and the dramatic application of red sandstone and white marble.

39 TAJ MAHAL

built in the 8th and 9th centuries, Borobudur Temple in Java, Indonesia, is the largest Buddhist monument in the world

THE RISE OF JAPANESE CASTLE ARCHITECTURE
1573-1615

During the Azuchi-Momoyama Period, when fiefdoms began to unify after years of civil war, castles replaced temples as the architectural focus in Japan. Instead of timber construction, stone now dominated, transforming the appearance of cities. Interiors became more ornate, brightening the dark spaces of the stone castles.

28 HIMEJI CASTLE

the Royal Exhibition Building was the first building in Australia to be declared a UNESCO World Heritage Site

Göbekli Tepe

The world's oldest known religious site, a complex of temples with intricate relief carvings and inscriptions

The contents of the mound known as Göbekli Tepe ("Potbelly Hill") in southeastern Turkey lay hidden until 1994, when German archaeologist Klaus Schmidt uncovered the first of the T-shaped stone pillars characteristic of the temples there. Further excavation revealed a number of circles of these stones, dating from as early as 10000 BCE, assumed to have a ritual or religious purpose. The pillars range from 10 ft (3 m) to 20 ft (6 m) in height and are set in sockets cut into the polished rock floor, forming a circle around two larger central pillars. It is thought that they were designed to support a roof.

Decorative sculpture

Just as remarkable as the stone structures are the sophisticated, decorative carvings on the individual stones. Cut into the flat surface of many of the pillars, these relief sculptures depict a variety of birds and other animals, including wild game caught by the hunter-gatherer people of the time, as well as some abstract designs and the occasional stylized human figure.

△ **BALIKLIGÖL STATUE**
A life-size limestone statue dating from around 9500 BCE was found at an excavation in nearby Balıklıgöl and is now on display in the Museum of Sanlıurfa. It is the oldest known Asian statue.

△ **COLUMNS AND PILLARS**
This image shows one of the circular temples excavated at Göbekli Tepe, with its characteristic central column. In the foreground is a pillar decorated with relief sculptures.

FUSION OF STYLES

The synthesis of local and Classical architecture is most apparent in temples such as that dedicated to the Semitic god Bel. From the outside, it appeared conventionally Classical, if unusually asymmetrical and topped with atypical roof terraces. The details, however, were local to Palmyra: sculptures on the columns and beams depicted local gods and acts of worship. The interior, which included shrines dedicated to indigenous gods, was also decorated in typical local style.

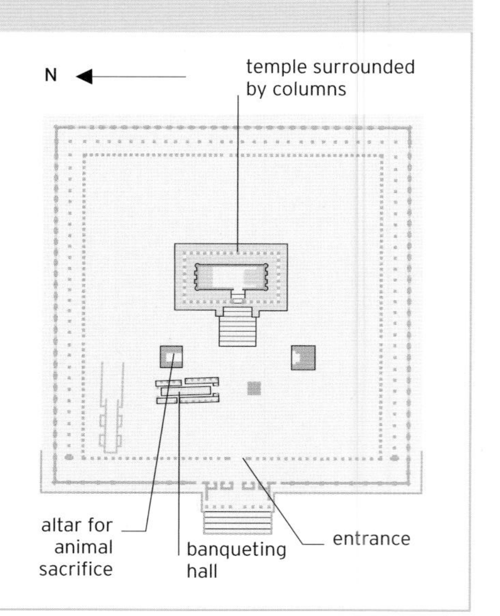

Palmyra

An ancient caravan oasis, which became a vital crossroads on the trade route from China, India, and Persia to the Roman Empire

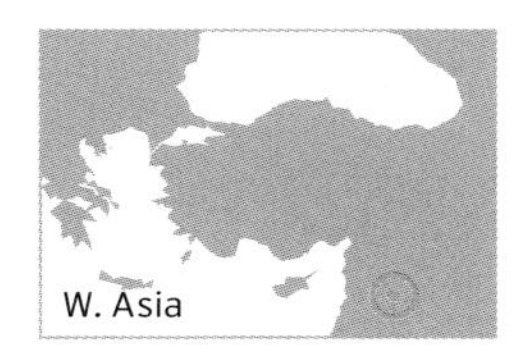

△ **ROMAN THEATER**
The 2nd-century theater stands in a semicircular piazza off the main colonnaded street. It has been a focal point of the city since its restoration in the mid-20th century.

The city of Palmyra was built on the site of a remote oasis in the Syrian desert, which had been known for millennia as a stopping place for caravans. It had its heyday, however, when it was part of the Roman province of Syria in the 1st and 2nd centuries CE and became an important trading center.

A meeting of cultures

The local population was Semitic with a well-established cultural tradition, and they were given a large degree of autonomy under Roman rule. This resulted in a synthesis of cultures, reflected in the architecture of the city. A wide colonnaded avenue in the Classical style ran through the center of the city, flanked by residential streets. Within the city walls, there were also the public buildings typical of a Roman town, alongside temples to the local deities. To the west of the city were the necropolises to house the dead.

Despite destructive onslaughts through the ages, many of the finest structures had survived into the 21st century, but during the Syrian Civil War, some buildings were damaged, and in 2015, Islamic State militants tragically demolished many of the remaining monuments.

Ephesus

An important port in Greek and Roman times, and later a place of Christian pilgrimage

Situated on the Ionian coast of present-day Turkey, Ephesus was first settled by the Greeks in about 1200 BCE and became one of the most prosperous cities of ancient Greece. After a period of Persian occupation from the 6th century BCE, it was recaptured by one of Alexander the Great's generals in 334 BCE and effectively refounded on a new site to the west to bring it nearer to the receding coastline.

The regenerated Ephesus boasted many fine buildings, including the famous Temple of Artemis, but sadly excavation has uncovered only fragments of the Hellenistic city. However, Ephesus underwent further rebuilding after it was acquired by the Roman empire, and much of the Roman city has survived. The splendor of its architecture was symbolic of Roman wealth and colonial supremacy. With a sophisticated infrastructure of streets and aqueducts, Roman Ephesus contained many fine public buildings, including a theater, the Library of Celsus, and the Temple of Hadrian.

▷ **ROMAN EPHESUS**
The best-preserved buildings of Ephesus date from the period of Roman rule, from the 2nd century CE. This magnificent facade of the Library of Celsus is an example of such well-preserved remains.

THE TEMPLE OF ARTEMIS

The third temple built on this site, built in 560–546 BCE, was the largest of its kind ever constructed–larger even than the Athenian Parthenon–and hailed as one of the Seven Wonders of the World. It was about 377 ft (115 m) long and 180 ft (55 m) wide, with a double row of columns, each about 60 ft (18 m) high and 4 ft (1.2 m) in diameter.

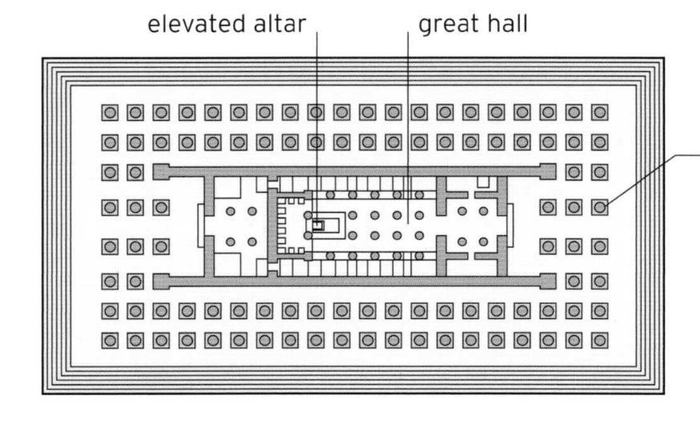

Paul the Apostle preached in the **theater** while **visiting** Ephesus

Mohenjo-Daro

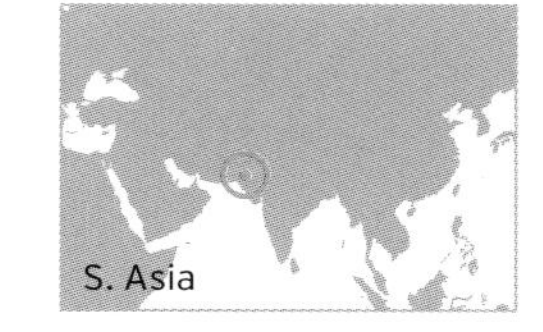

One of the earliest cities in the world and a major center of the ancient Indus Valley civilization

Located in the middle of the floodplain of the Indus Valley in present-day Sindh, Pakistan, Mohenjo-Daro dates back to the 3rd millennium BCE. It flourished at much the same time as the civilizations of ancient Egypt, Mesopotamia, Minoan Crete, and Norte Chico. The remains of the city are evidence of a system of urban planning: the unbaked brick buildings are laid out to a strict grid pattern, with a citadel on a promontory above the lower residential area.

The major structures of the city are within the citadel and include a large residential structure, assembly halls, and a central marketplace. Of particular interest are the public baths and wells, which were serviced by a sophisticated irrigation and drainage system. The central areas were heavily fortified, and the city was also protected by fortifications and guard towers to the south and west.

◁ **SOPHISTICATED ARTIFACTS**
This pottery container is just one of several artifacts excavated at Mohenjo-Daro that provide evidence of a highly developed civilization with a flourishing culture.

Ziggurat of Ur

A distinctive pyramidlike structure dominating the ruins of the ancient city of Ur

Ziggurats are a form of temple pyramids peculiar to the ancient civilizations of what is now Iraq and Iran. Unlike Egyptian pyramids, they are tiered structures with flights of steps between the different levels. One of the largest and best-preserved of these is at the site of the Neo-Sumerian city of Ur, near Nasiriyah in present-day Iraq, built in the 21st century BCE.

Structures of worship

The Ziggurat of Ur was constructed as a temple to the Moon god Nanna. Its original height is estimated to have been 70–100 ft (21–31 m), making the shrine at its top a highly visible landmark. By the 6th century BCE, however, its mud-brick structure had largely crumbled, and it was restored by the Neo-Babylonian King Nabonidus using a protective outer layer of baked bricks set in bitumen. It was further restored in the 20th century, with the addition of a reconstruction of the facade and main staircase, which suffered damage during the 1991 Gulf War.

▽ **COLOSSAL TEMPLE**
This image shows the restored facade of the ziggurat, with its main stairway leading up to the level on which the shrine of the Moon god Nanna would have stood.

AL-KHAZNEH
The massive facade of Al-Khazneh, meticulously sculpted into the pink rock face, provides a breathtaking vista when entering Petra through the narrow canyon known as the Siq.

CITY PLAN

Visitors to Petra usually approach the city via the Siq, its main entrance, passing a number of tombs cut into its steep walls before arriving at the famous view of Al-Khazneh at the end of the gorge. The major sites are found on either side of the valley, across which the colonnaded main thoroughfare leads to the Monastery.

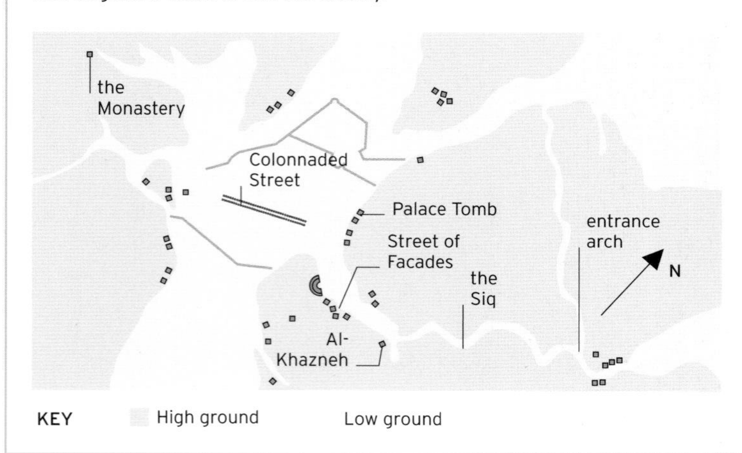

△ **PALACE TOMB**
The largest of the row of four Royal Tombs, the Palace Tomb is carved in imitation of the contemporary Roman style and is believed to have been inspired by Nero's Golden House in Rome.

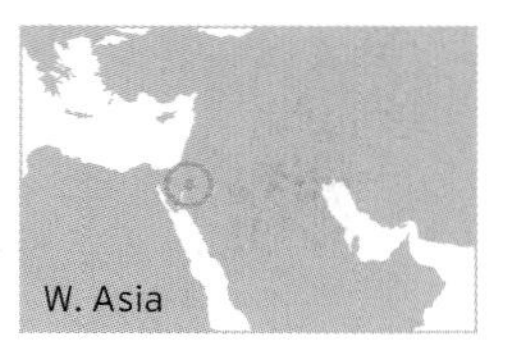

Petra

The "Rose-red City," capital of the Nabataean Kingdom, famous for the tombs and temples carved into its sandstone cliffs

The nomadic Nabataean people settled in the desert mountains of present-day southern Jordan more than 10,000 years ago, founding a settlement, Raqmu, that was to become a trading center for the region. Trade with the Greek and Roman empires brought prosperity, and the growing city became known as Petra, Greek for "stone."

Carved into the rock

Petra's location in steep-sided mountain gorges led the Nabataeans to develop a distinctive architecture, carving tombs and temples that mimicked the Greco-Roman Classical style but also included some characteristically Nabataean elements. The finest examples of this are the famous Al-Khazneh ("the Treasury," although it is in fact a mausoleum) at the entrance to the city and the Monastery, Al-Deir (actually a hilltop temple). Also of note is the row of four Royal Tombs—the Tomb of Urns, Silk Tomb, Corinthian Tomb, and Palace Tomb—each featuring a different Classical genre.

The city also has a huge theater hewn from the rock and a small number of free-standing buildings, such as the Qasr-al-Bint, the main temple of the city. Several of the soft sandstone buildings have suffered from erosion, and even in the best preserved ones the details have lost clarity.

At the height of its prosperity in the **1st century CE**, Petra had an estimated **population of 20,000**

◁ **ROCK-CUT TOMBS**
The tombs cut into the walls of the gorge leading into Petra are comparatively simple in design, but the multicolored, striated sandstone creates spectacular interiors, which were originally also decorated with frescoes.

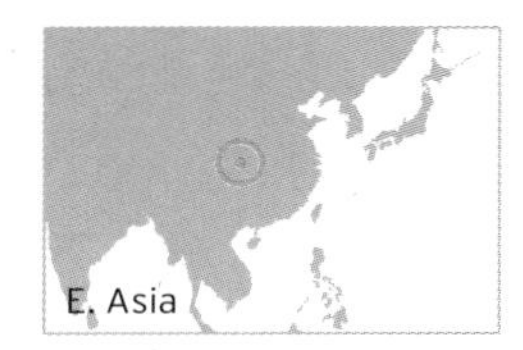

Terracotta Army

The most spectacular archaeological find of modern times, opening a unique window on China's past

Discovered by local farmers near Mount Li in China's Shaanxi province in 1974, the massed life-size figures known as the Terracotta Army are more than 2,000 years old. They form part of the mausoleum of Qin Shi Huang (259–210 BCE), who styled himself the "First Emperor" of China. Obsessed with death from an early age, Qin Shi Huang began creating the burial complex in 246 BCE, when he was only 13 years old and ruler of Qin, one of the seven major states into which China was then divided. From 221 BCE, with all of China united under his iron rule, he was able to mobilize massive resources for the mausoleum project.

Buried ranks

The Terracotta Army—consisting of at least 7,000 soldiers; 130 horse-drawn chariots; and 150 cavalry horses—guarded one side of the extensive imperial burial site. Most of the figures represent peasant foot soldiers arrayed in orderly ranks. Crossbowmen are shown kneeling and horsemen astride their mounts. As befits their status, the officer figures are taller and wear more elaborate armor. The real iron or bronze weapons they carried in their terracotta hands have been looted or disintegrated over time.

MASS PRODUCTION

The terracotta figures were created in workshops using mass-production techniques. The different parts were created in standard molds, fired separately, and then assembled. There were 10 basic molds for heads, with facial features added by hand to give them individuality.

figures held real iron or bronze weapons

painted and lacquered surface has degraded over time

Around **700,000 workers** were **employed** to help build the **emperor's mausoleum**

◁ **BATTLE FORMATION**
Three vaults were found, with figures all facing east, toward the enemies of the Qin. Two vaults were on the flanks and the third was the command post at the rear.

▷ **ETERNAL GUARD**
The figures of the soldiers are 5 ft 10 in (1.8 m) tall on average. Their headgear was an indication of their status and rank.

Ajanta Caves

A treasure trove of paintings and sculptures that are among the finest examples of Buddhist art

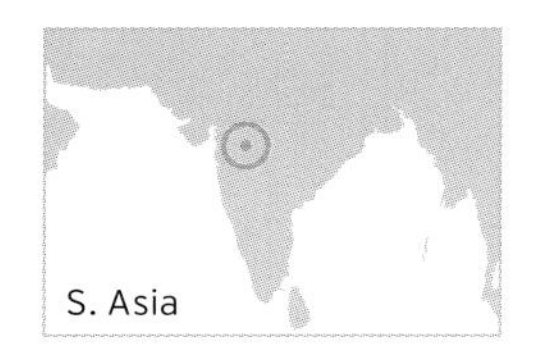

Despite their Hindu faith, the Gupta kings of ancient India were tolerant of other religions. Their period of rule (c.320–550 CE) saw a tremendous flourishing of philosophy, the sciences, art, and architecture. This included the creation of a number of significant early Buddhist sculptures and sites, the most famous of which are arguably the magnificent caves at Ajanta in western India.

Buddhist art

The 29 Buddhist cave temples and monasteries in Ajanta were carved from a basalt hillside intermittently between the 2nd century BCE and the 7th century CE. They contain sophisticated examples of Buddhist sacred art, including impressive representations of the Buddha and breathtaking paintings and murals that tell his life story, often in a rich, intense palette. It is generally believed that Cave 10 is the earliest cave sanctuary in the complex; it contains a *chaitya*—a prayer hall with a stupa (dome-shaped shrine) at one end. But most of the caves at Ajanta were for monastic residence and often featured an elaborate facade, with a central courtyard and adjoining cells.

The caves were abandoned for centuries. In 1819, a British hunting party stumbled upon them by chance—they soon became the subject of increasing global interest.

PHASES OF EXCAVATION

Ajanta's caves are laid out in an immense arc that follows the path of the Wagora River. Evidence suggests they were excavated in two main phases: an early phase around the 1st and 2nd centuries BCE; and a later phase, when most of the caves were dug, during the peak of India's golden age under Gupta rule, in the 5th century CE.

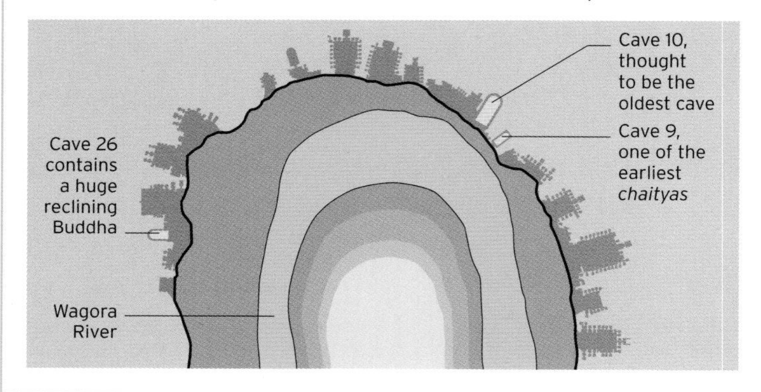

△ ANCIENT MURAL
This detail from a 5th-century CE mural in Cave 1 at Ajanta illustrates an episode from the *Jataka* tales–stories about the previous lives of the Buddha.

▷ ROCK-CUT ARCHITECTURE
Two magnificent, robust, and highly ornamented rock-cut pillars at Cave 26 testify to the astonishing artistic skill and craftsmanship of ancient India.

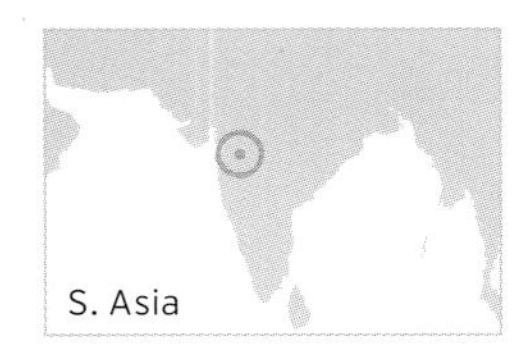

Ellora Caves

A cave complex in which three Eastern religions co-exist, famous for its sculptures and massive rock-cut Kailasa temple

India is home to three of the world's religions: Buddhism, Hinduism, and Jainism. The three traditions are brought together at the vast Ellora Cave complex in Maharashtra, western India. The site comprises 34 hand-carved, rock-cut temple caves built between the 7th and 11th centuries CE. Spread over 1.2 miles (2 km), they are most famous for the massive sculptures of deities.

The centerpiece of the complex is the spectacular 8th-century Kailasa temple, dedicated to the Hindu god Shiva. Off the courtyard are galleries filled with carvings and sculptural friezes, which narrate tales from the sacred Hindu texts, the *Ramayana* and the *Mahabharata*.

▽ **CAREFULLY CARVED**
The massive Kailasa temple, one of India's largest and most elaborate rock-cut temples, was excavated from the top downward, creating a magnificent free-standing structure.

The **Kailasa temple** is believed to have been excavated **out of a single rock**

The Hanging Temple

A precariously positioned temple that represents three of China's principal religions and schools of thought

Hengshan mountain, in Shanxi province, is the northernmost of China's sacred Five Great Mountains; clinging precariously to its side is the legendary Hanging Temple. Seeming to defy gravity, the structure is suspended some 246 ft (75 m) from the ground, with its 40 rooms secured by wooden crossbeams wedged into the cliff-face. The temple's pagodalike structures are linked to each other and the various halls by a maze of ramps, staircases, rickety passages, and walkways. The complex is thought to have been constructed during the Northern Wei dynasty (386–534/535 CE), although significant renovations and expansions were undertaken in the Ming (1368–1644) and Qing (1644–1912) dynasties.

In addition to its precipitous location, the wooden temple complex is notable for representing three of China's principal religions or schools of thought side by side: Buddhism, Daoism, and Confucianism, whose founders (Siddhartha Gautama, Laozi, and Confucius, respectively) are depicted in harmonious cohabitation in statues in the temple's northern Sanjiao Hall.

△ **BALANCING ACT**
Suspended on the side of Hengshan mountain above Jinlong Canyon, the temple buildings are secured by long, thin wooden poles that have been fastened into the near-vertical cliff-face.

Persepolis

The ceremonial capital of the First Persian Empire, now a cluster of ruined palaces

Located in modern-day Iran, the remains at Persepolis conjure up the glory of the Achaemenid kings. This great dynasty ruled the First Persian Empire in c.550–330 BCE, while Persepolis itself was founded by Darius I in c.518 BCE.

Persepolis seems to have been used only during certain festivals. Its main features—a massive gateway, the palatial Audience Hall, and the gigantic Throne Hall ("the Hundred-Column Hall")—were meant to impress. The sculpted reliefs of foreign dignitaries bearing gifts for the king, and the presence of a substantial treasury, have led to suggestions that tribute money was paid here. The mud-brick walls and roofs of Persepolis are gone, but many of the stone columns are still standing, often with elaborate, animal-shaped capitals.

▽ GATE OF ALL NATIONS
Part-man, part-bull, part-bird, this powerful figure is a *lamassu*—a guardian deity, which stood at the Gate of All Nations, a portal through which visitors to the royal palaces would have passed.

▷ TILED FACADE
The blue-and-white coloring of the İznik tiles (made in İznik in western Anatolia) is one of the highlights of the building's exterior. Added in the 16th century, the tiles survived until the 1960s, when most were replaced with replicas.

▽ GILDED DOME
The 66-ft- (20-m-) wide dome was originally made of gold but was later replaced with copper and then aluminum. It was re-covered with gold leaf under King Hussein of Jordan (r.1952–1999).

Dome of the Rock

An important Islamic shrine with a sumptuous interior decorated with mosaics and tilework

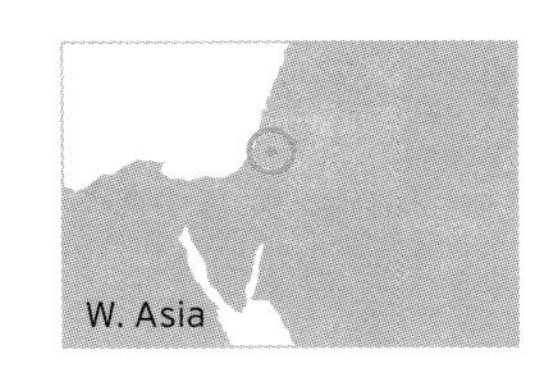

Situated on the eastern side of the Old City of Jerusalem, the Dome of the Rock was completed in 691–692 CE by the Umayyad Caliph Abd al-Malik to protect and celebrate the great rock—known as the Foundation Stone—at its core. The shrine has huge religious significance for Muslims—it is the place where their prophet Muhammad began his momentous Night Journey of 621—and for Jews, who identify it as the site of their original Temple.

Intricate decoration

The building has a Byzantine-influenced, octagonal structure that was common in the eastern Mediterranean (see pp.252–253). Inside, two octagonal arcades create a double walkway around the great rock. The two walkways recall the circular movement of pilgrims around the Kaaba in Mecca (see pp.282–283). Above the arches, much of the fine, original mosaic decoration has survived. The ornamental motifs, inspired by Byzantine and Sasanian models, are mostly based on plant forms, coiled around elaborate vases and jewels. The mosaics also feature inscriptions in kufic script, proclaiming messages of the Muslim faith.

The decorative elements of the building have been altered many times over the centuries. The most spectacular additions were executed in the Ottoman period during the reign of Suleyman the Magnificent (r.1520–1566). He commissioned the superb İznik tilework, which covers much of the exterior. Around 45,000 tiles were used. Suleyman's architects also redesigned the dome's 52 windows, filling the place with light.

COLONNADES

The internal layout of the building was designed to focus attention on the great rock. The wooden dome is supported on a circular arcade of 12 columns and 4 piers. Around the rock, there is a double ambulatory (walkway), divided by an outer circuit of 16 columns and 8 piers.

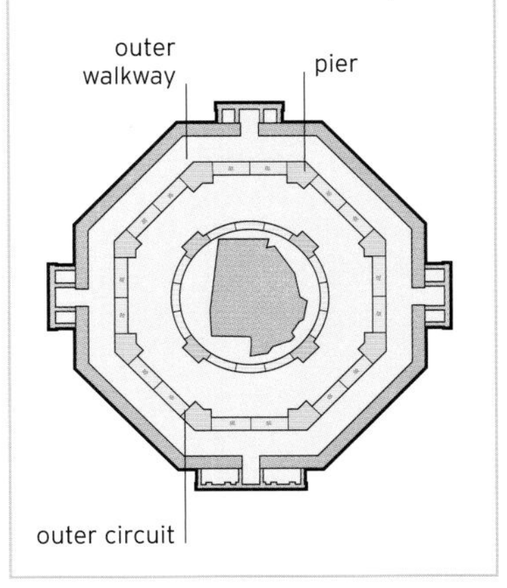

Architectural styles

EARLY ISLAMIC

Initially influenced by existing architectural traditions—notably the Classical and Byzantine—early Islamic architecture soon developed its own characteristics, which would lead to the creation of some of the world's greatest and most enduring monuments.

△ **GEOMETRIC PATTERNS**
Intricate geometric patterns are typically formed from intersecting circles and squares, with multiple overlays building up the complexity.

The mosque is the defining building of Islamic architecture. Primarily a place of worship, it is also a social and cultural center, and in many instances a place of education for all ages. The heart of the mosque is the prayer hall, a typically large space often supported by rows of columns and sometimes topped by a dome. Outside, mosques usually include a large courtyard with colonnades on each side, sometimes accompanied by a fountain, as well as a minaret—a tower clearly visible from the surrounding area and from where the faithful are called to prayer.

The Islamic faith prohibits depictions of human and animal forms in religious art and architecture, and as a result, Islamic culture evolved highly sophisticated decorative languages of geometry and pattern. These are sometimes combined with calligraphic inscriptions from the Qur'an. In addition to surface decoration, many Islamic buildings are ornamented by muqarnas—subdivisions of the underside of a dome or vault that create a honeycomb effect, as seen most dramatically in the Alhambra in Granada, Spain.

calligraphy is often the dominant decorative element

Arabic script has an aesthetic appeal

△ **CALLIGRAPHY**
Calligraphy is central to Islamic culture. Although linked to the Qur'an, from which it emerged, Islamic calligraphy is not limited to religious texts but is also used in cultural forms such as poetry.

octagonal shape may be based on polygonal Byzantine shrines

tiles are usually in blues, greens, and turquoise

four doors facing the cardinal points lead to the interior

▲ **DOME OF THE ROCK**
One of the oldest extant works of Islamic architecture, the Dome of the Rock is a shrine located on the Temple Mount in Old Jerusalem. Its gilded dome borrows heavily from Byzantine architecture, although its geometric decoration is characteristically Islamic.

MINARETS

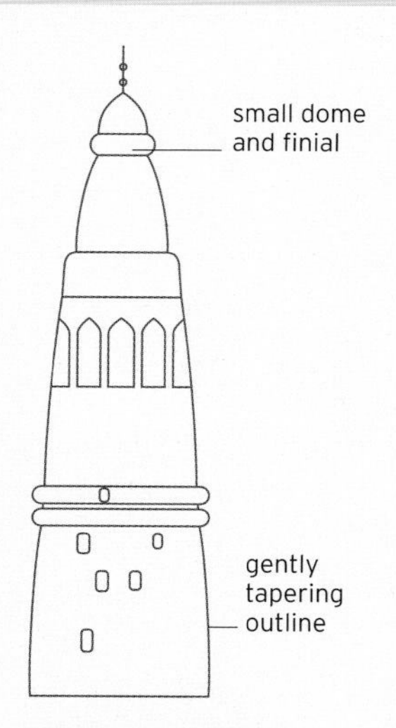

△ **YEMENI MINARET**
In this form, the different levels of the minaret are integrated into one single tapering form topped by a small dome.

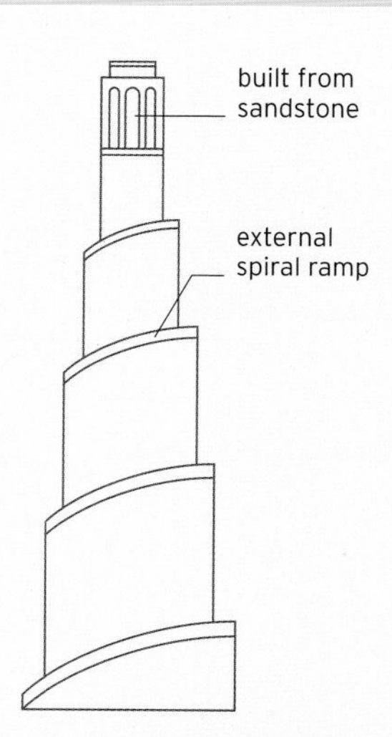

△ **AL MUTAWAKIL MOSQUE**
The spiraling conical form is unusual; it might reflect the importance of the mosque, which was one of the largest in the world when built.

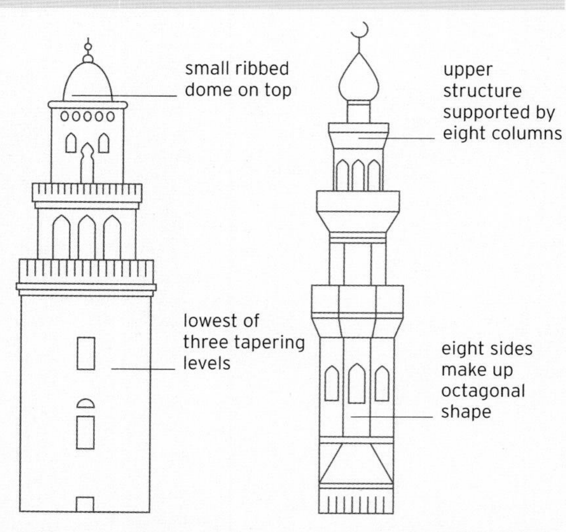

△ **KAIROUAN MOSQUE**
With a square base and squat in appearance, this simple minaret design proved highly influential in mosque design.

△ **SULTAN HASAN MOSQUE**
Standing in the old city of Cairo, the three octagonal minarets are tightly integrated into the structure and design of the mosque.

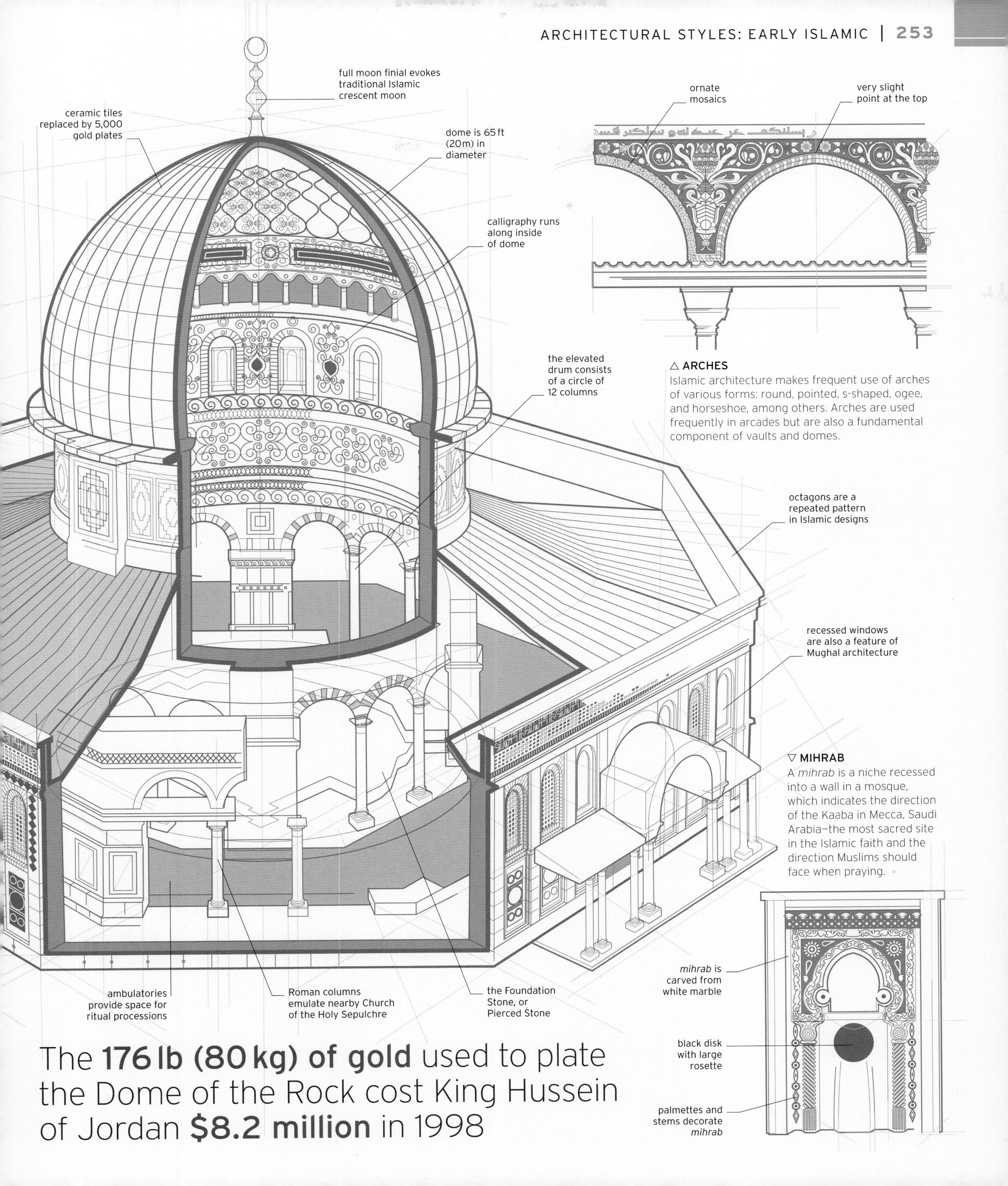

△ ARCHES

Islamic architecture makes frequent use of arches of various forms: round, pointed, s-shaped, ogee, and horseshoe, among others. Arches are used frequently in arcades but are also a fundamental component of vaults and domes.

▽ MIHRAB

A *mihrab* is a niche recessed into a wall in a mosque, which indicates the direction of the Kaaba in Mecca, Saudi Arabia—the most sacred site in the Islamic faith and the direction Muslims should face when praying.

The **176 lb (80 kg) of gold** used to plate the Dome of the Rock cost King Hussein of Jordan **$8.2 million** in 1998

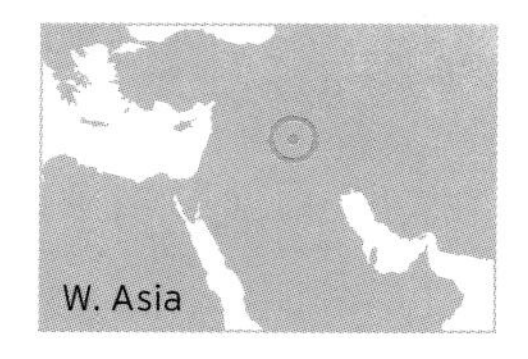

Great Mosque of Samarra

One of the world's largest mosques, and a spectacular example of early Islamic architecture and ingenuity

◁ **MALWIYA TOWER**
Standing some 171 ft (52 m) high, the sandstone minaret with its ascending spiral ramp would have been the tallest building in the region at the time of its construction.

Samarra, on the east bank of Iraq's Tigris River, north of Baghdad, was the 9th-century capital of the powerful Abbasid caliphate (the Muslim political and religious state). The town is especially famous for its Great Mosque, with its spectacular spiral minaret. Constructed between c.848 and 852 CE—during the Islamic golden age—by the caliph al-Mutawakkil, it is considered an important example of early Islamic architecture.

Snail shell tower

The mosque originally extended over some 1.8 million square ft (170,000 square m) but was damaged by Mongol armies during their invasion of Iraq in 1278. All that now remains of the mosque are its outer walls and the magnificent Malwiya ("snail shell") Tower—the undisputed highlight of the ancient site.

Combining science and artistry, the skillfully engineered tower is notable for its distinctive conical, or spiral, shape. It served as a forceful reminder of the power of the caliphate and the presence of Islam. In 2005, it became the subject of global attention after insurgents, having discovered that US troops were using it as a lookout post, bombed part of it.

LAYOUT AND FEATURES

The mosque is rectangular in shape and enclosed by a 33-ft- (10-m-) high brick wall, which includes 44 semicircular towers. It also featured 16 entrance gates, walls lined with dark-blue glass mosaics, 17 aisles, and a courtyard surrounded by an arcade. The minaret sits next to the mosque and was originally connected to it by a bridge.

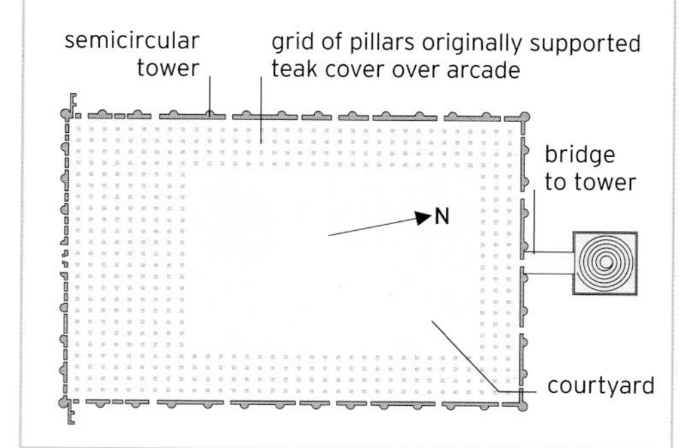

Leshan Giant Buddha

A masterpiece of engineering that is the world's largest stone Buddha

The Giant Buddha, carved out of a cliff at Mount Lingyun, in Leshan, in China's Sichuan province, presides over the intersection of three rivers: the Min, Qingyi, and Dadu. The sculpture was located there not simply to promote Buddhism but, according to legend, to pacify the river gods and help calm the treacherous waters that endangered human life and the smooth passage of ships. Appropriately enough, it depicts the bodhisattva ("enlightened being") Maitreya, a "future buddha," who was believed to rescue those in danger.

Larger than life

Construction of the monument was initiated by Hai Tong, a Chinese monk, in 713 CE; it was completed some 90 years later, after his death. Standing 233 ft (71 m) high, it is an astonishing feat of engineering and a magnificent example of Buddhist art: seated in tranquil pose and with his hands on his knees, Maitreya has a head that is approximately 46 ft (14 m) high and 33 ft (10 m) wide; his ears are about 23 ft (7 m) long; and his hair forms 1,021 exquisitely carved coils. It is claimed that the tremendous volume of stone that was cut from the Lingyun cliff-face and deposited in the river during the building of the giant statue caused a permanent change in the local river currents, making the surrounding waters calmer and far safer.

An ingenious drainage system was constructed inside the Giant Buddha when it was built to prevent erosion from rainfall. The statue was designated a World Heritage Site in 1996.

△ **GIANT STEPS**
The Buddha's feet—like most of the statue—are carved from stone. Only the ears are made of different material—wood covered with clay.

▽ **MOUNTAIN VIEW**
The spectacular Giant Buddha, carved out of the cliff-face at Leshan, gazes over the water toward Mount Emei, one of China's four sacred Buddhist mountains.

If upright, the statue would be **roughly** the **height** of the **Statue of Liberty**

Borobudur

A massive temple in the form of a mandala, with exquisite carvings that narrate the Buddha's teachings and life story

Built high on a hill in central Java, Indonesia, during the Shailendra dynasty (750–850 CE), the temple at Borobudur is dedicated to the Buddha. The world's largest and most elaborate Buddhist temple, it was buried under volcanic ash in around 1000 CE and rediscovered in 1814. Following extensive restoration in the 20th century, it was designated a World Heritage Site in 1991. The monument is an architectural expression of a three-dimensional mandala—a sacred pattern, symbolic of the Universe, which is contemplated in meditation.

Built up from a square base (representing Earth), Borobudur consists of 10 mounting terraces topped by a dome (representing Heaven). At the top of the structure is a large central stupa, dedicated to Vairochana, the Great Sun Buddha.

Borobudur has 504 statues of the Buddha and is decorated with 2,612 exquisitely carved relief panels, many of which narrate aspects of Buddhist scriptures and the Buddha's life and teachings.

◁ **PERFORATED STUPAS**
The three circular layers of the temple feature 72 separate perforated, domed stupas, each containing a statue of the meditating Buddha.

Borobudur was built from **2,000,000 cubic ft (57,000 cubic m)** of **volcanic stone blocks**

THREE-DIMENSIONAL MANDALA

Borobudur was built to replicate a mandala, its sides facing the four cardinal directions of the compass. Its design guides devotees to walk clockwise, along walkways that encircle the central cosmic axis, to the summit. The ascent symbolizes a spiritual journey from the realm of earthly desires toward enlightenment.

central stupa

base encasement

KEY

Open terrace with stupas

Square terrace

Broad terrace

△ **RELIEF PANEL**
Among Borobudur's most spectacular features are its decorative reliefs, which inspire pilgrims on their journey to the top of the temple. The main image in this carved panel is of the seated Buddha.

STATUE IN A STUPA
An atmospheric view from one of the upper levels of Borobudur shows a statue of the Buddha inside one of the site's many distinctive perforated stupas.

Vardzia

A multitiered complex of caves carved into the mountainside to accommodate a monastery and the Church of the Dormition

In the 12th century, when Georgia was threatened by Mongol invaders, King Giorgi III (1156–1184) began excavating a fortified monastery within the rocks of the Erusheti Mountain at Vardzia in southern Georgia. The project was completed by his daughter, Tamar the Great. The interconnected caves and tunnels went deep into the mountain over a 1,640-ft (500-m) stretch of the rock face and were supplied with water via a sophisticated irrigation system.

The invisible fortress

When completed, this underground citadel extended to 13 levels and contained some 6,000 apartments, a throne room, and a large church at the center of the complex. The Church of the Dormition was built on the orders of Queen Tamar when she succeeded the throne after her father's death. Hewn into the rock, the church is more remarkable for its mural decorations than its architecture. Paintings on the walls of the church, regarded as some of the finest examples of mural painting of the Georgian golden age (1089–1221), depict scenes from the life of Christ and of the saints, as well as portraits of Tamar and her father.

The greatest defense of the Vardzia monastery was its invisibility: its only entrance was hidden from view, and the system of caves was entirely within the mountain.

FRESCO-SECCO MURAL PAINTING

The murals in the church are painted using a fresco-secco technique, which differs from buon-fresco painting, as the colors are applied to a dry, finished surface rather than to wet plaster. Hence, the pigments form a superficial layer and do not seep into the plaster as they do in a true fresco.

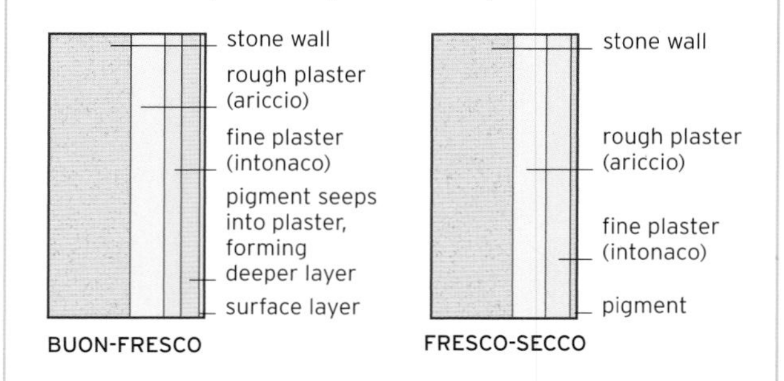

▷ **IN PLAIN SIGHT**
Once hidden within the mountain, the interior of the monastery was exposed by a rockfall caused by a severe earthquake in 1283. The entrance to the church and the honeycomb of caves and tunnels are now clearly visible on the hillside.

Dambulla Cave Temples

Sri Lanka's largest cave temple complex, containing numerous magnificent Buddha statues and decorated with Buddhist paintings

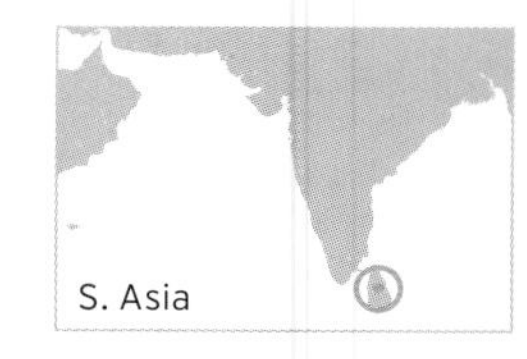

At the foot of a large rock outcrop near Dambulla, in the hills of central Sri Lanka, is a series of five caves, which were transformed into Buddhist temples over a long period from the 3rd century BCE to the mid-13th century CE. The site, now one of the country's main tourist attractions, is famed for its lavish mural decorations and the collection of more than 150 statues of the Buddha.

Colossal caves

The largest of the caves—Maharaja Lena ("Cave of the Great Kings")—contains 56 Buddha statues, as well as statues of the gods Saman and Vishnu and of the kings Vattagamani Abhaya and Nissanka Malla. A spring reputed to have healing powers trickles into the cave through the ceiling, which is painted with scenes from the life of the Buddha. Although considerably smaller, the neighboring Devaraja Lena ("Cave of the Divine Kings") houses an impressive reclining Buddha statue carved into the rock. To the other side of the Maharaja cave is the Maha Alut Vihara ("Great New Temple"), which takes its name from the comparatively recent 18th-century wall and ceiling paintings of the Buddhist revival period.

TEMPLE ENTRANCES

Although often referred to as a single temple, there are in fact five discrete cave temples in the Dambulla complex. All are approached through a courtyard, and the first three (the major Devaraja, Maharaja, and Maha Alut temples) via doorways in a verandah added in the 1930s. The two lesser temples have separate entrances at the far end of the courtyard.

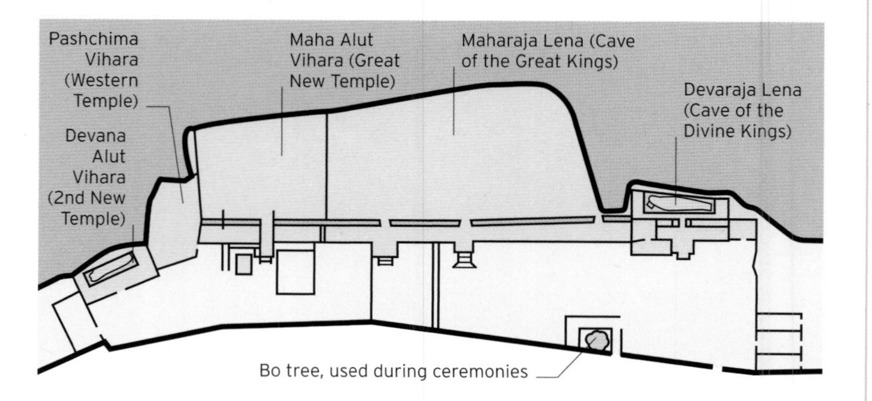

▽ CAVE OF THE DIVINE KINGS
The first temple, the Devaraja Lena, is occupied by an enormous painted statue of the reclining Buddha carved out of the rock, running the length of the cave.

▷ SITTING BUDDHAS
A collection of 40 sitting Buddha statues, along with 16 standing Buddhas, line the walls of the largest of the cave temples, the Maharaja Lena.

The **mural paintings** in the cave temples of Dambulla **cover** an area of **22,600 square ft (2,100 square m)**

Sacred City of Kandy

The former capital of the last independent kingdom in Sri Lanka and a Buddhist pilgrimage site

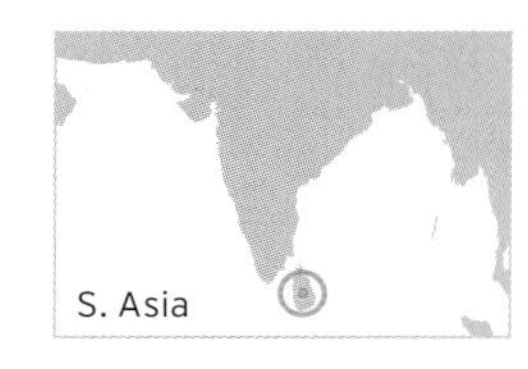

The picturesque sacred city of Kandy became prominent toward the end of the 15th century as the capital of Sri Lanka. Built on a plateau surrounded by hills in the country's Central Province, it was the last capital of the Sinhala kings until it was taken over by the British in 1815.

Cultural capital

The old city of Kandy is notable for its historic buildings; outstanding architecture; and numerous Buddhist shrines, temples, and manuscripts. It is most famous, however, for the Temple of the Tooth, or Dalada Maligawa—originally part of the royal palace complex—which is believed to house the supposed left upper canine tooth of the Buddha, one of Buddhism's most venerated relics. The tooth relic is stored in a chamber, encased within layered gold caskets. According to legend, the tooth was brought to the island in the 4th century CE. Construction of the extant ornate temple, whose function was to enshrine the relic, and its associated buildings was undertaken in phases from the early 18th century. The city of Kandy, also known as Senkadagalapura, was designated a World Heritage Site in 1988. It was and remains an important place of pilgrimage for Buddhists around the world.

△ **TEMPLE OF THE TOOTH**
Drummers perform in the Hewisi Mandapaya, the courtyard in front of the main shrine where rituals are held to venerate the tooth relic.

▽ **TEMPLE COMPLEX**
The octagonal pavilion, Pattirippuwa, was built in the early 19th century by the last king of Kandy. The low, white walls in the temple complex are perforated with small carved openings.

△ **PURPLE HEAVEN PALACE**
Set amidst breathtaking scenery, the Purple Heaven Palace is the best-preserved palace at Wudang and an outstanding example of Ming architecture.

△ **MONASTERY VILLAGE**
Wudang remains a major Taoist center that attracts many pilgrims and practitioners of tai chi. The monastery village shown here provides a place of rest for adherents.

Ancient Buildings of Wudang Mountains

A vast, ancient architectural complex in a spectacular location and one of the world's most important Taoist centers

The complex on the Wudang Mountains in central China forms an important part of the country's religious and cultural landscape. While some buildings on the site date to the reign of the Tang dynasty Emperor Taizong (r.629–649 CE), the area was most fully developed as a complex during the Ming dynasty (14th–17th centuries), with expansions in the late 17th and early 19th centuries.

Blending into the landscape

The vast site features spectacular examples of Chinese art and architecture, including numerous palaces, monasteries, and temples, all of which harmonize with the breathtaking natural environment, with its 72 peaks and various valleys, streams, caves, and ponds. Among the most important and impressive of the Wudang structures is the wooden Purple Heaven Palace, or Zixiao Palace, constructed during the 12th century, rebuilt in the 15th century, and expanded in the 19th century. Other notable buildings at Wudang include the Golden Shrine and the Ancient Bronze Shrine, both early 14th-century buildings in bronze; the Nanyang Palace (12th–13th centuries); and the Fuzhen Temple (15th and 17th centuries).

MOUNTAIN PALACE

The Purple Heaven Palace is spread over about 73,800 square ft (6,850 square m) and comprises 182 rooms laid out symmetrically. It rises from the Dragon and Tiger Hall, through the Great Hall (built for the worship of the deity Zhenwu), to the Parents' Hall. The Great Hall is among the most distinctive wooden structures at Wudang.

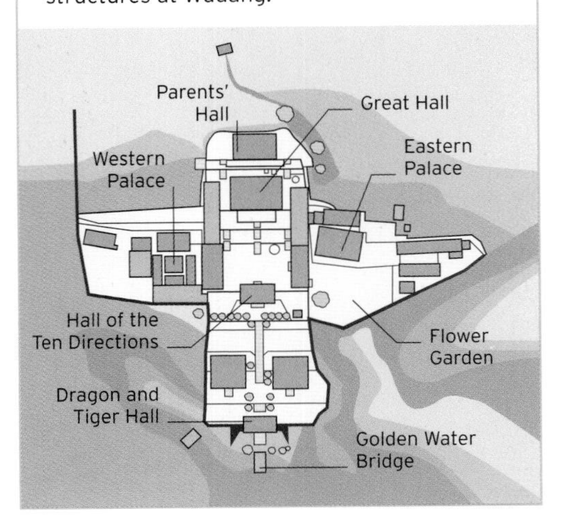

Candi Prambanan

A massive religious compound that reflects the craftsmanship and architectural skill of ancient Hindu culture

△ **TEMPLES OF THE HINDU DEITIES**
The temple of Shiva (center, foreground) is flanked by the smaller temples of Vishnu and Brahma. Behind them are lesser shrines dedicated to Garuda, Nandi, and Hamsa, the *vahanas* (animal vehicles) of these major deities.

Candi Prambanan in Yogyakarta, Java, Indonesia, is the country's largest and most impressive Hindu temple complex. It is dedicated to the *Trimurti* (trinity), the three major deities at the top of the Hindu pantheon: Brahma (the creator), Vishnu (the preserver), and Shiva (the destroyer).

Construction of the complex began around the middle of the 9th century CE, during the reigns of the Sailendra and Sanjaya dynasties. Like the nearby Buddhist temple of Borobudur (see pp.256–257), the plan of Prambanan is of a mandala, or cosmic diagram, with the soaring spires characteristic of Hindu architecture here representing the mythical Mount Meru, home of gods.

Temple complex

Prambanan has two principal areas, separated by a wall: an inner area, containing the main temples and associated shrines; and an area outside this containing 224 smaller temples (many of which are damaged). The complex centers on the Shiva temple, in the inner area—at 154 ft (47 m) high, it is the site's largest temple. It is flanked by the smaller temples of Brahma and Vishnu. The walls of these three main temples are adorned with magnificent stone carvings illustrating scenes from the Hindu epic the *Ramayana*.

◁ **GUARDIAN STATUE**
A fearsome 9th-century *dvarapala*, or temple protector, with bulging eyes and a moustache, keeps watch outside the Prambanan temple complex.

THE SHIVA TEMPLE

The Shiva temple has five main rooms. Like most Hindu architecture, it was conceived as a symbol of the Hindu cosmos: the base of the temple (*bhurloka*) represents the human realm; the central section (*bhuvarloka*) is the realm of sages and truth seekers; and the top of the temple, the holiest section (*svarloka*), represents the realm of the gods.

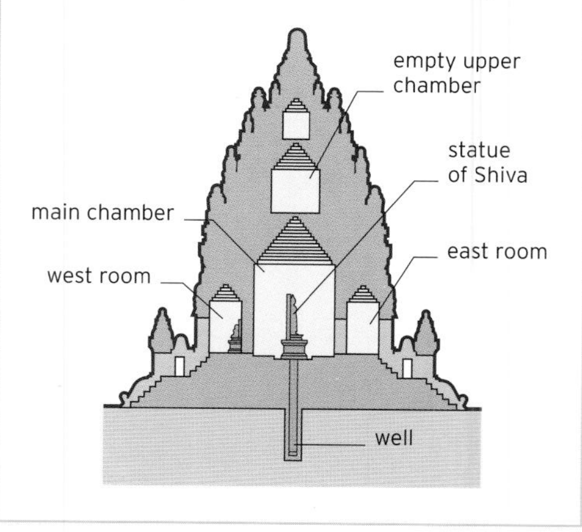

Chand Baori

A masterpiece of ancient Indian architecture and one of the world's largest stepwells

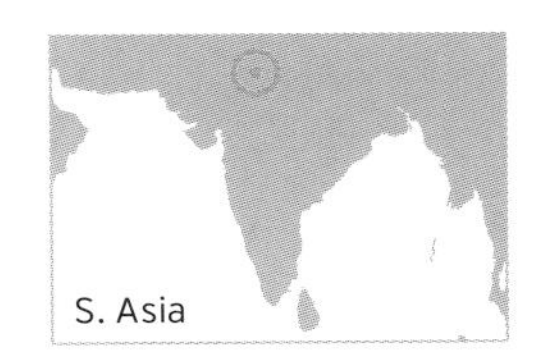

Chand Baori is a stepwell—a well or pond in which the water is accessed by steps—in the village of Abhaneri, Rajasthan. Constructed between the 8th and 9th centuries CE, it was designed not simply to collect and conserve water from the water table deep underground, but also—given its proximity to the picturesque 7th–8th-century Harshat Mata Temple—as an object of aesthetic beauty, a meeting space for the local community (including the royal family), and a shelter from the heat of the day.

Design and precision

Considered an artistic and architectural masterpiece, Chand Baori is one of the world's largest stepwells: it has 13 stories and 3,500 mazelike steps, which become narrower as they descend below ground level. It is also notable for its extraordinary geometric precision and breathtaking symmetry. Magnificent stone sculptures and friezes adorn the building's various halls and columns.

Chand Baori **featured** in the film ***The Dark Knight Rises***

◁ **VIEW OF THE INTERIOR**
Hailed as a masterpiece of Rajasthani art and architecture, Chand Baori has hundreds of steps on three of its sides; on its fourth side, seen here, is a pavilion with arches, balconies, and pillars, some of which have intricate carvings.

▽ **THE STAIRWAYS**
An ingenious system of multileveled, intersecting steps is built into the sides of the walls at Chand Baori, providing access to what was once a subterranean water supply.

Temples of Bagan

More than 2,200 Buddhist temples and pagodas spread across the plains of the ancient kingdom of Bagan

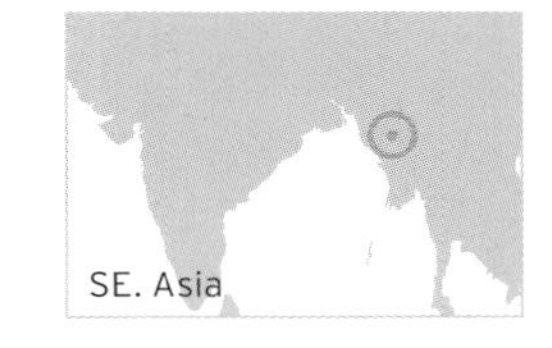

The kingdom of Bagan, or Pagan, on the plains of the Irrawaddy River in present-day Mandalay, Myanmar (formerly known as Burma), was the principal power in the region from the 9th to the 13th centuries. It established a distinctive Burmese culture, at the center of which was the practice of Theravada Buddhism. During this period, more than 10,000 Buddhist temples were erected on the Bagan plains, and although less than a quarter have survived, the panorama of pagodas rising from the landscape is an extraordinary spectacle.

Evolving style

The structures at Bagan provide a unique record of the development of Burmese temple architecture. Some are simple stupas, also known as pagodas—solid hemispherical shrines to contain relics—built in an essentially Indic style; later buildings, comprising temples topped with pagodas, show a transition to what is now recognized as a distinct Burmese style. Typically, the stupas evolved from a dome-shaped form to a more slender, bell-shaped structure while the temples beneath became more complex places of worship, with spacious, decorated interiors matching the sophistication of the exterior. These buildings generally had either one or four entrances, but a uniquely Bagan style of pentagonal temple also evolved.

Afterlife

Since the fall of the Bagan kingdom in the 13th century, few new temples have been built, and many were ruined by neglect and the effects of earthquakes. Unfortunately, in recent times, several of the remaining temples have been inappropriately restored, but a majority, including some of the finest, have been sensitively maintained.

◁ **VISTA OF PAGODAS**
The Bagan landscape is punctuated by elegant, conical or bell-shaped pagodas of countless Buddhist temples rising from the plains. Smaller temples are grouped around the more important temples.

△ **TEMPLE INTERIOR**
Beneath the pagodas, most of the distinctive Burmese temples have spacious interiors decorated with Buddhist artwork and containing statues of the Buddha.

THE *HTI*

A distinctive feature of almost all the temples in Myanmar is the *hti*, or "umbrella," a finial ornament on the spires of the pagodas. Typically, these spires are made of increasingly small concentric stone rings, with a *hti* at the tip. *Htis*, in a great variety of designs, are often adorned with bells and hanging decorations, and many are topped with precious stones. The *htis* of the Bagan temples, unlike the metal examples elsewhere in Myanmar, are made of stone plated with gold.

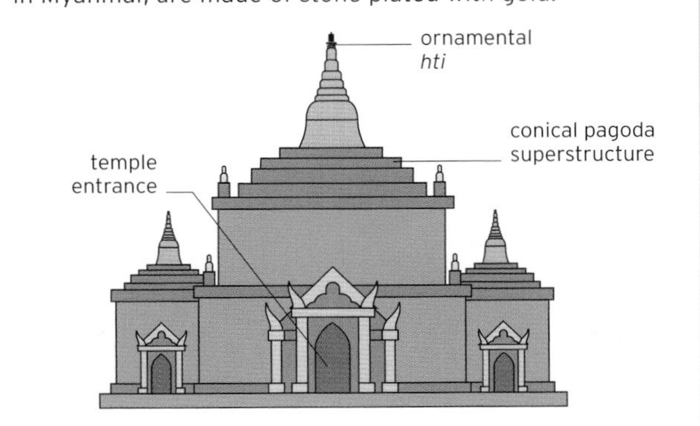

Bahla Fort

One of the finest examples of a medieval Islamic fortified oasis settlement

Bahla is one of four fortresses lying at the foot of the Jebel Akhdar highlands in Oman. Although evidence suggests that Bahla was settled and fortified from the first millennium BCE, the current fort was built by the Banu Nebhan tribe, which controlled much of Oman from the 12th to 15th centuries CE. The vast fort attests to the wealth of the Nebhani dynasty, accumulated through its dominance of the Middle Eastern trade in frankincense.

Bastion in the desert

Resting on sandstone foundations, 164 ft (50 m) above the oasis settlement that surrounds it, the fort is constructed in mud-brick. Inside the *Al-Qaabah* ("citadel"), there is a labyrinth of cells and chambers, with walls punctuated by openings through which boiling water or date juice could be poured on attackers. The outer walls, 8 miles (13 km) long, enclose the entire oasis settlement, including the Friday Mosque, family compounds, audience halls, bath houses, and the remains of a semicovered market. After restoration, the fort was opened to the public in 2012.

▷ **TYPICALLY OMANI**
Bahla Fort's round towers and smooth walls, punctuated by small windows, are typical of the more than 500 forts, castles, and lookout towers found across Oman. Around 20 of these structures have been meticulously restored.

▷ **OASIS CITADEL**
The fort rises above the surrounding mud-brick settlement and palm grove of the oasis, which is irrigated by a system of wells and underground channels that bring water from distant springs.

Bahla was the **center** of the **Ibadi school of Islam**, founded 20 years after Prophet Muhammad's death

△ **TRADITIONAL CONSTRUCTION**
Thick, mud-brick walls provide niches for a library beneath a ceiling constructed from split date-palm logs overlaid with palm frond matting.

Minaret of Jam

The world's second tallest brick minaret, famed for its intricate brickwork and inscriptions

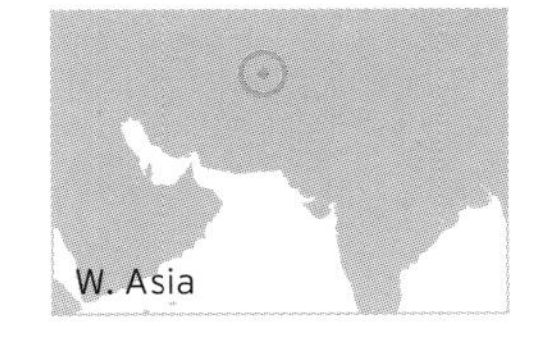

The minaret of Jam has stood in a remote valley nestled between mountains at the edge of the ancient kingdom of Ghor in western Afghanistan for more than 800 years. It was erected by the Ghurid ruler Ghiyath al-Din, probably to commemorate an important military conquest in 1173. It is thought to have been part of Firuzkuh, the summer capital of the Ghurid Empire, which was destroyed by the Mongols under Ghengis Khan in the early 13th century.

Elaborate brickwork

Standing 213 ft (65 m) high, the minaret was built entirely of red brick. It was decorated with high-relief terracotta tiles, arranged in geometric patterns of pentagons, hexagons, and diamonds, and a number of epigraphic bands in both kufic and naskh calligraphy. The inscriptions are 5–10 ft (1.5–3 m) in height and contain verses from the Qur'an and historical notes, including increasingly elaborate versions of the name and titles of Ghiyath al-Din and the name of the minaret's architect, Ali ibn Ibrahim of Nishapur.

The decoration of the minaret represents the high point of a technique typical of Afghan architecture first seen in the late 10th century. Frequent floods have damaged the base of the minaret, leaving it leaning and in danger of collapse, and it has lost more than one-fifth of its decorative brickwork.

TWO TIERS

The minaret has two levels. In the lower tier, two spiral staircases—a double-helix type not seen in Europe until the Renaissance—lead through six vaulted chambers to a balcony. In the upper tier, the stairs wind around a central void spanned by six vaults resting on four internal buttresses.

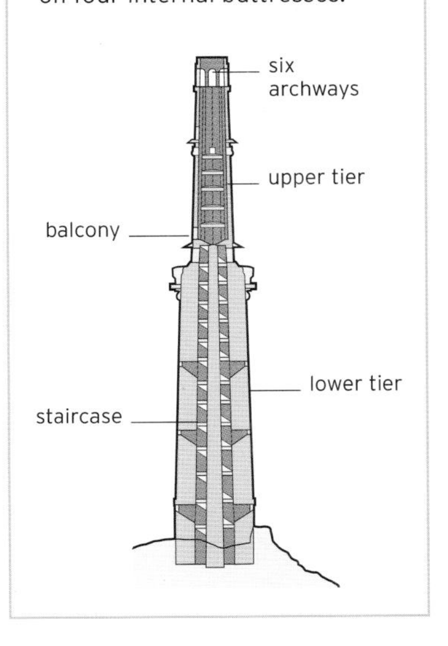

The minaret was **forgotten** after the 13th century, and **rediscovered** only in **1886**

△ **ELABORATE INSCRIPTIONS**
Five bands of decorated inscription encircle the minaret. The text is glazed in turquoise blue—the earliest known use of this feature in Ghurid architecture.

▷ **ENDANGERED SITE**
By 2014, a combination of floods and illegal excavations had left the minaret tilting at an angle of 3.47 degrees and in danger of collapse.

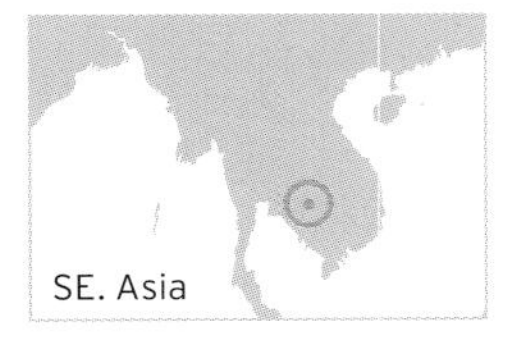

Angkor Wat

A masterpiece of Khmer art and architecture, and the world's largest religious structure

The temple of Angkor Wat (meaning "city temple") in Siem Reap, northwestern Cambodia, is the largest and one of the most impressive archaeological sites in the world, covering an area of some 0.6 square miles (1.6 square km). The temple was at the heart of a large, sprawling city, its parts linked by a network of roads and canals. Construction of the site began during the reign of King Suryavarman II (1113–c.1150), under the mighty Khmer Empire, which at the time ruled over much of mainland Southeast Asia. Vast sums were spent on glorifying the country's rulers by means of lavish building projects—none so awe-inspiring and ambitious as that of Angkor Wat.

An astonishing example of Khmer art and craftsmanship, the temple was conceived as a place of worship and protection and an endorsement of the king's authority and his claim to political office. It is dedicated to the great Hindu god Vishnu, the preserver who protects and sustains the world, and with whom Suryavarman II identified himself specifically.

Depiction of myths

The temple features 13,000 square ft (1,200 square m) of exquisitely carved bas-reliefs, including the visual narration of eight major stories in Hindu myth. The highlight of these is generally thought to be the depiction of the Churning of the Ocean of Milk on the east gallery of the temple. The magnificent carving shows gods and demons churning the ocean to extract *soma*, the elixir of immortality.

Angkor Wat went into decline in the 14th century and was sacked by a Thai army in 1431. It was listed as a World Heritage Site in 1992.

The **construction** of the **temple** is thought to have involved around **300,000 workers**

△ **CHURNING THE OCEAN OF MILK**
This image shows a detail from one of Angkor Wat's most spectacular wall carvings. It depicts the Hindu creation myth, the Churning of the Ocean of Milk.

◁ **TEMPLE COMPLEX**
The central tower of the temple complex rises to a height of 213 ft (65 m) and is flanked by four smaller towers and a series of enclosure walls, which have corners at angles of precisely 90 degrees.

A SACRED MICROCOSM

Angkor Wat conforms to the Hindu concept of a sacred microcosm. Its five spectacular towers symbolize the peaks of mythical Mount Meru–the dwelling place of the gods for both Hindus and Buddhists–and its enormous moat (now dry), more than 590 ft (180 m) wide, represents the cosmic ocean that was thought to encircle Mount Meru.

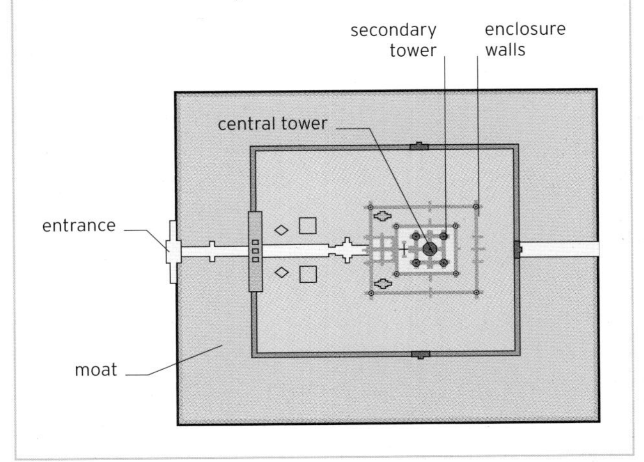

Monastery of Geghard

A medieval monastery with chapels and tombs carved into the cliffs of a spectacular gorge

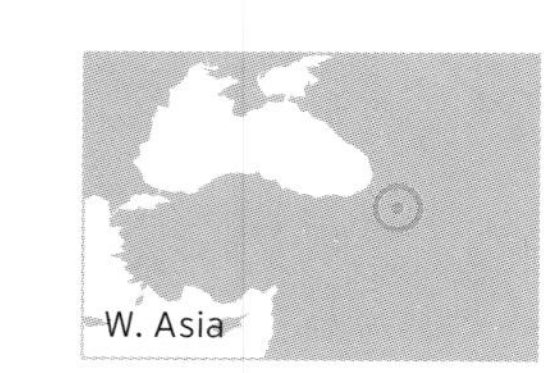

△ **INSIDE A ROCK-CUT CHAMBER**
Dramatic carvings on the walls of the *zhamatoun* are illuminated by sunlight from a decorated aperture in the ceiling.

Geghardavank, the "Monastery of the Spear," gets its name from one of the relics once housed there—the spear believed to have wounded Christ on the cross. The monastery is situated high on the cliffside of the Azat River gorge in Kotayk, Armenia. It was for several centuries a simple cave temple, until it was developed into a complex of churches and tombs in the 13th century.

Cut into the rock

Within the boundary of a high defensive wall is an imposing church, the Katoghike Chapel. On its western side is a large *gavit*, a square entrance lobby, which is partially carved into the rock. An entrance from the *gavit* leads into the mountainside to the series of chambers cut into the rock for which Geghard is best known. The complex includes tombs and vestries notable for columns cut from the solid rock. The walls of these chapels and tombs are decorated with elaborate geometric and floral relief carvings and the distinctive Armenian crosses known as *khachkars*. The whole complex represents a high point in medieval Armenian art and architecture.

▷ **THE KATOGHIKE CHAPEL**
The main church is in classically Armenian style with a floor plan of an equal-armed cross on a square base. It is topped by a dome on a cylindrical drum.

THE MONASTERY COMPLEX

Apart from the Chapel of St. Gregory the Illuminator, outside the walls of the complex, the principal churches of Geghard are all accessed via the *gavit*. The main Katoghike Chapel is free-standing but abuts the cliffside; the *gavit* cuts into the mountain, leading to the Avazan (meaning "basin"), the two *zhamatoun* containing tombs, and a second rock-cut church.

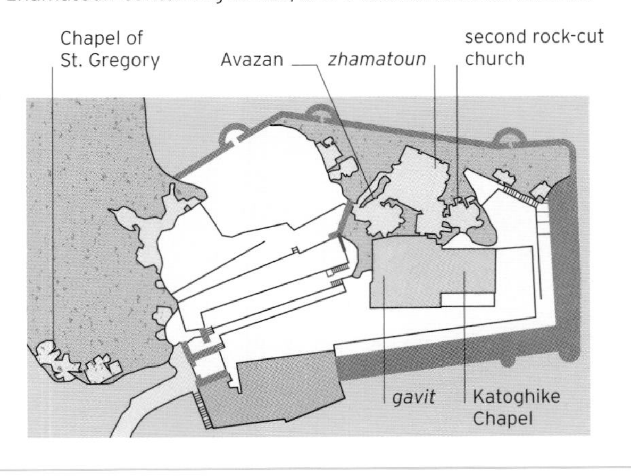

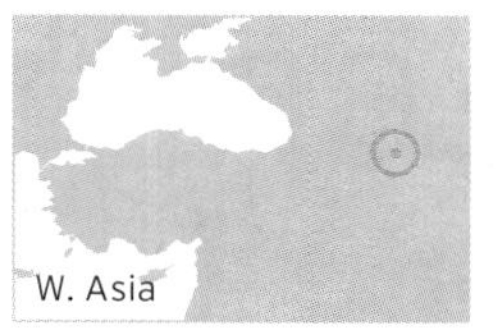

Palace of the Shirvanshahs

The luxurious residence built for the shahs of Shirvan on top of a hill in the ancient walled city of Baku

In the 15th century, Shah Khalilullah of Shirvan moved his capital to the port of Baku, on the Caspian Sea in present-day Azerbaijan, where he had a palace complex built. Within its walls, the complex occupies three levels. The main palace is at the summit of the hill, set in a courtyard, which forms the uppermost level; this level also contains the mausoleum of a court scholar, Seyid Yahya Bakuvi. To the side of the palace garden is an octagonal building enclosed by arcades, the *Divankhana*. On the second, lower level, the Shah's mosque and royal burial vaults occupy another courtyard; the lowest level consists of a terrace containing sunken bath-houses.

Constructed in the first half of the 15th century, with a grand eastern portal known as Murad's Gate added to the wall in 1585, the Palace of the Shirvanshahs is the focal point of the Icheri Sheher, the Inner City of Baku. It is described in its UNESCO World Heritage Site citation as "one of the pearls of Azerbaijan's architecture."

△ **CARVED INSCRIPTIONS**
Many of the buildings of the palace complex are decorated with stone friezes carved with historical and religious inscriptions in relief.

△ **PALACE IN THE EVENING SUN**
The domed rooftop of the palace is visible behind the hexagonal dome of the royal mausoleum and the domes and minaret of the palace mosque. All of these structures are made of blocks of the same honey-colored stone.

Himeji Castle

A masterpiece of wood-and-stone construction and one of Japan's last surviving feudal castles from the 16th and 17th centuries

Known to many as White Egret Castle or White Heron Castle because it is said to resemble a bird taking flight, Himeji Castle was built by Ikeda Terumasa in 1601–1609. Ikeda was appointed governor of the western provinces by Tokugawa Ieyasu (whose dynasty ruled as Shoguns over Japan until the late 19th century), and his castle was part of the Shogun's peacekeeping plan to build one castle per province.

Designed for defense

Ikeda Terumasa replaced the existing three-story keep with a five-story castle that in fact hides six floors and a basement (see pp.294–295). He added three further keeps, linked by two-story covered corridors in a formation known as *renritsu tenshu* ("connected towers"). There was only one access point to the keeps, and the outward-facing walls of the complex were protected with gun and arrow ports and stone drop windows (wooden structures with openings at the bottom through which rocks could be thrown). The castle was surrounded by a spiral-shaped moat, within which were a series of fortified gateways, and a labyrinth of courtyards and steep passageways designed to confuse attackers. An almost impregnable castle, its tiers of roofs and decorative gables were also a beautiful expression of traditional Japanese style. The castle somehow survived the extensive Allied bombing of Japan in World War II, becoming one of Japan's National Treasures in 1931 and one of the country's first two UNESCO World Heritage Sites in 1993.

THE *NAWABARI* OF HIMEJI

Many warlords believed that a castle's *nawabari*–its layout and defensive system–determined its fate. As this plan shows, Himeji Castle had a complex *nawabari* design with a three-layer spiral defense system. This created a series of descents and switchbacks to bewilder enemies and conceal the keep's entrance.

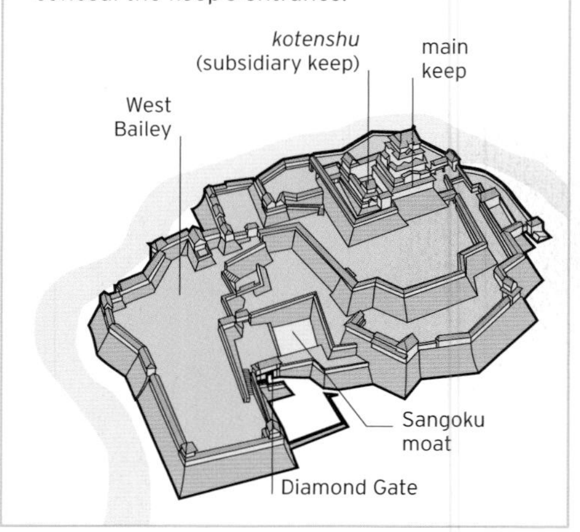

The **complex defenses** of Himeji were **never tested**, as the castle was **never attacked**

▽ IMPENETRABLE BEAUTY
The graceful eaves, elegant multiple-tiered roofs, and clean white walls of the five-story main keep at Himeji make it a prototypical example of Japanese architecture of the 17th century. Stone drop windows can be seen around the lowest story.

Gyeongbokgung Palace

Destroyed and rebuilt several times, a potent symbol of the pride and resilience of the Korean people

△ **RISEN FROM THE ASHES**
The reconstructed Heungnyemun Gate complex leads through to the two-story Geunjeongjeon (Throne Hall), one of the few buildings to survive the 20th-century destruction.

Also known as the Northern Palace, the Gyeongbokgung Palace is the largest of five palaces in Seoul that were home to the kings of Korea's ruling Joseon dynasty (1392–1910). Built in 1395, it was the main palace of the kings until it was destroyed during the Japanese invasion in the 1590s.

Surviving examples

In the 19th century, the regent, Heungseon Daewongun, rebuilt the palace, creating a vast complex of over 330 buildings covering 0.2 square miles (0.6 square km). This was largely destroyed during the Japanese occupation of Korea (1910–1945). However, several 19th-century buildings survived, providing beautiful examples of traditional Korean stone-and-wood architecture and decoration. They include the ornate two-story Geunjeongjeon (Imperial Throne Hall), the Gyeonghoeru (Royal Banqueting Hall), and Jagyeongjeon Hall (Quarters of the King's mother).

▷ **SYMBOLIC COLORS**
In traditional Korean building decoration (*Dancheong*), the paintwork protects the wood from the elements and the colors symbolize the compass points, the seasons, and both real and mythical creatures.

Architectural styles
CHINESE

For most of its history, Chinese architecture has been entirely tied to Chinese civilization—a common set of beliefs, writing, and aesthetic sensibilities that held sway for thousands of years across a vast area of East Asia.

Chinese architecture has proved influential across East Asia, most notably in Korea and Japan. Although stone and brick are used, it generally relies on timber construction in a post-and-lintel system.

Symmetry is a highly important aspect of Chinese architecture, with many buildings and complexes designed to be symmetrical along two axes. Buildings operate essentially as pavilions, with imperial structures aligned on an axis, and those of lesser importance to the side or in a courtyard. As a result, and despite the obvious importance of the pitched roof, the overall impression is usually horizontal rather than vertical.

With the opening up of the Chinese economy and society in the 1980s, there was an influx of Western styles and building technologies. But today, many architects are looking again at the forms and principles of traditional Chinese architecture.

▶ HALL OF SUPREME HARMONY, FORBIDDEN CITY
Constructed in the Ming dynasty during the early 15th century, the Forbidden City complex (see pp.276–277) contains the emperor's residence, as well as numerous ceremonial spaces, such as the Hall of Supreme Harmony. Historically, Chinese buildings typically follow codified construction methods and standards set out in 1103, and feature symmetrical layouts along a central axis, pitched roofs, mythological motifs, and color coding to signify rank.

dragon head waterspouts drain terrace of any rain

three-tiered marble precinct is 26 ft (8 m) high

copper stoves burn incense

broad staircase with shallow steps allows space for large processions to Hall of Supreme Harmony

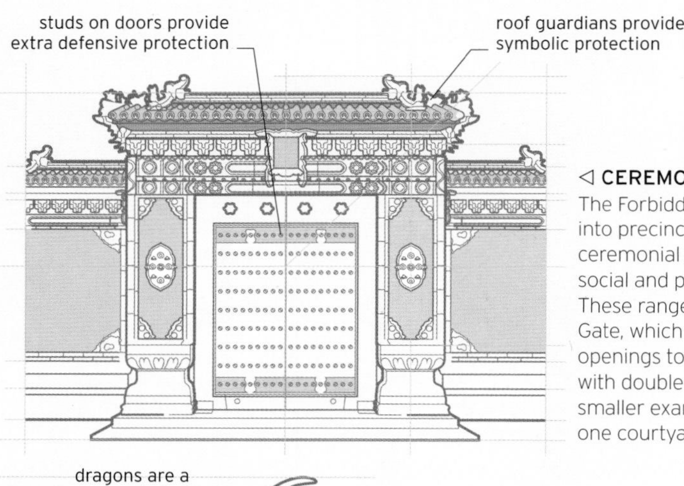

◁ CEREMONIAL GATEWAY
The Forbidden City is separated into precincts accessed by ceremonial gateways marking social and physical boundaries. These range from the Meridian Gate, which is actually five openings topped by buildings with double-tiered roofs, to smaller examples connecting one courtyard to another.

dragons are a symbol of imperial power

number of statuettes reflects high status of building

◁ ROOF GUARDIAN
Chinese architecture makes frequent use of the hipped roof form, in which all the sides of the roof slope down to the walls. In imperial architecture, the ridges are typically ornamented with mythical creatures, including the imperial dragon.

Beijing was the **most populated city in the world** from around **1450** until **the early 19th century**

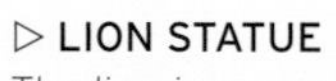

▷ LION STATUE
The lion is a common Chinese architectural ornament. Renowned as guardians, lion statues carved from stone were often placed outside notable buildings as a sign of prestige or wealth.

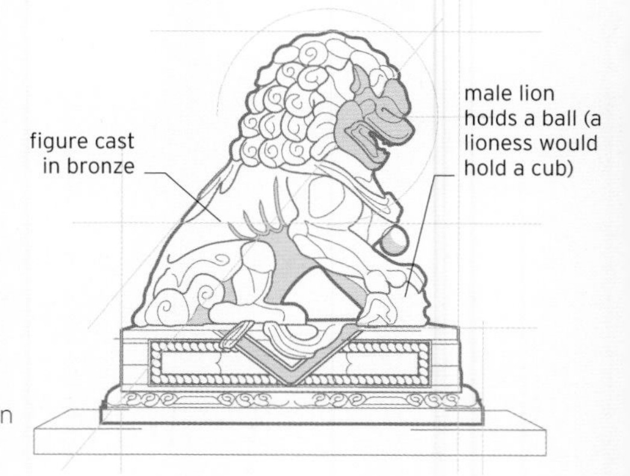

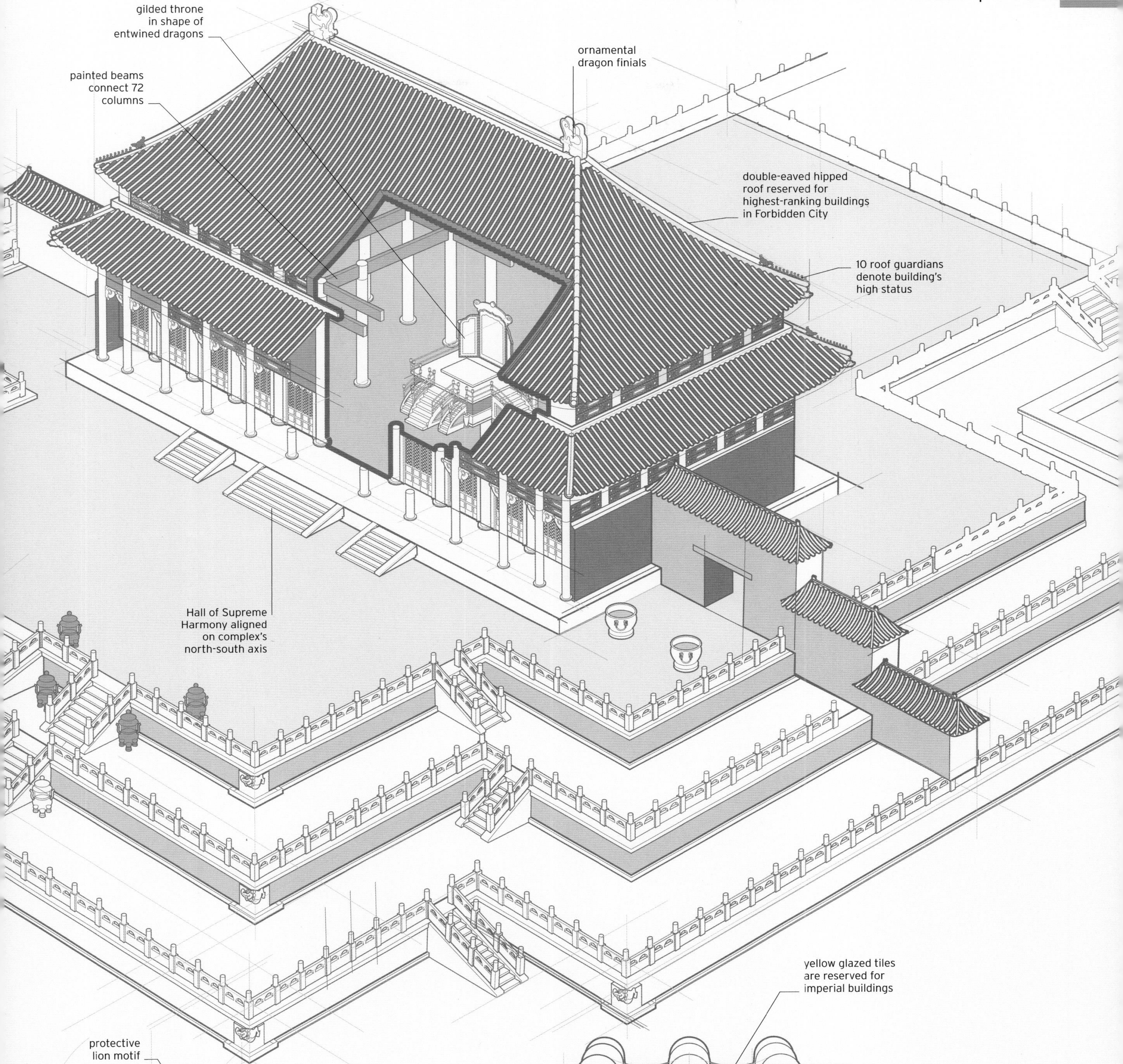

gilded throne in shape of entwined dragons

painted beams connect 72 columns

ornamental dragon finials

double-eaved hipped roof reserved for highest-ranking buildings in Forbidden City

10 roof guardians denote building's high status

Hall of Supreme Harmony aligned on complex's north-south axis

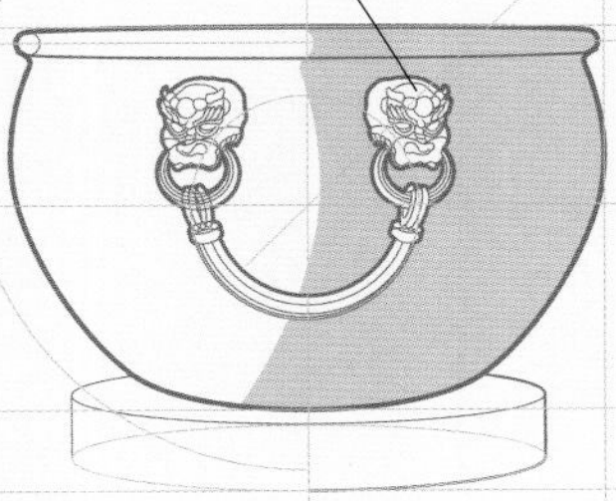

yellow glazed tiles are reserved for imperial buildings

protective lion motif

◁ CAULDRON
The Forbidden City contains 20 sizeable cauldrons made of bronze placed along the precinct. It has been suggested that, as well as being ornamental, the cauldrons carried water that could be used by palace firefighters in the event of a fire.

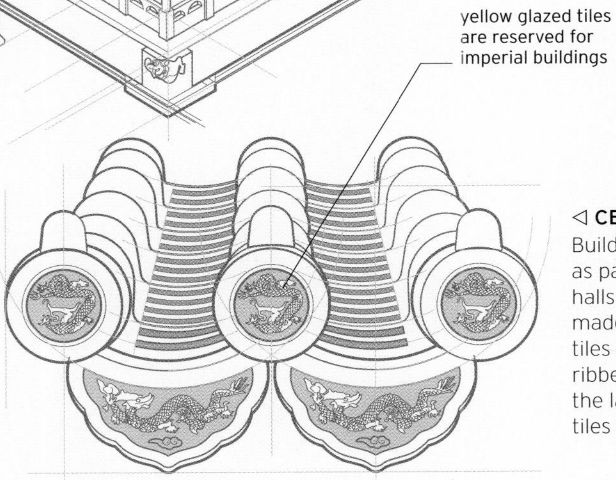

◁ CERAMIC ROOF TILE
Buildings of distinction, such as palaces and ceremonial halls in the Forbidden City, made use of glazed ceramic tiles to cover their roofs. The ribbed effect is created by the laying of semicircular tiles over flat ones.

PALATIAL CENTER OF THE CITY
Enclosed by a massive wall, the Forbidden City is a vast imperial complex, symmetrical in design and layout. It includes magnificent landscaped gardens, massive courtyards, and numerous palaces, as well as religious and government buildings.

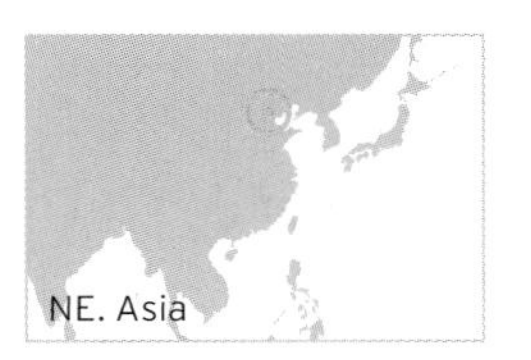

Forbidden City

An iconic Chinese landmark and the world's largest, best-preserved, and most spectacular collection of wooden buildings

Home to the rulers of China's last two dynasties (the Ming and the Qing, 1368–1912), the Forbidden City in Beijing was built as a display of imperial power and is now one of the world's most important cultural centers. Constructed in c.1406–1420 and rebuilt over the centuries, the palace covers an area of 7,750,000 square ft (720,000 square m) and contains around 800 buildings (see pp.274–275). Most of the structures are wooden, with pillars crafted from huge logs. The precinct is surrounded by a vast wall 33 ft (10 m) high.

Design and feng shui

The location, architecture, and layout of the complex followed the principles of feng shui. One of its most important features is its north–south orientation; many of the principal buildings face south, in homage to the Sun. Symmetrical in structure, the complex has an inner court—the imperial family's domestic space—and an outer court for state affairs. These areas include palaces, halls, towers, and pavilions, often with distinctive red walls and yellow roof tiles. Walls, railings, pillars, and roofs, with their characteristic upturned, overhanging eaves, are magnificently decorated. Among the many notable buildings are the Palace of Heavenly Purity and the Hall of Supreme Harmony—at 210 by 121 ft (64 by 37 m), the latter is one of the largest structures on the site.

◁ **GUARDIAN LION**
A magnificent gilded bronze lion with an ornamented sphere beneath his paw stands guard as a representative of imperial power and dignity in the Forbidden City.

The Forbidden City has **survived** more than **200 earthquakes**

△ **ELABORATE ROOFS**
The brightly colored rooftops in the Forbidden City have supporting beams and distinctive overhanging upturned eaves, beneath which are interlocking wooden roof brackets, or *dougong* (see panel, right).

THE CITY'S ROOFS

The elaborate roofs of the Forbidden City are supported by an ingenious system of wooden brackets, which can be assembled without using nails or glue. They provide a strong, flexible structure that allows the buildings to withstand regular earthquakes. Wooden blocks known as *dou* are placed onto the building's columns; further brackets (*gong*) are slotted into the *dou*. Multiple units of *dougong* below are compressed by the weight of the roof, making a robust structure.

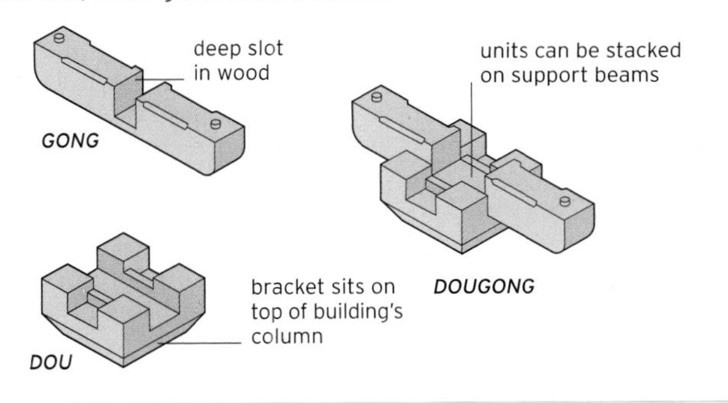

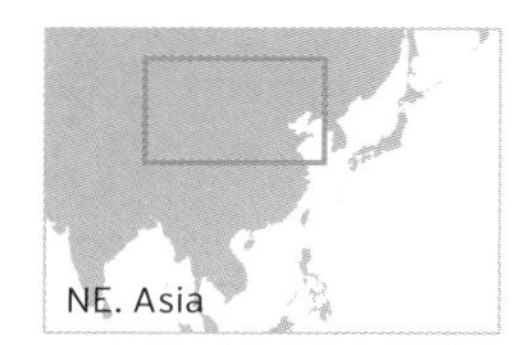

Great Wall of China

The longest man-made structure ever built, begun by China's first emperor and extended over thousands of years

Created to defend the Chinese empire from various ethnic tribes, including Mongolian nomads, to the north, the Great Wall of China is in fact a series of walls built between the 3rd century BCE and the 17th century CE. Altogether, the walls span 13,170 miles (21,196 km), traversing coastland, deserts, and rugged mountains. The most extensive section is 3,510 miles (5,650 km) long and was built by the Ming dynasty between 1368 and 1644. Around a quarter of the Ming wall consists of natural defenses, such as rivers and mountain ridges, while the rest is made from materials that were available locally, which range from dressed stone and kiln-fired bricks to rammed earth and sun-dried bricks, some held together with sticky rice flour.

Deathly wall

The Great Wall is an average of 21.3 ft (6.5 m) wide at the base and 23–26 ft (7–8 m) high, with some sections reaching 46 ft (14 m). An estimated 1 million of the convicts, peasants, slaves, guards, and citizens employed to build it died in the process, and bones found under parts of the wall suggest that many were buried on site. The wall provided protection from invasion for many centuries, but in 1644, the Manchus broke through and established themselves in Peking (now Beijing), bringing an end to the Ming dynasty's rule. The Great Wall of China was made a UNESCO World Heritage Site in 1987. However, sections in the north are still disappearing as a result of desertification and changing land use.

SPANNING THE CENTURIES

The series of walls that comprise the Great Wall of China include the preimperial walls linked together by the first Qin emperor Shihuangdi (r.220–210 BCE). The Han (202 BCE–9 CE) and Jin (1115–1234 CE) dynasties extended the Wall, and the Ming dynasty (1368–1644) built the brick and stonework Wall.

KEY
Pre-Qin walls
Western Han Great Wall
Ming Great Wall
Qin Great Wall
Jin Great Wall

Tianjin
Beijing

The **wall** functioned partly as a **border**, allowing China **to tax goods** transported along the **Silk Road**

▷ **FRAGILE FORTIFICATION**
Assaulted by the elements, watchtowers and sections of the walls are crumbling away on the unrestored Jiankou section of the Great Wall near Beijing, built by the Ming dynasty in 1368–1644.

Hampi

A treasure trove of sacred and secular architecture, hailed as the world's largest open-air museum

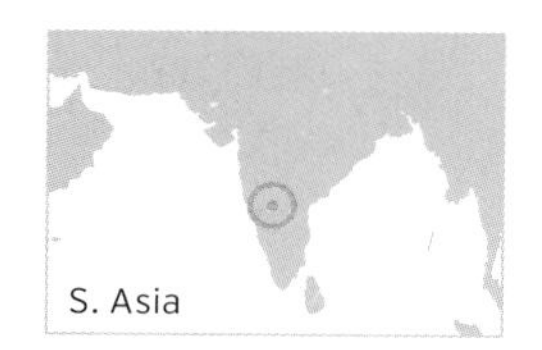

△ COURTESAN'S STREET
This temple street is the main approach to the exquisite Achyutaraya Temple. Lined with intricately carved pillars, the street was one of the liveliest marketplaces of the Vijayanagar empire.

Hampi in Karnataka, India, rose to prominence as the capital of the Hindu Vijayanagar empire (1336–1646). By the early 16th century, it was one of the largest and most prosperous cities in the world. Spanning several centuries and covering around 16 square miles (41 square km), the World Heritage Site includes some 1,600 ruins, from spectacular temples, palaces, and shrines to roads and irrigation systems. Local granite, burnt bricks, and lime mortar were among the primary building materials.

Architectural riches

Hampi's architectural treasures include the pre-Vijayanagar, 7th-century Virupaksha (Shiva) Temple, the tower of which soars to 160ft (49m). The centerpiece of the Vijayanagar period, however, is the 15th-century Vittala Temple complex, with its magnificent sculpture and huge pillars and courtyard. The Vijayanagar architects are renowned not only for intricate carvings and monolithic pillars but also for their massive *gopurams* (gateway towers), which mark the entrances to some of the temples.

△ STONE CARVINGS
Hampi is renowned for the outstanding stone carvings and reliefs that adorn its buildings, often depicting hunting scenes, warriors, and animals.

ROCK CUTTING

Many of Hampi's buildings are made from local granite. To cut slabs of rock for use in buildings, it is believed that a series of holes was made in the surface of a rock. Wooden pegs were then inserted into the holes and saturated with water. The pegs gradually expanded, causing the boulder to crack.

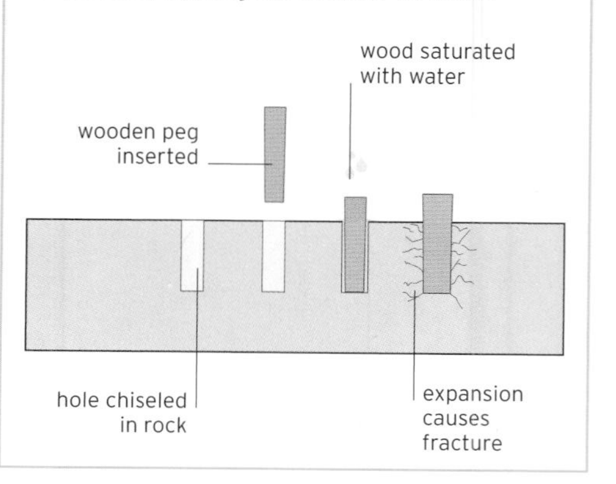

Mehrangarh Fort

A formidable 15th-century fort that towers in splendor above the city of Jodhpur

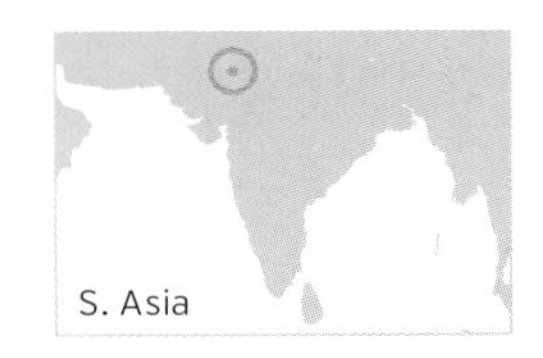

One of India's largest, most spectacular, and best-preserved forts, Mehrangarh ("citadel of the Sun" in Sanskrit), in Jodhpur, Rajasthan, was built by Jodhpur's founder, Rao Jodha (1416–1489). Following his decision to move the capital from nearby Mandore to a more secure location, construction of the fort began in the early 1460s, although it was modified and expanded considerably by successive generations. Situated on top of a red-sandstone cliff some 410 ft (125 m) above the now-sprawling city, the majestic structure continues to dominate the skyline.

Giant defenses

Enclosed by formidable walls 118 ft (36 m) high and 69 ft (21 m) wide, the fortress covers an area of around 860,000 square ft (80,000 square m). It boasts seven gateways and contains several opulent palaces, apartments, and reception rooms constructed or embellished by rulers at different periods. These are notable for, among other things, their intricate stonework, carved balconies, elegant courtyards and arches, and beautiful latticed *jali* screens in pink or yellow sandstone. The *mahal*, or palaces, often comprise a single, highly ornamented room, as in the Moti Mahal ("pearl palace") or the Phool Mahal ("flower palace"). Watchtowers, massive ramparts, and a series of strategically placed cannons along the battlements are also distinctive features of the site.

At the southern end of Mehrangarh Fort is the 15th-century temple dedicated to the Hindu goddess Chamunda Devi, revered by Rao Jodha. It houses an idol of the goddess, which he transported from his former capital at Mandore in 1460.

△ **MOTI MAHAL**
This 16th-century throne room of the Moti Mahal is notable for its vibrant stained-glass windows, alcoves, octagonal throne, and opulent gilt ceiling adorned with mirrors and crushed shells.

One gateway of the fort **bears the imprint** of **cannonballs** fired in battle

▽ **CITADEL OF THE SUN**
Built high on a red-sandstone cliff, the formidable Mehrangarh Fort dominates the landscape, towering above the modern city of Jodhpur and its distinctive blue houses.

Palitana Temples

Hundreds of Jain temples and shrines spread across a mountain site in western India

The Shatrunjaya hills outside the town of Palitana in Gujarat are sacred to the Jain religion, which has millions of followers across India. Especially associated with Rishabhanatha, the first Jain *tirthankar* ("spiritual leader"), the site has become a "city of temples," offering a crowded vista of domes, towers, and spires. Estimates of the number of sacred buildings on the hills vary, but the best count is probably 108 full-size temples and 872 smaller shrines. The first temples were built in the 11th century, but the majority date from the 17th century.

Temples of the sacred hills

The temples and shrines are made of marble, with exquisitely carved pillars. Their classic form is a square, with four doors opening toward the four points of the compass. The buildings house statues of gods, often decorated with gold and precious stones. The Adishwara temple, the largest in the complex, is dedicated to Rishabhanatha. Considered one of the finest religious buildings in India, it is decorated internally with images of dragons and lions.

Today, worshippers performing devotional acts mingle with tourists visiting the temples. All must leave by nightfall, because no one—not even Jain priests or monks—is allowed to spend the night on the hills. Visitors need to have a reasonable level of fitness, as reaching the Adishwara temple involves a climb up steep pathways with more than 3,500 steps.

△ **SCULPTURAL RELIEF**
Stone carvings are one of the major decorative features of the Palitana temples. They illustrate a variety of scenes from Jain mythology.

△ **CROWDED HILLSIDE**
Packed close together on the slope of a hill outside Palitana, the hundreds of ornate Jain temples and shrines were designed to make a dazzling impression on worshippers. The religious monuments stand within a walled enclosure.

△ **PILGRIMS AT MECCA**
An aerial view shows the Great Mosque flooded with Muslim pilgrims during the annual *hajj*. The central courtyard surrounds the Kaaba, the holiest shrine in the Islamic world.

THE KAABA

The Kaaba is a cube standing at the hub of the mosque's courtyard. The Black Stone embedded in the building's eastern corner, of unknown origin, is especially revered. The gold-embroidered silk *Kiswa* draping the exterior walls is changed annually. Pilgrims have to walk around the Kaaba seven times during *hajj*.

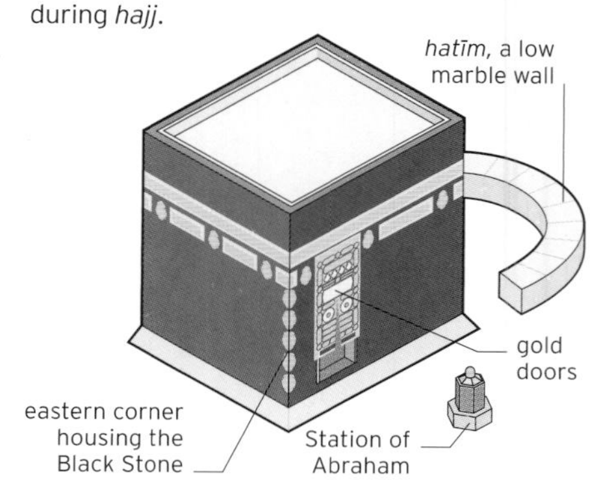

Great Mosque of Mecca

The largest mosque in the world, at Mecca in Saudi Arabia, and the focus of the annual hajj *pilgrimage performed by Muslims from all parts of the globe*

The site of the Great Mosque, also known as the Haram Mosque, was sacred ground long before the founding of Islam. Qur'anic tradition attributes the creation of the Kaaba—the granite cuboid structure around which the mosque is built—to the Biblical figure Abraham. It was a focal point of polytheistic worship for Arab tribes until 630 CE, when the Prophet Muhammad adopted it for the exclusive use of his monotheistic faith. The Kaaba became the holiest site in Islam, toward which all Muslims face when praying, and the object of the *hajj*—a pilgrimage that every Muslim is urged to make at least once in a lifetime. The mosque's courtyard houses not only the Kaaba but also the Station of Abraham—a stone bearing the imprint of Abraham's foot—and the sacred spring known as the Zamzam Well.

Expansion and modernization

The first mosque had been built around the Kaaba by the 8th century, but none of the original building remains. The oldest extant parts of the mosque's structure date from 1571, when it was rebuilt by the Ottoman Turkish architect Sinan. Most of the current building is far more recent, constructed in the 20th or 21st centuries. The need to cope with the ever larger number of pilgrims has led to massive expansion and modernization of the mosque since the 1950s. It now covers an area of some 4.3 million square ft (400,000 square m) and has escalators, pedestrian tunnels, and air conditioning. The mosque can now hold more than 800,000 pilgrims at a time. Admittance to the city of Mecca and its mosque is strictly forbidden to non-Muslims.

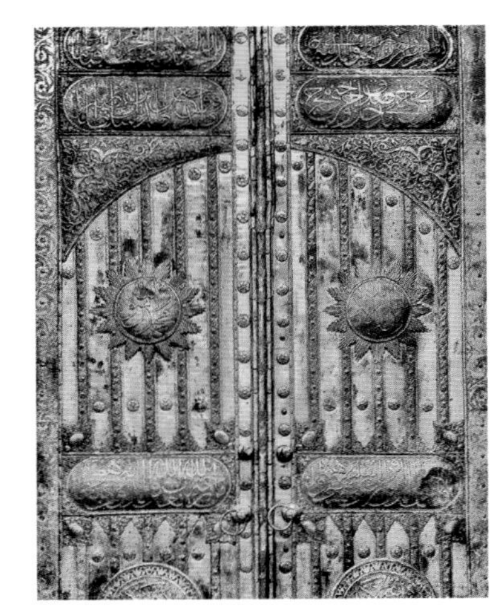

△ GOLD DOORS
The current gold entrance doors of the Kaaba were made in 1979. They are opened twice a year for the ceremonial "cleaning of the Kaaba."

Almost **3 million Muslims** make the pilgrimage to Mecca **every year**

Architectural styles

LATER ISLAMIC

Even as it has assimilated other styles and traditions, Islamic architecture remains closely tied to a distinctive vocabulary of forms and decoration informed by the traditions and principles of Islam.

As Islam spread from the Middle East throughout the world, new variations of Islamic architecture emerged, influenced by local traditions. One of the most distinctive was the Mughal empire's architecture. At its height during the 17th and 18th centuries, the Mughal empire controlled much of the Indian subcontinent. Architecturally, it drew from Persian, Indian, and Islamic traditions and is marked out by its use of large onion domes, impressive gateways, slender towers, and delicate ornamentation. The style was arguably at its apex under Shah Jahan, who built the Jama Masjid mosque in Delhi; the Shalimar Gardens in Lahore; and Mughal architecture's most famous building, the Taj Mahal.

Today, Islamic architecture continues to incorporate new forms and technologies while remaining closely informed by the style's rich traditions, heritage, and underlying basis in the Islamic faith.

△ **ARCHED PORTAL**
Arched portals—essentially large gateways or large arched openings—are a common feature of Mughal architecture. As in this example from the Taj Mahal, they often incorporate a variety of decorative schemes rendered in plaster, inlaid stone, or paint.

▷ **PERFORATED SCREEN**
Perforated screens of varying degrees of sophistication are a recurring feature of Islamic architecture. Often highly ornate with complex patterns or with motifs derived from plants, they also have a practical function in helping to keep interior spaces cool.

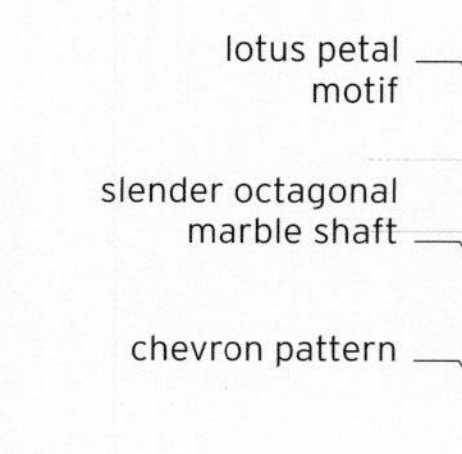

▷ **PINNACLES**
The iconic dome of the Taj Mahal is visually supported by a number of pinnacles that extend upward from the perimeter walls. The chevron pattern of the octagonal shaft is topped by a lotus bud form and a golden finial.

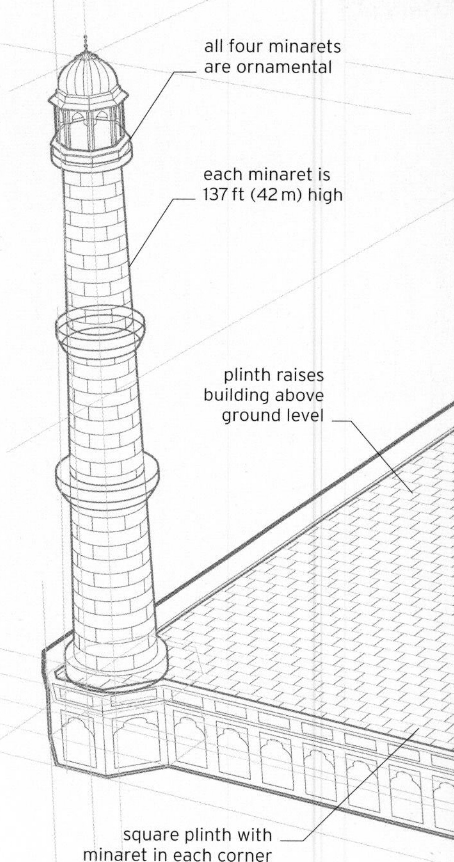

▲ **TAJ MAHAL**
The Taj Mahal (see pp.290–291) was built in the mid-17th century by Shah Jahan as a tomb for his favorite wife, Mumtaz Mahal. Constructed from white marble and richly ornamented, the tomb is at the center of a complex also containing a mosque and long reflecting pool.

ARCHITECTURAL EMBELLISHMENTS

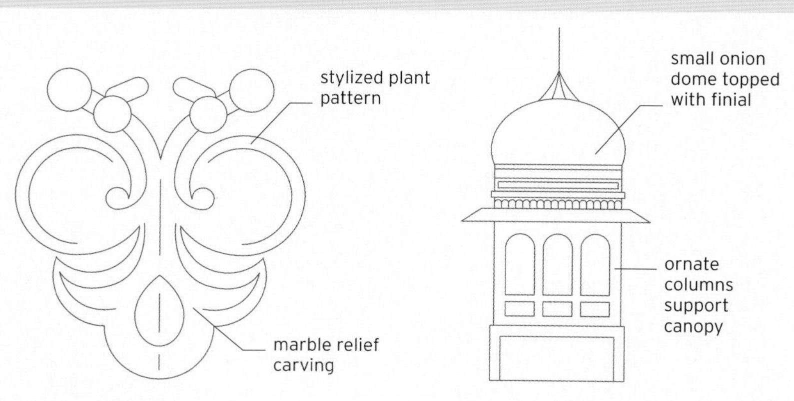

△ **PLANT-FORM RELIEF**
Plant forms are a frequent source of inspiration for Islamic craftsmen, with plant motifs of varying degrees of complexity appearing in Islamic architecture of all periods.

△ **CHATRIS**
A chatri is a pavilion-type structure consisting of a small dome resting on four or more columns. They are used as independent structures and on the corners of roofs.

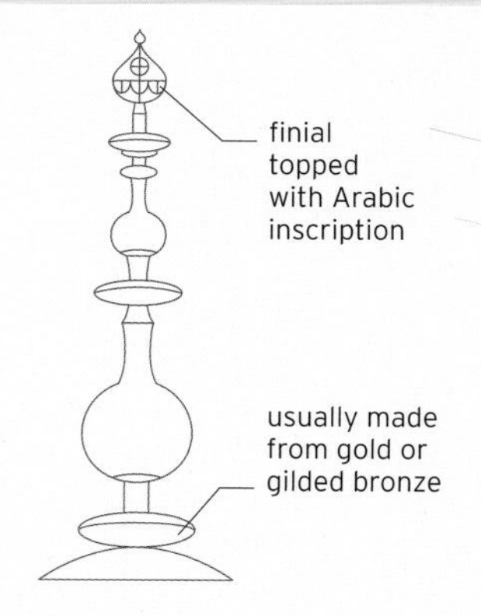

△ **FINIAL**
An architectural device that caps many minarets and domes, Islamic finials incorporate many decorative elements. Common motifs include a crescent emblem and a tulip shape.

It is estimated that the **construction** of the Taj Mahal involved the work of over **20,000 craftsmen and laborers**

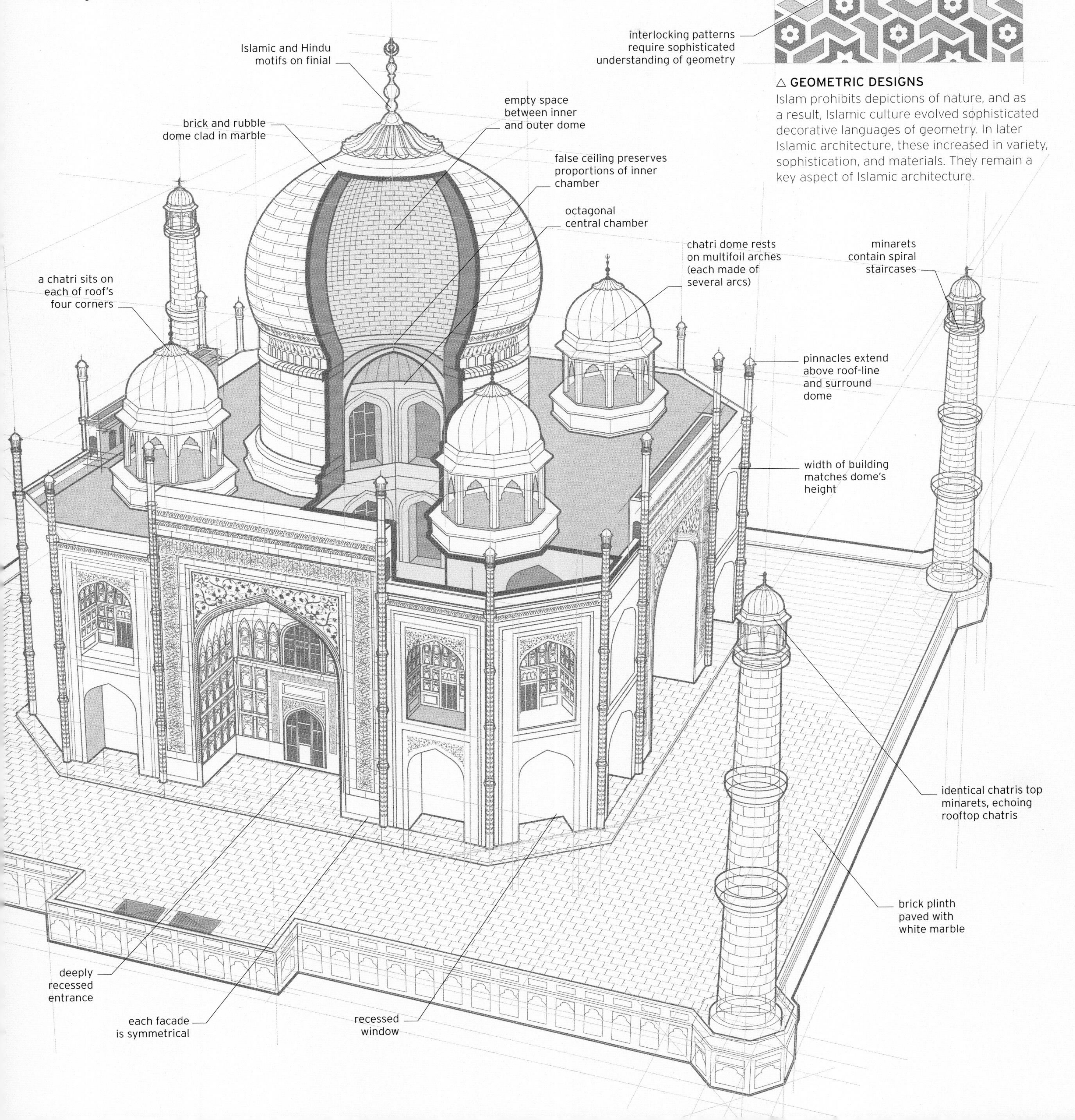

△ **GEOMETRIC DESIGNS**
Islam prohibits depictions of nature, and as a result, Islamic culture evolved sophisticated decorative languages of geometry. In later Islamic architecture, these increased in variety, sophistication, and materials. They remain a key aspect of Islamic architecture.

Registan Square

A monumental public square in the Uzbek city of Samarkand on the historic Silk Road

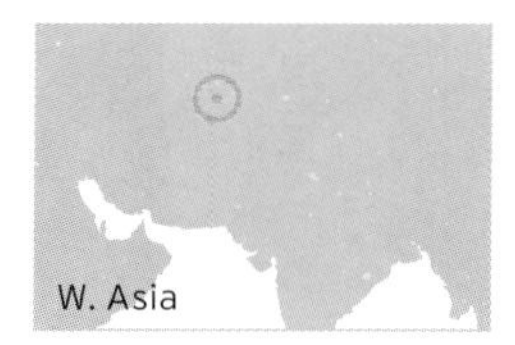

Five centuries ago, Registan Square was the center of commercial life in one of the world's wealthiest cities, Samarkand. The city had grown rich on account of its location on the Silk Road trade route between China and the West. In the 14th century, it had also become the capital of the empire created by the warrior Timur the Lame. It was Timur's grandson Ulugh Beg who, between 1417 and 1420, built the first of the three *madrasas*—Muslim schools—that now dominate the square. The Ulugh Beg *madrasa* developed into one of the most important centers of learning in the Islamic world. The other two were built in the 17th century, completing an architectural ensemble of breathtaking splendor.

A square of *madrasas*

The Ulugh Beg *madrasa,* and the Sher-Dor *madrasa* built to mirror it on the opposite side of the square, have vast portals flanked by free-standing minarets. The Tilya-Kori *madrasa* has a lower facade with minarets attached. Behind each of the portals were lecture halls, student dormitories, and a mosque. Samarkand is now one of the major tourist attractions of Uzbekistan.

△ TILYA-KORI *MADRASA*
The two minarets of Tilya-Kori *madrasa* are capped with turquoise glazed domes, matching the main dome that dominates the building. Completed in 1660, Tilya-Kori was the last of the three *madrasas* built on Registan Square.

△ GLAZED SPLENDOR
The facade of the 17th-century Sher-Dor *madrasa* is magnificently decorated with glazed tiles and bricks. Its mosaics of tigers pursuing deer breach the Muslim custom prohibiting representation of living creatures.

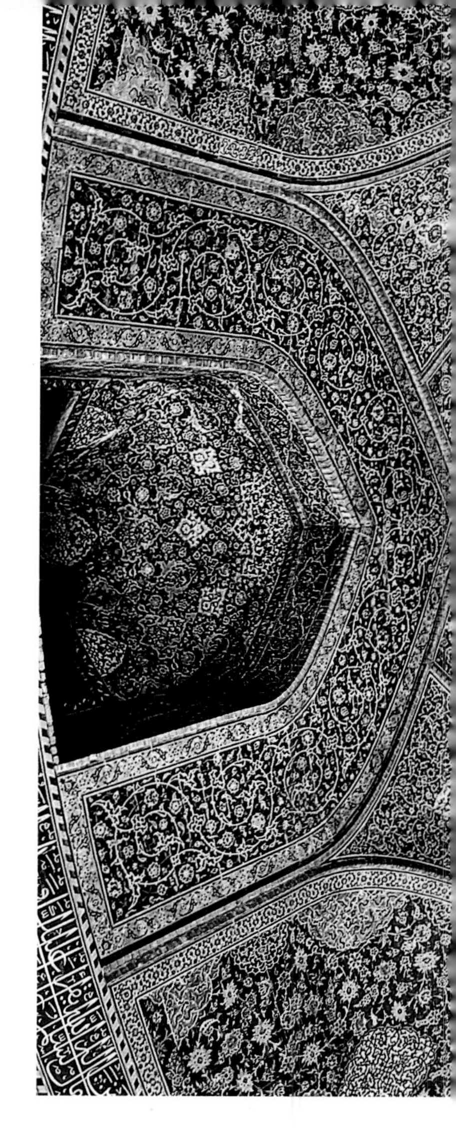

△ FLORAL CEILING
The interior of the Shah Mosque is lined with mosaics of hand-painted seven-color tiles. The technique of coloring tiles in several colors was perfected in Safavid Iran.

△ ELABORATE DECORATION
"Stalactite" tilework is a distinctive feature of Iranian Islamic architecture. This fine example embellishes the main entrance to the Shah Mosque.

Shah Mosque

A masterpiece of Islamic architecture reflecting the splendor of Safavid Iran

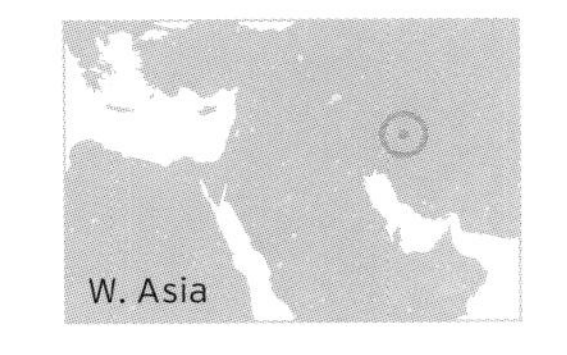

In 1598, Shah Abbas I, the greatest ruler of Iran's Shi'ite Safavid dynasty (1501–1722), decided to take Isfahan as his new capital and had the city rebuilt to reflect the secular and spiritual power of his expanding empire. The principal result was the creation of the magnificent Naqsh-e Jahan Square, bordered by the Ali Qapu royal palace; the Sheikh Lotfollah Mosque; and, above all, the immense Shah Mosque (now known in Iran as the Imam Mosque).

Color and proportion

Work on the Shah Mosque began in 1611 and was completed some 20 years later. Attributed to architect Ali Akbar Isfahani, it is a vast structure incorporating two *madrasas* and a separate winter mosque. The whole building is sumptuously decorated with polychrome tiles, with turquoise and dark blue being predominant. The focus of the mosque is the inner courtyard, built around a pool and dominated by four *iwans*, vaulted spaces open on one side. The walls of the *iwans* are embellished with tiles bearing floral motifs, and the ceilings are fine examples of the "stalactite" style known as *muqarnas*. The mosque is topped by a double-shelled dome, the outer shell rising to a height of 174 ft (53 m) and the inner shell reaching 125 ft (38 m) above the mosque floor. The exterior surface of the dome, covered in brilliant turquoise tiles, reflects sunlight to dazzling effect.

The Shah Mosque is very much a living center of worship for Iranian Muslims today, but non-Muslim visitors are also warmly welcomed. A major tourist attraction, the entirety of Naqsh-e Jahan Square, including the mosque, was designated a UNESCO World Heritage Site in 1979.

Some **18 million bricks** and **475,000 tiles** were used in the **construction** of the **Shah Mosque**

COMPLEX LAYOUT

The Shah Mosque is oriented toward Mecca, to the southwest of Isfahan, which places it at an angle to Naqsh-e Jahan Square, which is built on a north-south axis. Thus, the entrance from the colonnaded facade on the square twists through 45 degrees to the *iwan* leading into the central courtyard. The domed main prayer hall is flanked by two *madrasas*.

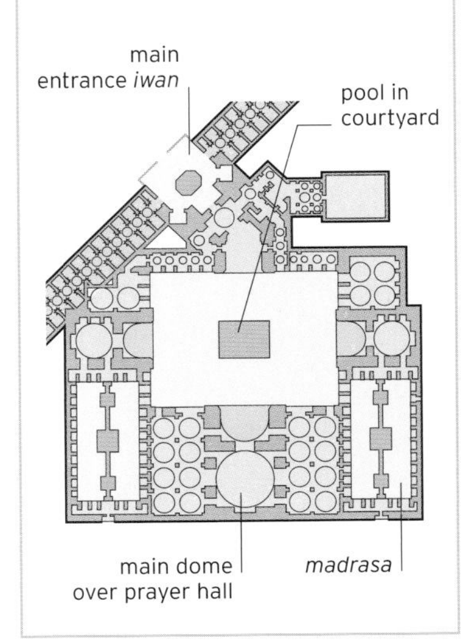

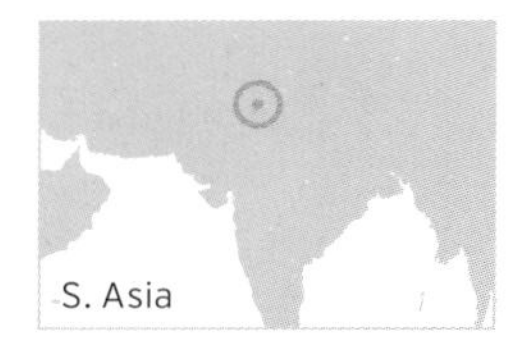

Golden Temple of Amritsar

The chief house of worship for Sikhs, lavishly decorated with gold and surrounded by a shimmering pool of "holy nectar"

The Golden Temple (or *Harmandir Sahib,* meaning "temple of god") is located in northern India's Punjab, where the Sikh religion originated. The construction work was carried out by Guru Arjan, the fifth Sikh guru, and the *Gurdwara* (the Sikh name for a place of assembly and worship) was completed by 1604.

Extravagance and humility

Nothing remains of the early versions of the temple because the shrine was repeatedly attacked by hostile Muslim armies. The present building dates mainly from the 18th century, when it was rebuilt. However, its most sumptuous adornments were added in the following century by Maharaja Ranjit Singh (r.1792–1804), who covered the entire roof area in gold foil, giving the temple its "Golden" nickname. The most striking feature of the interior is the ornate, jeweled canopy in the main hall, under which the *Guru Granth Sahib*—the Sikhs' sacred text—is placed on display.

In keeping with the Sikhs' sense of humility, the temple authorities claim to run the largest *langar* (community kitchen) in the world, offering free food to up to 100,000 people of all faiths every day.

◁ **COVERED IN GOLD**
This is one of the several glittering turrets that adorn the facade of the shrine. Made of pure gold, their harmonious design echoes the shape of the temple's dome.

The **roof** of the temple was **gilded** with **1,650 lb (750 kg)** of **gold**

△ **REMEMBERING THE FOUNDER**
This painting within the temple shows Guru Nanak (center right), the founder of the Sikh religion. His teachings are contained in the sacred text, the *Guru Granth Sahib*.

▽ **OPEN TO ALL**
Crowds of pilgrims and sightseers move slowly along the long marble causeway that leads to the entrance of the temple, which welcomes many thousands of visitors every day.

THE TEMPLE COMPLEX

The complex contains several structures. The temple itself is surrounded by the *Amrit Sarovar* ("pool of holy nectar"), the water in which is said to have curative powers. Other buildings include the Clock Tower, which serves as one of the main entrances; the *langar* (the community kitchen); the *Akal Takht*, where the Sikhs' official committee is based; the Sikh library; and a number of smaller shrines.

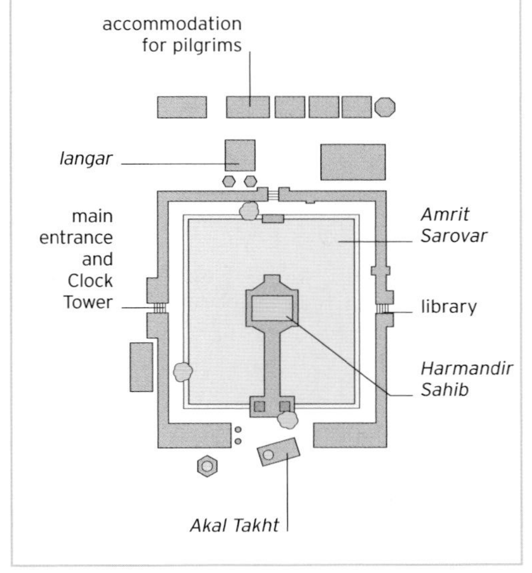

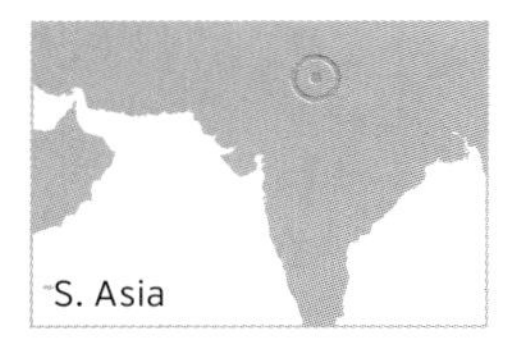

Taj Mahal

An icon of India, a symbol of love and a monument to the wealth and power of the Mughal Empire

The Taj Mahal in Agra was built in the 17th century by Mughal emperor Shah Jahan (1592–1666). The white marble edifice was to be a mausoleum for his favorite wife, Mumtaz Mahal, who died in childbirth in 1631. The grief-stricken emperor ordered work to begin in the same year, but it took 22 years to complete, employing more than 20,000 workers—including masons and craftsmen from Persia, the Ottoman Empire, and Europe—and over 1,000 elephants.

Lavish decoration

The complex of buildings sits within a walled rectangle. At the south end are the entry gates, forecourt, and former bazaar; in the middle are the gardens; and at the north is the mausoleum itself, which stands on a 22-ft- (6.5-m-) high platform, with a 137-ft- (42-m-) high minaret at each corner. It is flanked to the east by a mosque and to the west by the *mehmankhana* (guest house), both built in red sandstone. The mausoleum was built from milky marble from Makrana in Rajasthan and appears to change color from pink to white to gold depending on the time of day. It is exquisitely carved and decorated with 28 different types of gems, such as turquoise from Tibet and jade from China.

△ **MINARET FINIAL** The four minarets tilt slightly outward to avoid damaging the mausoleum should they collapse.

Famed Indian poet **Rabindranath Tagore** called the **Taj Majal** "a **tear of marble** ... on the cheek of time"

SYMMETRICAL DESIGN

The Taj Mahal complex is a work of perfect symmetry, broken only by the tombs of Mumtaz Mahal and Shah Jahan. The mausoleum is 187 ft (57 m) high–from the platform to the top of the dome–and is just as wide. Its layout is based on the *Hasht-bihisht* (eight paradises) plan often used by the Mughals. The four *iwans* (ceremonial halls) lead through decorated rooms to four octagonal chambers and a central, domed chamber.

KEY
- Iwan
- Octagonal chamber
- Central chamber

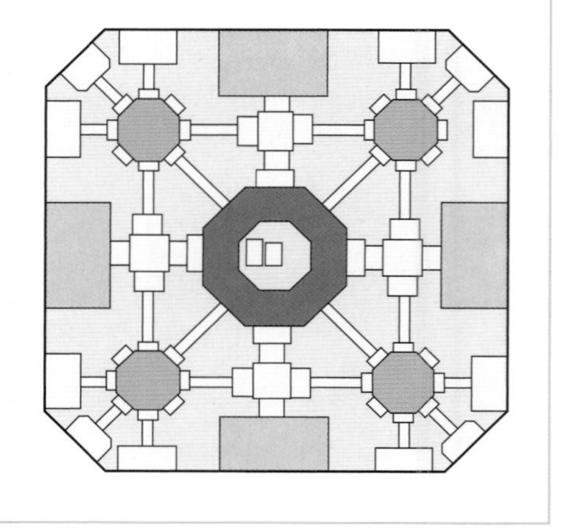

△ **THE MAUSOLEUM**
Built on the banks of the Yamuna River and set on its raised platform, the Taj Mahal's central mausoleum is strikingly regular and symmetrical in its proportions and the arrangement of its arches and domes.

△ **INTRICATE INLAYS**
The Taj Mahal is decorated with gemstone inlays that feature the plant forms, arabesques, and geometric patterns typical of Islamic art.

◁ **LOVERS' TOMBS**
The mausoleum's central chamber contains the cenotaphs of Shah Jahan and Mumtaz; their bodies lie in a lower level.

Toshogu Shrine

A beautiful 17th-century Shinto shrine dedicated to the first shogun of Japan's Edo Period

One of over 150,000 Shinto shrines in Japan, the Toshogu Shrine was established in 1617. It was dedicated to Tokugawa Ieyasu, the founder of the Tokugawa shogunate, which ruled Japan from 1600 to 1868—an unprecedented era of peace known as the Edo Period.

The shrine was enlarged to its current size in 1636, when 55 exquisitely carved buildings were completed in just 1 year and 5 months by over 400,000 carpenters. Its expansion cost about $350 million (¥40 billion) in today's money. A series of paths and stairways wind their way through the forest and the shrine's buildings to the bronze urn containing Ieyasu's remains.

Tradition in design

Intricately carved and richly painted and lacquered, the buildings and statues at the shrine provide a kind of visual dictionary of Japanese philosophy and spirituality that features, among other things, the three wise monkeys, lions, peacocks, cranes, and the five elements—earth, water, fire, wind, and ether.

The Toshogu Shrine has been well maintained and regularly renovated from the beginning. Most of the shrine's buildings were included in the Shrines and Temples of Nikko as a UNESCO World Heritage Site in 1999; five have been designated National Treasures of Japan and others are among Japan's Important Cultural Properties.

THE ISHIDORII GATE

A *torii* ("bird abode") is a traditional Japanese gate usually found at the entrance to a Shinto shrine. It marks the move from mundane to sacred. The Ishidorii Gate, the first *torii* at the Toshogu shrine, is a stone Myojin-type *torii*, characterized by its curved lintels, as shown in this typical example.

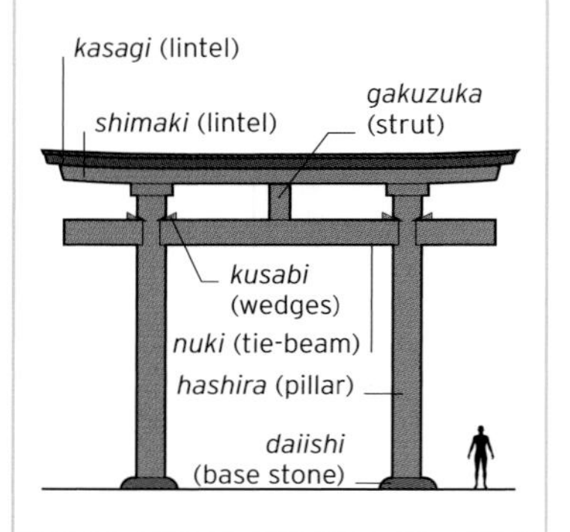

▽ **INTRICATE ENTRANCE**
The Karamon Gate is one of several elaborate portals at the shrine; its facade includes carvings of Kyoyu and Soho, legendary Chinese sages.

EARTHQUAKE RESISTANCE

The palace's construction helps it withstand earthquakes, which are frequent in the Himalayas. Its outer stone walls taper from an average of 16 ft (5 m) thick at the base to 10 ft (3 m) at the top; this lowers the wall's center of gravity, making it more stable. The walls and internal timber frame create a structure that is both rigid and flexible enough to withstand vibration.

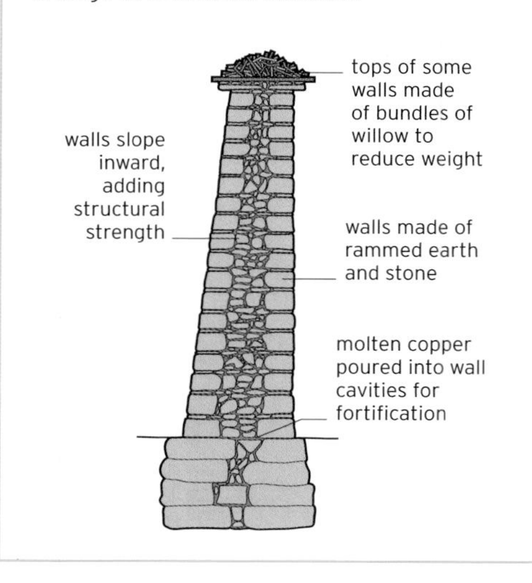

Potala Palace

The world's highest palace, home to Tibet's Dalai Lamas for over 300 years, and a symbol of Tibetan unity

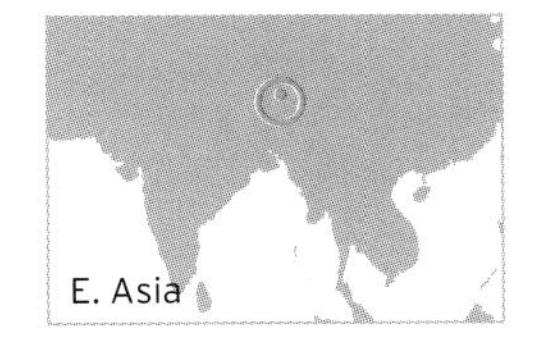

△ **FORMIDABLE PALACE**
The Potala Palace combines traditional Tibetan architecture with Indian and Chinese aesthetics. Covering 3,875,000 square ft (360,000 square m), it rises 660 ft (200 m) against a hillside.

A complex of buildings made up of the White and Red palaces and surrounding structures, the Potala Palace is situated 12,100 ft (3,700 m) above sea level on Red Mountain in Lhasa. The White Palace was built by the 5th Dalai Lama in 1645–1648 and served as his living quarters, home to over 100 monks, and the ceremonial, political, and administrative center of Tibet. The Red Palace, added in 1690–1693, was reserved for religious observances, meditation, and prayers. It also contained the gold-plated tomb of the 5th Dalai Lama. The uppermost terraces housed the palace's golden-roofed temples.

The palace grew to include more than 1,000 rooms arranged over 13 stories; 10,000 shrines; and 200,000 statues. Two other notable sites lie close to the palace: Norbulingka, the Dalai Lama's former summer palace, built in the 18th century; and the 7th-century Jokhang Temple Monastery.

The palace has been **home** to **10 Dalai Lamas** over **317 years**

▽ **HOLY TOMBS**
Gilded stupas containing Buddhist relics line the route to the Potala Palace, whose golden roofs are just visible above its monumental facade.

Architectural styles

JAPANESE

Japanese architecture is characterized by the consistency of its exteriors—timber construction supporting large gabled roofs—and the flexibility of its interior spaces, which are separated by moveable screens, allowing spaces to change depending on needs.

typically ceramic in composition

water to extinguish fire believed to be released through mouth

▷ SHACHIHOKO
A *shachihoko* is a mythical Japanese creature fusing the head of a tiger and the tail of a fish. *Shachihoko* ornaments are often found on roofs because they were said to guard against fire.

From 1633 until 1853, the policy of *sakoku* made it illegal for any Japanese person to leave the country and any foreigner to enter. This ensured that up until this point, Japanese architecture—as well as broader culture and society—existed in seclusion, developing in its own way according to its own recurring ideas and underlying principles. Shinto shrines and Buddhist temples became its most recognizable building types.

The result of this centuries-long isolation was the striking consistency of Japanese architecture in both design and materials. Stone is rarely used—only significantly in the bases of some temples and castles—with wood forming the starting point of a post-and-lintel construction method. Walls are not load-bearing and are usually thin, sometimes paper thin. The large gabled roofs are the main focus for architectural display, with ornamentation usually issuing from the structural elements.

With the opening up of Japan to international trade at the end of the 19th century, the native tradition began to be supplemented by Western styles. In the 20th century, Japan embraced Modernism, particularly after World War II, with the internationally significant Metabolist movement, which fused high technology with traditional Japanese architectural principles.

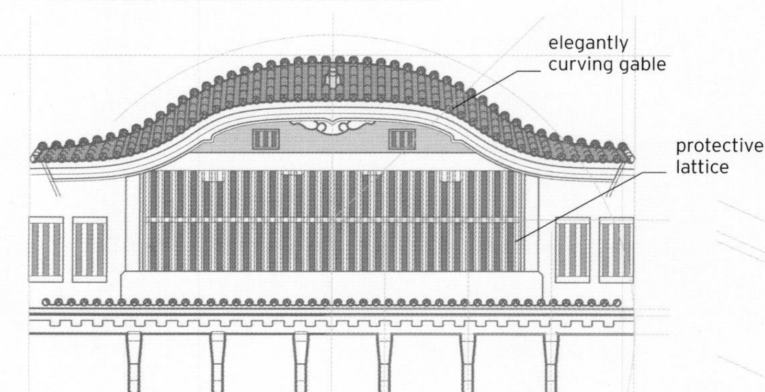

△ LATTICE WINDOWS
Lattice windows appear throughout Japanese architecture, usually in the form of internal *shoji* screens. At Himeji Castle, the latticework is much tighter, reflecting its exposed position and role as a defensive building.

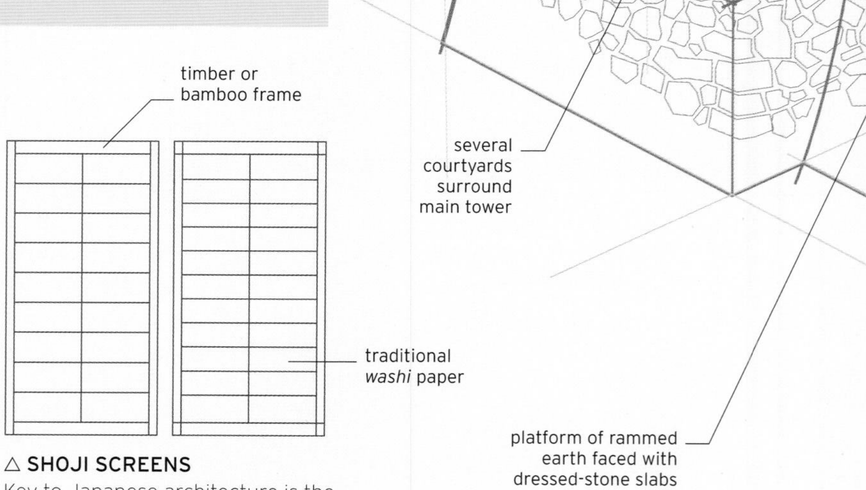

▶ HIMEJI CASTLE
Himeji Castle (see p.272) is regarded as one of the finest surviving Japanese castles. It comprises a stone base–a rare example of the use of this material in Japanese architecture–and a superstructure of timber tiered roofs.

TIMBER AND PAPER STRUCTURES

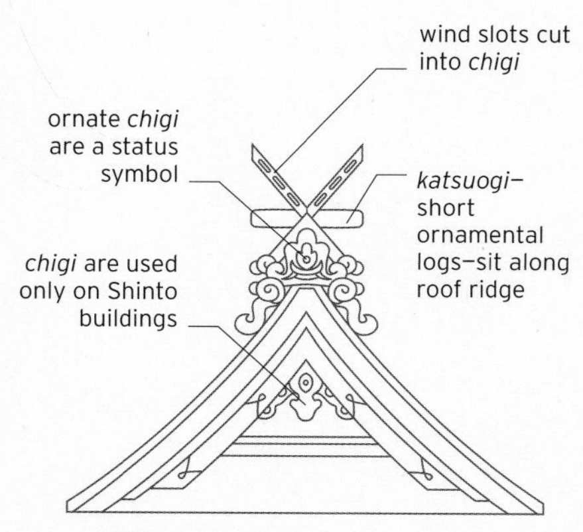

△ CHIGI FINIALS
Chigi are finials of two angled planks of timber projecting from the apex of gable ends of Shinto temples. Some are integrated into the roof structure, while others stand separate.

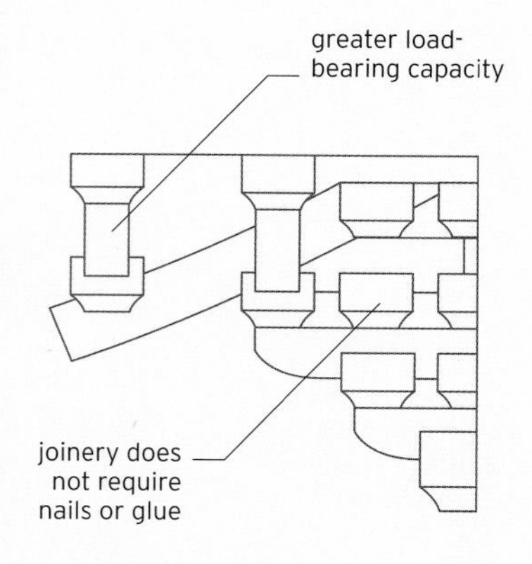

△ TIMBER JOINTS
Japanese craftsmen developed a distinctive language of woodworking joints with different traditions of carpentry focused on temples, domestic architecture, and interiors.

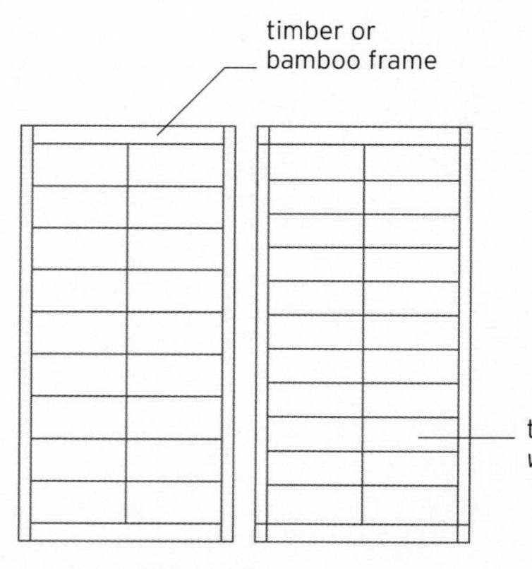

△ SHOJI SCREENS
Key to Japanese architecture is the division of internal spaces with sliding screens. Usually comprising a timber frame and paper, they allow rooms to be opened and closed.

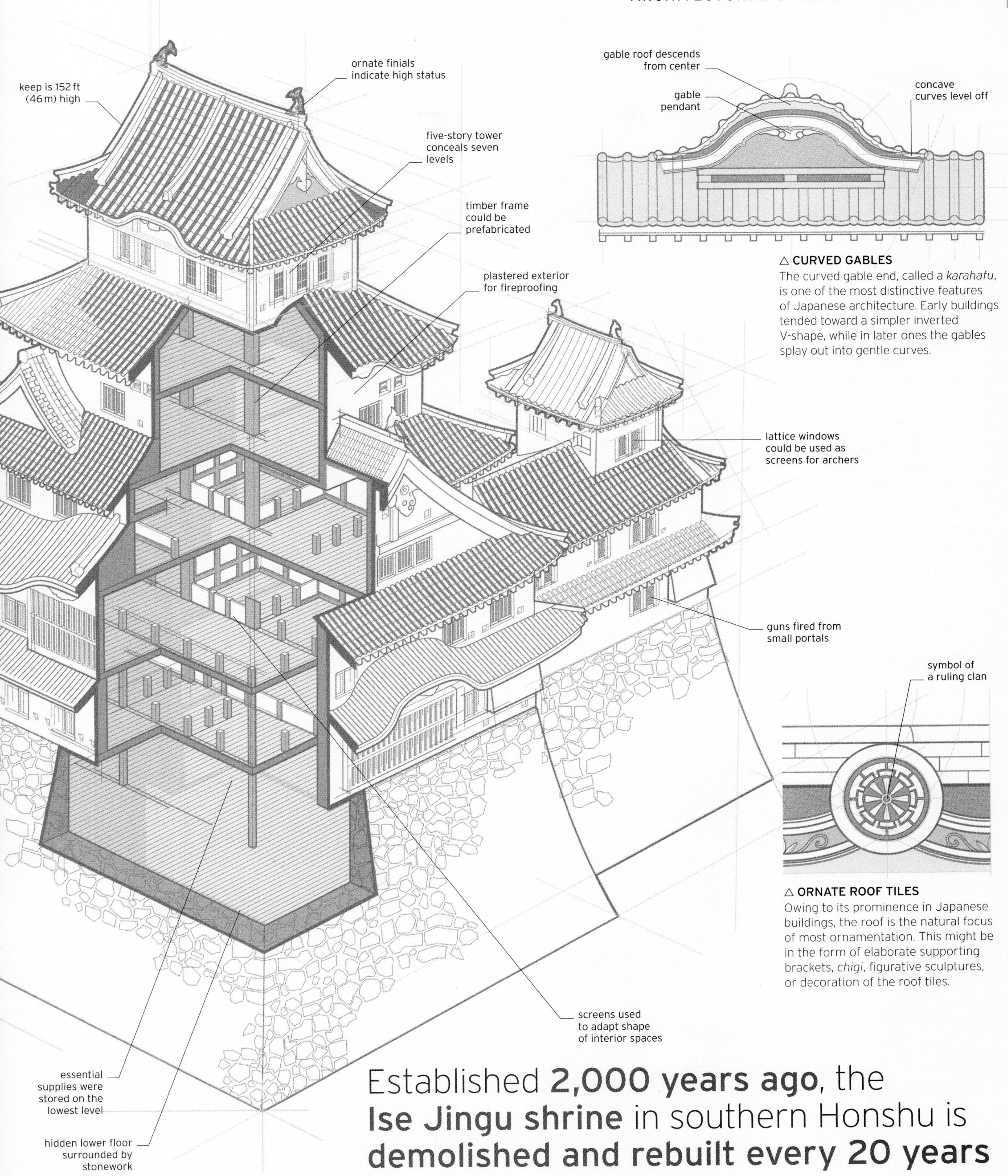

△ **CURVED GABLES**
The curved gable end, called a *karahafu*, is one of the most distinctive features of Japanese architecture. Early buildings tended toward a simpler inverted V-shape, while in later ones the gables splay out into gentle curves.

△ **ORNATE ROOF TILES**
Owing to its prominence in Japanese buildings, the roof is the natural focus of most ornamentation. This might be in the form of elaborate supporting brackets, *chigi*, figurative sculptures, or decoration of the roof tiles.

Established **2,000 years ago**, the **Ise Jingu shrine** in southern Honshu is **demolished and rebuilt every 20 years**

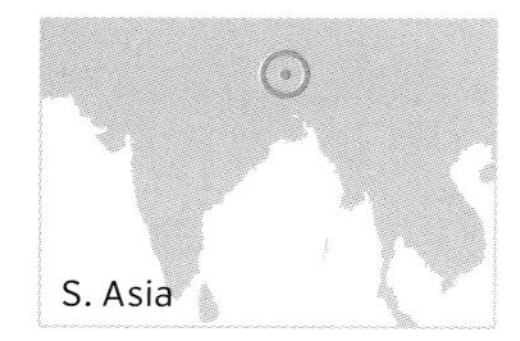

Paro Taktsang

Sacred Buddhist temples built on a precipitous mountainside in the Himalayas

Located in the kingdom of Bhutan in the eastern Himalayas, the Buddhist monastery of Paro Taktsang, widely known as the Tiger's Nest, clings to a cliffside 3,000 ft (900 m) above the Paro valley. The site is associated with Guru Rinpoche, also known as Padmasambhava, the revered master who is credited with introducing Tantric Buddhism to the area in the 8th century CE. Legend asserts that Rinpoche was carried to Taktsang from Tibet on the back of a tiger—hence the name Tiger's Nest. He meditated in a cave on the mountainside and, over the following centuries, many other Buddhist holy men imitated his example.

A home for the sacred

The monastery on the site was built in 1692 by Bhutan's ruler Gyalse Tenzin Rabgye. There are four temples and eight sacred caves, linked by vertiginous stepped paths and wooden bridges. The temples are in traditional Buddhist style, with plain white exterior walls and golden roofs. Some of the caves house religious statues and sacred scriptures. The monastery is still occupied by Buddhist monks today but it has also become a tourist destination. Paro Taktsang is the site of an annual Tsechu festival at which masked dancers celebrate Guru Rinpoche.

△ **PRECARIOUS PERCH**
The precarious location of Paro Taktsang makes the temples inaccessible for emergency assistance, and it burned down completely in a fire in 1998. It was restored to the form seen here in 2005.

△ **GOLDEN SPIRES**
The Temple of the Emerald Buddha and its surrounding buildings are a spectacular sight with the brightly colored tiles of their multilayered roofs and their glistening golden towers.

WAT PHRA KAEW

The walls of the Grand Palace enclose a total area of 2.35 million square ft (218,400 square m), within which the Temple of the Emerald Buddha (*Wat Phra Kaew*) stands in its own walled complex. The statue of the Emerald Buddha—carved from a single block of jade rather than emerald and dressed in gold clothing—is placed within the ordination hall, which is the main building in the complex.

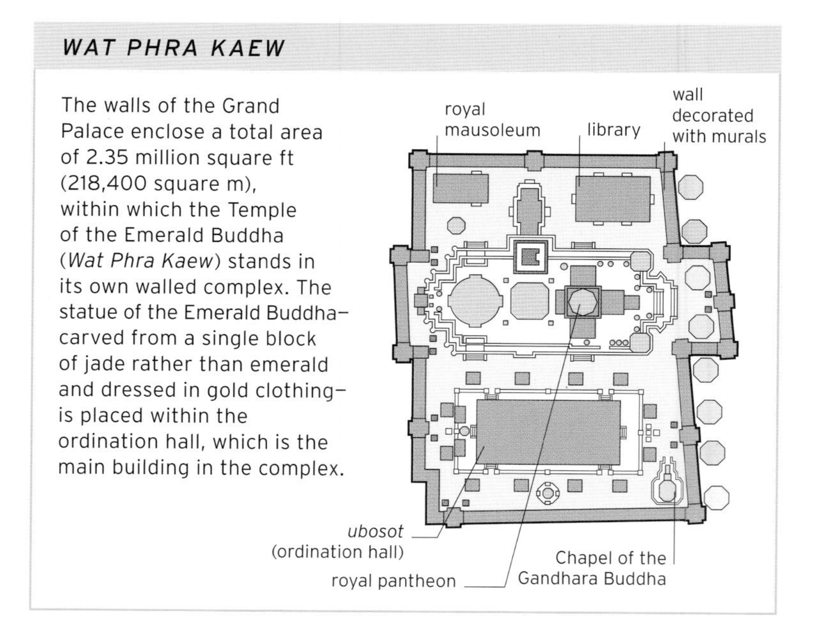

The Grand Palace

An historic Thai royal palace in Bangkok that houses the revered Emerald Buddha

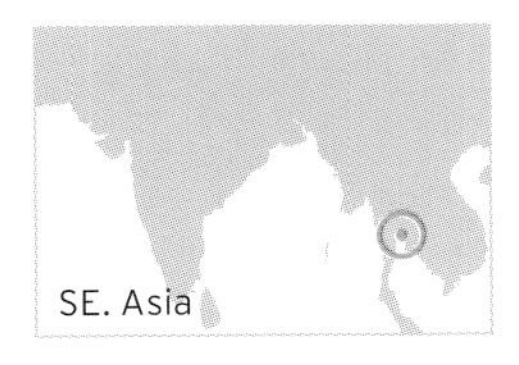

The Grand Palace in Bangkok is a complex of more than a hundred buildings. It was founded in 1782, when the first king of Thailand's reigning Chakri dynasty, Rama I, established his capital on the east side of the Chao Phraya River. A royal residence, an administrative center, and a site to hold ceremonies, the new palace became a "city within a city," teeming with thousands of officials, soldiers, servants, and concubines. Built over 200 years, the palace's state apartments, halls, and temples exhibit an eclectic mix of traditional Thai architecture and later foreign-influenced styles.

Temple of the Emerald Buddha

Among the many brightly colored, ornate palace buildings, the most famous is *Wat Phra Kaew,* also known as the Temple of the Emerald Buddha. Thailand's most sacred temple, it is named for the small green statue it houses, revered as the protector of the country's safety and prosperity.

The Grand Palace was occupied until 1932, by which time the royal family and government offices had dispersed to other locations. It is now chiefly a tourist attraction.

Only the **King of Thailand** is **allowed** to **touch** the **Emerald Buddha** in the **palace temple**

△ **GUARDIANS OF THE PALACE**
Statues of Yaksha, mythical giant warriors, guard the walls and doors of the Temple of the Emerald Buddha and other buildings in the Grand Palace.

Pink Mosque

A 19th-century mosque that, through its use of light and color, is among the most stunning examples of Islamic art in the world

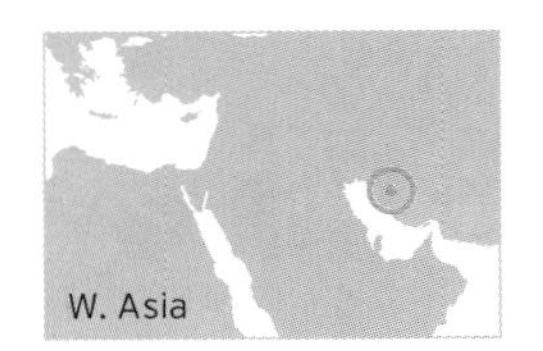

The Nasir al-Mulk Mosque is also known as the Pink Mosque, the Mosque of Colors, and the Rainbow Mosque. It was built in Shiraz, Iran, on the orders of Mirza Hasan 'Ali Nasir al-Mulk, a member of the Qajar dynasty, which ruled Persia (Iran) from 1789 to 1925. Begun in 1876, the mosque was completed by 1888. Its layout is traditional for Islamic, and specifically Persian, architecture—two prayer halls face one another across a courtyard and are joined by a vaulted hall, known as an *iwan*, topped by two low minarets. The courtyard has a rectangular pool, which both cools the air and symbolizes ritual purification.

Inner beauty

The mosque's external decoration of intricately patterned pink, yellow, and blue ceramic tiles gives only a hint of the beauty of its interior. Determined to create a place of worship that reflected the relationship between Heaven and Earth, the mosque's two designers—Mohammad Hasan-e-Memar and Mohammad Reza Kashi-Saz-e-Sirazi—turned the interior into a space filled with jewel-like light and color. As the morning Sun hits the stained-glass facade of the winter prayer hall, a kaleidoscope of color spills onto its inner walls, which are covered from floor to ceiling in tiled, stuccoed, and painted designs. The three-dimensional and highly decorated *muqarnas* (see panel, below) and *panj kaseh-l* (the "five concave shapes") that adorn the ceiling domes create patterns of light and shade that add to the mosque's vibrant beauty. This ebullient use of color is balanced by the simplicity of the building itself and the expert use of symmetry, repetitions, and geometry, typical of Islamic architecture, so that the mosque never feels overwhelming but remains a joyful and peaceful place.

The extensive use of **glazed, rose-colored tiles** gave the **mosque its name**

△ ISLAMIC SYMMETRY
The ritual pool in the courtyard of the Pink Mosque leads to the *iwan* with clearly visible *muqarnas*. The *iwan* links the prayer halls to the left and right.

◁ KALEIDOSCOPIC INTERIORS
Light floods through the mosque's stained-glass windows, casting colorful patterns on the woven carpets and the carved, tiled interior of the prayer hall.

MUQARNAS

The Pink Mosque features many examples of *muqarnas*, three-dimensional tiers of carved and molded stucco cells, pendants, and brackets. *Muqarnas* are used to decorate vaults, cornices, and other parts of buildings throughout the Islamic world. Seen from below, light plays over the smooth, sculpted, but regularly composed surface to dazzling effect.

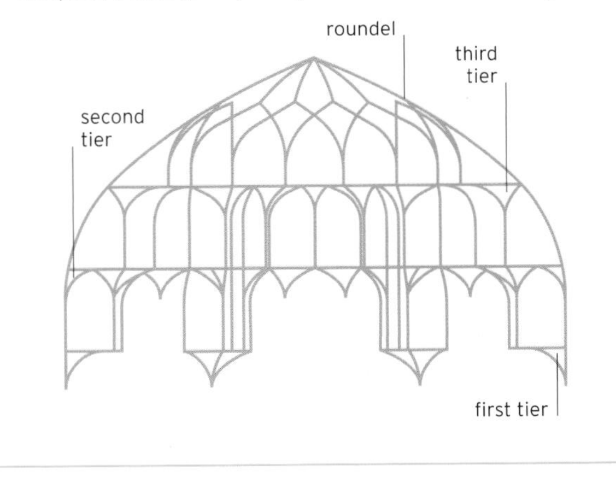

Royal Exhibition Building

A palace of culture, which signaled the arrival of Melbourne as a vital metropolis

As the self-governing Australian states of Victoria and New South Wales grew richer after the gold booms of the 1850s, rivalry between their respective capitals of Melbourne and Sydney intensified. Both wished to organize exhibitions modeled on the international exhibitions of Europe, a craze begun with the Great Exhibition in London in 1851. Sydney won the race and opened its fair in October 1879, but concentrated mainly on agriculture. Melbourne therefore decided to launch its more international-looking fair after Sydney's had ended, and the building opened on May 29, 1880.

△ **THE GREAT DOME**
The dome at the center of the building was built from cast iron and rendered stone. It rises 223 ft (68 m) high and is 60 ft (18 m) wide.

Exhibitions, parliaments, and sports

The exhibition was held in a vast building designed in cruciform style by the architect Joseph Reed. Its eclectic exterior included a dome modeled on that of Florence Cathedral (see pp.140–141) and elements of Byzantine, Romanesque, and Renaissance styles. After the exhibition closed on April 30, 1881, the building was used as the temporary home for the first Parliament of Australia in May 1901, and later as a venue for basketball and other sports during the 1956 Olympic Games. The first building in Australia to be listed as a World Heritage Site by UNESCO, the building is currently used as a commercial exhibition venue.

▽ **MAIN HALL**
The four branches of the main hall meet under the dome. High windows shed light on the opulent interior, covered with murals that represent the virtues of a new Australia.

1.3 million people visited the building during the **exhibition of 1880–1881**

BUILDING THE BRIDGE

Once the two abutment towers were built, cranes were used to construct the main arch. When the two sides of the arch met in the middle, vertical hangars and horizontal crossbeams for the roadway were put in place from the center to the sides.

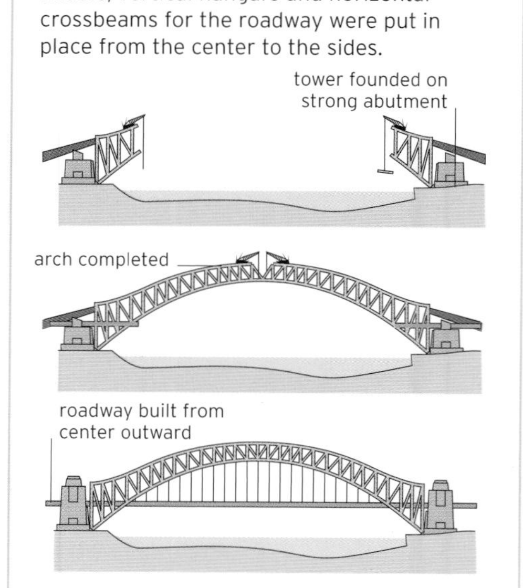

FRAMING THE HARBOR
Known locally as the "coat hanger," the bridge is illuminated at night and forms an iconic part of Sydney's panorama. The bridge regularly features in spectacular light shows in Sydney and is the centerpiece of Sydney's New Year's fireworks displays.

Sydney Harbour Bridge

A giant steel arch forming the centerpiece of one of the world's most dramatic harbors

The idea to build a bridge across Sydney Harbour connecting Sydney to the North Shore was first discussed in 1815, but it was not until the early 20th century that the proposal was taken seriously. In 1914, John Bradfield was appointed Chief Engineer of Sydney Harbour Bridge and Metropolitan Railway Construction. His work on the project later earned him the nickname "father of the bridge," such was his determination to see it built. Under his direction, the government approved the project in 1922 and design tenders were invited worldwide. In 1924, the British engineering firm Dorman Long & Co. Ltd. from Middlesbrough won the contract.

The winning contract

Well known as the builders of the Tyne Bridge in Newcastle, the winning firm proposed a similar arch bridge in Sydney. An arch was cheaper than the alternative cantilever and suspension bridges proposed, and also provided great rigidity to carry the expected heavy loads. Construction began on July 23, 1923, and was finished on January 19, 1932. The completed bridge was 3,770 ft (1,149 m) long and 440 ft (134 m) high. Its total width of 160 ft (49 m) allowed for six lanes of traffic, two lanes of trams, a path for pedestrians, a dedicated path for cyclists, and two railway tracks.

△ **BRIDGING THE GAP**
This photograph from August 1930 shows the two arms of the arch gradually coming together in the center.

Temple of the Golden Pavilion

An exquisite, gold-coated Zen Buddhist temple in Japan's ancient imperial capital

Reflected in the waters of the Mirror Pond (Kyōko-chi), the Golden Pavilion (Kinkaku-ji) is one of the most visited tourist sites in Kyoto. It was originally built as a private residence for Shogun Ashikaga Yoshimitsu (1358–1408), the most powerful man in Japan. He withdrew to the pavilion in 1398 after announcing his retirement from all official positions. On his death, as instructed in his will, the building was turned into a temple for Zen Buddhist monks.

Reconstructed treasure

Nothing of the original building has survived. In 1950, the temple was burned down by a 21-year-old monk—an event that formed the subject of a famous novel by Yukio Mishima, *The Temple of the Golden Pavilion* (1956). A new building has since been reconstructed, as close as possible to the original in every detail. The islands on the pond and the surrounding gardens, which form an essential part of the concept of the pavilion, are also scrupulously maintained.

CONTRASTING STYLES

The pavilion's three floors are in different architectural styles. The ground floor, the Chamber of Dharma Waters, harks back to the classical simplicity of Japan's Heian era. The second floor, the Tower of Sound Waves, is in the style of a samurai palace. Only the top floor, the Cupola of the Ultimate, is distinctively Zen in architecture.

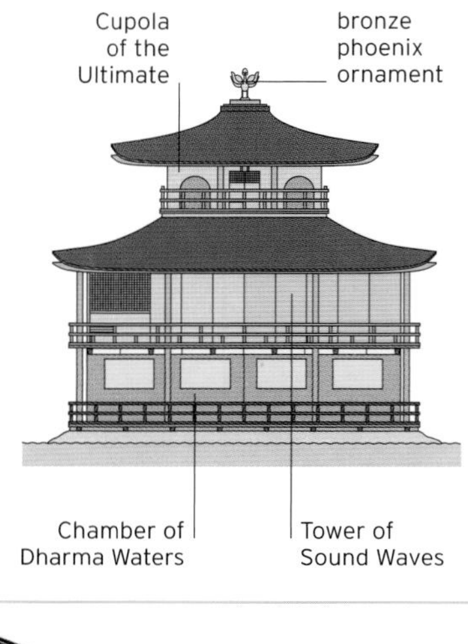

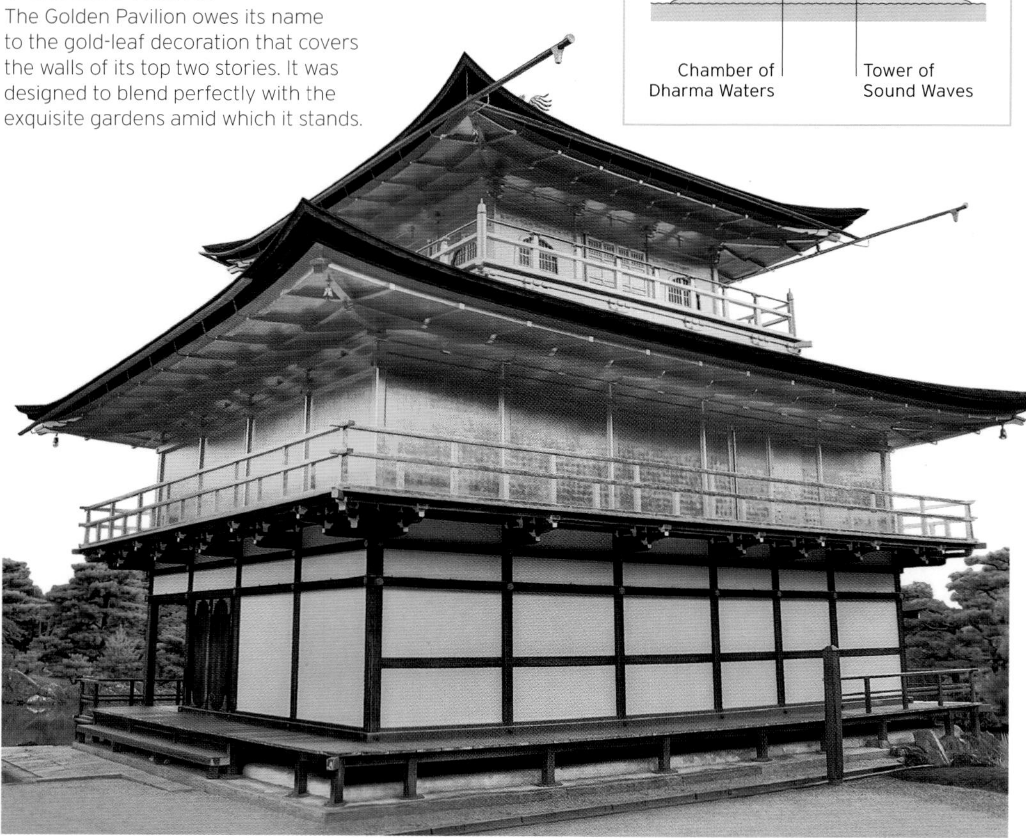

▽ **GLOWING TEMPLE**
The Golden Pavilion owes its name to the gold-leaf decoration that covers the walls of its top two stories. It was designed to blend perfectly with the exquisite gardens amid which it stands.

IMAGINATIVE STRUCTURE

The 27 petals of the Lotus Temple are arranged in three concentric rings. The outer nine petals arch outward over each of the nine entrances. The second ring surrounds the outer hall, and the innermost nine petals enfold the prayer hall. The glass-and-steel roof of the domed interior floods the hall with natural light. The temple is about 131 ft (40 m) tall at its highest point. The marble-covered petals are supported by massive concrete ribs, which are visible inside the hall but not from outside.

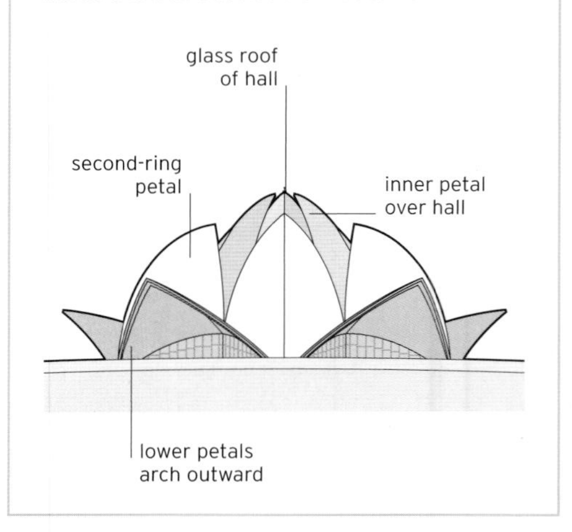

△ **GREEK MARBLE CLADDING**
Cleaners work hard to keep the marble surface of one of the temple's petals pristine white. The stone, brought from Greece, is of the same kind as that used to build famous ancient Greek temples such as the Parthenon (see pp.98–99).

▷ **PERFECT SYMMETRY**
An aerial view of the Lotus Temple shows its rigorous geometry, with the nine surrounding ponds completing the effect. The tips of the inner ring of petals do not meet, leaving a space for natural light to penetrate the interior.

Lotus Temple

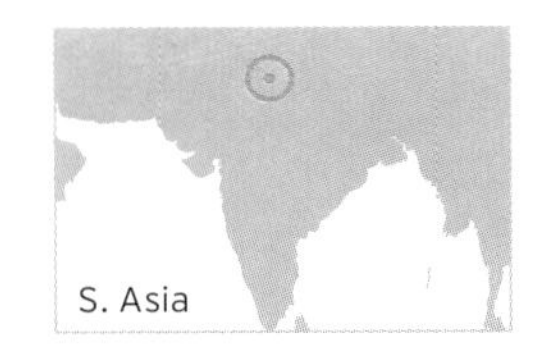

A spectacular Bahá'í House of Worship in Delhi that is designed to imitate the petals of a sacred flower

The basic tenet of the Bahá'í faith is the unity of all religions. The Lotus Temple, a *mashriqu'l-adhkár* ("house of worship") built in New Delhi in 1986, gives visual expression to this belief by adopting the form of a lotus, an aquatic flower recognized as a symbol of purity and immortality in Hinduism, Buddhism, Jainism, and Islam.

A flower about to bloom

Architect Fariborz Sahba, an Iranian living in North America, created an innovative building enclosed by 27 petals. Each petal is clad in dazzling marble brought from Mount Pentelicus in Greece. Reflecting the Bahá'í belief in the sacred properties of the number nine, the petals are arranged to create a building with nine sides and nine entrances. There are nine pools around the temple, which make the half-open lotus blossom appear as if it is floating on water. The domed inner space of the temple can hold 2,500 people. The building is open to people of all religious faiths. By the 21st century, the Lotus Temple has come to rival the Taj Mahal (see pp.290–291) as north India's most popular attraction.

△ **LANDSCAPED SURROUNDINGS**
The temple is built on a raised platform amid a wide expanse of landscaped gardens that extend over an area of 1.2 million square ft (110,000 square m).

▽ **AN AUSTRALIAN ICON**
The unique silhouette of Sydney Opera House's sail-shaped arches is instantly recognizable and has become a symbol of modern Australian culture.

The **roof** is made of **2,194 precast concrete sections** held together by **218 miles (350 km)** of steel cable

Sydney Opera House

A masterpiece of 20th-century design and engineering, with its roofs rising like sails above Sydney Harbour

Occupying the peninsula of Bennelong Point, Sydney Opera House, with its distinctive, sail-like roofs, gives the impression of floating on the waters of Sydney Harbour. More than merely an opera house, it is a complex of performance and exhibition venues. It includes a 2,679-seat Concert Hall, the Joan Sutherland Opera Theatre, a smaller Drama Theatre, studio and recording facilities, and an outdoor forecourt performance area.

Shells of concrete

The most striking feature of the opera house is the pioneering Expressionist design, conceived by Danish architect Jørn Utzon (1918–2008). His innovative idea of a group of parabolic, precast concrete shells to form a series of curved roof arches led him to win a design competition for the site in 1957. Construction began 2 years later and was completed in 1973.

The opera house has a binary structure: two sets of interlocking parabolic arches form the roofs of the two main halls, which diverge from a single entrance foyer. These arches are made of concrete, but faced with off-white glazed tiles in a chevron pattern, to form the distinctive shells. The two main buildings sit among a series of terraced pedestrian concourses, accessed via the Monumental Steps from the shoreside forecourt.

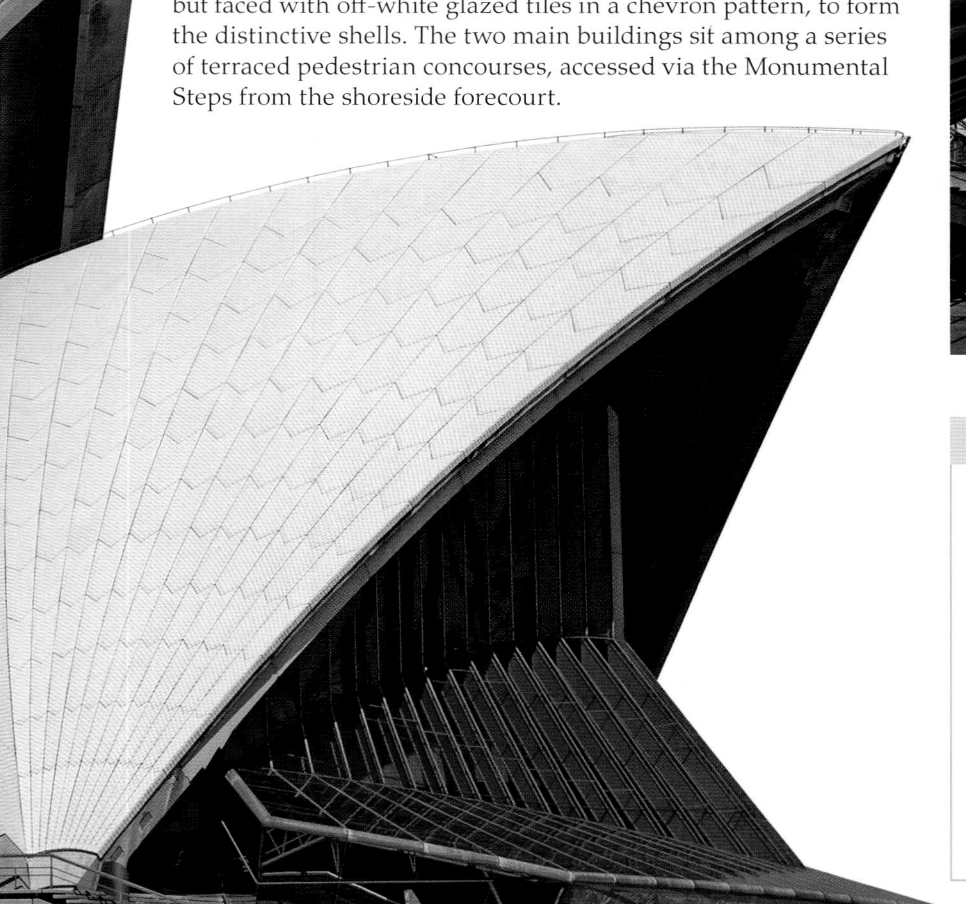

△ **PRESTIGIOUS AND MAJESTIC**
No less impressive than its exterior is the fully equipped Concert Hall, the largest performance venue at the opera house. It houses the world's largest mechanical action grand organ, just above the stage.

◁ **LIGHT-FILLED SPACES**
Looking out over Sydney Harbour, the foyers at the front of the two sets of arches are filled with natural light from the tall, metal-framed glass frontage.

THE "SPHERICAL SOLUTION"

Utzon's original design did not specify the exact curvature of the roof shells nor how they could be constructed. Various methods of prefabricating individual concrete panels were considered, but in 1963 Utzon and his engineers had a eureka moment, realizing that all the shells could be formed from sections of a sphere with a radius of 247 ft (75 m).

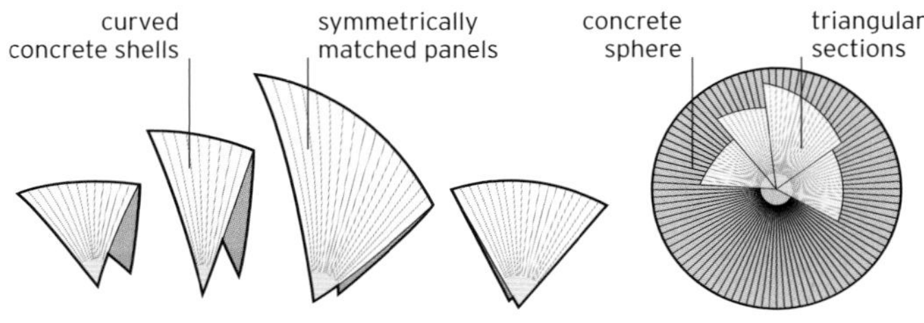

National Assembly Building of Dhaka

A striking Modernist parliament building and a remarkable statement of power and authority in a newborn nation

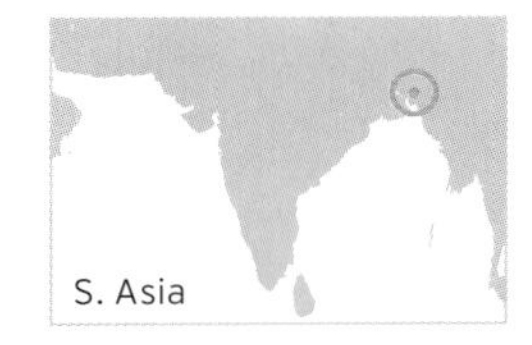

In 1962, with the decision to make East Pakistan's Dhaka the country's second capital, alongside Islamabad in West Pakistan, came the need for a space to house the National Assembly Building, or Jatiya Sangsad Bhaban. The government turned to Muzharul Islam (1923–2012), a leading Bengali architect, planner, and champion of Modernism. Islam opted to work with his former tutor at Yale University, Louis Kahn.

Kahn drew on the monumental architecture of the region, abstracting it through the prism of Modernism to create a building of bold geometric regularity. At the center of the complex is the assembly chamber, a domed amphitheater, and the library. The complex also includes a lake, gardens, and residences for members of parliament. The vast exterior walls of the main building (*Bhaban*) are deeply recessed, with porticos made from rough concrete inlaid with bands of white marble. Construction began in 1961 but was halted during East Pakistan's War of Independence in 1971 and not completed until 1982, by which time Dhaka was the capital of newly independent Bangladesh.

THE *BHABAN*

The *Bhaban* is actually a square that has been manipulated to form an octagon. It consists of nine individual blocks: the peripheral eight are used for committee meetings and house offices rising to 110 ft (34 m), and the central octagonal block, rising to 155 ft (47 m), is used for the national assembly.

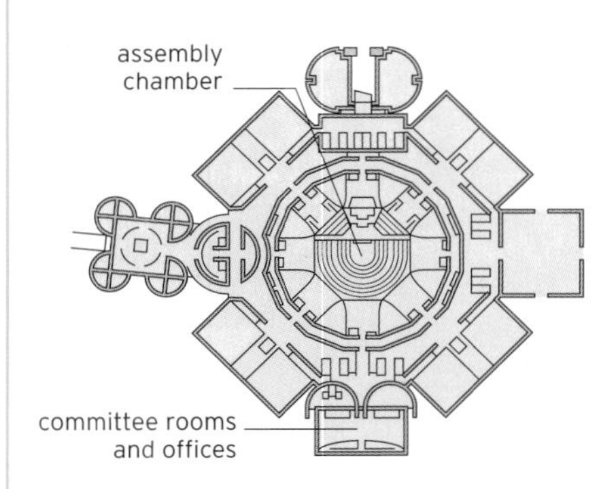

△ **POWER, PRESENCE, AND LIGHT**
The architect Louis Kahn flooded the main parliament building with light that filters between the solid columns and sheer walls.

▷ **AN ISLAND PARLIAMENT**
The *Bhaban* sits in an artificial lake that captures the riverine beauty of Bangladesh, a country situated in the Bengal delta, the largest delta system on Earth.

△ **WHITE VISTA**
Most buildings in the temple are coated with white-painted plaster to symbolize the purity of the Buddha. Mirror fragments on the exterior represent the spreading of Buddhist wisdom.

▽ **DESIRING HANDS**
In front of the bridge of "the cycle of rebirth" are hundreds of outstretched hands representing souls suffering in hell. From here, the only way forward is into the main *ubosot*.

Wat Rong Khun

A privately owned temple, art exhibit, and center of Buddhist learning and meditation

S. Asia

The original Buddhist temple of Wat Rong Khun, in the northernmost Chiang Rai province of Thailand, fell into disrepair during the late 20th century. A local artist, Chalermchai Kositpipat (1955–), decided to rebuild the temple from scratch, funding the work with his own money. To date, he has spent 1,080 million Thai baht (around $34 million) on the ongoing project, which is not expected to be completed until 2070 at the earliest. The temple opened to the public in 1997; admission is free for Thais, and only small offerings are accepted because Chalermchai refuses to be influenced by big donors.

White and gold

When completed, the temple—popularly known as the White Temple—will consist of nine buildings, including the existing *ubosot*, or prayer room, where ordinations take place; a meditation hall; a hall of relics; an art gallery; and living quarters for monks. The architecture is elaborate and ornate, much of it including elements from classic Thai buildings. While the *ubosot* is pure white with fragments of mirrored glass embedded in its exterior and represents the mind, the building housing the restrooms is golden, symbolizing how people focus on worldly desires.

Unusually for a Buddhist temple, the design, while based on classic Thai architecture, also includes depictions of western idols such as Michael Jackson, Freddy Krueger, and Neo from *The Matrix*, as well as controversial images of nuclear warfare, 9/11, and oil wells.

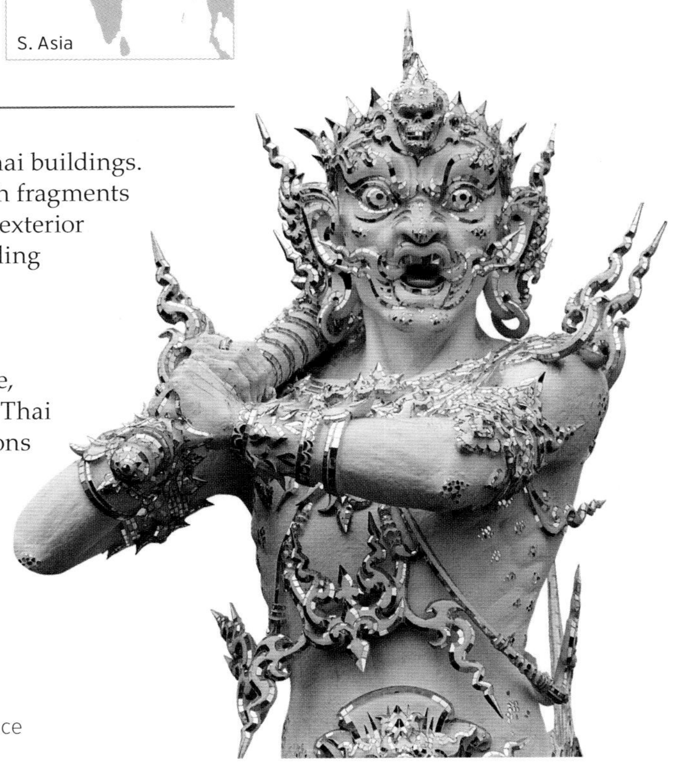

▷ **GUARDING THE TEMPLE**
Fierce guardian statues embedded with small mirror mosaics protect the entrance to the White Temple.

Akashi Kaikyo Bridge

An extraordinary feat of Japanese engineering, the world's longest suspension bridge

Opened on April 5, 1998, the Akashi Kaikyo Bridge—also called the Akashi Strait Bridge—carries a six-lane highway from Kobe city on Honshu Island to Iwaya on Awaji Island. The bridge is part of a larger system—the Honshu–Shikoku bridge project—which links the islands of Honshu and Shikoku across Japan's Inland Sea. The bridge is not only the world's longest suspension bridge, extending a total of 12,831 ft (3,911 m), but also one of the tallest, its two towers reaching 975 ft (297 m) into the sky.

Overcoming adversity

Spanning the Akashi Strait—a busy shipping route—was a considerable engineering challenge, not least because the area regularly experiences some of the world's worst storms and is seismically unstable. Indeed, while the bridge was under construction, an earthquake forced two of its towers 3 ft (1 m) farther apart. The designers used an innovative system of steel-truss girders and devices called tuned mass dampers (TMDs) to counteract the effects of wind and earthquakes. The triangular steel girders make the bridge rigid but also allow wind to pass through, while the TMDs swing in the opposite direction to the wind, balancing the bridge and cancelling sway. The bridge is capable of expanding and contracting by several feet. It can also withstand 180-mph (290-kph) winds and earthquakes of magnitude 8.5.

△ **VIEW FROM BELOW**
A steel-and-glass tunnel under a section of the bridge provides visitors with an unusual view of the scale of the construction.

▽ **LONGEST SUSPENSION BRIDGE**
Triangular steel girders provide the Akashi Kaikyo Bridge with the stability needed to span such an astonishing distance across the treacherous strait between the Honshu and Awaji islands.

△ **REFLECTED BEAUTY**
Pools of water surround the mosque's courtyard, reflecting the gilded and painted columns and elegant ceilings of its arcades. The mosque is illuminated at night in accordance with the phases of the Moon.

HYPOSTYLE MOSQUE

The Sheikh Zayed Mosque is a model hypostyle mosque (in which the prayer hall is made up of rows of columns). It has a main hall with 96 pillars, two open prayer halls, and a colonnaded courtyard. Four minarets, each 348 ft (106 m) high, sit at the corners of the courtyard.

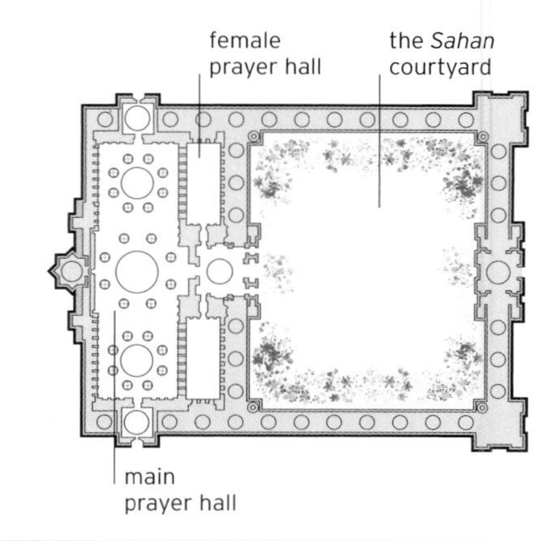

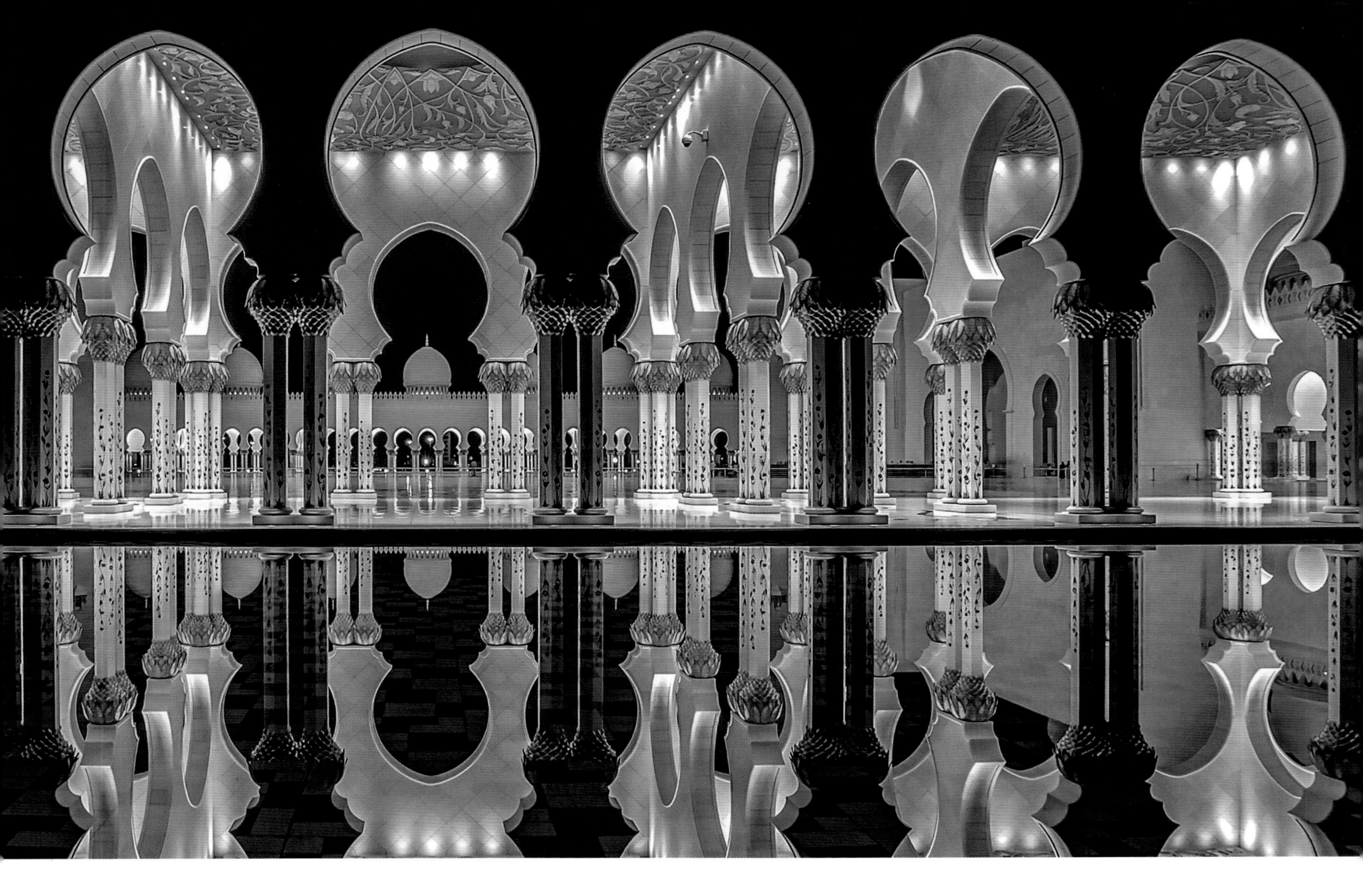

Sheikh Zayed Mosque

The largest mosque in the United Arab Emirates and the third largest in the world

In 1996, work began on the vast Sheikh Zayed Mosque, the brainchild of the first president of the UAE, Sheikh Zayed bin Sultan Al Nahyan. Construction lasted 11 years and was completed in two stages, in which a reinforced concrete shell was built and then clad in white marble and decoration. The resulting complex measures 430,000 square ft (40,000 square m) and can welcome 55,000 worshippers.

Combining cultures

The mosque's design is a fusion of Moorish, Arabic, and Mughal traditions and reflects the lunar cycle that underpins the Islamic calendar—its white marble changes from cold white to blue as the Moon wanes. More than 30 types of marble—brought from Italy, Macedonia, India, and China—were used in its construction, along with thousands of semiprecious stones, including lapis lazuli, amethyst, onyx, pearl, and aventurine. Floral decoration abounds, notably in the mosaic pavement of the *Sahan* courtyard, which extends over 187,000 square ft (17,400 square m), and on the almost 12,000 columns of the mosque as a whole. The mosque has several record-breaking features—the world's largest hand-knotted carpet, the largest crystal chandelier, and the largest Moroccan dome.

△ **MARBLE MASTERPIECE**
The mosque dwarfs the buildings that surround it. Its extravagance is reflected in its towering minarets, the multiplicity of its domes and columns, and the huge marble courtyard.

Burj Khalifa

An extraordinary feat of engineering, the world's tallest building and holder of several other world records

△ **PIERCING THE SKIES**
The Burj Khalifa dwarfs the other buildings in downtown Dubai, holding world records for the highest number of floors (163), highest occupied floor, highest observation deck, and tallest service elevator, among others.

Begun in 2004 and completed in 2009, the Burj Khalifa reaches over 2,716 ft (828 m) into the sky over Dubai. It is both the tallest building and the tallest free-standing structure in the world.

Deep foundations

The tower is built on a 12-ft- (3.7-m-) thick concrete slab supported by piles 5 ft (1.5 m) wide and 141 ft (43 m) deep. Its hexagonal central core is surrounded by a spiraling, Y-shaped plan with buttressing walls in its three arms—a design that can withstand the stresses of torsion and shearing to which the building is subjected. The building's wings decrease in size at the upper levels, a feature that breaks up the wind flow, which could cause the building to sway. The tower is clad in nearly 26,000 hand-cut reflective glass panels, designed to withstand Dubai's high temperatures, and topped by a telescopic steel spire, which was raised to its full height of more than 700 ft (213 m) by a hydraulic pump.

The Burj Khalifa is home to a hotel; restaurant; swimming pools; offices; and 900 residential apartments, including, of course, the world's highest.

△ **LOOKING DOWN**
The extraordinary view from the top of the tower shows the Burj Lake, the site of the 900-ft- (274-m-) long Dubai Fountain—the largest performing fountain in the world.

RECORD-BREAKING TOWER

On opening in 2010, the Burj Khalifa became the tallest building in the world, surpassing Taipei 101 in Taiwan—the previous holder of the title. It also beat the KVLY-TV mast in the US to become the tallest man-made structure in the world by some 650 ft (200 m), and beat Toronto's CN Tower (see p.53) to become the tallest free-standing structure in the world.

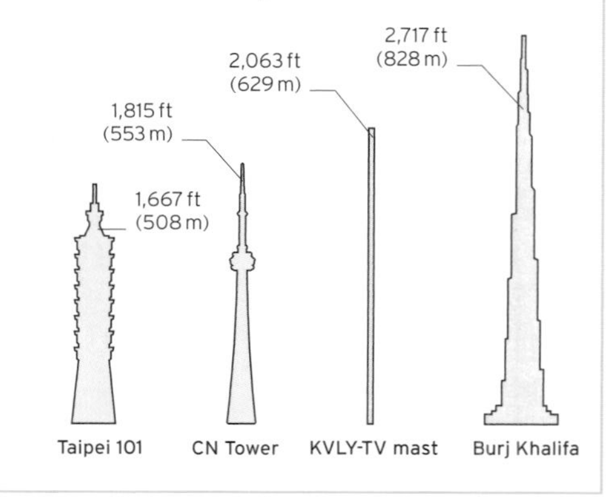

Gardens by the Bay

A stunning horticultural park that combines natural beauty with innovative green technology and architecture

Situated on 10 million square ft (1 million square m) of reclaimed land in the Central Region of Singapore, Gardens by the Bay is an award-winning horticultural park. It was created following an international design competition held in 2006 as part of Singapore's plans to transform its "Garden City" into a "City in a Garden" and build an internationally recognizable landmark.

Going green

The park is divided into three gardens: Bay South, Bay East, and Bay Central. Opened in 2012, Bay South is the largest and most developed garden; it spans 6 million square ft (540,000 square m), its layout inspired by Singapore's national flower, the "Miss Joaquim" orchid. Although it contains several distinct areas, the garden is defined by 18 massive "supertrees" and vast, climate-controlled biomes. The supertrees are 82–164 ft (25–50 m) high, and several are connected by a walkway that is 420 ft (128 m) long. Not only are the trees home to 162,900 plants and over 200 species, but they are also a valuable part of the development's sustainable energy-management system. Some are fitted with solar panels, and some harvest rainwater that both generates energy through a turbine and is used in the greenhouse cooling systems or for irrigation. Others serve as air intake and exhaust towers for the air handling systems. And, of course, they also provide shade for the park's visitors.

Cooled conservatories

Bay South's two climate-controlled biomes cover a combined area of 215,000 square ft (20,000 square m). The Flower Dome is the world's largest columnless glasshouse, while the Cloud Forest has its own 138-ft- (42-m-) high mountain, clad in tropical plants, encircled by a walkway, and housing the world's largest indoor waterfall. Gardens by the Bay are still under development, but they are already one of Singapore's most popular attractions.

Four themed "Heritage Gardens" **reflect Singapore's various cultural groups**—Indian, Chinese, Malay, and Colonial

△ **GLOWING ARBORETUM**
Illuminated by neon lights, the supertrees show off their own beauty, as well as that of the ferns, vines, orchids, and bromeliads that climb their sides.

◁ **HARDWORKING TREES**
Form and function combine in the elegant branches of the supertrees, which also help mask vents and rain-collection systems hidden in the treetops.

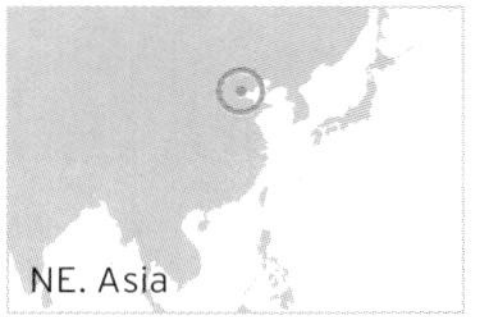

Tianjin Binhai Library

A spectacular, futuristic modern space that has become a Chinese architectural icon

The Binhai Library, in Tianjin, near Beijing, is a tribute to Modernism. The unorthodox space was designed by the Dutch architecture firm MVRDV and the Tianjin Urban Planning and Design Institute. Constructed in just 3 years, its doors opened to the public in 2017.

The sleek, five-story library covers 362,750 square ft (33,700 square m). It forms part of a cluster of cultural and educational buildings that were created at around the same time. The main entrance leads into a magnificent central atrium, which is for circulating, sitting, and reading. The first and second floors include reading rooms, book stacks, and relaxation areas; the upper floors contain meeting rooms, offices, computer and audio rooms, and two rooftop patios.

Spherical core

At the center of the space-age atrium is a gigantic sphere, or orb. Inside the orb is an auditorium, with seating for 100 people; above the orb, circular roof lighting helps to maximize illumination. Around the atrium are breathtaking, floor-to-ceiling shelves of books. However, some of the terraced shelves double as seats and stairs, and many of the "books" on the shelves are in fact imitations, created from embossed aluminum plates to resemble the real thing.

THE EYE OF BINHAI

The architectural highlight of the library is the spherical atrium, which comprises a large open space with a luminous orb at its center that is surrounded by shelves. A circular skylight above the orb, combined with louvres and glass windows at ground level, allows natural light to penetrate the interior. The orb is referred to as the Eye. To add to the effect, the shelving curves around the orb, its contours mimicking the outline of an eye. An oval shape punched through the exterior walls also creates a resemblance to an iris when the building is viewed from outside.

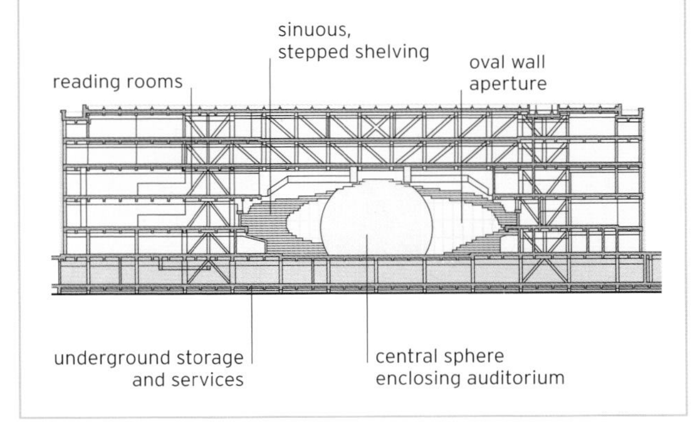

The library has the **capacity to hold** more than **1 million books** in its collection

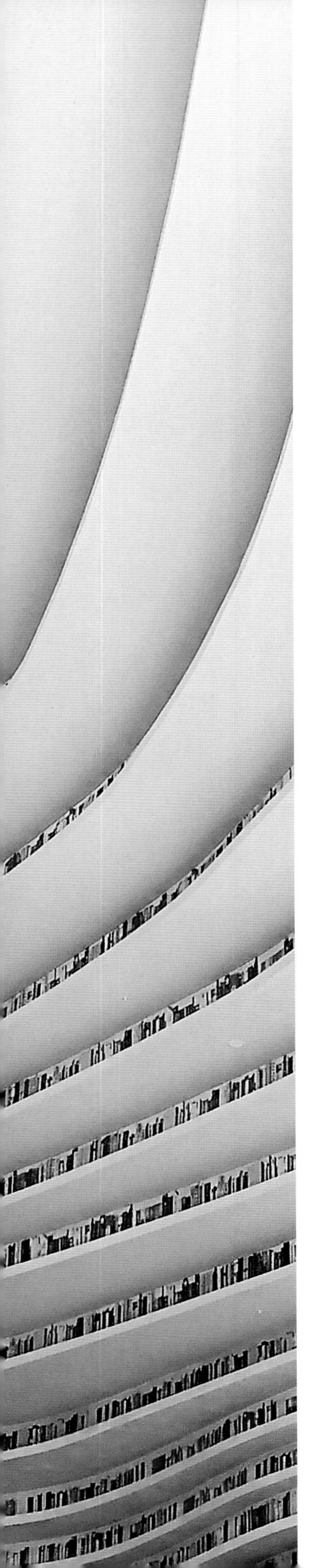

◁ **UNDULATING SHELVES**
Floor-to-ceiling shelving in the main atrium flows around the central sphere, creating a spectacular sculptural effect and dissolving any clear distinctions between walls, floor, and ceiling.

▷ **LOOKING IN**
The interior and exterior of the library were ingeniously designed to represent a giant eye, with a glowing white iris at its center.

Directory of other sites & Glossary

The sites featured in this section include not just highly visible structures built on a grand scale but also smaller constructions, down to the size of individual houses, as well as hidden places such as catacombs. They have been arranged first by continent and then by date of completion.

High vantage
At more than 2,600 ft (800 m) tall, the Burj Khalifa is the tallest structure ever built. Supported by buttresses, its concrete-and-glass towers look down on the trading hub of Dubai, one of the world's fastest-growing cities.

North America

Cahokia

Location Illinois

Cahokia is the largest and most important pre-Columbian archaeological site in North America. It was the administrative, trading, and religious center of the Mississippian Culture (800–1600 CE), which extended across the southeastern US. At its height, in the 13th century, Cahokia covered more than 6 square miles (16 square km) and supported a population of 40,000. Its most distinctive feature was a series of devotional earth mounds, or temple pyramids, of which 80 survive. The largest, Monks Mound, is the largest pyramid in North America and the biggest prehistoric earthen structure in the Americas. Its base extends over 657,000 square ft (61,000 square m)—comparable in size to that of the Great Pyramid of Giza (see pp.208–209). Built between 900 and 1200, it rises in four terraces, or steps, to a height of 98 ft (30 m). Its flat-topped summit was once the site of a wooden temple.

San Xavier del Bac

Location Tucson, Arizona

One of the finest examples of Spanish colonial architecture in the US, San Xavier del Bac was built by Franciscan missionaries between 1783 and 1797. Designed by Ignacio Gaona, the mission displays a synthesis of Moorish, Baroque, and native styles. Its brilliant white stucco exterior—which gives it the name the Dove of the Desert—is dominated by two massive, pierced square towers. These frame a three-story brick facade, featuring elaborate sculptures topped by a broken pediment. The 52-ft- (16-m-) high white dome sits over the transept. The imposing interior is richly decorated with multi-colored sculptures, paintings, and carvings featuring Spanish and Native American motifs.

Trinity Church

Location New York, New York

Trinity Church is the third church built on its site on Wall Street, lower Manhattan. Constructed out of locally sourced brownstone between 1839 and 1846 by British-born architect Richard Upjohn, it was the tallest building in the US for 23 years. It is also the earliest example of Gothic Revival architecture in the country. An elegant, soaring building, the Trinity Church is dominated by a spire that rises 281 ft (86 m) over the main facade. The vertical accent is echoed in the nine slender turrets that stud each side of the roof of the nave. Prominent features of the church include the large stained-glass windows that bathe the interior in multicolored light and its three sets of bronze doors, which are modeled on those of the Florentine Baptistery (see pp.140–141).

Monadnock Building

Location Chicago, Illinois

Built between 1891 and 1893, the 16-story Monadnock Building, designed by Daniel Burham and

▼ THE BRIGHT AND ELEGANT EXTERIOR OF SAN XAVIER DEL BAC

John Root, was on completion the world's largest office block. It was also one of the earliest examples of what would become the signature of downtown urban America: the skyscraper. It is constructed almost exclusively in dark, almost purple bricks. The designers deliberately avoided any form of decoration beyond the bay windows that surge upward, almost like half-columns, to give the building a sort of undulating elegance. Although modern in style, structurally it looked backward, being supported by masonry rather than constructed around an iron frame. Its 6-ft- (2-m-) thick ground-floor walls would not have been out of place in a medieval castle.

Wainwright Building

Location St. Louis, Missouri

Completed in 1891, the Wainwright Building was an example of a new sort of immense urban architecture. It also looked back to the past: the building, faced with vivid terracotta, was conceived by its designers Dankmar Adler and Louis Sullivan as a sort of huge Classical column with base, shaft, and capital. It does feature decorative elements, notably the elaborate panels under the windows and the mass of curling foliage under its boldly projecting cornice. However, it also looked decisively ahead. Not only was it built around a steel core, but it was intended to create and house a series of identical cells that were functional; efficient; and, above all, modern.

Pennsylvania Station

Location New York, New York

Pennsylvania Station—completed in 1910 to the designs of the firm of McKim, Mead, and White—was a late example of Classical architecture in the modern world, partly influenced by Rome and partly by Beaux Arts Paris. Intended to stun and amaze, it had a vast multicolumned facade, accented at each end by templelike pediments, that gave way to an immense central concourse and deliberately echoed the marbled public glories of ancient Rome. Beyond them was a series of platforms lit by large windows. A belief that the site could be used more profitably saw the building controversially demolished 50 years later.

Woolworth Building

Location New York, New York

The Woolworth Building is arguably the most magnificent pre-World War I building in the US. A Neo-Gothic skyscraper, it consists of an immense 29-story base from which surges an even taller 30-story central tower, topped by huge pinnacles and lanterns. It was completed in 1913, and at 792 ft (241 m) high, it was the world's tallest building until 1930. It also boasted the most complex elevator system yet built. Its architect was the prolific Cass Gilbert. Its presiding genius was F. W. Woolworth, a man intent on creating a building that would celebrate his commercial supremacy.

▲ THE CENTRAL TOWER OF THE WOOLWORTH BUILDING RISING ABOVE THE LOWER TOWERS

▲ TRAYLIKE BALCONIES OF THE FALLINGWATER HOUSE

Fallingwater

Location Mill Run, Pennsylvania

Fallingwater is Frank Lloyd Wright's masterpiece and one of several supreme statements he made as an architect. Completed in 1938, it is a private house built not just on the side of a rocky hillside but into it. The merging of building and site, each reinforcing the other, remains a tour de force. The house itself is unapologetically modern—sparse, spare, and angular—but it achieves the remarkable feat of seeming to grow from its site. Its most obvious technical achievement is the use of a series of apparently unsupported protruding horizontal exterior surfaces. Inside, the rooms flow seamlessly together. Its name comes from the river that cascades through and under the house.

San Simeon

Location California

Also known as Hearst Castle, San Simeon was built by architect Julia Morgan between 1919 and 1947 as a home for the newspaper magnate William Randolph Hearst. It is a lavish fantasy vision, on a gargantuan scale, of what a multimillionaire's palace should resemble. The rocky site overlooking the Pacific covers 390 square miles (1,000 square km), while the elaborately landscaped gardens extend over 549,000 square ft (51,000 square m). The style of the four principal buildings—the Casa Grande, where Hearst lived, and three guest houses—is broadly a fusion of Spanish Baroque and Moorish. The interiors are no less lavish and were largely conceived as suitable settings for Hearst's enormous art collection. They feature complete fireplaces shipped from Europe and Italian Renaissance ceilings.

Rockefeller Center

Location New York, New York

The Rockefeller Center is Manhattan's most celebrated complex of office blocks and corporate headquarters—a monument to the virtues of capitalism. The site, between 5th and 6th Avenues, covers some 1 million square ft (89,000 square m). It consists of 19 separate buildings linked by walkways and plazas, which are studded with statues. It is also home to an open-air ice rink. It was built in two stages: the first 14 buildings, all Art Deco in style, were constructed in 1931–1939; a further five buildings, in International Style—typically involving simple geometry and vast expanses of glass—were built in 1958–1974. The most famous building in the complex is Radio City Music Hall but the most commanding is that at its heart, the 850-ft- (260-m-) high, 66-storey RCA Tower.

Eames House

Location Los Angeles, California

Named for its designers and owners, Charles and Ray Eames, the Eames House was a product of the Case Study Houses program, the brainchild of John Entenza, the publisher of *Arts & Architecture* magazine. Its goal was to exploit the boom in the US house-building industry after World War II to champion Modernism. All the houses were to be economical and functional and to be made of mass-produced materials. Thirty-six designs were produced, and 26 were built. The Eames House, completed in 1949, sits on a wooded site of 54,000 square ft (5,000 square m). It consists of two simple rectangular buildings—the house and a studio—with their external walls divided into repeating colored rectangles. Internally, the most dramatic space is the double-height living room.

GM Technical Center

Location Michigan

The General Motors Technical Center, built between 1953 and 1955, is the first undisputed masterwork of the Finnish American architect Eero Saarinen. It is a corporate campus arranged around two lakes on a landscaped site of 0.5 square miles (1.3 square km). The buildings are long, low, and flat-roofed, dominated by steel and glass. However, changes of scale aided by splashes of vivid color, often from glazed bricks, add an almost theatrical air to the monumentlike structures. The futuristic feel is epitomized by the sleek steel of the 180-ft- (55-m-) diameter Design Dome and the almost spaceshiplike 140-ft- (43-m-) tall water tower that rises from the larger of the lakes.

Crown Hall

Location Chicago, Illinois

The famous maxim of the German-born architect Ludwig Mies van der Rohe, "less is more," is encapsulated by his Crown Hall at the Illinois Institute of Technology, built in 1956. It is a glass-and-steel rectangular box—220 ft (67 m) long, 120 ft (36 m) wide, and 18 ft (5.5 m) high—of absolute simplicity and minimal structure. The principal facade consists of six central glass bays created by vertical black steel supports. On either side, there are a further eight bays. The flat roof is suspended from four large steel girders that span the building's width. Everything else is glass—sandblasted for the lower 8 ft (2.5 m), but clear for the remaining 10 ft (3 m). The result is a brilliantly lit, entirely unobstructed interior.

Case Study House No. 21

Location Los Angeles, California

The most celebrated result of the Case Study Houses program (see also the Eames House, left) is the house known simply as No. 21, or the Bailey House, after its first owners, Walter and Mary Bailey. Designed by Pierre Koenig and completed in 1958, it is perhaps the purest statement of domestic Modernism ever built. It is not just functionally efficient; it is simple to the point of austerity. The floor plan is L-shaped. The walls are painted a uniform white. The south-facing front consists entirely of windows separated by steel supports. It is also water-cooled, with water pumped up to the roof being channeled to flow down the building. After a series of unsympathetic alterations, in 1997, Koenig restored the house to its original understated elegance.

▶ THE WATER TOWER AT THE GM TECHNICAL CENTER

▲ THE BLACK-AND-WHITE UNDERSIDE OF THE SHARP CENTRE FOR DESIGN

Seagram Building

Location New York, New York

German American architect Ludwig Mies van der Rohe's Seagram Building, completed in 1958, was the most influential skyscraper of the second half of the 20th century. Standing 516 ft (157 m) tall, it is a minimalist masterpiece—slim and entirely uniform—with no trace of the massive masonry of its predecessors. Constructed of a steel frame from which hang dark-tinted glass walls, the building has a series of bronze beams running vertically up its 38 stories to accentuate the underlying structure. The interior is lavish, featuring marble and more of the bronze and rich brown glass, complete with travertine lobby and elegant elevators. In front of the building sits a large, open granite plaza with two rectangular fountains, forming an urban landscape that allows the building space and light.

Sharp Centre for Design

Location Toronto, Canada

The Sharp Centre for Design, completed in 2004 to designs by Alsop Architects, is an extension to the Ontario College of Art and Design. One of the most immediately striking structures of the 21st century, it is a rectangular, two-story, steel-frame box, 275 ft (84 m) long, 101 ft (31 m) wide, and 30 ft (9 m) high. Its aluminum walls are painted white, to which seemingly random black squares have been added. Held 85 ft (26 m) off the ground by six pairs of angled, multicolored steel columns, it sits directly above the otherwise unexceptional brick buildings of the preexisting college. It is linked to a box, painted dark gray, containing elevators and stairs.

Aqua Tower

Location Chicago, Illinois

Opened in 2010, the Aqua Tower, designed by Studio Gang Architects, is an 82-story, 876-ft- (267-m-) tall skyscraper close to the shores of Lake Michigan. Its most obvious feature is its series of alternately projecting and receding balconies. These give the building an organic quality, with echoes of naturally weathered rock formations. As a result, the building seems to be in a state of permanently shimmering movement—almost as if it is alive. The effect is greatly heightened where the balconies disappear entirely, revealing areas of blue glass that look like vertical pools of water. It is a multiuse building, with shops; offices; a hotel; and, from the 19th story upward, 739 apartments. The roof is given over to a vast, intricately landscaped garden.

Central and South America

Tiwanaku

Location Province of Ingavi, Bolivia

Once the heart of Tiwanaku Culture, Tiwanaku was the capital of a pre-Hispanic empire that ruled over part of the southern Andes. The city flourished between 400 and 900 CE and had a strong civic character. A series of buildings underline the importance of the political and religious power within the city; these include the Palace of Putuni and Kantatallita, the semi-subterranean Temple, Kalasasaya's Temple, Pumapumku's Pyramid, and Akapana's Pyramid—arguably Tiwanaku's most imposing structure. Much of the architecture was made out of carefully carved ashlar (finely cut stone) masonry, while the urban plan boasts a complex system of underground drainage that controlled the flow of rainwater.

▶ ENGRAVED DEITY STATUE AT THE ENTRANCE TO TIWANAKU'S TEMPLE COURT

Temple of the Sun

Location Trujillo, Peru

Built by the ancient civilization of Moche, also known as Mochica, this giant temple is one half of a pair of large-scale religious structures nestled in the Moche river valley, just outside the city of Trijullo in northern Peru (the other being the Temple of the Moon). The stepped, pyramidal temple, built from adobe brick, spans an impressive 1,115 ft (340 m) and is 135 ft (41 m) high. Sitting within a site that was densely occupied between the 1st and 8th centuries CE, this monument was, spatially and socially, the heart of this Moche urban center.

Pyramid of the Magician

Location Uxmal, Mexico

The Pyramid of the Magician is unusual, in that it is set on an oval footprint and has rounded edges. Located in Mexico's ancient Maya city of Uxmal in Yucatán, the dramatic, 91-ft- (27.6-m-) high structure is the tallest building in the entire archaeological site. It was completed in about 900 CE. Uxmal is representative of the region's Puuc architectural style, and the pyramid is no exception, featuring stone decorations and a concrete core. Two steep and massive staircases lead to the top of the pyramid on the east and west sides. The structure contains a series of five nested temples resulting from successive building phases.

Governor's Palace

Location Uxmal, Mexico

The Governor's Palace at Uxmal is one of Mexico's most instantly recognizable pre-Columbian structures. Elevated on an 26-ft- (8-m-) high platform that can be reached by a dramatic stairway with three landings, this administrative center was built in the 10th century. Despite the palace's imposing setting and impressive size—it is about 330 ft (100 m) in length—its most eye-catching feature is arguably the facade's intricate relief decorations, crowned by a mosaic frieze. The decorations include an abundance of glyphs representing Venus, and the palace itself is aligned toward the point where the planet rises in the southernmost part of its cycle.

Casa de los Azulejos

Location Mexico City, Mexico

This ornate 18th-century palace, situated in what was one of the most desirable streets in the center of Mexico's capital, Francisco I. Madero Avenue, was built by the Count of the Valley of Orizaba. As its name implies, the facades of Casa de los Azulejos ("House of the Tiles") are covered in beautiful white and blue tiles in the distinctive style of the country's state of Pueblo. It is considered a fine example of the Baroque architecture of New Spain, with intricate decorations inside and out, as well as an open courtyard with a fountain at its heart.

▼ WHITE AND BLUE TILING OF THE CASA DE LOS AZULEJOS IN MEXICO CITY

Metropolitan Cathedral

Location Mexico City, Mexico

One of Mexico City's most iconic buildings, the Metropolitan Cathedral sits right on the Zócalo—the city's central square. Built in stages, between 1573 and 1813, the dramatic structure is one of Latin America's largest and oldest churches. The design, by Spanish architect Claudio de Arciniega, includes Gothic, Baroque, Neo-Classical, and Neo-Renaissance influences. The Baroque front facing the square features portals with inset marble panels and bas-reliefs but is dominated by two huge bell towers. A key highlight of the interior is the gilded Altar of the Kings, built in 1718–1737 by Spanish architect and sculptor Jerónimo de Balbás.

Iglesia de la Merced

Location Quito, Ecuador

Quito, the capital of Ecuador, has the best preserved historical center in Latin America. The imposing Basílica de Nuestra Señora de la Merced, consecrated in 1747, is a fine example of its monumental buildings. It was constructed to replace an older church that was destroyed by an earthquake. It is a beautiful but austere building with a plain white exterior. It features five domes and a tall, square tower in one corner. The church features a mix of Inca and Moorish influences—its pagan motifs include images of the Sun and the Moon. Legend has it that its tower is possessed by the devil.

▲ CONTRASTING DOME, BOWL, AND TOWERS OF THE NATIONAL CONGRESS BUILDING

Teatro Colón

Location Buenos Aires, Argentina

Inaugurated in May 1908, with a performance of Italian composer Giuseppe Verdi's *Aida*, Teatro Colón is Buenos Aires's main opera house. The building was initially designed by Italian architect Francesco Tamburini but was later completed by Italian Victor Meano and Belgian Jules Dormal. Sitting right in the heart of the city, the majestic structure is generally Neo-Classical in style—its ordered facade features an array of columns and triangular pediments. However, its character is better described as "eclectic," as it displays a mixture of architectural motifs. The theater seats 2,478 people (with standing room for another 500) and is renowned for its excellent acoustics. A series of restoration works was completed in 2010.

Church of St. Francis

Location Belo Horizonte, Brazil

This striking church, set by the Pampulha Lake in Belo Horizonte, the state capital of Minas Gerais, is part of a group of buildings—the UNESCO World Heritage Site of the Pampulha Modern Ensemble—that also includes a casino, restaurant, dance hall, and yacht club. The complex was commissioned by the city's then-prefect, Juscelino Kubitschek, in 1940 and was designed by Oscar Niemeyer. It is considered to be one of the first significant interpretations of modernism in Brazil. The church's controversial design features an undulating reinforced concrete roof over a T-shaped plan. The distinctive tiled mural on the northern side depicts scenes from the life of St. Francis by the Brazilian painter Cândido Portinari.

Ministry of Education and Health, Rio de Janeiro

Location Rio de Janeiro, Brazil

Considered to be one of Brazil's most important buildings of the 20th century, the Ministry of Education and Health—also known as the Gustavo Capanema Palace—was the work of a team of avant-garde architects, including Brazilians Lúcio Costa and Oscar Niemeyer and Swiss-French modernist Le Corbusier. Built between 1937 and 1943, the 15-story, concrete office block features a striking grid facade with a distinctive system of movable louvres that provide shade within. The ground level is raised on supporting pillars, or pilotis (see p.193), which gives the building a sense of lightness.

National Congress Building, Brasilia

Location Brasilia, Brazil

Occupying one of Brasilia's most prominent locations, the National Congress Building represents Brazilian Modernism through its geometric shapes and striking composition in glass and concrete. Oscar Niemeyer designed it with a long, flat roof, upon which sit two distinct formations—a dome, placed over the Senate chamber, and a bowl, containing the Chamber of Deputies. A long ramp leads up to the main entrance and a pair of twin towers rises beyond, housing further office spaces. The project was inaugurated in 1960, along with Brazil's new capital city.

▲ THE QUEEN'S MEGARON (SUITE) AT THE PALACE OF KING MINOS

Europe

Palace of King Minos

Location Crete, Greece

From about 2800 BCE, the first distinctively Greek culture, the Minoan, flourished on the island of Crete. During its heyday, it was a vibrant, organized society. It was epitomized by the palace of the mythical King Minos at Knossos, built perhaps after 2000 BCE—the high point of the Minoan civilization. What remains is largely a reconstruction, built by British archaeologist Arthur Evans in the 1920s. The mostly two-story complex, built around a central courtyard, sprawls over 21,500 square ft (2,000 square m) and has more than 1,300 rooms—some ceremonial, most more utilitarian. The Minoan taste for vivid colors is evident in the reconstructed frescoes, teeming with animal life.

Capitoline Temple of Jupiter

Location Rome, Italy

The Capitoline Temple of Jupiter, dedicated in 509 BCE, at the very start of the Republican era, was the first of the great temples of Rome. It was built on an imposing stone podium 203 ft (62 m) long; 174 ft (53 m) wide; and, importantly, 13 ft (4 m) high, although only parts of this structure have survived. The temple represented a transplanted Greek tradition filtered through Etruscan practices, evident above all in its use of a high base and a deep and commanding porch, with three rows of six columns. Unlike later, more characteristic Roman temples, the columns continued along the side walls. The building was rebuilt with progressively more sumptuous materials several times, notably in 83 BCE. The original consisted mostly of wood.

Statue of Zeus

Location Olympia, Greece

The Statue of Zeus, which stood in the Temple of Zeus, built in the 5th century BCE at Olympia, was one of the Seven Wonders of the World. Completed around 435 BCE, it was the work of Phidias, the most celebrated of all ancient Greek sculptors. No trace of the statue remains today, and there are no visual representations other than those on a few Roman coins. Even the circumstances of its loss are unknown. However, there are descriptions by contemporary writers that report a seated figure of massive size—43 ft (13 m) tall—and extraordinary opulence, constructed of ivory, gold, ebony, and glass, with jewels studding Zeus's robes and his immense throne. In one hand, he clasped a staff; in the other, he held a gold-and-ivory figure of Nike, the Greek goddess of victory.

Erechtheion

Location Athens, Greece

Built in 421–405 BCE, the little temple of the Erechtheion is on the Acropolis, to the north of the Parthenon (see pp.98–99). It was extensively damaged during its life, and what little remains is the result of rebuilding. It is unusual in that its plan is asymmetrical, with three separate porticos; however, this reflects its role as a shrine to three objects of devotion, including King Erechtheus of the Archaic Greek era, after whom the temple is named. Its most famous features are the caryatids that support the south porch. These six slim and precisely carved female figures act as columns, their robes echoing the fluting of the slender shafts of the Ionic columns on the other two porticos. All six are reconstructions.

Theatre of Epidaurus

Location Epidaurus, Greece

By the time the open-air Theatre of Epidaurus was built in the mid-4th century BCE, the tradition of Greek drama pioneered in Athens in the 5th century had ended. Nevertheless, the building is not just the most complete example of an ancient Greek theater; it is also remarkable for its simplicity, size, and functionality. For example, its acoustics are thought to be near perfect. Originally, it had three elements: the immense semicircular marble auditorium; the circular stage, or orchestra, at its foot; and behind that a stage building, or skene. Only the first two survive. The auditorium, which seats more than 13,000 in 55 rows, has a diameter of 387 ft (118 m) and the orchestra has a diameter of 67 ft (20.5 m). It remains in regular use.

The Colossus of Rhodes

Location Rhodes, Greece

The Colossus of Rhodes, representing the Sun god Helios, was one of the Seven Wonders of

the World. Unfortunately, it was destroyed by an earthquake in 226 BCE. According to legend, the statue was an immense figure that straddled the entrance to the harbor at Rhodes. It is thought to have been 108 ft (33 m) high, which would make it by far the largest statue of the Classical world. However, it remains an enigma. Almost all that can be said with any certainty is that the figure was created in 292–280 BCE, and was clad in bronze. It was the work of Chares of Lindos, whose teacher, Lysippus, had been responsible for an 58-ft- (18-m-) high bronze statue of Hercules in Taranto in southern Italy.

Pompeii

Location Pompeii, Italy

The city of Pompeii, along with its neighbor on the Bay of Naples, Herculaneum, was buried deep below layers of ash and rock by an eruption of Mount Vesuvius in 79 CE. Miraculously, it was largely preserved intact and was rediscovered in the 18th century. Presenting a moment frozen in time, its excavation transformed our understanding of Roman life. Much of what is known of Roman painting comes from the frescoes discovered here. Pompeii's forum, temples, and huge amphitheater attest to the priorities of Roman society. Of the many houses, some are sumptuous, with colonnaded courtyards and gardens, but most are humble, giving a glimpse of the normal Romans' daily life. The layout of the city is largely a regular grid plan based on two principal streets running northeast, with blocks of buildings covering some 38,000 square ft (3,530 square m).

▼ FORUM BATH IN THE ANCIENT ROMAN CITY OF POMPEII

Stoa of Attalos

Location Athens, Greece

The modern version of the Stoa of Attalos, built in the 1950s, is a precise reconstruction of the original 2nd-century BCE structure destroyed in 267 CE. It formed one side of the agora, or marketplace, that was a prime focus of Athenian public life. The stoa is a substantial, two-story structure, measuring 377 ft (115 m) long and 66 ft (20 m) wide. Constructed of marble and limestone, it consists of a ground-floor colonnade fronted by fluted Doric columns and a smaller, upper colonnade with similarly fluted three-quarter Ionic columns. The precision of the stoa's craftsmanship highlights how Greek urban planning was increasingly directed toward ordered, composed public spaces.

▼ DORIC AND IONIC COLUMNS AT THE STOA OF ATTALOS

Ostia's Insulae

Location Ostia, Italy

During Rome's early Imperial period (1st century CE), the growing importance of Ostia, on the mouth of the Tiber, as Rome's port led to a rapid increase in its population, which in turn put great pressure on housing. As a result, there was a reduction in the number of private houses being built, and the creation of high-rise tenement blocks, or *insulae*—literally "islands" (the term also means a city block)—increased. Ostia has some of the best examples of this new type of domestic architecture. *Insulae* were often cheaply built of unadorned bricks and concrete, with poor sanitation. Because of the risk of fires, the least desirable rooms were those located on the upper floors. Fears over safety led to height restrictions—first to 70 ft (21 m), later to 60 ft (18 m). Nonetheless, *insulae* with as many as six stories were common.

Maison Carrée

Location Nîmes, France

Built around 19 BCE, the Maison Carrée ("square house") is the most complete surviving Roman temple. It is raised on a substantial 10-ft- (3-m-) high podium and is reached by steps at the main west end flanked by the podium's arms. The deep portico, which is fronted by six fluted Corinthian columns, with two more on each side, is topped by a commanding pediment. Both sides of the walled, windowless temple feature eight Corinthian half-columns. An elaborate frieze, with tendrils and acanthus decorations, runs around the building. Like all the carvings on the limestone temple, it is precise and rich. The length of the building, 87 ft (26.5 m), is almost exactly twice its width, 44 ft (13.5 m).

Pyramid of Cestius

Location Rome, Italy

The Pyramid of Cestius—in effect an imperial Roman–Egyptian pyramid—is among Rome's most surprising structures. It was built in about 15 BCE, as the tomb of senator Gaius Cestius. Its form is explained by a sudden taste for Egyptian art after the Roman conquest of Egypt in 31 BCE; its unusual steepness suggests that it was modeled on similar Nubian pyramids. The base of each side is 95 ft (29 m) long, and it rises 121 ft (37 m). It is built of concrete and brick, which is faced with marble. Contemporary Latin inscriptions on two walls explain whose tomb it is and who built it. In the late 3rd century CE, the pyramid was incorporated into new city walls, of which it still forms a part.

Temple of Mars Ultor

Location Rome, Italy

Other than a few steps, three battered columns, and the stubs of four more columns, little remains of the Temple of Mars Ultor—"Mars the Avenger." It was completed in 2 BCE on the orders of the Emperor Augustus to commemorate his victory at the Battle of Philippi in 42 BCE over the assassins of his great-uncle Julius Caesar. Raised on a platform reached by steps, the temple's portico had rows of 59-ft- (18-m-) high Corinthian columns on three sides, above which sat a massive carved pediment. The temple dominated the huge Forum of Augustus. Together, they had been one of the most imposing sights of imperial Rome, reinforcing Augustus's claim, "I found Rome a city of bricks and left it a city of marble."

▲ BAS-RELIEF ON TRAJAN'S COLUMN CELEBRATING ROMAN VICTORY IN DACIA

Trajan's Bridge

Location Alcántara, Spain

Trajan's Bridge, named after the Spanish-born Emperor Trajan who ordered its construction, was completed in 105 CE. With a total length of 623 ft (190 m), the bridge spans the Tagus River on what is now the Spanish–Portuguese border. Much like Pont du Gard (see pp.100–101), it exemplifies the Roman genius for engineering, which also produced impressive architecture. It has six huge stone-built arches—the most characteristic of Roman engineering solutions. The vast masonry of the two central arches rises 157 ft (48 m) above the river. Straddling the bridge's center is the structure's only nonfunctional element—a severe, castellated triumphal arch.

Trajan's Column

Location Rome, Italy

The Roman practice of creating imposing public monuments to celebrate military triumphs was given new form in Trajan's Column, built in 112 CE. Made of a series of huge marble drums, the column is 115 ft (35 m) high and originally featured an immense, gilded statue of Trajan at its summit. It celebrates the victories of Trajan in Dacia, in the region of modern-day Romania, in two campaigns between 101 and 106 CE. It is decorated with a continuous spiral of carving, in relief, climbing up and around the column, forming a dense narrative of Trajan's conquests. It features around 2,500 individual figures, and, if unraveled, it would be more than 600 ft (183 m) long.

Hadrian's Villa

Location Tivoli, Italy

Roman opulence is nowhere more obvious than at the vast villa built by the Emperor Hadrian in Tivoli, east of Rome, in the 2nd century CE. The entire complex, including the villa, covered an area twice the size of Pompeii (see p.325). Today, only fragments remain of what was once a dazzling array of formal and informal buildings, with sprawling gardens, lakes, pools, fountains, and myriad statues—all heavily inspired by ancient Greece. There is little of the rigid formality associated with the great buildings of Rome. Instead, the villa incorporates a mixture of different styles, features, and innovations.

Hadrian's Mausoleum

Location Rome, Italy

Hadrian's Mausoleum, which from 590 CE has been known as the Castel Sant'Angelo, illustrates the changing fate of many buildings that survived the fall of Rome in the 5th century. It was built by Hadrian, who also built the bridge over the Tiber that leads to it, as a mausoleum for him and his family. It was completed in 139 CE, the year after his death. In 401, it became a fortress; in 1277, it was taken over by the Vatican, and it became a museum in 1901. The mausoleum is an immense, circular concrete structure—203 ft (62 m) in diameter and 68 ft (21 m) high—but its exterior was originally clad in marble.

Roman Baths, Aquae Sulis

Location Bath, England

The Romans built public baths in almost every part of their empire. At Bath (Aquae Sulis to the Romans), they had the additional advantage of natural hot-water springs, eliminating the need for the type of heating normally required for the *caldarium* and *tepidarium*, the hot and warm baths. As elsewhere, the role of such baths was as much social as hygienic—a focal point of Roman public life. The centerpiece, then and now, is the Great Bath, with its distinctive green-gray water. It was built from about 70–80 CE and successively rebuilt and enlarged. Originally vaulted, today it lies open to the sky. The stern colonnade that surrounds it would once have been filled with statues.

▲ COLONNADE SURROUNDING THE WATERS OF THE GREAT BATH AT AQUAE SULIS

Baths of Diocletian

Location Rome, Italy

The imperial baths of Rome were among the city's greatest splendors. The most opulent—those of Emperor Diocletian—were the last to be built, completed in about 306 CE. Covering 1.3–1.5 million square ft (120,000–140,000 square m), the baths consisted of a series of vast marbled interiors housing hot rooms, cool rooms, sweating rooms, and plunge pools, made possible by a system of heating—worked by an army of slaves—as elaborate as the aqueducts that delivered water to the city. A magnificent part of the otherwise ruined building was recreated by Michelangelo as the Church of Santa Maria degli Angeli in 1563–1564.

Basilica of Constantine

Location Rome, Italy

Even for Imperial Rome, the Basilica of Constantine, or Basilica Nova, was overwhelming in its size and richness. Today, all that remains are the immense concrete barrel vaults of one of the aisles. The rest of the building—stripped of its treasures—was destroyed by earthquakes in the 9th and 14th centuries. The basilica was built between 307 and 312 CE to house a variety of administrative functions. Its plan was comparatively simple. A groin-vaulted nave—265 ft (80 m) long, 83 ft (25 m) wide, and 115 ft (35 m) high—was flanked by slightly smaller, barrel-vaulted aisles. At one end, an apse housed a similarly oversized statue of Emperor Constantine. Because they were not traditionally associated with Roman temple designs, basilicas of this type were adopted for most Roman Christian churches.

Arch of Constantine

Location Rome, Italy

The triumphal arch, built to commemorate military victories, remains the most immediately distinctive form of Roman public monument. Such arches sprouted across the empire, and in Rome alone, there were more than 50. The Arch of Constantine, built between

▲ THE THREE ARCHWAYS OF THE ARCH OF CONSTANTINE

312 and 315 CE, was the last to be erected and is one of only three that survive. At 85 ft (25.9 m) wide and 69 ft (21 m) high, it is the largest of Rome's triumphal arches. It has three archways—the central one larger than the other two—flanked by four projecting columns on raised bases. Carved reliefs, simply lifted from earlier structures, and an inscription over the central arch, originally in bronze, attest to the emperor's martial valor.

St. George's Church

Location Salonika, Greece

There was always an uncomfortable transition in building types between those of pagan Rome and those of the Rome decreed Christian by Constantine after 313 CE. It was an awkwardness reinforced by the division of the empire in 395 CE into western and eastern halves, which in the east gave rise to an increasingly distinctive Byzantine architecture. St. George's Church reflects the first phases of this process. It is an extension, completed in about 395 CE, to an earlier, circular structure. To this was added a portico, a commanding apse, and an even more commanding dome, 80 ft (24.5 m) in diameter. The church has been stripped of almost all of its original decoration, leaving behind its massive, arched masonry, some 20 ft (6 m) thick, and some surviving later Byzantine mosaics.

Sant' Apollinare in Classe

Location Ravenna, Italy

The 6th-century Sant' Apollinare in Classe is among the first mature early Christian basilicas, consisting of a nave, aisles to either side, and an apse at its east end. The 183-ft-(56-m-) high nave is exactly twice the height of the aisles and is separated from them by 24 huge columns. However, the true glory of the building is one of the most characteristic forms of Byzantine-influenced church decoration: the shimmering mosaics of the apse. Predominantly gold, green, and blue, they follow a complex iconographical scheme, celebrating not just the divinity of Christ but also of the 1st-century CE martyr St. Apollinaris. He is shown calling his flock of 12 sheep—a reference to the Apostles—against a vivid, flower-strewn landscape.

Skellig Michael

Location County Kerry, Ireland

Skellig Michael is a bleak and rocky island that rises steeply from the Atlantic 7 miles (12 km) off the southwest coast of Ireland. From around the early 6th century CE to about 1200, it was the site of a tiny Augustinian monastery. The monks lived in six dry-stone beehive cells. These austere structures have survived, as have the monastery's two tiny stone-built chapels, both of which resemble upturned boats. The larger of the two measures just 14 ft (4.3 m) long and 12 ft (3.6 m) wide. The site also includes a similarly diminutive cemetery and the remains of a kitchen garden. Upward of 100 gaunt stone crosses are studded across the island. As Rome collapsed, it was in such remote places that Christian learning clung to survival in northern Europe.

◀ MONASTIC BEEHIVE HUTS ON THE ISLAND OF SKELLIG MICHAEL

Theotokos, Hosios Lukas

Location Phocide, Greece

The later 10th-century church of the Theotokos is one of two churches at the Greek Orthodox monastery of Hosios Lukas. Its significance lies in the fact that it was the first church to be built in what would become the dominant plan of Byzantine ecclesiastical architecture: the cross-in-square, or crossed dome. Within an overall rectangular floor plan, a square is formed in the center of the building by four piers, and from it project four identical arms at right angles. As is typical, a dome rises from the central space. Smaller domes over the arms were a characteristic later development. The church's design would prove immensely influential. Its symmetry not only satisfied the Byzantine near veneration of geometry but also lent itself to the complex hierarchical decorative schemes, almost always in mosaic, that reinforced Byzantine liturgical practices.

Cathedral of Santa Sophia

Location Novgorod, Russia

The influence of the Byzantine cross-in-square plan (see Theotokos, Hosios Lukas, above) is reflected in Novgorod's mid-11th-century Orthodox Cathedral of Santa Sophia. Its five onion domes—the central one gilded, the rest silvered—rise to a height of 125 ft (38 m) over the heart of the church. However, its asymmetry, small windows, and vertical emphasis suggest an influence from the massive Romanesque cathedrals simultaneously appearing across parts of Western Europe. It is the oldest existing church in Russia, which was first reached by missionaries in only 933 CE. The church has remained mostly unchanged externally since it was built, and its stern appearance is emphasized by its plain white walls. However, much of the decoration in the dimly gleaming interior is from the late 19th century.

Ely Cathedral

Location Ely, England

Ely Cathedral is remarkable for its huge bulk, most memorably seen in silhouette, which reveals a long, low building with various towers and spires. It contains prime examples of almost every major period of the Gothic style in England, as well as an 11th-century Norman nave and transepts with sturdy, round-headed arches. The West Tower over the main entrance is a similarly bulky Norman structure. By contrast, the 14th-century Lady Chapel is a riot of exuberant carving. However, the most distinctive feature is the huge eight-sided tower, or octagon, over the crossing, built in 1322–1340. From it climbs a second, wooden octagon.

Maria Laach Abbey

Location Rhineland, Germany

By the end of the 11th century, economic revival, an assertive Catholic Church, and the continuing rediscovery of the engineering feats of the Romans made possible a series of vast, vaulted stone-built cathedrals in Western Europe. In early medieval terms, they were extraordinary technical triumphs. The Romanesque Benedictine abbey-church of Maria Laach, completed after 1093, is a prime example. Its most obvious feature is not just its size and uniformity, but the impact of its bold masses—above all its imposing towers, three at each end, with the square tower at the west end being the most dominant. Boldly projecting apses reinforce the sense of massive masonry. The heavily shadowed interior is as daunting as it is impressive.

▲ THE ROUND STONE CHURCH CAPPED BY A CONICAL WOODEN ROOF AT OSTERLARS

Osterlars Church

Location Bornholm Island, Denmark

Like Norway's "stave" churches (see p.128 and p.332), the stone church at Osterlars, built on the Baltic island of Bornholm in c.1150, belongs to no recognizable tradition of European architecture. Instead, it represents the response of local Scandinavian traditions to Christianity, which by the mid-11th century was spreading across the former Viking world. The building, rough-hewn under its conical wooden roof, is circular. Its near-windowless interior boasts a remarkable column—hollow in its center, pierced by a series of round arches—that spreads outward to support the roof. A band of frescoes, probably dating from the 14th century, encircles it. They include a vivid representation of the Last Judgment, in which hell is represented by a dragon.

Borgund Church

Location Sogne Fjord, Norway

The exterior of the late 12th-century Borgund Church is an example of how the requirements of Christian worship were fused with existing Scandinavian building types to produce a unique architecture. Constructed almost entirely of wood, Borgund Church's most obvious features are the heaped-up, weathered shingle roofs (arranged over six levels) that culminate in a tapering spire. Also noteworthy are the elaborately carved dragons' heads that project from the upper levels. The dim and sparse two-story interior follows a conventional design: the nave, marked by 12 sturdy piers, is derived from Byzantine Greek cross models. Beyond it, a substantial square chancel gives on to an apse. An internal ambulatory surrounds the entire building.

Cloth Hall

Location Ypres, Belgium

The burgeoning prosperity of the Low Countries in the high Middle Ages (c.1000–1350) gave rise to a sense of civic pride that was expressed in a series of imposing public buildings. As "temples to trade," they were consciously intended to rival even the largest and most opulent cathedrals. The largest of all was the 13th-century Cloth Hall in Ypres. The principal facade, of three stories, is 433 ft (132 m) long. At its center, there is a vast square bell tower, turreted at its corners and topped by a lantern. Each end of the facade is marked by smaller but imposing spires. The entire structure, apart from the steeply pitched roof, is smothered in a wealth of decorative detail.

Siena Cathedral

Location Siena, Italy

Siena Cathedral, built after 1226, underlines how the absolutes of French Gothic architecture were rarely echoed in medieval Italy. Superficially, there are similarities with the Gothic style; these are most obvious on the main facade, articulated by three commanding portals and a series of sculptured figures, and the large central rose window above. However, the differences are striking. For example, the hexagonal crossing is covered by a distinctly non-Gothic dome, which is topped by an

▼ SHINGLE ROOFS OF BORGUND CHURCH

incongruous Baroque, 17th-century lantern. Similarly, black-and-white horizontal marble stripes run around the entire building. They are conspicuously present on the six-story, square bell tower, next to the south transept. This banding is continued throughout the vividly rich interior, which is also notable for its intricate marble floor.

Sainte-Chapelle

Location Paris, France

Sainte-Chapelle, completed in 1248, is the purest expression of the Gothic style in France. In essence, the chapel is no more than a delicately ribbed framework supported by flying buttresses, and its walls are its stunning stained-glass windows. The result is a uniform, vertiginous space. It was built as a chapel but has become an oversized reliquary, housing some of the most venerated relics in Christendom: a fragment of the true cross and the crown of thorns, acquired by the pious Louis IX. Where the interior is not gilded, it is painted. The deep-blue vaulted ceiling is covered with gold *fleur-de-lys*, and a needlelike *flèche* (spire) soars from the pitch of the roof. The church was heavily damaged during the French Revolution but restored in the mid-19th century.

San Pablo

Location Valladolid, Spain

The church of San Pablo in Valladolid was built between 1445 and 1468, although its most notable feature, the facade, was completed in 1550. It is an extreme example of what has come to be known as Isabeline architecture, named after Isabella I who, with her husband Ferdinand II, expelled the last of the Moors from Spain in 1492. If in style it sits somewhere between the last vestiges of the Gothic and the Classicism of the early Renaissance—laced with Flemish and Moorish influences—its most obvious characteristic is its extreme elaboration of detail. Two rough-hewn square towers frame San Pablo's pedimented facade, which bursts with sculptured decorative features, including figures, coats of arms, tracery, and finials.

▲ BRICK-BUILT GOTHIC STRUCTURE AT MALBORK CASTLE

Malbork Castle

Location Malbork, Poland

Malbork Castle on the Nogat River, built from the late 13th century onward, was the headquarters of the Teutonic Knights. This was a crusading order that, from 1233, had been given papal permission to convert Europe's Baltic states, chiefly Prussia, to Christianity. By 1309, when the Teutonic Grand Master established himself at Malbork, the Teutonic Knights controlled a substantial Christian Baltic state. In due course, the castle—built throughout in red brick—would become the world's largest, sprawling over 3.4 million square ft (320,000 square m). In addition to immensely powerful defensive walls and towers, the castle contains a series of sumptuous Gothic residences—notably the Palace of the Grand Master, a lavish, four-story brick-built structure completed in 1399.

Albi Cathedral

Location Albi, France

Albi Cathedral, begun in 1282, seems more like a fortress than a place of worship. This is no accident. It was built following the 12th-century Cathar heresy—a major challenge to the Roman Catholic Church—in Languedoc, southwest France. The brick-built cathedral was constructed as an absolute reminder of the reach of the Church, albeit filtered through local styles. In place of the soaring cathedrals of northern France, with flying buttresses, huge windows, and upward-reaching spires, Albi Cathedral has buttresses that merely bulge from the walls; its windows are little more than vertical slits; and the thick-set tower at the west end is more Romanesque than Gothic.

Conwy Castle

Location Conwy, Wales

Conwy Castle, in north Wales, was one of four castles built by Edward I of England to consolidate his conquest of Wales between 1277 and 1307. By any measure, it is a brutal, brooding structure. It was built on a more or less rectangular plan, its layout dictated by the narrow promontory on which it sits overlooking the Conwy River. Its limestone walls are 15 ft (4.6 m) thick, and at regular intervals along them are eight massive, circular battlemented towers, each 70 ft (21 m) high. There is a single entrance that originally featured a drawbridge over a deep moat. Within a century, however, gunpowder would render even such a seemingly impregnable fortress obsolete, and by the late 17th century, it had fallen into ruin.

Carcassonne

Location Aude, France

Seen from afar, Carcassonne appears as the model fortified medieval city: perched on a low hill, enclosed within a complete circuit of walls, and topped by towers and spires. In fact, much of the city today is a reconstruction, the brainchild of Eugène Viollet-le-Duc, 19th-century France's most industrious champion of the past. Medieval Carcassonne was a major fortification, reaching its peak in the 13th century under Louis IX, when it was crucial to the defense of France against the Spanish kingdom of Aragon. With the consolidation of the French monarchy in the mid-15th century, the city's significance inevitably faded. It was revived by Viollet-le-Duc in 1853, when he began the city's remarkable, if not always accurate, reconstruction.

◀ THE TORRE DEL MANGIA TOWER SEEN FROM THE COURTYARD OF PALAZZO PUBBLICO

Westminster Hall

Location London, England

Built for William II from 1097, Westminster Hall is the oldest surviving part of the Palace of Westminster (see p.178). Measuring 240 ft (73 m) long and 67 ft (20 m) wide, it was once the largest hall in Europe. However, the roof is its most remarkable feature. Its design is the most impressive example of a uniquely English contribution to Gothic style—the hammer beam. The roof was constructed from 1393 on the orders of Richard II. It consists of a series of huge oak timbers projecting inward, from which pointed wooden arches spring out to span the hall. It is not merely technically remarkable, but it is also very decorative; for example, the projecting timbers are carved with oversized angels.

Sant'Antonio

Location Padua, Italy

The idiosyncrasies of Italian Gothic style are nowhere more obvious than in the vast, late 12th-century basilica of Sant'Antonio, which combines elements from at least three different and apparently incompatible styles—French, Lombard, and Byzantine. For example, the church's roof is crowded with domes, rising from huge drums, interspersed with a series of slim towers, all with lanterns. Unusually, both the transepts and the nave are marked by gables, and those on the nave are separated by hefty buttresses. The similarly gabled facade is more coherent: the entrance is flanked by pointed arches containing attenuated windows under an elegant Venetian arcade. The interior contains a riotous assembly of decorative styles, from the Gothic to the Baroque.

Palazzo Pubblico

Location Siena, Italy

Siena's Palazzo Pubblico, completed by 1344, is among the most impressive medieval structures in Italy. The palace forms an eloquent, self-confident statement of civic pride, built at a time when the rival city-states of Italy continually sought to outdo each other. Built out of bricks above a limestone ground floor, it comprises two distinct elements. The imposing main body of the building, slightly concave though symmetrical, features a commanding four-story central block, its crenellated summit flanked by two three-story blocks. The central block has four bays, and the flanking blocks have three. To the right, a bell tower, the Torre del Mangia, soars 335 ft (102 m) high. Its elaborate limestone summit flares outward and is crowned by a further crenellated marble tower.

Ca d'Oro

Location Venice, Italy

The fantasy Gothic style of 15th-century Venice—a mixture of Gothic, Byzantine, and Arabic—is precisely captured by perhaps the most sumptuous residential building on the Grand Canal—the Ca d'Oro ("House of Gold"). The building was never completed: a three-bay wing, matching that on the right, was planned for its left-hand side. Nevertheless, what was built is remarkable, and would have been even more so when it was sheathed in gold leaf—hence the name. Constructed over three stories and a mezzanine, it features central bays flanked by windows with projecting arcaded balconies. The building has three loggias—the largest is on the ground floor, and the upper two feature interlinked, pointed arches decorated with tracery. Across the top of the facade runs a line of spires of rhythmically varying height.

Coca Castle

Location Coca, Spain

Coca Castle, in central Spain, is one of the last and largest of Europe's medieval castles. Built from 1453, mostly using brick, it is also a striking example of Mudéjar architecture, which developed in Spain during and after the Christian reconquest. It saw buildings that essentially followed established Latin Christian forms but were executed with Moorish details. In the case of Coca Castle, this meant that the overall plan of the castle—with a massive moat; tower-studded outer walls; and, at its heart, a heavily defended strong point, or keep—echoed the form of castle-building developed during the Crusades. By contrast, the clusters of polygonal corner towers, sharply pointed at their bases and fluted at their summits, are completely Moorish.

▲ POLYGONAL AND SEMICIRCULAR TOWERS OF COCA CASTLE

▲ RICHLY TILED WALL AT THE CHINLI KIOSK, TOPKAPI PALACE

Chinli Kiosk, Topkapi Palace

Location Istanbul, Turkey

The Chinli Kiosk is one of the oldest buildings in the Topkapi Palace, the sprawling complex built from 1473 on the orders of Mehmet II, the conqueror of what was Byzantine Constantinople. The administrative and imperial home of the Ottoman sultans, the Chinli Kiosk is a pleasure palace—delicate, airy, and instantly memorable for the glazed tiles with complex geometrical patterns that cover much of the exterior. It is also a typically Persian building, encircled by elegant arcades and polygonal columns. The building has a cruciform plan, with a shallow dome over its center. The overall impact is one of sophistication and worldliness. It would immensely influence subsequent Ottoman architecture.

St. Martin

Location Landshut, Germany

The late-Gothic St. Martin in Bavaria, completed in 1498, is what is known as a hall church—its nave and aisles are the same height. Also, it has no transepts. The effect is the creation of a single, soaring interior, emphasized by plain polygonal columns that line the vertiginous nave and the pencil-thin windows in the aisles that flood light into the building. The stern elegance this creates is reinforced by the sparse rib-vaulting—the ribs are picked out in a somber, flat purple-red. The exterior features are arranged in ordered, repeating patterns below a steep red roof. Its most dramatic feature, dominating the church and town alike, is the slender brick tower over the main entrance, 428 ft (131 m) high.

Tempietto of San Pietro in Montorio

Location Rome, Italy

The importance of Donato Bramante's Tempietto, a small tomb set in the courtyard of the church of San Pietro in Montorio, vastly outweighs its diminutive size. Built in 1502, it is arguably the most influential building of the Renaissance. Its Classicism, combined with a sense of harmony, makes it the first architectural masterpiece of the Renaissance. It is circular, raised on three shallow steps, with 16 Doric columns supporting an entablature over which there is a balconied walkway. From this rises a high drum, with alternating round- and square-headed niches, above which sits the supreme architectural symbol of the Renaissance—a dome (see pp.146–147).

Segovia Cathedral

Location Segovia, Spain

Segovia's vast cathedral, built in 1525–1577, represents a last stand of the Gothic style in Western Europe. Externally, the most notable features are the elaborate finials that rise from almost every point. Flying buttresses spring from many of the finials. The austere bell tower rises 290 ft (88 m) to one side of the main entrance. There is a similar sense of the vertical inside, the most notable feature being complex, decorative fan-vaulting of the nave and aisles. Out of style with the rest of the building, both the bell tower and the crossing are crowned with Baroque domes, added in 1614 and 1630 respectively, each carrying a lantern.

Suleymaniye Mosque

Location Istanbul, Turkey

The Suleymaniye Mosque was built in 1550–1557 on the orders of the greatest Ottoman sultan, Suleiman the Magnificent. Intended to rival Hagia Sophia (see pp.108–109), it is, in equal measure, a deliberate statement of Ottoman might and Ottoman piety. It is an immense, brooding structure, dominated by a shallow dome, 174 ft (53 m) high. Smaller domes bulge around the building, and there are four thin minarets. The interior, under its glittering, golden dome, consists of a single space flanked on two sides by half-domes and on the other two sides by arches with rows of round-headed windows. The arches of the piers supporting the dome are

picked out in alternating red and white. The mosque is reached via a stately, arcaded courtyard.

Villa Farnese

Location Caprarola, Italy

The Villa Farnese was originally planned as a fortress, but only its five-sided foundations were constructed. In 1556, it was decided instead to build an imposing villa on those foundations. The result, designed by Giacomo da Vignola, is among the largest and most important Italian villas of the 16th century. It is a monumental and supremely confident building. Its base is heavily rusticated; its two, many-windowed upper stories are symmetrical and precise; and it is capped by a boldly projecting cornice. At the villa's heart lies a circular courtyard formed by pairs of Ionic columns. Perhaps its most striking features are the theatrical stairways that lead to the building. The first, curving pair give way to two further flights that dramatically double back on each other.

▼ CIRCULAR STAIRCASE WITH DOUBLE COLUMNS IN THE VILLA FARNESE

Town Hall, Antwerp

Location Antwerp, Belgium

As the influence of the Renaissance spread across Europe in the 16th century, at least initially, Classicism was frequently adopted alongside existing architectural styles. Antwerp's town hall, completed in 1564, is an excellent example of the resulting hybrid—Flemish fused with Classical. Set on a rusticated base, it features nine bays on each side of three central bays that project slightly forward. Each of the central section's five stories, which burst through the roof-line, is conceived as a series of triumphal arches: three on the first two floors, two on the third, and one each on the final two. The upper floor is crowned by a pedimented gable. Slim obelisks rise from each side of the third level.

▲ FOUNTAINS IN THE GARDENS OF VILLA D'ESTE

Palace of Charles V

Location Granada, Spain

The Holy Roman Emperor Charles V was the most powerful ruler in Europe, commanding a vast accumulation of territories across the continent, as well as in the Americas. In 1527, he commissioned a new palace in Granada. However, when work stopped in 1568, it was still incomplete. Nevertheless, it is a remarkable building—majestic and idiosyncratic. Its principal facade of only two stories consists of 18 bays flanking a wider entrance bay, with a pedimented door below and an arched window above. The ground floor is very heavily rusticated, while the upper floor is crowded with pilasters and windows. The astonishing courtyard features a vast and austere, circular, two-story colonnade—Doric below, Ionic above—with arcades on both levels.

Villa d'Este

Location Tivoli, Italy

The significance of the Villa d'Este, created for the hugely wealthy Cardinal Ippolito II d'Este (1509–1572), lies as much in its gardens as in the villa itself. In the 16th century, for the first time since the Roman empire, a concerted effort was made to produce landscaped gardens on a major scale. The goal was to create a kind of idealized landscape, alive with allusions to the Classical past, filled with statues that were intended to delight and surprise. Above all, fountains—symbols of abundance—expressed this new vision. They abound at the Villa d'Este. The result was a kind of intellectually exuberant recasting of nature, spilling downward terrace after terrace.

Selimiye Mosque

Location Edirne, Turkey

Completed in 1574, the Selimiye Mosque was designed by Sinan, the architect of the Suleymaniye Mosque in Istanbul (see p.336). He declared it his finest work. Again, bold masses, built up layer by layer, are deployed to startling effect, culminating in a vast, shallow dome that rears 138 ft (42 m) over the city. Four 272-ft- (83-m-) high minarets frame the entire structure. The breathtaking interior is simultaneously monumental and seemingly weightless. The dome is carried on eight vast piers that rise into huge arches. Outside and below them are further arches. Throughout, there is a rhythmically repeating series of small, round-headed windows through which light floods. The dome itself is a wonder of vividly precise decoration.

El Escorial

Location Central Spain

The Escorial, more properly El Real Monasterio de San Lorenzo de el Escorial, completed in 1584, was built by Philip II of Spain as the palace of the kings of Spain. It is a daunting structure, a reflection of the power of not just Europe's richest ruler but also, after the Reformation, the Catholic Church's most implacable champion. At its heart is a church dedicated to St. Laurence, after whom the Escorial is named. The gridlike floor plan is claimed to represent the grill over which St. Laurence was said to have been martyred in the 3rd century CE. With its 4,000 rooms and 15 miles (24 km) of passages, the Escorial is a building that enshrines scale over any more obvious architectural virtues.

Château de Fontainebleau

Location Fontainbleau, France

Fontainebleau was the favorite of the energetic King François I's vast architectural creations. Situated in the forest of Fontainebleau, southeast of Paris, from the 12th century it was the site of a royal hunting lodge. But from 1528, under the direction of Gilles Le Breton, François utterly transformed it. Although it has been added to by later French monarchs, today's building is essentially his. Its plan is nothing if not irregular. It features an asymmetric assemblage of wings, courtyards, and connecting blocks, all under steeply pitched roofs punctuated by turrets, towers, and chimneys. Inside, it is lavish to a fault, with the Galerie de François I containing the first major cycle of Renaissance frescoes in France. It sits in immense, precisely landscaped gardens.

Rialto Bridge

Location Venice, Italy

There are only four bridges over Venice's Grand Canal. The oldest and most celebrated of these is the Rialto. It was completed in 1591, to designs by the aptly named Antonio da Ponte. It was the first stone bridge over the canal, replacing a number of wooden bridges that were prone to collapsing. It represented a significant engineering challenge, not so much because of its 157-ft (48-m) span but because of its substantial 26-ft (8-m) width—the result of the need for two rows of shops on the bridge to generate revenues for its upkeep. At its center is a pedimented triumphal arch flanked on each side by six arches. Elegant balustrades form pedestrian passages on both sides of the bridge.

Bridge of Sighs

Location Venice, Italy

The limestone Bridge of Sighs—a name coined only in the early 19th century, by Lord Byron—links the courtroom of the Doge's Palace to the heavily rusticated New Prison. It was built in 1600 by Antonio Contino. It may be small, spanning only 36 ft (11 m), but it remains among Venice's most instantly recognizable sights. It is supported by a shallow arch, the shape of which is echoed by the pediment that tops the bridge, on which are curving, wavelike volutes. Two square windows with stone grills on each side of the bridge are flanked by diminutive banded pilasters. Eleven slightly menacing carved stone heads line the lower arch.

▶ THE ARCHED AND ENCLOSED STRUCTURE OF THE BRIDGE OF SIGHS

Church of San Giorgio Maggiore

Location Venice, Italy

The glory of Andrea Palladio's Church of San Giorgio Maggiore, built between 1566 and 1610, is its commanding facade, which towers over the wide waters of St. Mark's Canal facing the Doge's Palace (see pp.144–145). It presents an extremely satisfying solution to a perennial problem facing Renaissance architects: how to reconcile a Classical temple front with the liturgical demands of a Christian church, with a high nave and lower aisles. Palladio's answer was to use two pediments: a high central one for the nave, superimposed on a lower, broken pediment for the aisles. The four three-quarter composite columns of the central pediment are raised on vast bases. Five similarly oversized sculptures adorn the roof-line.

Basilica Palladiana

Location Vicenza, Italy

The grandeur that Andrea Palladio so effortlessly achieved in his creations is typified by the Basilica Palladiana. Built in stages between 1549 and 1617, although entirely to Palladio's designs, it forms a screen around Vicenza's 15th-century Palace of Justice. Arranged over two stories, it consciously aims to recreate the nobility of a Roman public building. Both stories—the lower Doric and the upper Ionic—consist of nine arched bays separated by 10 half-columns. The arches are supported by smaller free-standing columns set in slightly from the wall to create what are sometimes called Serlian or Palladian windows. Balustrades run between the openings on the upper story and along the length of a roof-line studded with statues. The roof is memorably copper clad.

Onze Lieve Vrouwekerk

Location Scherpenheuvel, Belgium

Onze Lieve Vrouwekerk, or the Church of Our Lady, built between 1609 and 1627, was the first fully Classical building in what was then the Spanish Netherlands. It is an imposing, seven-sided structure. A Palladian, arched motif forms the main entrance, flanked by niches between pilasters. The upper story features a boldly broken pediment. The church's dominant feature is its vast, spreading dome, studded with little metal stars and buttressed by oversized volutes separated by an arched window. At its summit is an elaborate domed lantern. The interior of the church is richly decorated. Its prime focus is a miracle-working statue of the Virgin Mary—discovered in the 13th century—for which the church was built.

The Exchange

Location Copenhagen, Denmark

The Exchange, or Børsen, built between 1625 and 1640, is potent proof of the determination of Christian IV of Denmark to transform the Danish capital. Made of bricks but with seemingly endless

decorative stonework details, the building is long and low—only two stories high—but has vertical elements provided by nine voluted gables that project boldly up and outward. The facades at both ends feature similarly sumptuous gabling, arranged over five bands. But the Exchange's most distinctive feature is the copper-sheathed spire, 184 ft (56 m) high, which rests on an open-sided lantern in the center of the steeply pitched copper roof. It comprises four entwined dragons' tails that spiral exotically upward.

Les Invalides

Location Paris, France

Les Invalides was begun by Louis XIV in 1670 as part hospital, part home for the veteran soldiers of France. Among the most ambitious urban projects of the Baroque period, it is a vast complex of 17 stately courtyards in the heart of Paris. Its most imposing feature is the church of the Dôme des Invalides—from 1840, the site of Napoleon's tomb—designed by Jules Hardouin-Mansart. Its two-story facade, separated by a pronounced entablature, features clusters of columns under a projecting pediment. From this rises a double-drum, with pairs of projecting columns below and volutes above. It supports the most dramatic dome in France, glistening with gold ribs and trophies of arms, from which a slim, gold-and-green spire thrusts 351 ft (107 m) above the city.

▼ THE SPRAWLING COMPLEX OF LES INVALIDES

Zwinger

Location Dresden, Germany

The Zwinger palace—built between 1710 and 1728, then rebuilt after World War II—is perhaps the most outlandish structure in Europe. In its different guises as a courtly arena, pleasure garden, art gallery, and theater, it exudes exuberance. It was built for the Saxon Elector Augustus the Strong by his court architect, Matthäus Daniel Pöpplemann. It is essentially an arcaded courtyard, with four fountains at its center, formed by low galleries interrupted by larger buildings. Two sides are apsed. In the center of three of its sides are entry pavilions of startling originality—part Baroque, part Rococo. The whole building is festooned with architectural and sculptural decorations.

Abbey Church

Location Weltenberg, Germany

The unremarkable exterior of the Abbey Church at Weltenberg, on the banks of the Danube, promises little. By contrast, the interior marks the first great example of one of the Rococo's most distinctive qualities—the fusion of painting, sculpture, and architecture—in an ingenious spatial plan. Built in 1717–1721, the church is the work of the Asam brothers, Egid Quirin and Cosmas Damian. Despite the lavishness of the oval nave, with its paired columns, the main focus is on the gilt and silvered equestrian statue of St. George, over the high altar, slaying a dragon. From the hidden light sources to the twisting barley-sugar columns that frame the altar, the impact is compellingly theatrical.

The Spanish Steps

Location Rome, Italy

The sinuous cascade of Rome's Spanish Steps, built between 1721 and 1727, is both the final flourish of the city's ambitious Baroque urban schemes and the best example of the period's dramatic use of steps as an architectural device. They sweep commandingly between the church of the Trinità dei Monti at their summit and the Piazza di Spagna at their foot—hence the name. Halfway up is a terrace; as the steps rise toward it, they narrow. As they pass the terrace, the steps widen before narrowing and then widening again to pass a further terrace at the summit. The obelisk in front of the church is a later addition, erected in 1786.

Melk Abbey

Location Melk, Austria

Completed in 1736, Jakob Prandtauer's late-Baroque Benedictine abbey and church at Melk takes full advantage of its setting on a rocky spur above the River Danube. Long and low under its steep-pitched red roofs, it rises to a glorious crown in the church itself. This sits in an apsed courtyard formed by the projecting wings of the monastery, over which two highly ornate towers rise under onion domes picked out in gold and green. Behind them is a double dome on a high drum. The impact is heightened by the color scheme of white-and-ocher stucco. The slightly later interior of the church is a symphony of deep-red marble and gold under soaring frescoes.

▲ ORNATE DETAILING ON A SPIRAL STAIRCASE AT MELK ABBEY

Amalienburg

Location Munich, Germany

If the Rococo was essentially a decorative rather than architectural style, the Amalienburg, built between 1734 and 1739 on the grounds of the Nymphenburg Palace on Munich's outskirts, is its supreme expression. Externally, the Amalienburg is a modest, one-story structure. Internally, it explodes in a riot of precise decoration. The dazzling highlight is the circular Hall of Mirrors at the heart of the building. It largely consists of a series of round-headed mirrors around which—and sometimes over which—a profusion of silver-gilt stucco tumbles, featuring animals, cornucopias, cherubs, plants, and musical instruments. Where it strays onto the cool blue of the domed ceiling, it metamorphoses into a series of exotic landscapes.

New Cathedral

Location Salamanca, Spain

Salamanca's New Cathedral is a true hybrid. A late-Gothic cathedral of exceptional splendor, begun in 1513, it boasts an 18th-century Baroque dome over its crossing and a similarly domed bell tower. Adding to the incongruity is the fact that the city's original 12th-century Old Cathedral is not only still intact but effectively forms part of the southern wall of the New Cathedral. The most impressive elements of the building's 16th- and 17th-century Gothic features are the three richly carved ogee (S-shaped) arches on the main facade and the massive flying buttresses that line the nave. Over it all, on its tall drum with double three-quarter columns and high, round-headed windows, rises the authentically Baroque dome.

Frauenkirche

Location Dresden, Germany

However surprising it may be that Dresden's Lutheran Frauenkirche, completed in 1743, was built in a style so closely associated with Catholicism, it is an emphatically bold Baroque building on a towering scale. Other than the projecting semicircular chancel at the east end, it is centrally planned. The three remaining facades, all with large, round-headed windows, have broken pediments on giant pilasters. The corners are angled and topped by curved cornices from which rise ornate towers. Above them is a steep-sided dome crowned by a lantern under an onion dome. The echoing interior is brightly lit. The building was largely destroyed in World War II and rebuilt, using some of the original stones, between 1992 and 2005.

Rottenbuch Abbey

Location Rottenbuch, Germany

Rottenbuch Abbey in the foothills of the Bavarian Alps almost perfectly encapsulates the Bavarian affinity with the Rococo. Built between 1085 and 1125, the church itself is Romanesque and traditionally laid out with a nave, aisles, a crossing, and a chancel. In the mid-18th century, the interior was transformed by stuccoist Josef Schmuzer and fresco painter Matthäus Günther. The building's original Gothic architecture was smothered in stucco—mainly pink, white, and gold. Inside, Günther's illusionistic frescoes dramatically extend the real space of the building into imaginary heavens above.

Catedral Metropolitana de Santiago

Location Santiago, Spain

Structurally, the Romanesque origins of the Catedral Metropolitana de Santiago are unmistakable. However, decoratively, especially externally, the building displays 18th-century Churrigueresque detailing—a form of elaborate decoration popularized by the Churriguera family from Barcelona. The essential elements favored by the Churrigueresque are more or less Classical, but their impact is not, and any discordance generated by clashes between styles is simply ignored. There is an array of intricate detailing. The skyline of the main facade—a central lantern flanked by towering square towers, with volutes and turrets abounding—is particularly audacious.

▶ ACIBECHARÍA FACADE OF THE CATEDRAL METROPOLITANA DE SANTIAGO

Radcliffe Camera

Location Oxford, England

The rationality of 18th-century English Classicism—exemplified in architecture by simple geometric shapes—is precisely expressed by the circular Radcliffe Camera, designed by James Gibbs. Completed in 1748, it is simple in its conception and assured in its execution. Its 16-sided ground floor, with alternating pedimented doors and niches, supports a second story with pairs of giant three-quarter columns and, between them, two levels of windows—the upper arched and the lower square. A plain entablature and cornice above give way to a balustrade, punctuated by pairs of urns over the columns. Atop it all sits a strongly ribbed dome crowned by a lantern.

Wieskirche

Location Bavaria, Germany

No building better encapsulates the Bavarian Rococo than the Wieskirche, built between 1745 and 1754 by the Zimmerman brothers, Johann and Dominikus. Its exterior promises little; internally, it delights and astounds. It is the Rococo at its most uninhibited and self-confident. There is almost no part of its gleaming white interior that is not animated by elaborate stucco and spreading frescoes. However, there is also almost no part that does not follow a precise, liturgically dictated color scheme. If it is an interior that revels in decorative delights, it is no less one in which the religious intent is clear.

▼ GOLD-CLAD EXTERIOR OF THE CATHERINE PALACE

Catherine Palace

Location Tsarskoe Selo, Russia

The emergence of Russia, sparked by the energetic reforms of Peter the Great (r.1682–1725), was underlined by the construction of a series of vast imperial buildings in and around the new Russian capital of St. Petersburg. Among the most sumptuous and impressive is the Catherine Palace, built for the Empress Elizabeth, between 1752 and 1756. It was designed by Bartolomeo Rastrelli, the architect of the Winter Palace at St. Petersburg (see p.176). The building's extravagant magnificence is

encapsulated by the 978-ft- (298-m-) long principal facade. Other than the cluster of gilded onion domes at one end, it is a precisely symmetrical structure, with five projecting blocks separated by giant columns, the whole multicolored and gilded.

▲ THE IMPOSING INTERIOR OF THE PANTHÉON IN PARIS

Bom Jesus do Monte

Location Braga, Portugal

The significance of the church of Bom Jesus do Monte, built from 1784, lies less in the comparatively sober two-story church itself, its facade flanked by two bell towers, than in the extraordinary flight of 577 steps that leads to it. Their interlocking, zigzagging design, festooned with increasingly elaborate statuary and urns, is made even more memorable by a contrast of white stucco and gray, weathered stonework. However, they are more than simply an architectural tour de force; they are also packed with religious symbolism. At every intersection, there are fountains—those at the lower levels representing the senses; those higher up, the virtues. Similarly, each crossing point contains one of the 14 Stations of the Cross that document the Crucifixion of Jesus.

Catacombs of Paris

Location Paris, France

Also known as the Denfert Rochereau Ossuary, the Catacombs of Paris are located on Paris's Left Bank. There are perhaps 174 miles (280 km) of these tunnels, containing about 6–7 million bodies. They were not built as catacombs but instead were quarries, some dating back to Roman times, that were hurriedly pressed into service after 1786 to house the dead that could not be accommodated by the city's overflowing cemeteries. Only a few can be visited, but they make for a genuinely macabre experience, not least because in 1810 the inspector-general of the quarries ordered that the bones be precisely, almost decoratively stacked.

Panthéon, Paris

Location Paris, France

The Panthéon, in Paris's Latin Quarter, is among the city's most stately buildings. Designed by Jacques-Germain Soufflot and built between 1757 and 1790, it was originally intended as a church honoring the city's patron saint, St. Geneviève. But in 1791, the National Constituent Assembly, set up after the French Revolution, decreed it a mausoleum honoring the great and the good of France. Externally, its most dramatic features are its projecting central portico, its pediment carried by 22 Corinthian columns, and its high dome supported by a colonnaded drum. The interior is gloomily sepulchral, the result of its windows having been shuttered. However, the marbled grandeur of its Greek-cross plan is immediately evident.

Custom House

Location Dublin, Ireland

Although it was rebuilt after being damaged in 1921, during the Irish War of Independence, James Gandon's Custom House, originally completed in 1791, remains among the most elegant of the city's many 18th-century buildings. Its most memorable facade faces south over the waters of the River Liffey. It is focused on a projecting, pedimented central block flanked by pairs of recessed Doric columns. These flanking elements are repeated in two further three-bay pavilions at each end of the building. A slim, elegant limestone dome rises over the pediment, its drum articulated by pairs of columns—four projecting discreetly outward. The lightly rusticated north facade is no less soberly Classical.

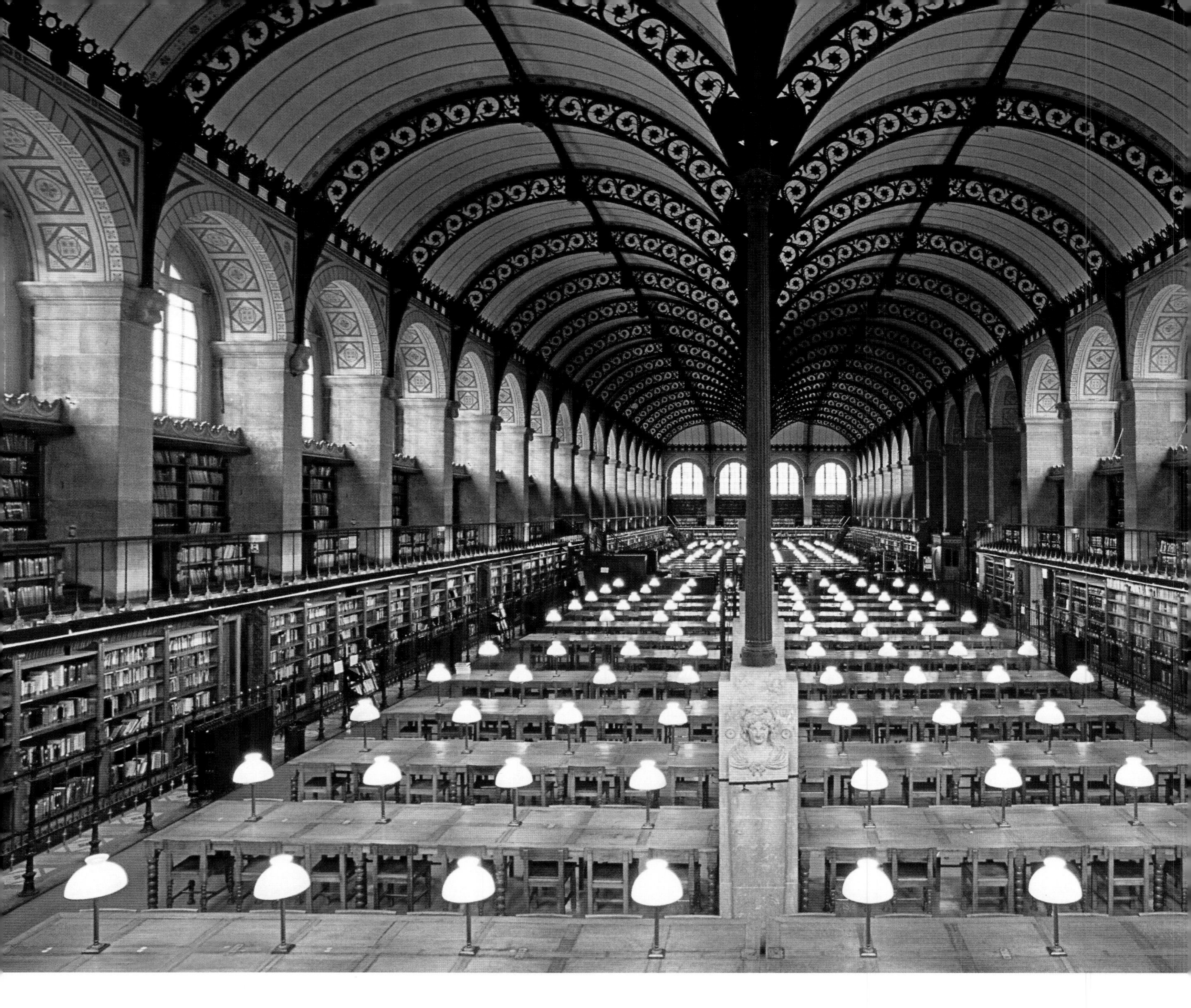

▲ THE READING ROOM OF THE LIBRARY OF SAINTE-GENEVIÈVE

The Palm House, Kew Gardens

Location London, England

Like its immediate predecessor, Joseph Paxton's great glass conservatory at Chatsworth House, the Palm House at Kew Gardens made clear the advantages of prefabrication in iron and glass. It also demonstrated how iron-framed buildings were poised to revolutionize architecture.

The Palm House was built between 1844 and 1848 by the otherwise Classically inclined British architect Decimus Burton and Irish iron engineer Richard Turner. The pair subsequently teamed up to build the Temperate House at Kew. Burton's influence is most obvious in the structure's bulging curves, above all in the high central vault—the whole forming a series of uninterrupted surfaces. The interior, with its slim iron columns, spiral staircases, and elevated walkways, allows itself the occasional decorative flourish.

Library of Sainte-Geneviève

Location Paris, France

Henri Labrouste, the architect of the Library of Sainte-Geneviève, completed in 1850, was among the pioneers of iron-framed construction. The restrained two-story exterior of the library, in effect a kind of Renaissance palazzo with rhythmically repeated arched windows, provides no hint of the technological advances of the vast reading room within. This room consists of two barrel vaults that run the length of the building, supported by 16 slim cast-iron columns on high stone bases. From them spring elegantly curved transverse iron ribs. The architectural language may have looked backward, but the building's use of iron looked decisively forward. Labrouste's later National Library in Paris would be an even more emphatic assertion of the transforming possibilities offered by industrial technology.

Crystal Palace

Location London, England

Joseph Paxton's Crystal Palace, as it was instantly dubbed, was the single most important building of the first phase of the Industrial Revolution of the 19th century. It was the setting for the Great Exhibition of 1851, the brainchild of Prince Albert, intended to showcase Great Britain's myriad technological advances. This called for an exceptional setting, and it got it. The Crystal Palace was 1,851 ft (564 m) long and rose to 128 ft (39 m). Constructed entirely of iron and glass, it was prefabricated and erected on site in Hyde Park. Its design was simple: a central flat-roofed aisle flanked by two lower flat-roofed aisles. At its center, a vaulted transept large enough to contain mature trees cut across them. It was a dazzling demonstration of Britain's industrial prowess. Relocated to Penge Common, South London, after the exhibition, it was destroyed by fire in 1936.

Lutheran Cathedral

Location Helsinki, Finland

Helsinki's bold Neo-Classical cathedral dominates the heart of the city. It was designed by a German, Carl Ludvig Engel, but finished, in 1852, by a second, Ernst Lohrmann, after Engel's death in 1840. Stark and simple, it comprises a central block with four identical six-columned, pedimented porticos on each facade. It sits on what amounts to a gigantic granite podium approached by steps. Above the central block towers a hefty green dome on an elongated, two-story drum. Four smaller drums, added by Lohrmann, are at the roof corners. The grandeur of the building is enhanced by its setting in Engel's spaciously harmonious Senate Square. The dazzlingly white, sparse interior is flooded with light.

St. Isaac's Cathedral

Location St. Petersburg, Russia

Although it is on a vastly greater scale than Helsinki's Lutheran Cathedral (see above), St. Isaac's Cathedral is essentially the same in plan: a Greek cross with four identical porticoed facades having pediments borne by giant columns. A dome, flanked by four smaller domes, dominates the structure. However, its execution could hardly be more different from the Lutheran Cathedral. Apart from its impressive size—the building can hold 14,000 people—this comes down to materials. St. Isaac's, the work of French architect August Montferrand, completed in 1858, is ostentatious. Its domes are gilded, and its columns are made of red granite. Forty-eight bronzelike groups of sculptures stud the roof lines. The interior is more lavish still and glints with colored marble and semiprecious stones.

Clifton Suspension Bridge

Location Avon Gorge, England

There is no structure that more emphatically proclaims the technological advances of mid-Victorian Britain than the Clifton Suspension Bridge. Construction began in 1831, but the bridge was not completed until 1864. It was the engineering wonder of its day. Measuring 1,350 ft (412 m) long, it dramatically spanned the rocky Avon Gorge 250 ft (76 m) below. Its architectural pretensions are made clear by the Egyptian motifs on the two vast splay-footed towers that support it. But it is as an example of engineering daring, in which Brunel excelled, that the bridge impresses. Although it was widely presumed that the bridge would collapse, it remains sturdily intact.

▼ THE IMPOSING TOWERS OF THE CLIFTON SUSPENSION BRIDGE

Turbine Building, Menier Chocolate Factory

Location Noisiel-sur-Marne, France

The Turbine Building at the Menier Chocolate Factory, east of Paris, is a purely functional building used to mill cocoa beans. Positioned on the River Marne, the mill's power was supplied by three water-driven turbines. It was designed by Jules Saulnier and completed in 1874. Its importance lies in the fact that although externally it is made of pale yellow bricks, its internal iron frame is tied together by diagonal bracing on the outside of the building. In other words, the bricks have no structural role at all. However, the brickwork is highly decorative, with a range of motifs based on flowers and cocoa beans. The impact is appealingly playful.

▲ THE PALAIS GARNIER'S AUDITORIUM, RICHLY DECORATED IN RED AND GOLD

Palais Garnier

Location Paris, France

The Palais Garnier is the home of the Paris Opéra, a ballet and opera company, and the building itself is often simply called the Opéra. Designed by Charles Garnier, it opened in 1875. It was both the most feted and the most controversial structure in the rebuilding of Paris initiated by Napoleon III (r.1852–1870). It was criticized as a vanity project on the part of the emperor, and for its immense cost. The mixing of Renaissance, Baroque, and Rococo styles and lavish decoration led many to brand it merely vulgar. Nevertheless, its almost heroic self-assertion—expressed most obviously in its huge, marbled foyer, dominated by a giant staircase that branches in two—is a potent assertion of Belle Epoque opulence.

Galleria Vittorio Emanuele II

Location Milan, Italy

Giuseppe Mengoni's towering Galleria Vittorio Emanuele II, built between 1865 and 1877, was always meant to be more than a vast covered public space. Its goal was to showcase a new, technologically potent Italy after the country's unification in 1861, while paying homage to the country's past. Its iron-and-glass tunnel-vaulted roofs and octagonal glass dome are clear indications of Italy's industrial prowess and presumed future. Simultaneously, its cross-shaped floor plan recalls its Christian past. No less symbolically, the longest arms run from the Piazza della Scala opera house to the Piazza del Duomo museum, reinforcing the link between the country's artistic and religious heritage. The architecture of the body of the structure is impressively, if ponderously, Classical.

Cardiff Castle

Location Cardiff, Wales

Constructed on Roman foundations, Cardiff Castle is a Norman building that, between 1868 and 1885, was wholly transformed by the most imaginative English Victorian Gothic revivalist, William Burgess. Burgess was employed by the enormously rich Marquess of Bute, whose great delight in the medieval world matched that of his architect. They proved a formidable pairing. The chief joy of Cardiff Castle lies in its intricate and astonishingly rich interiors, most of them vast—the banqueting hall and library in particular. The Arab room is a no less astounding expression of brilliantly vivid medieval fantasy, its ceiling a dizzy, overlapping maze of Moorish arches climbing upward. Imposing stone fireplaces, brilliantly colored, abound. Almost everywhere, there are sudden touches of precisely crafted whimsy.

Palácio da Pena

Location Sintra, Portugal

Progressively remodeled between 1842 and 1885 as a summer retreat for Portugal's royal family by the German architect Baron Ludwig von Eschwege, Palácio de Pena ranks high among the world's most unlikely buildings. Built on the remains of a late 15th-century monastery, it rises, domed, turreted, and castellated, on a steeply wooded crag over the elegant town of Sintra. It is part Rhineland castle, part Moorish stronghold, and part mere fantasy belonging to no obvious architectural type. The building's asymmetrical, blocky elements are vividly rendered in ochers, purples, and deep reds. Few buildings are more improbable or possess so clear a sense of romantic unreality. Its rich, dusky interiors conjure a feeling of equal astonishment.

▶ THE BRIGHTLY COLORED EXTERIOR OF PALÁCIO DA PENA

Rijksmuseum

Location Amsterdam, the Netherlands

One of the world's great galleries, the Rijksmuseum is home to about a million objects from every artistic field. Designed by Dutch architect Pierre Cuypers, it was built in 1876–1885. With its turrets and towers, gables and dormers, and boldly projecting masses with stone banding and molding over its deep-red bricks, it might almost be a hybrid late-Gothic-early-Renaissance Loire château. It has two main facades. The central bays of that facing the Singelgracht canal are animated by large, round-headed traceried windows. The opposite facade features a no less imposing entrance arch. The interior is organized around two substantial, glassed courtyards. From 2004 to 2012, the museum underwent a huge renovation program aimed at restoring Cuypers's lavish interiors.

Forth Bridge

Location Firth of Forth, Scotland

By any measure, Scotland's Forth Bridge is an impressive engineering feat. When it was opened in 1890, it was easily the longest bridge in the world. It spans a length of 8,094 ft (2,467 m), and its two central sections are each 1,709 ft (521 m) long. About 58,000 tons (53,000 tonnes) of steel were used, along with 640,000 cubic ft (18,000 cubic m) of granite for the 12 piers that support the steel structure. The bridge is thought to contain 6.5 million rivets. At the peak of its construction, it required 4,600 workers—of whom 57 were killed. It cut the journey time between London and Aberdeen by almost 5 hours.

Hôtel Tassel

Location Brussels, Belgium

If externally, Victor Horta's Hôtel Tassel generally follows Classical examples, albeit modified, the interior of the building was startlingly novel. A private house, built between 1893 and 1894, it is considered the first building in the Art Nouveau style, which dominated the arts between about 1890 and 1910. For the first time, here was a house that did not look back to historical models, but created an entirely new architectural language of its own. The building's unity was ensured by Horta having designed every detail. The most notable aspects of its revolutionary interior are the central light well, which floods the space with yellowish light, and the use of an open plan, with rooms and spaces flowing together.

Majolica House

Location Vienna, Austria

Compared to its weightily Classical neighbors, Otto Wagner's Majolica House, completed in 1899, is almost shockingly modern. The building itself—two stories of shops with four stories of apartments above them—is restrained, its windows regular and rectangular. By contrast, its decoration has a refined yet vivid Art Nouveau opulence. Ceramic tiles of red flowers are used across the building, their design increasing in complexity as they climb toward the roof, reaching a boldly projecting cornice, with a row of open-mouthed lions below it. The cornice itself is sheathed in elegantly attenuated acanthus leaves below striking red poppies. The building is framed on either side by cast-iron green balconies that jut from recessed vertical bays.

Town Hall, Copenhagen

Location Copenhagen, Denmark

Copenhagen's brick-built town hall, designed by Martin Nyrop and completed in 1905, was a potent vote of confidence in Denmark's future. However, architecturally, it

▲ THE MAIN ALTAR OF LIVERPOOL CATHEDRAL

looks backward to the civic buildings of the late Middle Ages—especially the Palazzo Pubblico in Siena (see p.335). The facade is animated at either end by impressive oriel windows (protruding bay windows). Over the principal entrance is a gilt figure of the 12th-century Bishop Absalom. The skyline is dominated by bold castellations. To one side is a similarly bold bell tower under a soaring spire, 348ft (106m) high. Throughout, the detailing is precise and lavish. The interior features a huge, glassed-in courtyard, with a handsome arcade in early Renaissance style running around three of its upper stories.

Liverpool Cathedral

Location Liverpool, England

Liverpool Cathedral, with its assured, bulky masses, is the last great building of the Gothic Revival in Britain and the largest cathedral in the UK. It was begun in 1903 and completed some 75 years later. It was the first building designed by its architect, Giles Gilbert Scott, but he proved an inspired choice. From the deeply recessed, soaring lancet windows over the main entrance to the great square tower that rears 280ft (85m) over the crossing, the building was conceived on the grandest of scales. The red sandstone exterior and lavish, shadowy interior complement each

other precisely. The result is a building that exudes confidence and manages to be simultaneously uplifting and brooding.

Glasgow School of Art

Location Glasgow, Scotland

Charles Rennie Mackintosh's Glasgow School of Art was built in two phases: between 1897 and 1899, and in 1907. It is the most original and important Art Nouveau building in Britain. While it found admirers across Europe, it had few in Britain. It is a building of astonishing complexity. Playful and stern, it is constricted by its site but is ever inventive. The building, especially its principal facade, seems little more than a kind of factory, with wide, repeated iron-framed windows. However, sudden details intrude and symmetry is cast aside. The same mixture of the dutiful and the subversive, the rational and the organic, spreads through the interior. In May 2014, the building was extensively damaged by fire. In June 2018, a second, more serious fire left its future in doubt.

Vittorio Emanuele II Monument

Location Rome, Italy

Although celebrating the reunification of Italy, the immense Vittorio Emanuele II Monument is a structure that looks decisively to the Roman and Renaissance past for its inspiration. It was designed by Giuseppe Sacconi. Built intermittently from 1885, it was finally completed in 1935. An equestrian statue of Vittorio Emanuele is placed in the middle of a vast, white-marbled stage set at the head of an imposing flight of steps. Behind him, a colonnade recalling St. Peter's Square (see pp.158–159) ends in two Roman temple fronts, topped by bronze chariots. Numerous other sculptural groups embody virtues such as work, liberty, sacrifice, and patriotism. Fountains to the left and right symbolize the Adriatic and Tyrrhenian Seas.

▶ THE BRONZE STATUE OF VITTORIO EMANUELE II SET UPON A MARBLED STAGE

City Hall, Stockholm

Location Stockholm, Sweden

The City Hall in Stockholm, designed by Swedish architect Ragnar Östberg, was built in 1911–1923. It belongs to no obvious architectural style, but draws elements from many. It is detailed, inside and out, and, although brick-built, its impact is anything but utilitarian. Its most dominant external feature is the immense, tapering tower on the southeast corner of the building. On top sits an open copper lantern under a squashed golden dome, capped by three crowns—a Swedish national emblem—at its apex. The arches that face the water on the main facade, with blank expanses of wall above, are reminiscent of the Doge's Palace in Venice (see pp.144–145). Inside, the building contains offices and a series of grand assembly rooms and halls, including the Blue Hall, which hosts the annual banquet for the awarding of Nobel Prizes.

Notre Dame du Raincy

Location Le Raincy, France

If ever a cathedral could be said to have been entirely modern while remaining Gothic in spirit, it is Notre Dame du Raincy, northeast of Paris. It was built in 1922–1923 by Auguste Perret as a memorial to the Battle of the Marne (September 6–12, 1914). The most interesting aspect of the building is that it uses undecorated, reinforced concrete. There is no architectural elaboration, merely a nave flanked by two aisles whose arches, supported on slender columns, are so shallow as to seem undulating. However, it is a structure that allows soaring walls that consist almost exclusively of glass. The exterior of this building is dominated by a tower that looks like a skyscraper.

▲ GERMAN SCULPTOR GEORG KOLBE'S SCULPTURE *ALBA* AT THE GERMAN PAVILION

Bauhaus

Location Dessau, Germany

The Bauhaus, the renowned design school in Germany, was founded by Walter Gropius in Weimar in 1919. It was moved to Dessau 6 years later, and it was here, in 1925–1926, that Gropius designed the school's buildings. They had all the elements of the Modernism that Gropius championed. The buildings were entirely faithful to his belief that truth—and hence worth—in architecture could only be the product of utility. Curves were banished, as was any decoration. Above all, in an industrial age, only industrial materials—steel, glass, and concrete—were used. Windows were regular and repeated; the complex of interlocking buildings formed a series of right angles. In 1932, the school moved to Berlin. Running afoul of the Nazis, it closed the following year.

German Pavilion

Location Barcelona, Spain

Walter Gropius's successor as head of the Bauhaus was the German architect Ludwig Mies van der Rohe. Although a believer in Gropius's ideas, the pavilion he designed as the official German contribution to the 1929 Barcelona International Exposition reflected his own characteristic use of opulent materials. By using marble in spaces that were geometrically pure, almost abstract, he created an understated luxuriance that remained functional. The German Pavilion contained little more than examples of Mies van der Rohe's Barcelona chair—an instant icon of modern design. This reflected the fact that while other countries had used buildings to house their exhibits, van der Rohe's building was itself the exhibit.

Villa Savoye

Location Poissy, France

Outside the Bauhaus, no architect more precisely reflected the stern intellectual demands of Modernism in the 1920s than the Swiss-French Le Corbusier. His *Vers Une Architecture* (1923) was a rallying call for a new architecture reflecting a socialist world. Henceforth, buildings were to be "machines for living," and his Villa Savoye, built in 1929, was a precise expression of this theory. Long, low, and starkly white, it is raised from the ground on plain concrete columns (see pp.192–193). All the living spaces, surrounded by rectangular windows, are located on the open-plan second floor, which overlooks a similarly stark roof garden. Encapsulating the Modernist style, Villa Savoye was one of the most influential 20th-century buildings.

Tugendhat House

Location Brno, Czech Republic

Mies van der Rohe's Tugendhat House, built between 1928 and 1930, was designed as a deliberately lavish domestic villa, built without regard to cost while also remaining true to the structural and architectural demands of Modernism. If its planning was stark, its materials were certainly not. The villa is built on a steeply sloping site, with the main entrance at street level. Stairs lead down to a sweeping living room, 80 ft (24 m) long by 40 ft (12 m) wide, and the garden front is given over to floor-to-ceiling windows. An onyx partition adds an air of obvious luxury. The sense of spaces flowing together is immediate. Further unity is provided by the decorative details, all designed by the architect.

Karl Marx Hof

Location Vienna, Austria

Vienna's Karl Marx Hof is an example of social planning given solid form. It was built in 1927–1930 as part of a sustained drive by Vienna's socialist ruling council to house the city's dispossessed. The result was 1,382 apartments of 322–644 square ft (30–60 square m) that could house more than 5,000 people. Karl Marx Hof is the longest residential building in the world, at 3,600 ft (1,100 m) from end to end. Architecturally, it is ponderous, an aspect reinforced by its unrelieved ocher and yellow coloring. Its centerpiece—four towers flanked by balconies, topped by tall masts—resembles a row of ocean liners.

Moscow Metro

Location Moscow, Russia

The building of Moscow's metro system was always intended to be more than a demonstration of Soviet engineering prowess. The goal was to create what were, in effect, a series of opulent and overwhelming underground palaces. The first line was opened in May 1935; by 1938, a further two had followed. Glistening with mosaics and marble statues of Soviet heroes—soldiers, workers, and collective-farm laborers—they were unlike stations in any other metro system in the world. Ploschad Revolutsii station has 72 such statues flanking deep-red marble arches. The undulating, arched walls of Novoslobodskaya station contain 32 stained-glass panels in intricate brass frames. Perhaps the most spectacular of all is Komsomolskaya station, from whose yellow vaulted ceiling, highlighted with white stucco, hang vast chandeliers.

Villa Mairea

Location Noormarkku, Finland

In the late 1930s, Finnish architect Alvar Aalto introduced a new concept to architecture: Modernism made organic. He encapsulated this in the Villa Mairea, built in 1938–1939. Aalto's clients were rich and encouraged him to experiment. The result was an L-shaped house that was plainly modern, with expansive rooms flowing together, but which also embraced traditional materials, above all stone and wood. This, in part was an attempt to make the house an extension of the woods that surround it. The villa sought to be formal and informal, restrained and inventive, in a way rivaled only by Frank Lloyd Wright's architecture (see p.318).

▶ THE YELLOW-AND-OCHER FACADE OF KARL MARX HOF

▲ MOSCOW STATE UNIVERSITY'S MAIN BUILDING DOMINATING THE CITY'S SKYLINE

Grundtvig's Church

Location Copenhagen, Denmark

Danish architect P. V. Jensen-Klint's Grundtvig's Church, built between 1921 and 1940, is an architectural oddity. More cathedral than church in its dimensions, it looms almost threateningly over the houses (also built by Jensen-Klint) that frame it. It is constructed entirely from brick, and its vast west front consists of three squat-arched entrances, over which a deeply striated triangle rises. Over this is an expanse of brick that climaxes in three giant, similarly striated triangular towers, their edges variations of the stepped gables of Danish medieval traditions. This brutal monumentality is continued not just in the awkwardly angular flying buttresses at the east end, but also in the soaring and stark interior. The few windows the architect included are pencil thin.

Moscow State University

Location Moscow, Russia

The main building of Moscow State University was one of Josef Stalin's Seven Sisters, a series of skyscrapers through which the Communist leader sought to redefine the Moscow skyline. Designed by Lev Vladimirovich Rudnev, the building is a gargantuan propagandist symbol of Russia's postwar resurgence. On completion in 1953, it was the tallest building in Europe. In style, it is a supremely over-the-top vision of totalitarian Baroque. The central tower, 787 ft (240 m) high, becomes more ornate as it climbs before culminating in a huge red star. On every side are similarly overblown, tower-studded wings. Much of the labor for its construction was provided by political prisoners. Its completion coincided almost exactly with Stalin's death.

Notre Dame du Haut

Location Ronchamp, France

After World War II, Le Corbusier was an architect reborn. His initial Modernist strictures gave way to an architecture more attuned to human needs. His attachment to concrete and steel never lessened. However, his new recognition—that architecture should reflect rather than direct the vagaries of humanity—produced a series of remarkable buildings. His church of Notre Dame du Haut, completed in 1955, is a supreme example. Sinuous, organic, and innovative, it depicts Modernism transformed and recast. It has no obvious predecessors. Its curving walls and vast enveloping roof appear suddenly conjured, as if the result of a vivid imagination.

Jespersen Office Building

Location Copenhagen, Denmark

Completed in 1955, Danish architect Arne Jacobsen's Jespersen Office Building is cool, calm, restrained, and precisely detailed. If its regular pattern of green-gray panels does

almost nothing to draw attention to itself, the building nevertheless achieves an understated elegance. It is also an astonishing piece of engineering. Raised off the ground, it is supported only by two narrow piers, to provide access to an apartment building on its far side. These piers not only support the entire building, but the cantilevered structure they carry means that every one of the building's seven floors is entirely open plan. The result is a structure of deceptive simplicity.

Pirelli Tower

Location Milan, Italy

The Pirelli Tower was designed by Italian architect Gio Ponti and engineer Pier Luigi. Built in 1955–1958, it dramatically expanded the meaning of skyscrapers. This was partly because, at 417 ft (127 m), it was in some way the tallest building in Italy, but also partly due to its shape. In contrast to the rectilinear high-rises being built in the US, the ends of the Pirelli Tower taper, forming, in effect, a wing-shaped building. It is a visual device of elegance, reinforced by a secondary roof that appears to float over the 32-story structure. The building inspired the design of the Pan Am building (now, MetLife Building) in New York, among others.

Olympic Stadium, Munich

Location Munich, Germany

Munich's Olympic Stadium, built for the 1972 games by German architects Günter Behnisch and Otto Frei, was the most revolutionary and instantly recognizable sports stadium built at the time. The architects used rigorous mathematics and computer-aided design to determine its structure and flowing shape. To a large extent, the aim was to avoid the kind of monumental, stone-built stadium used for the Berlin games in 1936. By extension, it was also intended that the 1972 games would be a means of demonstrating to the world that West Germany had put its Nazi past behind it. The solution was a stadium set into the ground rather than rising out of it. Its partial roof of rising and falling translucent, tentlike acrylic panels is supported by eight steel pylons and braced by 271 miles (436 km) of steel cabling. Although now transformed into a movie theater and performance venue, it remains a memorable location.

Lloyd's Building

Location London, England

When it opened in 1986, Richard Rogers's building for insurers Lloyd's instantly established itself as the most innovative building in London. The impact of this extraordinary, almost sculptural, steel structure has never lessened. Its most obvious feature is that all of its vital systems are placed on the outside; its three main towers are covered with pipes, ducts, vents, cabling, and glass-fronted elevators. Similarly, a series of cranes, used for cleaning, stud the roofs. Not only does this make repairs easier, it creates space inside. The most dominant internal feature is the glass-vaulted atrium rising the full 289-ft (88-m) height of the building, criss-crossed with escalators. There are 14 floors of office space around it.

▲ THE ZINC-CLAD EXTERIOR OF THE JEWISH MUSEUM

Jewish Museum

Location Berlin, Germany

Berlin's first Jewish Museum opened 6 days before Hitler came to power in 1933. Unsurprisingly, its end was sudden and absolute. In 1988, the city's government announced plans to build a new museum, which opened in 2001. The museum consists of two startlingly different buildings. The first, which is Baroque in style, was originally the Berlin Museum. The other, accessible only from the first, is a disconcerting and powerful new building with a fractured and tormented zinc exterior. Designed by Polish American architect Daniel Libeskind, the building is laid out in a snakelike zigzag pattern, and the interior is slashed through with what the architect called "voids"—narrow, disjointed, alienating corridors representing the continuity of Jewish life in Germany. These corridors lead progressively to the Garden of Exile, its sloping floor disconcertingly studded with towers, and the Holocaust Tower, an empty, echoing silo with no exhibits.

Tate Modern

Location London, England

Tate Modern is housed in what was formerly the Bankside Power Station, on the South Bank of the River Thames. Designed by British architect Giles Gilbert Scott and built in stages from 1947, it eventually closed in 1981. In 2000, triumphantly recast by Swiss architects Jacques Herzog and Pierre de Meuron, it reopened as an art gallery. It was an instant success, in part because the original building was so powerfully molded, but also because its new architects recognized that this industrial muscularity would act as the perfect foil for the museum's collection of 20th-century art. For example, what was once the Turbine Hall was reinvented as a vast display space. A further phase of redevelopment followed, culminating in 2016 with the opening of Herzog and de Meuron's Switch House, a dramatically contorted brick tower.

30 St. Mary Axe

Location London, England

There is no more instantly recognizable building in London than British architect Norman Foster's 30 St. Mary Axe, known informally as the Gherkin. Built in 2001, this gleaming, bulletlike structure, rising 590ft (180m), is criss-crossed by its strongly banded diagonal steel grid. If it belongs to no obvious architectural manner other than the startlingly new, it is nevertheless a supremely rational building. It has been praised for its energy efficiency, with a natural ventilation system acting between twin outer layers of glass to warm the building in winter and cool it in summer. It has a sophisticated computer-controlled system that opens windows automatically if

the temperature goes above 68°F (20°C). Above all, it represents a vision of architecture that suggests unlimited possibilities.

Oslo Opera House

Location Oslo, Norway

Oslo Opera House, designed by architectural firm Snøhetta, was built in 2003–2007. Its waterfront setting and contrasting planes and angles inevitably suggest a kind of Nordic iceberg—a point reinforced by the expanse of marble that slopes down, around, and across it. The vast plaza is truly a public space that actively encourages visitors to explore. Even the roof is accessible. It also acts as a dramatic setting for the glass-fronted lobby that rises from and across it like a gigantic dormer window. The detailing is exceptional. The looming bulk of the stage-tower is clad in burnished aluminum. The no less high-tech auditorium, under a vast oval chandelier, is faced with darkly stained oak.

National Museum of 21st Century Arts

Location Rome, Italy

The National Museum of 21st Century Arts—*Museo nazionale delle arti del XXI secolo* in Italian, MAXXI for short—was the work of the prolific Iraqi-born British architect Zaha Hadid. It was built in 1998–2010. Hadid conceived of the building as a river delta, its channels dividing and multiplying. Despite the massiveness of much of the concrete building—itself a reference to the ancient Roman use of concrete—it ebbs and flows in ways that always surprise. Corridors suddenly curve or straighten, resembling a surreal highway intersection. Materials take on a sort of dissolving, translucent quality. Running through much of the building are underlit black staircases.

Rolex Learning Center

Location Lausanne, Switzerland

The Rolex Learning Center, which opened in 2010, is part of the École Polytechnique Fédérale de Lausanne, or EPFL. Containing a library, teaching centers, and various public amenities, it was designed by Japanese architects Kazuyo Sejima and Ryue Nishizawa. Seen from above, it appears modest: a simple concrete rectangle of 400,000 square ft (37,000 square m), pierced by a series of round and oval openings. However, seen from the side, it reveals itself as a kind of magic carpet, in a permanent state of elegant, unforced flux. It is scarcely imposing—nowhere is it more than 30 ft (9 m) high—and its walls are made entirely of glass. Inside and out, it rises and falls in gentle rhythms, as if breathing.

▼ THE UNDULATING STRUCTURE OF THE ROLEX LEARNING CENTER

The Shard

Location London, England

No building as vast as Italian architect Renzo Piano's Shard—officially known as the London Bridge Tower—can fail to generate controversy. Constructed between 2009 and 2012, its 95 stories soar 1,020 ft (310 m) over London. Perhaps ironically, its universally used name was originally intended to denigrate it. The English Heritage Trust called the proposed building a "shard of glass through the heart of London." There is no doubting its audacity. An irregularly shaped, attenuated pyramid ends in eight separate razorlike points. Differences in the angling of the 11,000 panes of glass cause the building to change its glinting appearance as the light and seasons change. At night, lit from within, it appears otherworldly. Loved or loathed, it has transformed London's skyline.

Africa

Step Pyramid of Zoser

Location Sakkara, Egypt

The Step Pyramid of Zoser, located about 19 miles (30 km) south of Cairo, sits in the vast burial grounds of the ancient Egyptian capital of Memphis. The structure is both the world's earliest stone monument and Egypt's first pyramid. It was commissioned in around 2600 BCE by Pharaoh Zoser (or "Djoser") and was designed by the architect Imhotep. Reaching some 197 ft (60 m) high, the pyramid is made of small limestone blocks arranged in stacked layers that look like giant steps. Built around a *mastaba*—a type of tomb in which the pharaohs of ancient Egypt were buried—it was the centerpiece of a complex that included courtyards, shrines, a temple, and priests' dwellings.

Fortress of Buhen

Location Buhen, Egypt

Located on the west bank of the River Nile, in the region of Nubia (currently on the northern edge of Sudan), the Fortress of Buhen was built during the reigns of Senwosret I and Senwosret III in the 12th Dynasty of ancient Egypt (c.1990–1800 BCE). Stretching some 490 ft (150 m) along the river, the fortress covered an area of 140,000 square ft (13,000 square m). Its outer walls, made of stone and bricks, are up to 16 ft (5 m) thick and 33 ft (10 m) high. Inside the walls was a town with a population of more than 3,000 people. The complex also included a temple of Horus. The fortress became completely submerged during floods in the 20th century and is now part of Lake Nasser.

Temple of Queen Hatshepsut

Location Der El-Bahari, Egypt

Situated on the west bank of the Nile, near the Valley of the Kings and the cliffs of Der El-Bahari, the huge temple of Queen Hatshepsut is one of Egypt's best known architectural masterpieces. Featuring several colonnaded platforms arranged over different levels and linked by steps and ramps, this majestic monument was built from masonry and cut stone as a mortuary temple to honor one of ancient Egypt's first female pharaohs. Built between 1490 and 1460 BCE by architect and royal chancellor Senenmut, the structure is adorned with striking reliefs and pillared porticos.

Valley of the Kings

Location Luxor, Egypt

Located across the Nile from the ancient city of Thebes (modern Luxor),

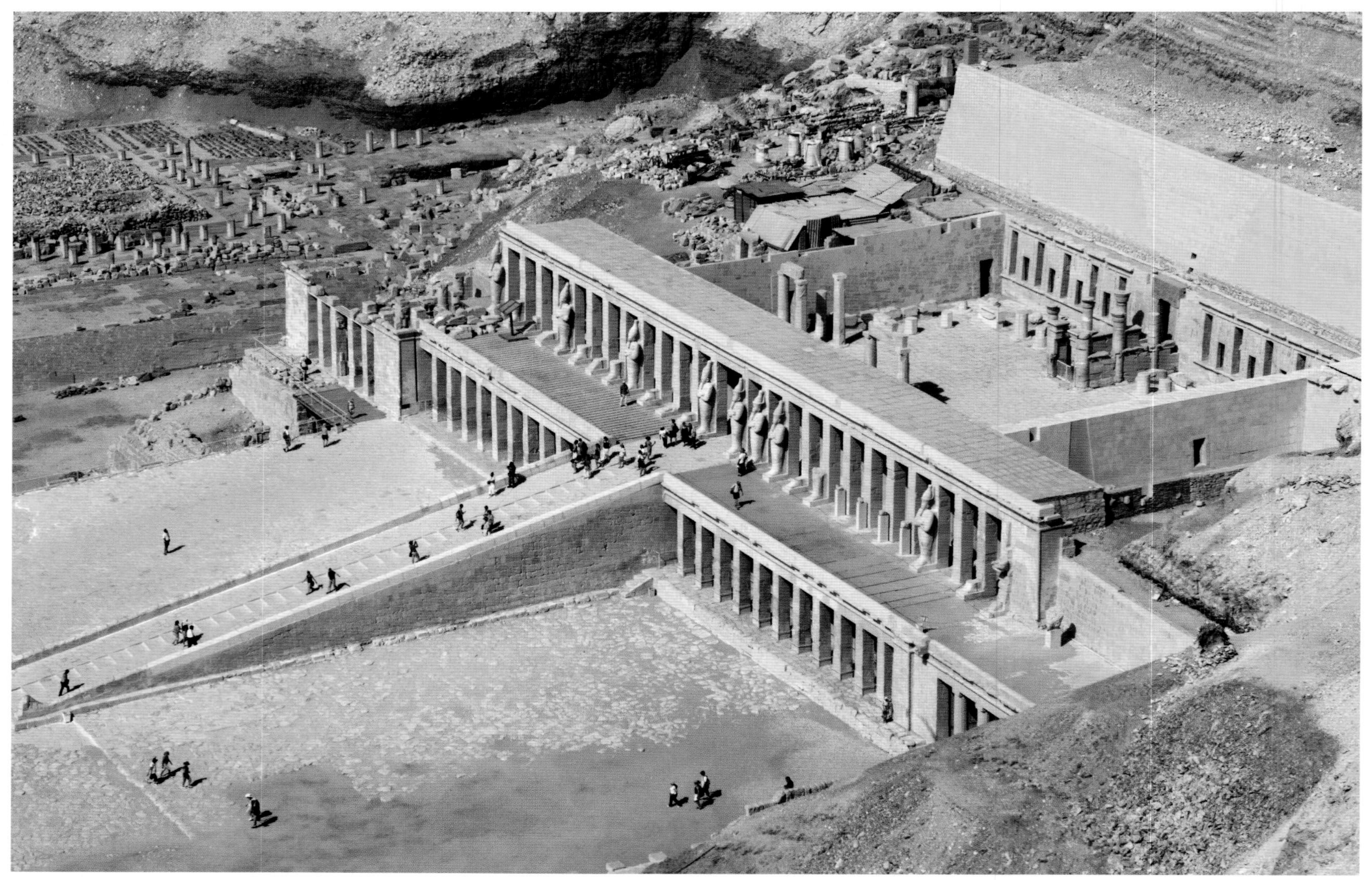

▲ THE TEMPLE OF QUEEN HATSHEPSUT NESTLED IN THE DER EL-BAHARI

the Valley of the Kings is a renowned burial site of pharaohs, featuring tombs dating from 1539 to 1075 BCE. Spread over a wide area, the complex contains a total of 64 known tombs that vary greatly in size and style. The structures were intended to be concealed within the hills to protect the pharaohs' remains and riches from looters. Generally, the interior plans of the tombs contain complex series of shafts and chambers, but only one corridor leads to the burial chamber, where a stone sarcophagus containing the royal mummy was laid. Nearly all of the tombs of the Valley of the Kings were looted in antiquity. Nevertheless, the site has been an important place of archaeological exploration since the 18th century and was made a UNESCO World Heritage Site in 1979.

Lighthouse of Alexandria

Location Alexandria, Egypt

Said to have risen to a height of up to 600 ft (183 m), the Lighthouse of Alexandria (also known as the Pharos of Alexandria) was a technological marvel, considered to be one of the Seven Wonders of antiquity. It was built in around 280 BCE, during the time of the Ptolemaic Kingdom, to a design by Greek architect Sostratus of Cnidus. Ancient descriptions suggest it was built from limestone blocks,in three tapering sections built on top of one another on a base some 98 ft (30 m) square. Unfortunately, a series of earthquakes damaged and eventually destroyed the lighthouse in the 13th century. However, in 1994, archaeologists discovered its remains in Alexandria's harbor, just off the small Pharos Island.

Temple of Isis

Location Agilkia Island, Egypt

The ancient island of Philae on the River Nile, in southern Egypt's Aswan region, is renowned for its wealth of archaeological treasures. Among them is the Temple of Isis, a complex constructed in stages and eventually completed during the reign of Ptolemy III Euergetes (r.246–221 BCE). The temple is dominated by two large pylons—monumental gateways consisting of two tapering towers—decorated with carvings depicting pharaohs and gods, including Isis. The temple was partially flooded by the building of the old Aswan Dam but reemerged after 1970, when a new dam was constructed. To save the rapidly deteriorating structure from water damage, archaeologists moved the temple's ruins to the nearby Agilkia Island.

Temple of Horus

Location Edfu, Egypt

The well-preserved Temple of Horus, on the west bank of the Nile, was built in the Ptolemaic Kingdom during the Hellenistic period, between 237 and 57 BCE. Dedicated to Horus, ancient Egypt's falcon-headed god, the structure is instantly recognizable by its huge pylons, about 118 ft (36 m) high, and vibrant external decorations that include carvings depicting the deity. Beyond the pylons is a courtyard, a series of halls, and an inner sanctuary surrounded by chapels. In one of the halls, known as the Court of Offerings, stand two granite sculptures of Horus. With its simple plan along a main axis, the building is considered a classic example of an Egyptian temple.

▶ PYLON DECORATED WITH DETAILED CARVINGS AT THE TEMPLE OF HORUS

Djinguereber Mosque

Location Timbuktu, Mali

The Djinguereber Mosque is the oldest and largest of Timbuktu's three great mosques. It was built in the 14th century, when the city was experiencing a long period of prosperity as an important West African trading post and center of Islamic culture. The mosque is made out of earth, used in mud bricks and rendering, and has been regularly maintained using traditional techniques. One of its most prominent features is its pyramidal minaret with protruding wooden beams. The structure, and its intricate facade, was rebuilt and enlarged between 1570 and 1583 and is considered a symbol of Timbuktu's golden age.

Madrasa of Sultan Qaytbay

Location Cairo, Egypt

The Madrasa of Sultan Qaytbay, which sits in Cairo's old Islamic quarter, is one of the city's best-preserved and most highly regarded examples of Mamluk architecture. Completed in 1474, in the Egyptian capital's Northern Cemetery, it is a building complex containing a variety of structures—a mosque, a mausoleum, a minaret, a Qur'anic school, and a public well. A square courtyard sits at the heart of the site. The mosque and mausoleum are beautifully adorned, full of rich carvings and patterns made of marble. The mausoleum also features an intricately decorated dome.

Tomb of Askia

Location Gao, Mali

The Tomb of Askia is part of a key building complex in the Malian city of Gao. It comprises a pyramidal tower, two flat-roofed mosques, a necropolis, and a white stone square. The 56-ft- (17-m-) high pyramid was created in 1495 by the Emperor Askia Mohamed, after he declared Gao the capital of the Songhai Empire and made Islam the empire's official religion. The tower—the tomb itself, where the emperor is thought to be buried—is an earthen structure made using the region's traditional mud-building techniques and featuring permanent wooden scaffolding. It also functions as a minaret. The monument has been severely damaged due to erosion and political conflicts. It was designated a UNESCO World Heritage Site in 2004.

◀ THE CENTRAL COURTYARD AND POOL OF THE ALI BEN YOUSSEF MADRASA

Ali Ben Youssef Madrasa

Location Marrakech, Morocco

Built in the 14th century, the Ali Ben Youssef Madrasa was enlarged and redecorated in 1570, becoming one of the most architecturally distinguished and largest Islamic colleges of the Maghreb region of North Africa. At one point, its 132 rooms housed up to 900 students. The structure, which is influenced by Andalusian architecture, is delicately decorated with cedar wood and mosaic carvings, colorful tiles, and stucco inscriptions. A large marble courtyard, 50 ft (15 m) wide and 65 ft (20 m) long, is bordered by open-air galleries on its east and west sides, and a pool sits at its center. The *madrasa*'s prayer room is capped by a heavily ornate, wooden dome. In the 1960s, the building was transformed into a museum.

Enda Mariam Orthodox Cathedral

Location Asmara, Eritrea

The architecture of this Eritrean Orthodox church reflects the distinct Modernist style of many of the buildings of Asmara, which was developed in the 1890s as an Italian colony. It was originally designed in 1920 by Ernesto Gallo, but the structure was replaced, in 1939, by the current cathedral, whose architect is unknown. The structure consists of a main body flanked by two imposing towers built in alternate layers of brick and stone, reflecting the local Aksumite architecture. The central facade features brightly colored murals.

African Renaissance Monument

Location Dakar, Senegal

Designed by the Senegalese architect Pierre Goudiaby and completed in 2010, the African Renaissance Monument was built to commemorate Senegal's 50 years of independence from France. It is a bronze statue that sits at the summit of one of the twin hills of Collines des Mamelles, overlooking the Atlantic. The project was spearheaded by the country's then-president Abdoulaye Wade and was conceived as a celebration of a new era for Africa. The dynamic composition, consisting of a man, a woman, and a child, can be approached by a dramatic series of steps. At 160 ft (49 m) high, it dominates the Dakar skyline and is thought to be the tallest statue in Africa. An observation deck, in the head of the male figure, offers spectacular views of the Atlantic Ocean.

▼ THE MAJESTIC STATUES OF THE AFRICAN RENAISSANCE MONUMENT

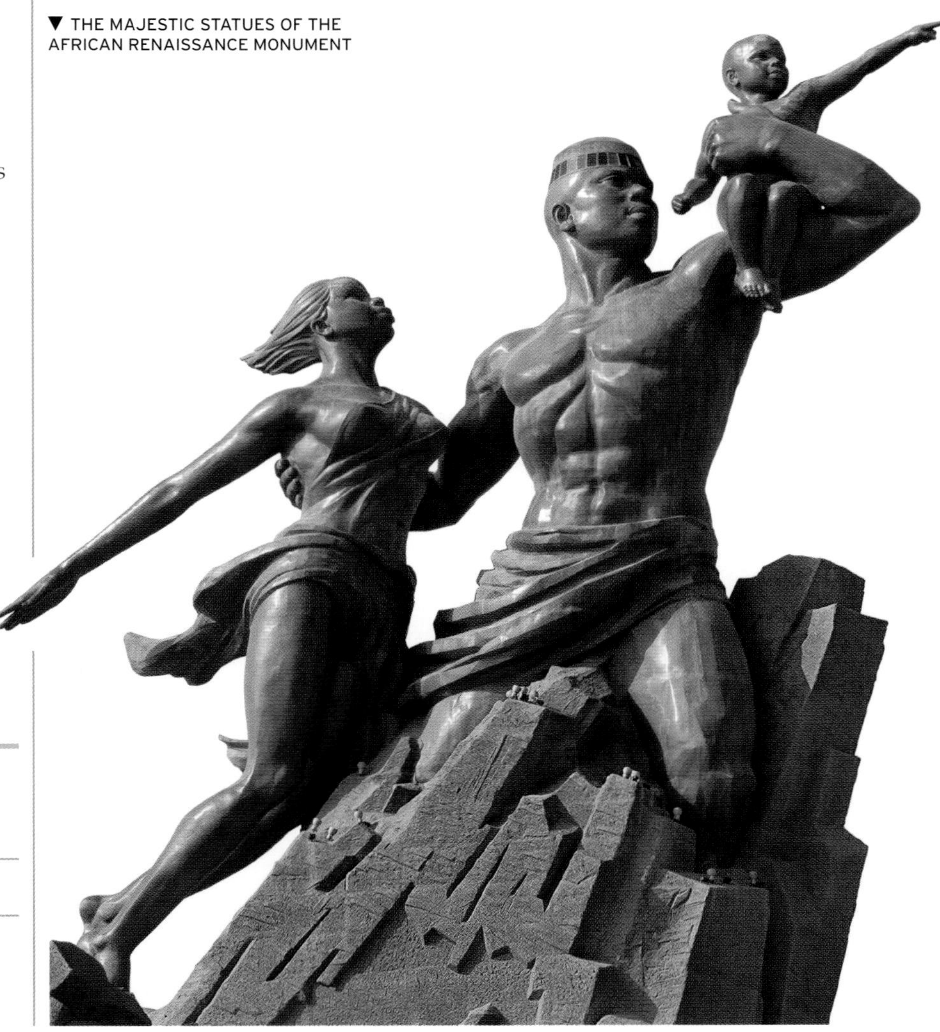

Asia and Australasia

Harappa

Location Punjab, Pakistan

Harappa, the first known city in south Asia, and one of the first in the world, was settled in around 3300 BCE. Stretching some 0.6 square miles (1.5 square km) along the Indus River Valley (in modern-day Pakistan), it is thought to have been home to around 24,000 people and been supported by local agriculture. The city was arranged in a grid pattern and included a distinctive citadel with crenellated walls, flat-roofed houses, granaries, halls, and sophisticated drainage and sanitation systems. Mud and baked brick were the primary building materials. Harappa was abandoned in around 1750 BCE, possibly as a result of rising sea levels and the silting up of rivers.

The Temple of Solomon

Location Jerusalem, Israel

The center of Jewish worship in ancient times, the Temple of Solomon was erected by King Solomon and stood from around 957 BCE until 587 CE, when it was destroyed by the Babylonians. The rectangular stone structure took 7 years to build and had three main areas: a porch, flanked by two huge bronze pillars; a main interior hall, overlaid with gold; and an even more sumptuous 30-ft- (9-m-) square gold-clad inner sanctum containing the Ark of the Covenant—a gold chest said to have housed two stone tablets of the Ten Commandments. The two main rooms were paneled in carved cedar wood, with floorboards made from cypress. The cost of the spectacular temple is said to have been exorbitant.

Nimrud

Location near Mosul, Iraq

The city of Nimrud (formerly Kalhu) was founded in the 13th century BCE but did not rise to prominence until the 9th century BCE, when the Assyrian king Ashurnasirpal II rebuilt and expanded it as his capital. By 879 BCE, supported by a vast labor force, he had created a city of some 1.4 square miles (3.6 square km), with enormous defensive walls, temples, shrines, irrigation canals, and a distinctive ziggurat. The most impressive building was his Northwest Palace, which covered almost 322,917 square ft (30,000 square m) and contained breathtaking examples of Assyrian art, including vibrant frescoes and colossal limestone bas-reliefs and monuments. Nimrud was extended by successive rulers but was virtually abandoned in the 7th century BCE, when the Assyrian Empire fell.

Citadel of Van

Location Van, Turkey

The Citadel of Van was a fortress town built by the Urartu Kingdom of ancient Armenia between the 9th and 7th centuries BCE. The imposing fortification was constructed on stepped terraces, high on a limestone promontory overlooking the Old City of Van (Tushpa), Lake Van, and the region's hugely fertile lands. The citadel's vast walls, made from mud-brick and unmortared basalt, were designed to facilitate control of regional territories rather than simply be a means of defense. Sardur Tower, built with huge limestone blocks in c.840–830 BCE, is thought to have been the first building on the site. Thereafter, the citadel was greatly expanded to include many impressive structures, including palaces, royal burial chambers, sanctuaries, canals, and inscribed stelae.

Palace of Sargon II

Location Khorsabad, Iraq

Constructed on a high terrace within the heavily fortified Assyrian capital city of Dur-Sharrukin (present-day Khorsabad), the palace of King Sargon II was dedicated in 706 BCE, the year before the king's death in battle. Two enormous limestone bulls almost 13 ft (4 m) high, with wings and human heads, guarded the entrance. The palace itself was also monumental in design, testimony to the might of the king. Built mainly from mud-brick, it included a spectacular entrance court, a state court, a throne room, and numerous halls and courtyards, some with pools. Among the decoration were bronze doors, carved ivory ceilings, and bas-reliefs on the walls depicting victories in battle. The royal enclosure included a multistoried ziggurat and temples dedicated to major deities.

▶ THE CITADEL OF VAN, LOCATED ON A LIMESTONE PROMONTORY

Nineveh

Location Mosul, Iraq

The splendor of the Assyrian city of Nineveh was renowned, and skillfully planned. Although founded as early as 6000 BCE, it was transformed into a glorious royal and imperial capital by King Sennacherib in the 7th century BCE. The king erected city walls, gates, temples, palaces, and beautiful public gardens and parks, as well as sophisticated irrigation systems, canals, and aqueducts. His own palace, which he hailed as "the palace without rival," contained around 80 rooms. He also showcased Assyrian art in all its magnificence. Particularly distinctive were immense statues of animals that guarded the entrances to buildings; inside, spectacular bas-relief panels glorified the triumphs of the empire. Nineveh was sacked and burned in around 612 BCE, when the Assyrian Empire fell to the Medes and the Babylonians.

Ishtar Gate

Location Babylon, Iraq

Constructed in around 575 BCE by King Nebuchadnezzer II, the Ishtar Gate became the eighth fortified gate in the magnificent city of Babylon. It was dedicated to Ishtar, the Mesopotamian goddess of fertility, love, and war. More than 39 ft (12 m) high and built from burned brick, the enormous double gate had a central arch that spanned 25 ft (7.5 m). Brilliant blue glazed tiles—depicting rows of bulls, lions, and dragons—decorated the facade. The gate was built at the start of the Processional Way, an avenue that is thought to have run southward for almost 0.6 miles (1 km).

► RECONSTRUCTED FACADE OF THE ISHTAR GATE DECORATED WITH ANIMAL FIGURES

▲ THE DOME-SHAPED STUPA INSIDE THE GREAT CHAITYA

Babylon

Location near Al-Hillah, Iraq

In the first half of the 6th century BCE, the ancient city of Babylon on the Euphrates River was spectacularly rebuilt by King Nebuchadnezzer II to become one of the most celebrated cities in the Near East. The city was laid out on a grid pattern, with the huge temple of Marduk at its heart. Its precincts also contained the great ziggurat of Babylon, which is thought to have been the prototype of the Tower of Babel. Nebuchadnezzer refurbished temples and built vast new city walls, canals, a moat, and a palace with 250 rooms. He is also thought to have created the Hanging Gardens of Babylon, one of the seven wonders of the ancient world. According to ancient sources, the terraces of the ziggurat rose to a considerable height and were maintained by an ingenious irrigation system. Babylon fell to the Persian king Cyrus the Great in 539 BCE, never to regain its former glory.

Pasargadae

Location near Shiraz, Iran

Pasargadae was founded as the dynastic capital of the Achaemenid Empire under King Cyrus the Great, who reigned between 559 and 530 BCE. Despite its location in the arid plains of southern Iran, Cyrus created wonderful lawns and flowerbeds, maintained by impressive water systems. Historians believe that his residential palace occupied an area of around 7,000 square ft (650 square m). Among its distinctive features was an imposing central hall with rows of massive columns, the bases of which still remain. However, arguably the most striking monument in the city is Cyrus's massive, austere limestone mausoleum, built in around 500 BCE. More than 33 ft (10 m) high, it is raised on a plinth and accessed by six steps.

Tomb of Darius I

Location Naqsh-e-Rustam, Iran

The tomb of the fourth Persian king of the Achaemenid Empire, Darius I, was cut from the cliffs at Naqsh-e-Rustam, near Persepolis, around 485 BCE. It is distinctive for its richly decorated facade, carved out of a rock in the shape of a cross. Various impressive reliefs feature on the cross, including that of a palace, resting on a throne-bench supported by 30 throne bearers. The figure of Darius also appears on a platform in front of a fire altar, gesturing to a winged figure above him, thought to be the deity Ahura Mazda. By

contrast, the interior of the tomb is unadorned and simple, comprising a few bare chambers. Three of Darius's successors also commissioned tombs at the site, but these cannot be identified with certainty from the accompanying inscriptions.

Temple of Artemis at Ephesus

Location Ephesus, Turkey

The temple of Artemis, the Greek goddess of the hunt and the Moon, at Ephesus (see p.242), in modern-day Turkey, originated in the 7th century BCE but was destroyed by floods and rebuilt from around 550 BCE by King Croesus of Lydia. His magnificent marble Ionic temple, measuring about 360 ft by 180 ft (110 m by 55 m), was surrounded by a double row of decorated columns—126 in all. In 356 BCE, it was deliberately burned down, but it was rebuilt in the same place to a similar design. Although the new building was slightly smaller, it was raised somewhat—giving it greater stature—and it also housed spectacular works of art. The reconstruction was impressive enough to be hailed as one of the seven wonders of the ancient world. The temple was destroyed by the Goths in 262 CE.

Mausoleum of Halicarnassus

Location Miletus, Turkey

Designed by the Greek architects Pythius and Satyros, the mausoleum of the Achaemenid ruler Mausolus was built on a hill above the city of Halicarnassus (now Bodrum), in c.353–350 BCE. A stairway, flanked by statues of lions, led to the podium supporting the marble tomb. The structure itself was virtually square, around 140 ft (43 m) high, surrounded by 36 Ionic columns, and adorned with artworks that included statues and sculptural reliefs. A 24-step pyramid at the top was crowned by a sculpture depicting a marble chariot. One of the seven wonders of the ancient world, the tomb was damaged by earthquakes between the 13th and 15th centuries CE.

Ruwanveliseya Dageba

Location Anuradhapura, Sri Lanka

The Ruwanveliseya is a gigantic Buddhist *dageba* (stupa, or domed reliquary shrine) built by King Dutugemunu in around 140 BCE, in the sacred city of Anuradhapura. The original brick structure would have been encircled by two staggered platforms. An outer boundary wall, also made of brick, featured statues of elephants facing outward, reputedly with tusks of real ivory. Today, following considerable restoration and renovation, the structure is gleaming white and stands at around 338 ft (103 m) tall—almost twice its original height. It is one of the country's most venerated stupas and an important place of pilgrimage for Buddhists.

Great Chaitya

Location Karli, India

The *chaitya* (Buddhist prayer hall) at Karli, Maharashtra, is one of a series of rock-cut cave shrines that were excavated between the 1st century BCE and the 1st century CE. Known as the Great Chaitya, it is distinctive not only as one of the best preserved and largest such halls of the period—it is 148 ft (45 m) long and 46 ft (14 m) high—but also for its impressive carvings and sculptures. The facade of the cave has a window and a huge column crowned by four lions. Inside, there is a vaulted ceiling, and the hall is flanked by two rows of columns with elaborate carvings, which lead to a circular ambulatory that houses a dome-shaped stupa.

▼ THE DRAMATIC CLIFF-CUT FACADE OF THE TOMB OF DARIUS I

Great Stupa, Sanchi

Location Sanchi, India

The Great Stupa at Sanchi was commissioned by the Mauryan king Ashoka in the 3rd century BCE. The unadorned, hemispherical brick burial mound housed a relic chamber thought to contain the Buddha's ashes. Substantially embellished over the centuries, it is one of Buddhism's most treasured monuments—for its distinctive art and architecture, as well as its sacred significance. By the 1st century BCE, its dome had been flattened slightly and its size increased to around 121 ft (37 m) wide and 56 ft (17 m) high. It was covered with stone slabs and encircled by a railing and a low outer wall. Four spectacular *toranas* (ceremonial gateways) were also added, featuring statues and exquisite relief carvings that include fertility deities, auspicious symbols, and scenes from the Buddha's life.

Ise Shrine

Location Ise, Japan

The Grand Shrine at Ise (formerly Uji-Yamada), in Mie prefecture, dates from the 3rd or 4th century CE and is the Shinto religion's most sacred shrine complex. Its many buildings are divided into two areas: an outer shrine, dedicated to Toyuke, the food goddess; and an inner shrine—the site's oldest and holiest place—dedicated to Amaterasu, the Sun goddess. The Ise Shrine is distinguished for its simplicity and harmony. Its buildings, made from thatch and unpainted cypress, are raised about 7 ft (2 m) off the ground on posts. Every 20 years since the 8th century, the buildings have been demolished and replaced by exact replicas; this is thought to revitalize the site and the deities themselves.

Lingyin Temple

Location Hangzhou, China

One of the oldest, largest, and most notable of China's ancient Buddhist temples, the Lingyin Temple, also known as the Temple of Souls, was constructed during the Eastern Jin dynasty (317–420 CE). It was significantly embellished in later dynasties, particularly in the 10th century, when it is said to have housed more than 3,000 monks. The site's numerous palaces, halls, pavilions, and grottoes include striking examples of Buddhist art and architecture. Notable are the elegant double-eaved Hall of the Heavenly Kings, with its paintings of dragons and phoenixes, and the 112-ft- (34-m-) high Hall of the Great Hero, with a huge statue of Sakyamuni Buddha carved from camphor wood and coated in gold leaf.

▼ *TORANA* AT THE GREAT STUPA AT SANCHI

Tāq Kasrā

Location Ctesiphon, Iraq

The ancient imperial capital of Ctesiphon is known for the remains of its palace, the residence of the kings of the Sasanian Empire (c.224–651 CE). The palace's throne room is thought to have been 112 ft (34 m) high, 80 ft (24 m) wide, and 161 ft (49 m) long. However, the site's most distinctive feature is the Tāq Kasrā, or Archway of Ctesiphon, a remarkable, majestic structure, which formed part of the palace. At 121 ft (37 m) high and spanning around 85 ft (26 m), it is the largest single-span, unreinforced brick-built arch in existence. Its date of construction is disputed—historians are divided between the 3rd and the 6th centuries.

Temple of Henan

Location Dengfeng, China

Also known as the Shaolin Temple, the Henan Temple on Mount Song, Henan, was one of the first centers for the translation of Buddhist scriptures in China. It also housed magnificent libraries and became a major monastery for Shaolin monks. However, it is best known as a principal site for the martial arts. Founded in around 495 CE, the Zen Buddhist temple has been destroyed and rebuilt many times. Covering around 620,000 square ft (57,600 square m), the complex includes—from different periods—an entrance gateway; a bell tower; a drum tower; around 15 halls; an outer wall; and distinctive artworks, including statues, murals, and sculptures. A series of stelae recount the temple's history.

▲ THE *KONDŌ* AND FIVE-STORIED PAGODA OF THE HŌRYŪ-JI TEMPLE COMPLEX

Yungang Grottoes

Location Datong City, Shanxi Province, China

A treasure trove of Buddhist cave art dating from around 460 to 520 CE, the Yungang Grottoes comprise 252 major caves containing around 51,000 statues. Stretching for about 0.6 miles (1 km), these magnificent cave sanctuaries, hollowed out of steep limestone cliffs, were probably influenced by the rock-cut traditions of India and Central Asia. Among the masterpieces at Yungang are gigantic Buddhist statues, including several of Chinese emperors; works depicting buddhas, *bodhisattvas*, and other sacred beings; and frescoes and vibrantly painted ceilings. The carvings inside the caves became more intricate and detailed over time.

Hōryū-ji Temple Complex

Location Ikaruga, Nara Prefecture, Japan

The earliest Buddhist monument in Japan, the temple complex of Hōryū-ji includes some of the world's oldest wooden buildings. The original temple (destroyed by fire in 670 CE) was constructed by Prince Shōtoku in 607 CE. Adapting Chinese style and temple layout to Japanese culture, and evolving over the centuries, the complex now comprises 48 buildings. Its masterpieces of wooden architecture include a five-story pagoda measuring over 105 ft (32 m) high, a two-story *kondō* (sanctuary hall), and the octagonal *Yumedono* (Hall of Dreams). Hōryū-ji contains more than 2,000 cultural artifacts and art treasures, including sculptures and murals, showing the influence of China, Korea, and India on the development of Japanese Buddhist art.

▲ ONE OF THE TWO MASSIVE BUDDHAS AT BAMIYAN

Wofo Temple

Location Beijing, China

The Wofo Temple—also known as the Wofosi Temple or the Temple of the Sleeping (or Recumbent) Buddha—was originally constructed between about 626 and 629 CE but has been rebuilt many times. As its name suggests, the temple is most famous for its statue of a Buddha reclining as he awaits his *nirvana*—the state of perfect enlightenment. Originally a sandalwood sculpture, it was replaced during the Yuan dynasty (1279–1368) by a colossal bronze version, more than 17 ft (5 m) long, which is said to have been cast from some 275 tons (250 tonnes) of bronze by several thousand workers.

Giant Buddhas, Bamiyan

Location Bamiyan, Afghanistan

The two soaring Buddhas carved into the sandstone foothills of the Hindu Kush mountains at Bamiyan, central Afghanistan, were the largest stone statues in existence until 2001, when they were destroyed by Taliban fighters, who declared them to be false gods. Probably created in the 6th to 7th centuries, the spectacular monuments—125 ft (38 m) and 180 ft (55 m) tall, respectively—flanked the numerous rock-cut Buddhist cave monasteries (3rd to 5th centuries) that stretch for around 4,265 ft (1,300 m) along the base of the ocher-colored cliff-face. It is thought that the folds in the Buddhas' robes may have been created by hanging thick ropes from the rocks and then covering them in plaster.

Great Mosque of Damascus

Location Damascus, Syria

The Great Mosque of Damascus, also known as the Umayyad Mosque, is the world's oldest surviving stone mosque. A spectacular example of early Islamic architecture, it was built between 705 and 715 CE by the Umayyad caliphate. It is rectangular in layout, measuring 515 ft by 328 ft (157 m by 100 m), and has two main parts: an open courtyard surrounded by arches and a prayer hall and sanctuary with three aisles formed by arches. The building includes notable features of mosque architecture, such as the *mihrab* (prayer niche), the *minbar* (pulpit), the semicircular (horseshoe) arch, the square minaret, and the *maqsūrah* (a box or screen near the *mihrab*).

Seokguram Grotto

Location Gyeongju, South Korea

Seokguram is a Buddhist cave temple in the form of an artificial grotto constructed of large granite blocks. Built between 751 and 774 CE, during the Unified Silla period (676–935 CE) of ancient Korea, it is located on the upper slopes of the holy Mount Tohamsan and looks eastward over the sea. It comprises an antechamber, a corridor, and a domed rotunda. Inside the inner chamber is a 11-ft- (3.5-m-) tall monumental statue of the Buddha, seated on a lotus pedestal and surrounded by an entourage of exquisitely carved reliefs, including bodhisattvas, disciples, and divine guardians. The grotto, with its sculptures, is an exceptional example

of the Buddhist art and architecture that flowered during the Silla Kingdom.

Pre Rup Temple

Location Siem Reap, Cambodia

The majestic temple-mountain of Pre Rup, consecrated to the Hindu god Shiva in around 961 CE, was the state temple of the Khmer King Rajendravarman II. The pyramid-shaped structure is square in layout and has two enclosing walls: an outer wall, measuring around 384 ft by 417 ft (117 m by 127 m), with *gopurams* (gateway towers), and an inner wall. Long halls, now derelict, line the inner enclosure. At the heart of the complex, five sanctuary towers are set on a square base—four at the corners of the square and the fifth, raised on an upper tier, at its center. Stairways on each side of the base, flanked by lion statues, lead to the upper levels. Brick and laterite (a red clay), the primary building materials used at Pre Rup, lend a warm, reddish hue to the compound.

Brihadisvara Temple

Location Thanjavur, India

Dedicated to the Hindu god Shiva, the Brihadisvara Temple is a masterpiece of southern Indian Dravidian architecture. Completed in about 1010 CE at the height of the Chola dynasty (10th–13th centuries). The huge complex is enclosed by a colonnaded *prakara* (veranda), with a 98-ft- (30-m-) high *gopuram* (gateway tower) to the east and smaller *gopurams* around the perimeter. Above the inner sanctum, its central tower, or *vimana*, rises to a height of around 200 ft (60 m) and, like the rest of the temple, is built entirely of granite and covered with intricate carvings. A massive statue of Nandi—the sacred bull ridden by Shiva—carved from a single piece of granite and weighing around 28 tons (25 tonnes), faces the sanctum. The temple is still in use.

▼ COURTYARD OF THE GREAT MOSQUE OF DAMASCUS

▲ THE ICONIC PHOENIX HALL AT BYŌDŌ-IN TEMPLE

Byōdō-in Temple

Location Kyoto, Japan

Originally built in 998 CE as a country villa, Byōdō-in Temple was converted into a Buddhist temple in around 1053. The focus of the complex is the elegant Phoenix Hall, added in 1053, which contains treasures of Japanese art, including a spectacular gold-coated wooden statue, almost 10 ft (3 m) tall, by the sculptor Jōchō. The distinctive wooden building itself is acknowledged as an exceptional example of architecture from the Heian period (794–1185). The Phoenix Hall comprises two symmetrical wings flanking a central hall, with a tail corridor—from the front, the overall shape, with its upturned eaves, recalls a bird with outstretched wings. Two bronze phoenixes face each other on the rooftop. The site also features a pond and a beautiful garden.

Ancestor Temple of Foshan

Location Foshan, China

The Ancestor Temple of Foshan (or Foshan Zumiao), built in 1078–1085, during China's Northern Song dynasty (960–1126), was originally devoted to the worship of ancestors. It was rebuilt in the 14th century as a Daoist temple—dedicated to the deity Beidi (or Xuanwu), who presides over the waters—with further significant rebuilding taking place in subsequent centuries. The complex covers some 27,000 square ft (2,500 square m) and now comprises the Ancestral Temple; an early 20th-century temple devoted to China's major educator and political reformer, Confucius; and two 21st-century temples dedicated to the martial arts. The complex contains notable carvings and sculptures and examples of Song-dynasty *dougong* (interlocking wooden roof brackets).

Ananda Temple

Location Bagan, Myanmar

Bagan's Ananda Temple—Ananda was a first cousin of the Buddha—is among the most impressive of Myanmar's Buddhist temples. Dating from the 11th or 12th century, it is dominated by its bulbous gilded spire, or *shikhar,* which rises 171 ft (52 m) from the center of the cruciform structure. Below it, on the second of the stepped terraces that support the *shikhar,* are four smaller domes at each corner of the building. The entire brick-built structure is wreathed in intricate ornamentation. Glazed terracotta plaques surround the four principal entrances. Inside, there are two concentric corridors, in effect ambulatories, surrounding the dark space at the heart of the building, in which stand four gilded Buddhas, 30 ft (9 m) high.

Chennakeshava Temple

Location Belur, India

One of the defining features of Hindu temples as they developed across India from the 5th century onward was their profusion of sculptured decoration. Buildings seemingly shimmered in a state of permanent motion. Belur's Chennakeshava Temple, begun in 1117 on the orders of the Hoysala ruler Vishnuvardhana, is a prime example of this exuberant sculptural delight. The level of detail is such that the figures more closely resemble highly wrought metalwork

than stone carvings, particularly those of the celestial dancers. The temple is dedicated to the four-armed Lord Vishnu. More than 100 ft (30 m) high, it originally boasted a tapering, curvilinear tower, or *shikhar*. The building is surrounded by four smaller temples.

Krak des Chevaliers

Location Homs, Syria

Krak des Chevaliers ("Castle of the Kurds") was rebuilt in 1144 by the Knights Hospitallers following the First Crusade. It is one of the most memorable and formidable of all medieval castles, and one of the best preserved. It was built on a natural, craglike stronghold, although the nature of the site dictated the slightly irregular layout. It occupies an area of 256,000 square ft (23,800 square m). Following a further stage of building in the 13th century, it became a concentric castle with two tower-studded defensive walls—one outer and one inner—that made it essentially impregnable. Any parts of the inner wall thought to be relatively vulnerable were reinforced by sloping sections, or "glacis," in parts more than 82 ft (25 m) thick.

Kalyan Minaret

Location Bukhara, Uzbekistan

Completed in 1127, the Kalyan Minaret, meaning "great minaret," is a structure of exceptional simplicity and remarkable elegance. It is a 154-ft- (47-m-) high, free-standing tower of golden bricks that tapers slightly as it climbs before flaring outward at its summit. The diameter of the base is 30 ft (9 m) and that of the summit is 20 ft (6 m). The body of the tower is banded with 12 rows of elegantly incised geometrical decorations. Just below the more elaborately carved rotunda at the summit is the sole touch of color: a glazed frieze of pale blue, with inscriptions from the Qur'an. Sixteen arches, from where the faithful are called to prayer, circle the rotunda. The minaret stands in the courtyard of the 16th-century blue-domed Kalyan Mosque.

Summer Palace

Location Beijing, China

Beijing's Summer Palace is more than a palace. Grouped around a man-made lake, it is a series of gardens, lakes, and ornamental buildings in a landscaped park that spreads over 1.1 square miles (2.9 square km). Although there had been an imperial garden in this area since the mid-12th century, it was in the 18th century, during the Qing dynasty, that the gardens took on their present form. They almost form a microcosm of subsequent Chinese history. A place of luxury, the Summer Palace was repeatedly looted and largely destroyed between 1860 and 1900. It was only in 1953 that restoration work began.

▶ INTRICATE DETAILING ON THE WALLS OF THE KALYAN MINARET

Jain Temples, Mount Abu

Location Rajasthan, India

There are five Jain temples in the high, wooded hill station that is Mount Abu. Collectively known as the Dilwara Temples, they were built between the 11th and 13th centuries. The most famed are the two oldest: the Vimala Vasahi, built in c.1030, and the Luna Vasahi, built in 1230. They are dedicated, respectively, to the first and the 22nd of the *tirthankaras,* the 24 most esteemed figures in Jainism. Inside and out, these domed temples burst with intricately carved details, including dancing figures, geometric patterns, and tiny spires projecting down as well as up. However overwhelming this remarkable decorative program appears, the impact is never oppressive. Both buildings are made of white marble and gleam with light.

Rukn-e-Alam

Location Multan, Pakistan

The Rukn-e-Alam, "Pillar of the World," is the tomb of an Islamic mystic. It was built between 1320 and 1324 by a local ruler, Tughluq I. This astonishing building, both sturdy and delicate, is octagonal in plan and built mostly of red brick. Eight inward-leaning, tapering buttresses support the massive base of the structure, and the whole is elaborately banded with incised patterns and turquoise tiles. At the summit of each buttress is a white dome. Above these is another even more opulently decorated octagon, studded with blue and turquoise glazed tiles and pierced by arches, each angle marked by further domes. Above all this, slightly set back, is a brilliantly white hemispherical dome that rises to a height of 115 ft (35 m).

Wuxianguan Temple

Location Guangzhou, China

The Taoist Wuxianguan Temple in Guangzhou, formerly Canton, in southern China, is also known as the Temple of the Five Immortals. It commemorates the arrival of five immortals from the heavens—three men and two women—said to have been riding goats and carrying food during a time of famine, possibly in the Zhou dynasty (c.1030–771 BCE). They brought with them "the gift of rice cultivation and the promise of bountiful harvests." Today's temple is a late 19th-century replica of a Ming dynasty building of 1378, the original having been destroyed by a fire. Its sharply upward-curling, yellow-and-green tiled eaves are alive with finials and statuary. A deep-red, sturdy, arched tower houses a Ming bell weighing 5.5 tons (5 tonnes).

Drum Tower of Xi'an

Location Xi'an, China

Xi'an's imposing Drum Tower is one of two almost identical structures built early in the Ming dynasty (1368–1644). The Drum Tower dates from 1380, and its sister building, the Bell Tower, from 1384. Both stand on substantial brick bases and are built of elaborately decorated brick and wood, with three eaves of glazed, dark green tiles. Their purpose was partly to assert Ming imperial might and partly to act as timekeepers—the bell was rung to signal a new day, and the drum was struck to signal its end. Although the original drum is no longer there, a new drum—the largest in China—was placed there in 1996. Twelve vivid red drums line the exterior of the north and south sides of the building's second floor.

▲ DETAILED MARBLE CARVINGS IN THE JAIN TEMPLES AT MOUNT ABU

Taimiao Ancestral Temple

Location Beijing, China

Although it lies just outside the Forbidden City (see pp.276–277), the complex of ordered buildings that make up the Taimiao Ancestral Temple, built from 1420, was sacred to both the Ming and Qing dynasties. As the name suggests, it was here that the ancestors of emperors and empresses were venerated. The complex consists of several separate structures, most of them dedicated to lesser nobles and courtiers. The most important and largest structure—reached by the complex's principal entrance, the Halberd Gate—is the Hall for Worship of Ancestors. This long, low building is twin-eaved and rests on a three-tiered base. Under its overhanging eaves are elaborate glazed green, blue, and cream tiles. Following the Chinese Communist Revolution in 1949, the complex became the Beijing Working People's Cultural Palace, where cultural events are held.

▲ A PAGODA AT THE HUMBLE ADMINISTRATOR'S GARDEN

Temple of Agriculture

Location Beijing, China

In a country periodically racked by famines, the importance of China's harvests loomed large. Dedicated to Xiannong, the father of agriculture, the Temple of Agriculture was built in 1420 to celebrate abundant harvests. It sits in a 1.2-square-mile (3-square-km) complex, complete with its own palace, the Qing Cheng Palace. It was to the temple's Jufu Hall that the emperor himself came every year at the spring equinox for the Tilling Ritual. The temple is a single-story building, much of it of a deep-red lacquer, with complex geometrical patterning highlighted in gold. Above the lower level is a blue-and-green frieze, filled with sumptuous details, and a sweeping, dragon-studded roof of green glazed tiles. Having fallen into almost complete dilapidation, the entire complex was painstakingly restored from 2001.

Humble Administrator's Garden

Location Suzhou, China

The central principle behind Chinese gardening has been to co-opt nature, so as to create an idealized, harmonious vision of the world. Suzhou, in southern China, was particularly famed for its gardens, and the Humble Administrator's Garden, laid out by a former court official in the early 16th century, is the largest and most celebrated of all. It comprises three sections—Eastern, Western, and Central—arranged around a large lake, making water the dominant element of the garden. Ponds and streams, rocks and bridges, as well as many pavilions and monuments, are artfully combined to create an apparent informality. The mood of the setting is contemplative, serene, and refined.

▲ THREE-EAVED STRUCTURE OF THE HALL OF PRAYER FOR GOOD HARVEST AT THE TEMPLE OF HEAVEN

Temple of Heaven

Location Beijing, China

The potency and reach of Ming China (1368–1644) are encapsulated by the buildings of the Temple of Heaven. Many of them were constructed under the original builder of the Forbidden City, the Emperor Yongle, in 1420. The complex covers a site of 1 square mile (2.6 square km). There are three principal structures, all arranged on an exact north–south axis, and all, as symbols of the sky, are circular. The site itself is rectilinear—a symbol of Earth. The Altar of Heaven, built in 1530, is an open platform made of three levels of white marble. The Imperial Vault of Heaven, which was rebuilt in 1752, is a single-eaved building with a blue-tiled roof. The three-eaved Hall of Prayer for Good Harvest is the most important of the buildings. Its interior is an expanse of dazzlingly detailed, symbolically charged decorations.

Tomb of Humayun

Location Delhi, India

The Tomb of Humayun, completed in 1566, commemorates the second of India's Mughal emperors. Heavily influenced by Persian architecture in style, the building displays the balance, refinement, and symmetry that would become the defining characteristics of Mughal architecture. Also influenced by Persia is the garden in which the tomb sits—it exudes courtly sophistication, with its precise planting and well-defined waterways. It was intended to be an earthly representation of the Gardens of Paradise. The three-story tomb consists of a multiarched base, above which rises a much larger, three-bay second story with angled corners. This is dominated by a central pointed arch with an apselike interior that leads on to further arches. On top of the building sits a white marble dome, 140 ft (42.5 m) high, flanked and fronted by kiosks.

Fatehpur Sikri

Location Uttar Pradesh, India

Akbar, the third and greatest of India's Mughal emperors, had an energetic interest in the arts, which received a huge boost under his rule. Among his ambitious projects was an entirely new capital, Fatehpur Sikri, constructed from 1569 and largely complete by 1585. However, by 1610, it had been abandoned, possibly because of inadequate water supplies or simply because Akbar had lost interest in it. Nevertheless, it remains among the most sumptuous and startling of Mughal achievements. Its red sandstone courtyards, palaces, arcades, towers, and domes are richly ornamented. Among the highlights are the vast Buland Darwaza, the principal entrance to the mosque courtyard; the mosque itself, the Jama Mosque; and the shimmering white marble tomb of the Sufi mystic Sheikh Salim Chishti.

Pavilion of Ali Qapu

Location Isfahan, Iran

The Pavilion or Palace of Ali Qapu (*Ali* meaning "exalted" or "great" and *Qapu* meaning "gate") was part of the ambitious works launched by the Safavid emperor Shah Abbas I. Built in the early 17th century, it overlooks the Meidan Emam, a public square larger than any in Europe. Externally, its most obvious feature is the lofty terrace, or *talar*, supported by tapering columns. However, the building's real glories are inside. The Ali Qapu had no real function, other than as a kind of pleasure palace, and this is reflected in its decorations. Produced under the direction of the shah's court painter, Reza Abbasi, they are supremely refined. For example, the Music Room on the sixth floor has decoration piled on soaring decoration.

Beehive Village

Location Aleppo, Syria

The conical beehive houses of northern Syria (see p.10) represent a remarkable survival of what was a vernacular house-building tradition, stretching back thousands of years. Though they appear crude, they are ideally adapted to local conditions, and being made only of mud and straw, they can be repaired more or less indefinitely. Their most obvious advantage is that their thick walls are ideal for absorbing fierce desert heat during the day and reradiating it during what are often bitterly cold nights. The heat that does enter the structures is able to escape from the small opening at the summit of the cones, guarded by a sort of tiny chimney that prevents rain from getting in.

Shibam

Location Hadhramaut, Yemen

Shibam, on the edge of the fearsome Rub' al Khali ("Empty Quarter") desert, is one of the world's most extraordinary walled cities. Its buildings are a group of mud-brick high-rises clustered tightly together. Most are between five and eleven stories, and a few exceed 98 ft (30 m). The city itself dates from the 3rd century CE, but most of the current buildings were built from the 16th century. Although they were painstakingly constructed, the buildings require almost constant repair and have all been rebuilt many times. They are simple, rectangular structures that taper slightly as they rise. Smaller, lighter bricks are used for the upper floors, and the windows are small and square. The walls are whitewashed to reflect solar radiation and reduce heat gain.

Tomb of Jahangir

Location Lahore, Pakistan

The son of Akbar the Great, Jahangir, the fourth Mughal emperor of India, ruled from 1605 to 1627. His tomb is a prime example of the severity, regularity, symmetry, and delicacy of Mughal architecture as it approached its peak in the 17th century. It is a long and low square building, with four kiosk-topped minarets at its corners. It sits in the center of an immense walled garden courtyard. While the external appearance is severe, internally it is a precise expression of the Mughal empire's capacity to use astonishing decorative detail in its dynastic monuments. The tomb itself, under a commanding dome over the exact center of the complex, rests under a single slab of marble. On the tomb are inscribed the 99 names of God.

▲ RED SANDSTONE COURTYARD OF FATEHPUR SIKRI

Kiyomizu-dera Temple

Location Kyoto, Japan

Kiyomizu-dera Temple is among the largest and most important of the many Buddhist temples in Kyoto, which, until 1868, was the imperial capital of Japan. It stands in an extensive complex of buildings on a rocky, wooded hillside overlooking the city. The temple was founded in 798 but most of the current buildings were rebuilt after the originals were destroyed in a fire in 1633. The main hall, or *hondo,* of the temple itself was built in honor of Kannon Bodhisattva, the Goddess of Mercy. The structure is cantilevered over a precipitous slope and supported by substantial timbers. Above this is a large terrace under sweeping roofs set at right angles to each other. The heart of the temple is aglow with brilliantly gilded images.

Shugakuin Imperial Villa

Location Kyoto, Japan

The chief glory of the Shugakuin Imperial Villa is not its architecture—which is harmonious, simple, and modest—but its gardens, which extend over the rocky, heavily wooded foot of Mount Hiei. There are three principal buildings: the Upper, Middle, and Lower villas. The Upper and Lower villas were built between 1655 and 1659, and the Middle villa in the 1680s (as a nunnery). Apart from the Middle villa, the gardens and buildings were the work of the former emperor Go-Mizunoo, who, in 1629, stood down in protest at the constricting power of the Tokugawa Shogunate (military government). The dominant feature of the overall design is a large man-made lake. Rock-strewn streams and ponds criss-cross the well-maintained mossy gardens. The mood throughout is contemplative and calm.

Citadel of Bam

Location Bam, Iran

In December 2003, the Citadel of Bam (or Arg-é Bam) on the high desert plateau of southeast Iran was largely destroyed by an earthquake. Until its sudden destruction, it had been the largest walled, mud-built city in the world, dominated by its fortress. However, what has since been rebuilt makes its former magnificence clear. The site covers 1,940,000 square ft (180,000 square m), and its walls are 6,000 ft (1,815 m) long. There had been a settlement at this natural oasis since the 6th century BCE, but the citadel's much later importance stemmed from its access to the Silk Road to the east and the Persian Gulf to the south. Its importance reached a peak under the Safavid dynasty, which came to an end in the early 18th century.

Shwedagon Pagoda

Location Yangon, Myanmar

Yangon's Shwedagon Pagoda may be the most instantly stupefying building in the world. The date of its construction is uncertain, but according to legend, it is up to 2,500 years old. The most sacred of Myanmar's Buddhist shrines, it is said to contain a collection of holy relics. It is a stupa—a moundlike structure, typically with a hemispherical dome, topped by a spire. The stupa was a major Buddhist contribution to the architecture of India and later of Southeast Asia. It is a design rich in symbolism, the base and summit representing Earth and the sky, respectively. Several stupas were built across what is today Myanmar. The exterior of the Shwedagon Pagoda is covered entirely in gold, and in the top of the spire—which rises to 325 ft (99 m)—are set thousands of precious stones, including a single 76-carat diamond. The main pagoda is surrounded by 64 smaller spires.

Genbaku Dome

Location Hiroshima, Japan

Just as the shattered tower of the 19th-century Kaiser Wilhelm Memorial Church in Berlin stands as a stark reminder of the city's devastation by Allied bombing during World War II, so do the ruins of the Genbaku Dome. Also called the Atomic Bomb Dome and the Hiroshima Peace Memorial, it stands as a memorial of the destruction of Hiroshima by the first atomic bomb, in August 1945. Opened in 1921 as the Product Exhibition Hall, it was the only building left standing in the center of the city after the bombing. The decision to preserve it in its ruined state, as the centerpiece of the Hiroshima Peace Memorial Park, was made in 1950. It has periodically been discreetly reinforced to ensure that it remains intact.

◀ THE NIOMON GATE OF KIYOMIZU-DERA TEMPLE, GUARDED BY TWO LION DOGS

▲ THE GOLD-PLATED DOMES OF THE SHWEDAGON PAGODA

Meiji Shrine

Location Tokyo, Japan

The Japanese Emperor Meiji, who at least nominally ruled Japan from 1867 to 1912, gave his name to the Meiji Restoration, a process that saw the sudden westernization of Japan after 1868. Although he was not responsible for the process, it was decided that he and his empress, Shoken, should be well commemorated. The Meiji Shrine, completed in 1921, is the city's largest Shinto shrine. It comprises Japanese cypress structures of great simplicity and elegance under sweeping copper roofs. They stand in a 7.5-million-square-ft (700,000-square-m) park planted with close to 120,000 trees. The shrine was destroyed in the bombing of Tokyo in March 1945, but it was painstakingly rebuilt in 1958.

Rashtrapati Bhavan

Location New Delhi, India

The decision, in 1911, that the capital of British India should move from Calcutta (now Kolkata) to what would be New Delhi resulted in the construction of a great architectural set piece that would house the country's ruler, the viceroy. Designed by British architect Edwin Lutyens, the Viceroy's House—now known as Rashtrapati Bhavan—was completed in 1929. It may have been an assertion of imperial might, but it was a notably sensitive one, marshaling Mughal, Hindu, and Buddhist elements alongside an otherwise Classical European design. The main building, which covers 218,000 square feet (20,200 square m), has four floors and 340 rooms, and is set in the expansive, landscaped Mughal Gardens. It is a deeply satisfying building, at once vast, harmonious, and stately. It is now the official residence of the President of India.

Chandigarh

Location India

The city of Chandigarh was created in the 1950s as the capital of the states of Punjab and Haryana. It was a monument not just to the optimism of a newly independent India, but also to the vision of the man who led the team that built it, the Swiss-French architect Le Corbusier. It was the great opportunity that this pioneer of modern architecture had been waiting for: the creation, from scratch, of an ideal city that was at once harmonious, rational, and bold. Chandigarh's three largest and most important buildings—the High Court of Justice, the Secretariat, and the Palace of Assembly, built between 1951 and 1962, all in raw concrete—are astonishingly original, with a sculptural dynamism that consistently surprises.

Jean-Marie Tjibaou Cultural Center

Location New Caledonia

The Jean-Marie Tjibaou Cultural Center in the former French colony of New Caledonia is not just a remarkable tribute to the Kanak people of the island but an act of contrition on the part of the French government for its treatment of them. It is named after a staunch campaigner for Kanak rights who was assassinated in 1989. It was designed by Italian architect Renzo Piano and completed in 1998. Essentially, it features a radical reworking of traditional Kanak conical huts on a much larger scale. A total of 10 pavilions, each up to 92 ft (28 m) high, are arranged in three groups, or "villages," along a central alley. The organic structures swell outward but are left only half-built. As a result, they have an almost dematerialized quality, greatly enhanced by their oceanside setting. They contain permanent displays of Kanak life and culture.

Beijing National Stadium

Location Beijing, China

Better known, for obvious reasons, as the Bird's Nest, this stadium was the principal venue of the 2008 Beijing Olympics. Designed by Swiss architects Jacques Herzog and Pierre de Meuron and several Chinese architects, it was completed in 2007. It is, in fact, two separate structures. A more or less conventional, albeit substantial, concrete stadium that seats more than 80,000 people is contained within an elaborate, seemingly random network of interlocking steel girders—part structural, part decorative—that rise at each end. These give the stadium a distinctive, saddle-shaped profile. Laid end to end, the girders would stretch 22.5 miles (36 km). It was a solution that proved not just cost-effective but architecturally dramatic, producing the most memorable Olympic stadium since that used at Munich in 1972.

▼ THE INTERLOCKING STEEL GIRDERS OF THE BEIJING NATIONAL STADIUM

Liyuan Library

Location Jiaojiehe, China

At first sight, the Liyuan Library, situated in the woods outside a small town 50 miles (80 km) north of Beijing, is unpromising: a dark, modest, single-story rectangle covering about 1,880 square ft (175 square m). However, closer inspection reveals something much more remarkable. The exterior is loosely sheathed with fruit-tree branches, arranged vertically. The frame of the fully glazed building is steel, but inside wood again dominates, with strong uprights crossed by beams—a pattern continued on the roof. The split-level, partly stepped floor is similarly made of wood. There are no chairs or tables, just rectangular cushions. Throughout, diffused light plays through the exterior cladding. Completed in 2011, Liyuan Library represents the simple made memorable.

Ryugyong Hotel

Location Pyongyang, North Korea

The Ryugyong Hotel is a vast pyramidal skyscraper that dominates the cityscape of Pyongyang, the capital of North Korea. Comprising 105 floors and reaching 1,080 ft (330 m) high, the building is the country's tallest structure. It consists of three wings that rise jaggedly at an angle of 75 degrees and are capped by a cone-shaped structure formed by a series of concentric rings. Its construction began in 1987 and was due to be finished by 1992. However, by 1992, it was still incomplete and work was suspended. Construction resumed in 2008 and continued until 2013, by which point the exterior was more or less complete. In 2018, a huge LED display was added to one side of the building.

Heydar Aliyev Center

Location Baku, Azerbaijan

The sweeping curves and flowing spaces that mark so much of Iraqi-British architect Zaha Hadid's work reach perhaps their ultimate expression in her Heydar Aliyev Center, completed in 2013. The huge building, 243 ft (74 m) tall at it highest, is a convention center, library, and museum—its forms melting and merging into one another. It is enclosed in white, almost wavelike shapes that reach rhythmically skyward and fold in on themselves. Openings are filled with huge expanses of bronze-tinted glass that glow dramatically at night. The interior is no less dynamically curvilinear, displaying a kind of fluid geometry with almost no straight lines. The building stands in absolute contrast to the rigid monumentality of Soviet- era Baku.

House for Trees

Location Ho Chi Minh City, Vietnam

In a city with almost no trees or parks left because of rapid urbanization, the House for Trees, designed by Vietnamese firm Vo Trong Nghia (VTN) Architects and completed in 2014, is an ingenious means of reintroducing greenery while producing a low-cost, simple-to-build house. In fact, the house consists of five separate, two-story structures, grouped irregularly around a courtyard, with their upper levels linked by open bridges. They are built of concrete that is patterned to resemble bamboo. The interior walls feature plain exposed brick. All sections have large, airy windows, but what most distinguishes them is that they are, in effect, huge flower pots. From their strengthened roofs rise substantial trees. They also function as storm-water basins during heavy rains, helping reduce flooding in the city's narrow streets. In one stroke, an otherwise unrelieved and polluted urban environment is suddenly transformed.

GLOSSARY

ABACUS ▼ A flat slab forming the topmost part of a column. Typically wider than the capital on which it sits, the abacus receives the weight of the structure above the column, usually an arch or architrave. *See also* arch, architrave, capital, column.

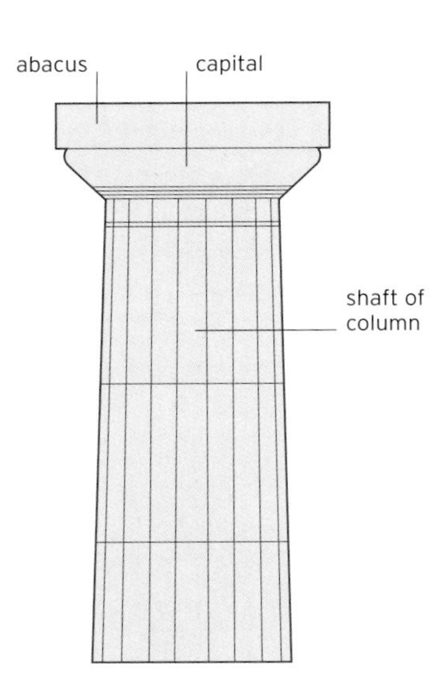

ABUTMENT ▼ The substructure on which a superstructure rests. Abutments are typically found at the ends of bridges or either side of arches, transferring horizontal and vertical loads to the foundation.

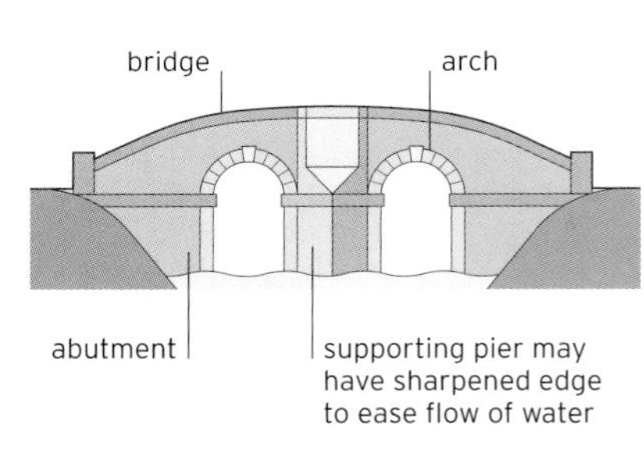

ACANTHUS A flowering plant native to warm regions, particularly the Mediterranean. In ancient Greek architecture, acanthus leaves often appear as an ornamental pattern on capitals of the Corinthian order. *See also* Corinthian, order.

ACROPOLIS The fortified part of an ancient Greek city, typically built on a hill. The most famous example is the Acropolis of Athens, site of the Parthenon.

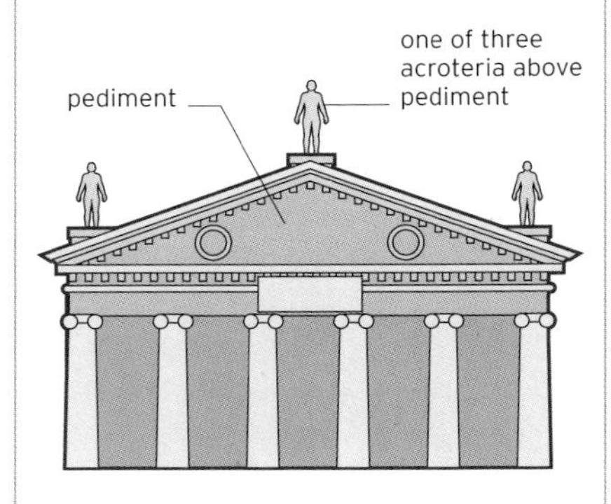

ACROTERION ▲ A decorative sculpture or ornament at the apex or ends of a pediment. Originally, acroteria were often petal-shaped ornaments; later they developed into statues or groups of statues.

ACRYLIC A type of plastic. It is often used in architecture as a substitute for glass in high-pressure, high-impact scenarios such as for skylights and underwater.

ADOBE Sun-dried mud-brick used for building. Adobe often contains chopped straw to reinforce the mixture of compressed earth and clay.

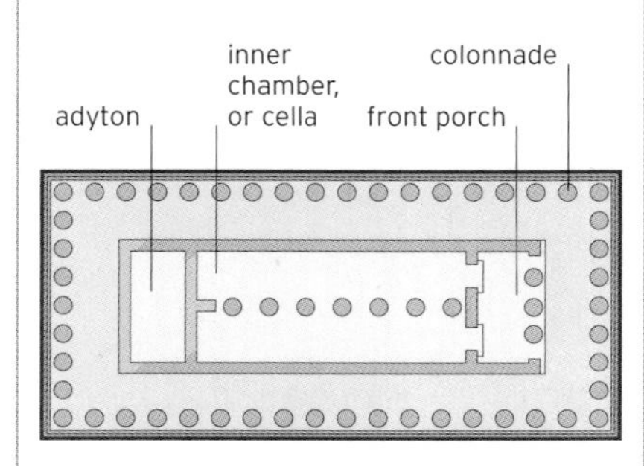

ADYTON ▲ The innermost sanctuary within a Greek or Roman temple. The adyton (*adytum* in Latin) often housed the cult image of the god, and entry was forbidden except to priests.

AESTHETIC Relating to the notion of beauty and good taste. The Aesthetic Movement was prominent in 19th-century European art, which valued beauty above all other qualities.

AGGREGATE Fragments or particles loosely compacted together. A coarse aggregate of sand, gravel, and crushed stone is an essential component of concrete. *See also* concrete.

AIR CONDITIONING A mechanical air-cooling system. Air conditioning can refer to any technology that modifies air in a closed environment by cooling, heating, dehumidifying, cleaning, or ventilating.

AISLE A lengthwise corridor or wing that typically runs parallel to the nave in a church. Aisles often have a roof at a lower level than the nave. An aisle may also be a passage between rows of seats in a building or the passage between shelving or storage such as in a supermarket or warehouse. *See also* church, nave.

ALABASTER A light-colored, translucent, soft stone. Alabaster has historically been favored by sculptors for small, decorative items because it is easy to work.

ALCOVE A small, arched recess in a room or external wall.

ALTAR A flat-topped table or block used as the focus for a religious ritual.

ALTAR-TOMB A raised tomb or funerary monument covering a tomb that resembles an altar. These tombs are not used as altars; they only resemble them.

AMBULATORY A round walkway encircling the altar in a cathedral or large church. The term can also refer to the covered passage around a cloister. *See also* altar, church, cloister.

AMPHITHEATER A round-shaped open-air theater. Developed by the ancient Romans, amphitheaters were free-standing buildings, circular or oval in plan, with tiered seating all around a central arena.

ANASTYLOSIS The rebuilding of historical monuments using only the original elements. Components must be placed back in their original positions, and new materials can only be introduced if they are necessary for stability.

ANG A feature in Chinese roof construction; this is a long bracket arm that functions like a lever.

ANTECHURCH The portico at the main entrance to a church, also known as a narthex. *See also* church, narthex, portico.

APEX STONE *See* keystone.

APSE A semicircular recess covered with a hemispherical vault. In a church or cathedral, there is often an apse behind the altar. *See also* altar, cathedral, church, vault.

AQUEDUCT A bridge for conveying a water course. Aqueducts usually span valleys or ravines. They are particularly associated with the Romans, who built many notable examples.

ARABESQUE ▼ A decorative style based on sinuous lines, often incorporating plant motifs. Arabesques are typically derived from Islamic art, hence the name.

ARCADE A covered passage with arches along one or both sides. In modern use, an arcade can also be a covered passageway of shops.

ARCH See panel, below.

ARCHITRAVE A beam that rests on the capitals of columns. In modern buildings, the architrave is the wooden molding that frames a door or window. *See also* capital, column, entablature.

ART DECO A decorative art style of the 1920s and 1930s. Taking its name from the International Exhibition of Modern Decorative and Industrial Arts, held in Paris in 1925, the style was characterized by bold, geometric lines and shapes; strong symmetry; and a modern, streamlined look.

ART NOUVEAU A decorative art style popular from approximately 1890 until World War I. Whimsical in nature, Art Nouveau employed undulating, asymmetrical shapes and lines, typically evoking the forms of flowers, plants, insects, and other natural objects.

ASHLAR Stone that has been cut and smoothed into regular-shaped blocks and laid in a bricklike fashion.

ARCH

An arch is a structure that is curved at the top and directs load around an opening, transferring it to pillars, piers, abutments, or walls on either side. Its construction depends on wedge-shaped blocks, called voussoirs, in which the upper part of the block is wider than the lower part. The central voussoir is called the keystone. During construction, the voussoirs require support from below until the keystone has been put in place to lock the arch.

The completed structure is self-supporting, held in compression by gravity and the weight of the structure above. This makes arches stable and efficient and capable of spanning larger distances and supporting greater loads than a horizontal beam. Different types of arches have evolved over time, while the techniques of arch-building have developed into many other structural forms, such as vaults and domes.

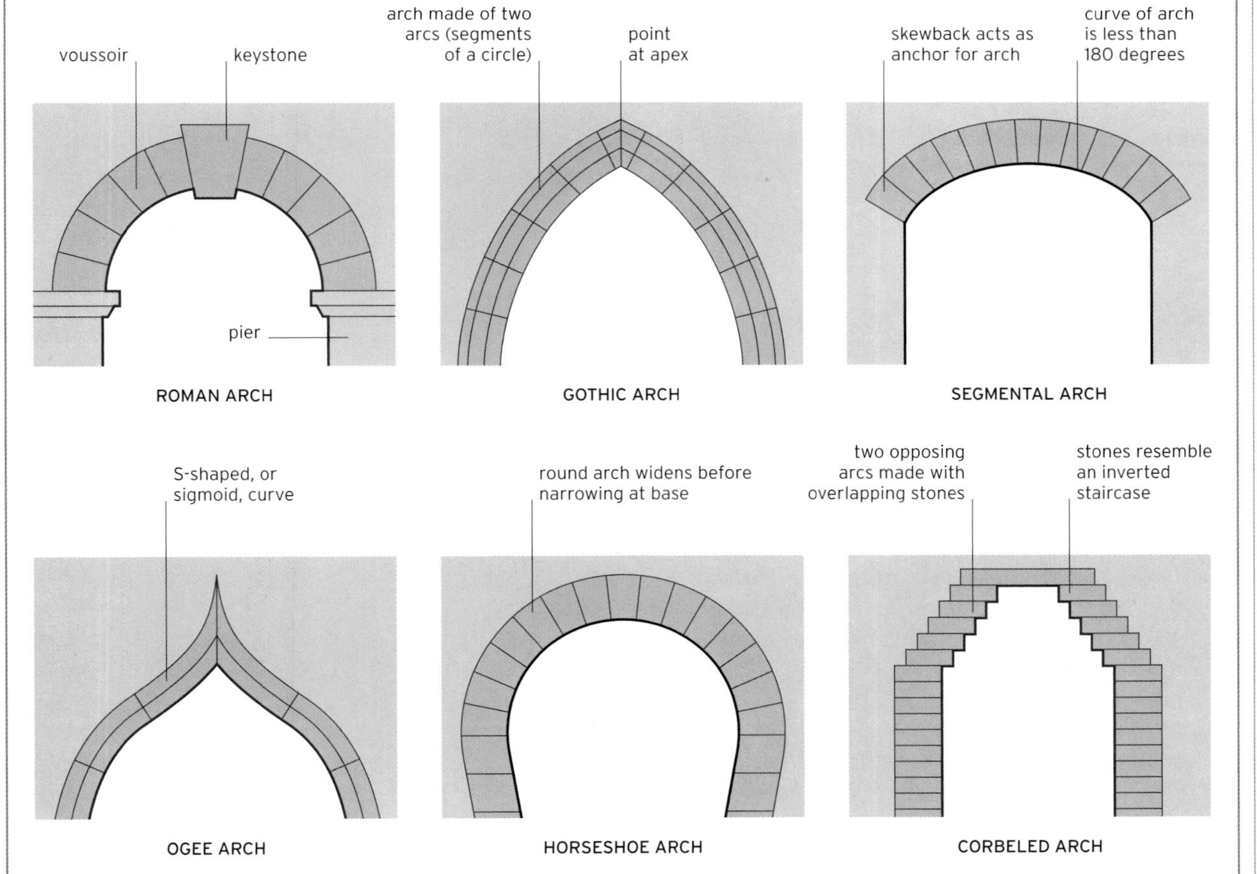

ATRIUM An open space in a building that extends up through several floors, often with a glass roof. Atriums are a common feature of hotels and shopping centers.

AUDITORIUM A place where people gather to watch a performance. In architectural terms, this is the part of a theater or concert hall where the audience sits, or a building used for events such as concerts.

AVANT GARDE Pioneering or experimental.

AVENUE A wide road with trees or tall buildings on either side.

AXIAL PLANE An arrangement of elements along an extended central axis.

AXIS A real or imaginary straight line to which elements on a plan relate, typically through symmetry.

AZULEJO A type of Spanish and Portuguese tiling. Colorful, glazed azulejo tiles are commonly used as decorative elements on the interiors and exteriors of churches, palaces, and public buildings.

B

BAILEY A courtyard in a castle or fortress enclosed by a wall. Larger structures might often have an outer (lower) bailey and an inner (upper) bailey. *See also* castle.

BALCONY A platform projecting from the outside wall of a building, with a low wall and railings or a balustrade for safety. *See also* balustrade.

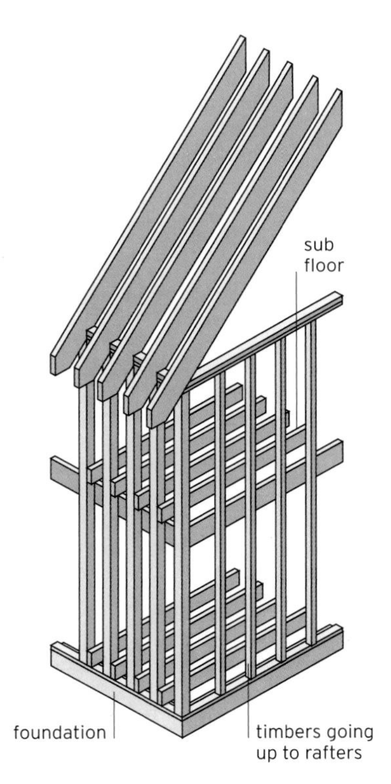

BALLOON FRAMING ▲ A type of timber construction used to build houses in which long timbers are used as supports extending from the foundations to the rafters. This contrasts with platform framing, in which each story is formed by floor-to-ceiling timbers topped by a floor deck that then becomes the platform for the next story.

BALUSTRADE A row of small columns topped by a rail, employed to prevent a person falling off a stairway, balcony, or terrace.

BAPTISTERY The area that surrounds the baptismal font in a church or cathedral. This may be incorporated within the body of the main structure, or it could be a separate building.

BARBICAN A fortified gate that protects the entrance to a city or castle. *See also* castle.

BARGEBOARD A board that runs under the slopes of a projecting gable roof. Historically, these were often decorated with carved or cut-out patterns. *See also* gable.

BARN An agricultural building, usually found on farms, used for storage purposes.

BAROQUE An architectural style originating in 17th-century Italy and pioneered by the Catholic Church, which looked to add dynamism and drama to the rules of Classical architecture. Floor plans incorporated ovals and curves, facades played with scale, and interiors employed theatrical effects through the manipulation of light, gilding, and extensive use of trompe l'oeil. *See also* Classical, trompe l'oeil.

BARREL VAULT A structure that acts like a series of arches placed one after another to make a semicylindrical ceiling, often seen in church or cathedral architecture. *See also* vault.

BARROW An elongated mound of earth or stones used in ancient times to cover a burial site.

BARTIZAN ▼ A small kiosk or tower protruding from a fortified wall. Bartizans are most commonly found at corners to provide better defensive visibility and are furnished with arrow slits.

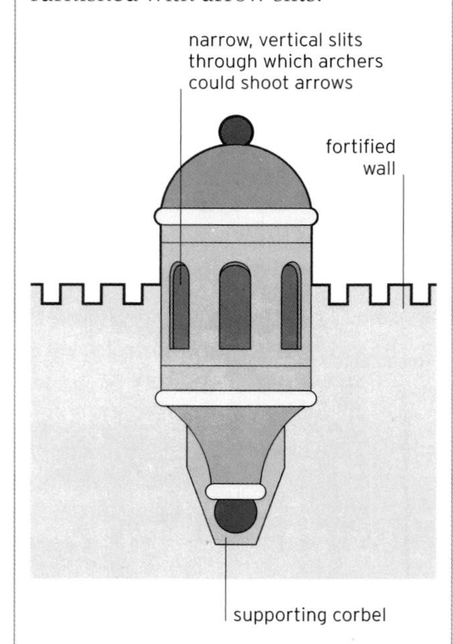

BAS-RELIEF A form of sculpture in which part of a flat material, such as wood or stone, is carved away to create a shallow, raised three-dimensional effect. It was a favored decorative element in the cultures of ancient Egypt and Mesopotamia. *See also* relief carving.

BASCULE ▼ A moveable bridge that uses a counterweight to help raise its deck. It is sometimes referred to as a drawbridge.

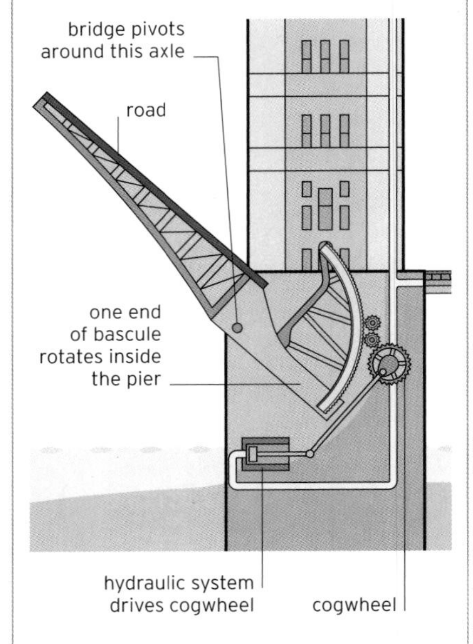

BASE The disk that forms the lowest part of a column, on which the cylindrical shaft sits. *See also* column.

BASEMENT The floor or floors of a building that are wholly or partially underground.

BASILICA Originally, an ancient Roman public building with a raised platform in an apse for the seating of officials. After the arrival of Christianity, the design was adopted for places of worship, featuring a central nave with a door at one end and an altar platform at the other end. *See also* altar, apse, nave.

BASTILLE A fortified tower or small fortress.

BASTION A structure that projects out from a castle or fortress wall. Bastions are most often angular in shape and situated at corners to give an improved range of defensive fire.

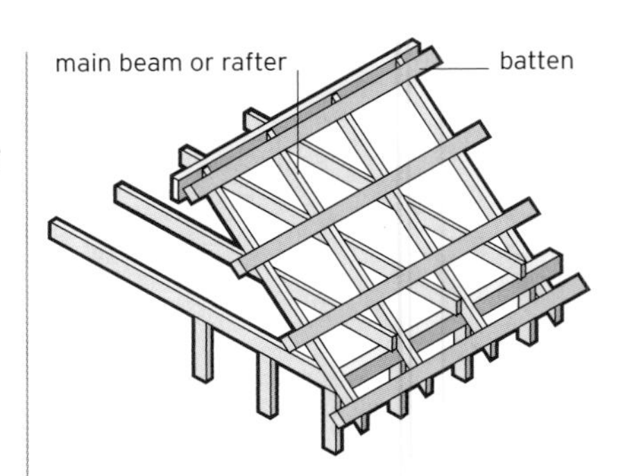

BATTEN ▲ Generally, a strip of timber, used as a secondary framework in construction, onto which a surface is fixed. For example, in roofing, battens are laid across the main structural beams and the tiling is laid on top of them.

BATTLEMENTS The crenellated top row of stones on a wall or tower. The alternating high parts (merlons) and gaps (crenels) provide cover to defending soldiers while allowing them to fire arrows. *See also* crenellation.

BAUHAUS A highly influential German art and design school operating from 1919 to 1933. It pioneered a Modernist and minimalist approach, characterized by the statement "form follows function." Utility was paramount, ornamentation was avoided, and there was an emphasis on uniting art and technology through the use of new materials and mass production.

BAY The space between architectural elements—for example, between columns or posts.

BAY WINDOW A window that projects outward from the wall of a building.

BAZAAR A Middle Eastern market, typically with a multitude of small shops and stalls in alleys, often with certain trades and types of goods grouped together. The alleys are sometimes roofed over, and historically there were gates so that the bazaar could be locked at night.

BEAM A horizontal structural component spanning an opening and carrying a load, such as a floor or wall. Traditionally, beams were wooden, but in modern construction they are more often concrete or steel.

BEAUX ARTS A grand style of architecture associated with the École des Beaux-Arts in Paris, which prevailed in France in the late 19th century. Buildings are typically large and elaborate with a symmetrical facade, their features derived from Classical Greek or Roman architecture. Interiors are similarly ostentatious, often featuring high, vaulted ceilings; central domes; and grand staircases. The style was popular in the US in the first decades of the 20th century. *See also* Classical.

BEEHIVE HOUSE A primitive circular structure with walls sloping inward as they rise to a rounded peak, like a cone. Common in ancient times, beehive houses are still found in some parts of the world today, including Syria.

BELFRY ▼ The upper part of a church tower where the bells hang. It is also sometimes used to refer to the entire tower.

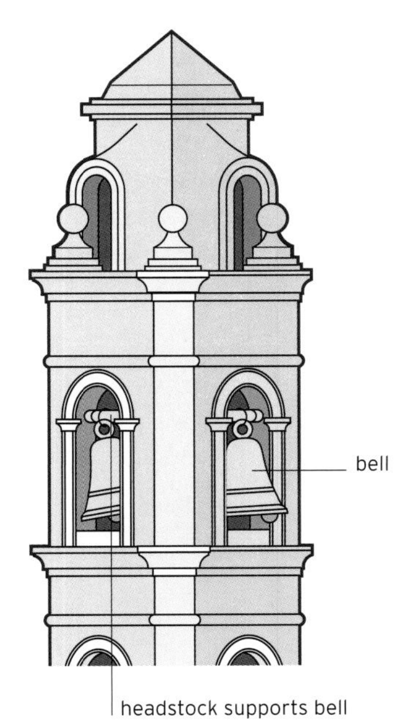

BELLCOTE A turret or small structure on a roof on which bells are hung.

BELVEDERE A structure built specifically to take advantage of an attractive view. It may be the upper part of a building or a free-standing element. The word comes from two Italian words, *bel* ("beautiful") and *vedere* ("view").

BERM A narrow shelf or path typically at the top or bottom of a slope. In a castle, the berm is the narrow area between the outer wall and its protective ditch or moat.

BOSS ▼ A carved ornament at the junction of stone vaulting ribs or roof beams. Bosses are often painted, and the subject of the carving is usually animals, birds, mythological beasts, or human faces. *See also* beam, rib vault.

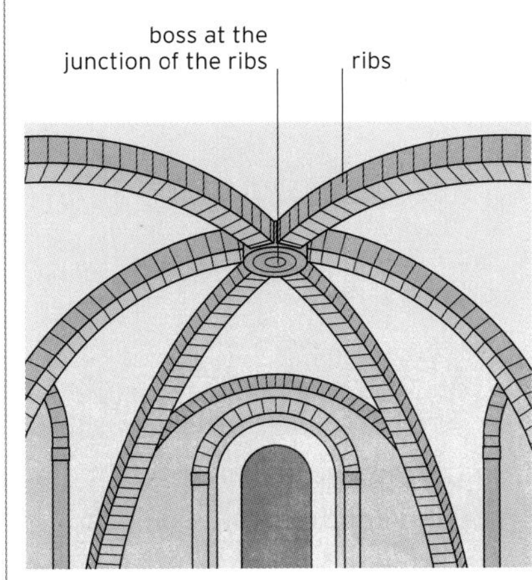

BOULEVARD A wide city road, usually in central or more affluent neighborhoods, with attractive landscaping and architecture.

BOX-FRAME A type of construction in which the external and internal concrete walls carry the weight of the floors and walls above. It is only suitable for buildings up to a height of five stories and where individual cells or rooms on each floor conform to a similar grid, so that they form a structural frame.

BRACE ▼ A supporting wire, rope, or beam that steadies or holds something else erect.

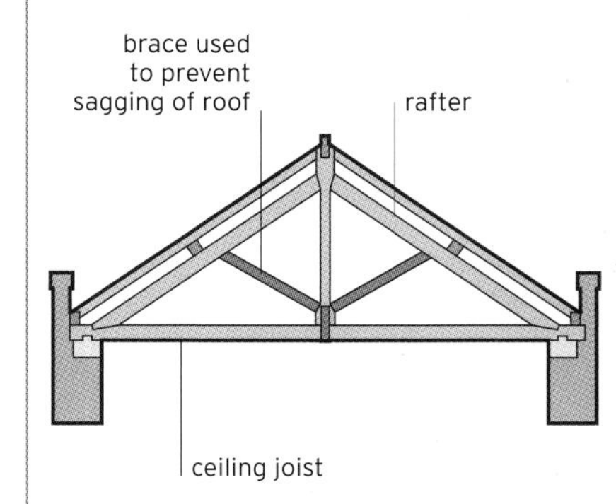

BRACKET An element that projects from a wall to support an arch, beam, shelf, balcony, or statue. Brackets are often decorated in the form of scrolls and can sometimes be entirely ornamental.

BREEZE BLOCK *See* cinder block.

BRISE-SOLEIL An external structure designed to provide protection from the Sun. It can take the form of baffles or lattices outside windows or can extend over the entire facade of a building.

BROCH A circular Iron Age stone structure found in Scotland. The exact function of brochs is a subject of debate, but many historians believe they were defensive in nature.

BRUTALISM An architectural style of the 1950s and 1960s characterized by monolithic blocklike forms and raw concrete construction.

BUDDHIST RAILING A stone barrier resembling a timber fence, with horizontal beams running through vertical posts.

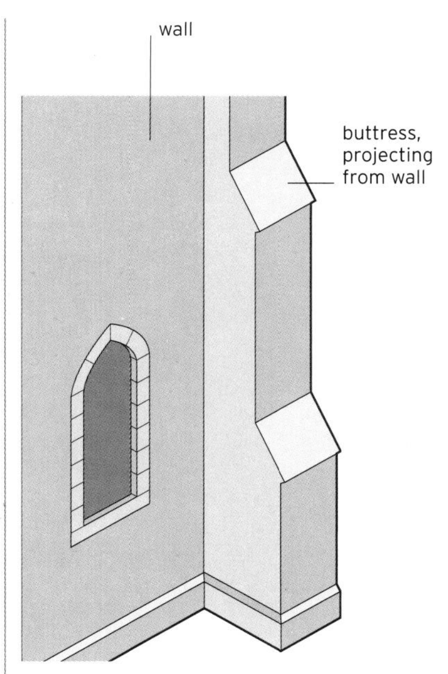

BUTTRESS ▲ A structure that supports a wall. *See also* flying buttress.

BYZANTINE Relating to the Byzantine empire (395–1453 CE). Byzantine architecture typically combines elements from ancient Rome and eastern Mediterranean countries such as Syria.

CABLE MOLDING A type of convex molding that resembles a twisted rope.

CAISSON A boxlike structure used to keep construction dry in areas submerged in water. Caissons are commonly employed when building bridge piers.

CAMBER A convex curvature built into a beam to compensate for any anticipated sag when loaded.

CAME A metallic strip used to hold small pieces of glass to make a large panel, as in a stained-glass window. Cames are typically made of lead and are in H-shaped sections, with U-shaped sections for the borders.

CANTILEVER

A beam, slab, or other structure supported only at one end, with a clear space underneath and the other end having no supporting columns or bracing, is called a cantilever. Vertical structures, such as towers or walls, can be given stability through cantilevers at their base.

cantilever retaining wall

cantilever footing attached to inside face of wall

beam is supported at one end by wall

beam transfers load to wall

CANTILEVER WALL

CANTILEVER BEAM

CAMPANILE A free-standing bell tower usually built beside or attached to a church. Campaniles are most often found in Italian architecture.

CAMPO SANTO A cemetery; the term is derived from the Italian and Spanish words for "holy field."

CAMPUS The land on which a group of college or university buildings is situated.

CANDI ▼ A Hindu or Buddhist shrine or temple in Indonesia.

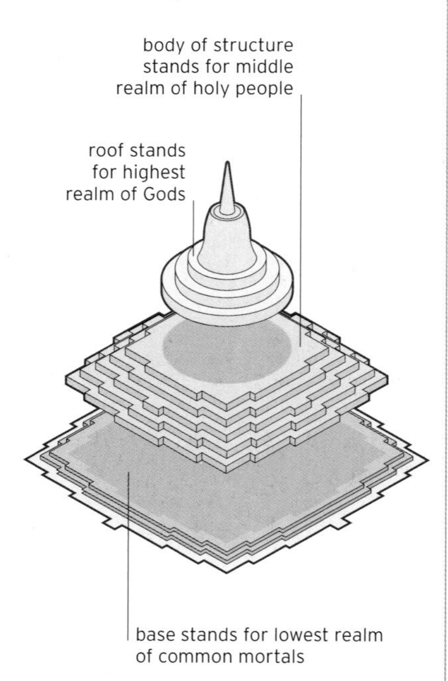

CANEPHORA A sculpture of a young woman carrying a basket of ceremonial offerings on her head. The sculpture may be free-standing but more often takes the form of a caryatid. *See also* caryatid.

CANOPY A projecting hood or cover over an altar or statue. In medieval times, this symbolized a divine or royal presence.

CANTILEVER See panel, above.

CAP The horizontal slab on top of a parapet. *See also* parapet.

CAPITAL The crowning element of a column that sits between the shaft and the abacus. The capital is usually decorated, and in the Classical tradition, its form belonged to one of three distinct orders: Doric, Ionian, and Corinthian. *See also* abacus, Classical, order.

CAPSTONE ▼ The top stone; for example, capstones may be used to top off a wall. It is distinct from a keystone, which sits at the apex of an arch. *See also* keystone.

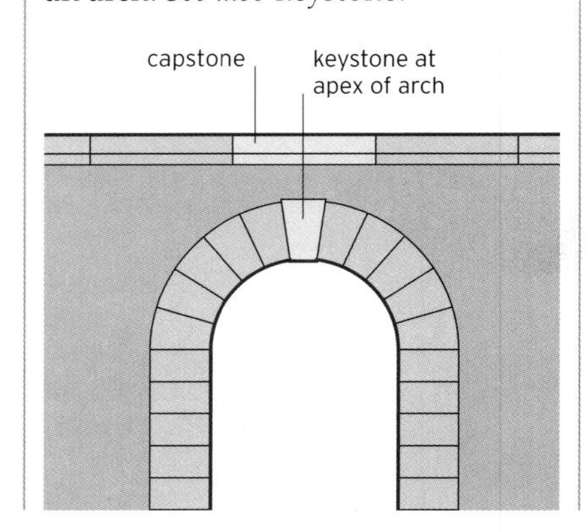

CARACOL A staircase in a helix or spiral form.

CARAVANSERAI A merchants' inn organized around a central courtyard with stables on the ground floor and accommodation on the upper floor. Caravanserais were a traditional type of building in Central Asia, in the Middle East, and across North Africa. In some places, they are also known as *wikala* or *fundouk*.

CARYATID A sculpted female figure that acts as a column supporting a structure on her head.

CASEMATE A small room in the wall of a fortress with openings from which guns could be fired.

CASEMENT A window frame that opens on side hinges, like a door.

CASTLE See panel, below.

CATACOMB A system of underground tombs.

CASTLE

A castle, or fortress, is a medieval stronghold, usually the residence of a king or a lord. Its stone fortifications provided protection for those within and a secure base from which to rule the surrounding territory. In its most basic form, the castle consisted of a central keep on a mound surrounded by a moat. More elaborate castles would have had a wall punctuated by defensive towers and sometimes even a series of concentric walls.

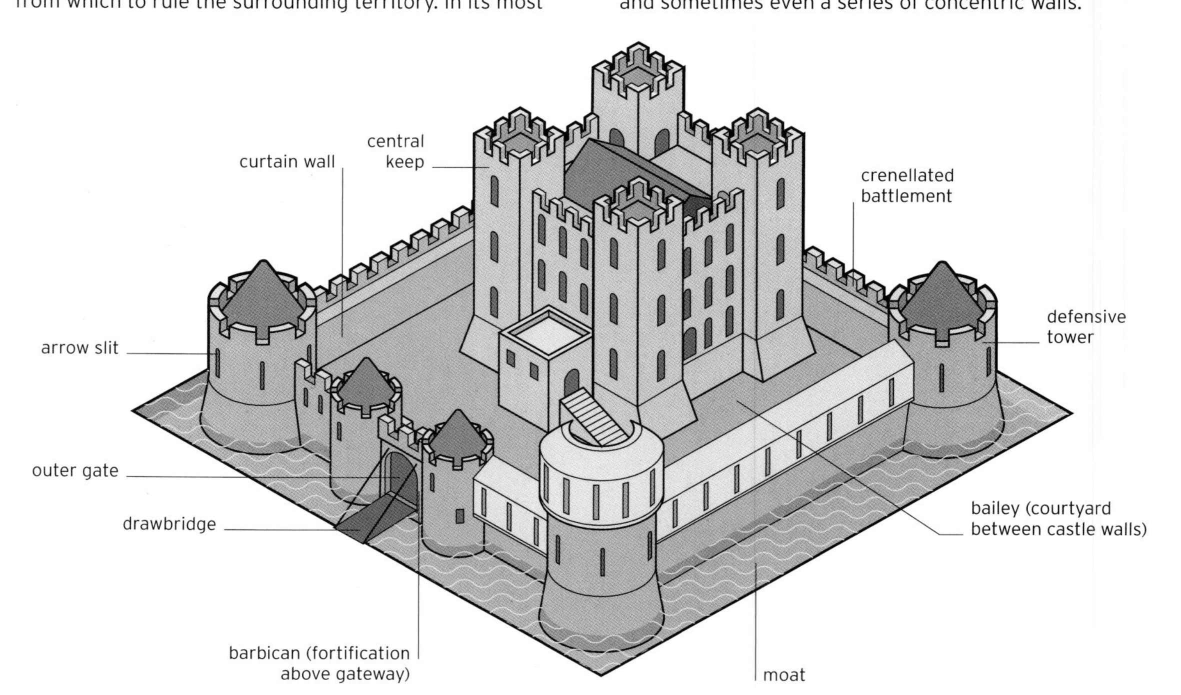

CHAITYA

A chaitya is a Buddhist shrine or prayer hall with a stupa at one end. The end of the hall is usually rounded to allow devotees to circumambulate the stupa. Such halls are typically high, long, and narrow and have a vaulted ceiling with prominent ribbing. Many early chaitya were carved out of rock, and the ribbing was decorative rather than structural.

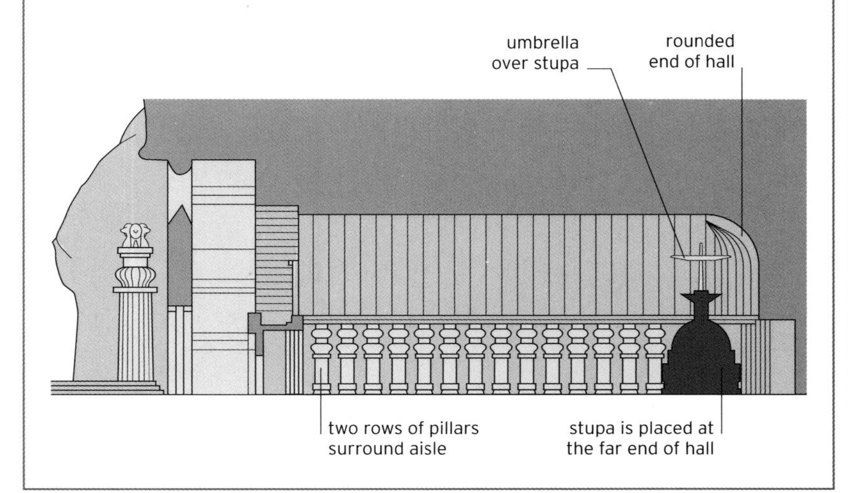

CATHEDRAL A Christian church that serves as the seat of a bishop and the central church for an administrative region, or diocese. Cathedrals tend to be larger than the average church. *See also* church.

CEILING The overhead surface covering a room.

CEILING JOIST *See* joist.

CELLA The inner chamber of a temple in Classical Roman architecture. It usually contained a statue of the deity to whom the temple was dedicated. *See also* adyton, Classical.

CEMENT A binding agent used to make concrete. Cement has been used since ancient times. Modern cement is made by heating limestone with other materials, such as clay, to high temperatures and then grinding the resulting material.

CENOTAPH A memorial monument erected in honor of a person or persons whose remains are elsewhere.

CENTERING The timber framing used to support an arch or dome while it is under construction.

CHAIR-RAIL A molding fixed horizontally around the walls of a room at roughly waist height. Originally, it was intended to protect the wall from being damaged by furniture, but in modern rooms, it is used for decorative effect. It is also known as a dado rail. *See also* dado.

CHAITYA See panel, above.

CHAITYA ARCH An ornamental arch used to decorate a facade, often placed around or above a doorway. The design of the arch is based on a cross-section of a chaitya and looks like an upturned ship's hull with a keel and ribbing. *See also* chaitya.

CHALET A timber house characteristic of Switzerland, Bavaria, and the Alpine regions, with a shallow, sloping roof and wide eaves. Originally, chalets were the dwellings of shepherds. *See also* eaves.

CHANCEL The area where the altar is located at the end of a nave. It is the area used by the clergy during worship. *See also* altar, church, nave.

CHANTRY CHAPEL An endowed chapel in which a priest was employed to sing regular masses for the chapel's founder. It was believed that such masses would speed the deceased's soul to Heaven. Chantry chapels were abolished during the Protestant Reformation (16th century).

CHAPEL A Christian place of worship that does not have a parish attached to it. Chapels can be free-standing but are more commonly part of a larger structure such as a cathedral, castle, college, palace, or hospital. *See also* cathedral.

CHARNEL HOUSE A vault or building in which human skeletal remains are kept.

CHATRI ▼ A small, domed pavilion supported by columns at each corner. *Chatri*s are used to adorn the corners of a roof or placed above the main entrance in Indian Mughal architecture. They are decorative and have no function. *See also* Mughal.

CHEQUER-WORK A design of alternating squares that resembles a chessboard.

CHEVRON ▼ An inverted V-shape. In Norman architecture, it is used to form zigzag patterns on arches and columns.

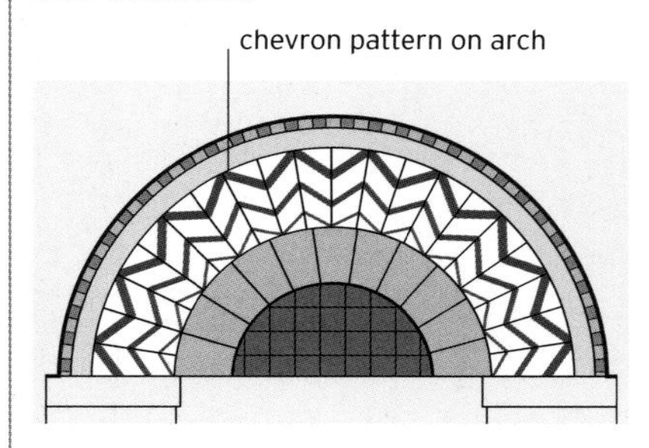

CHIGI ▼ Forked roof finials found in Japanese architecture. *See also* finial.

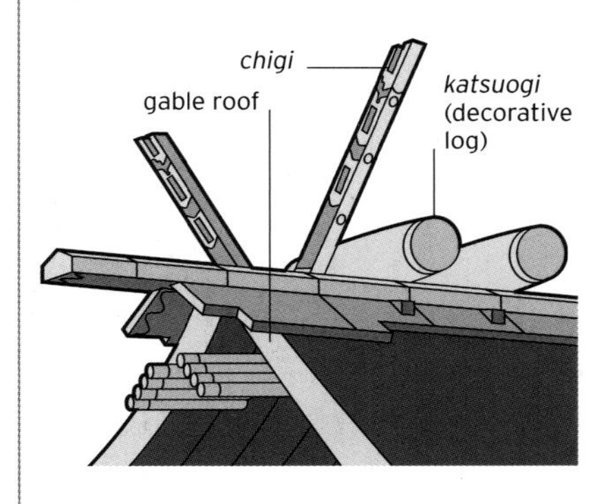

CHIMNEY-BREAST The part of a chimney that projects forward into a room with the fireplace set into its base.

CHIMNEY SHAFT A high chimney stack with only one flue (duct leading up to the roof).

CHIMNEY-STACK The part of a chimney that projects above the roof.

CHINOISERIE A European decorative style inspired by art and design from China, Japan, and other Asian countries. The name comes from the French word for Chinese, *chinois*.

CHOIR The part of a church or cathedral that provides seating for the clergy and church choir. Choirs are commonly located in the chancel. *See also* chancel, church.

CHULLPA An ancient funerary tower of the Aymara, an indigenous people of the Andes and Altiplano regions of South America.

CHURCH See panel, below.

CHURRIGUERESQUE A style of elaborate sculptural ornament typically applied to the facades of buildings. It was popular in the 17th–18th centuries and derives its name from the Spanish architect and sculptor José Benito Churriguera (1665–1725). Its extravagant, florid detailing is most commonly found in Spain and Mexico.

CINDER BLOCK A large brick made from cement and ashes. Cinder blocks are called breeze blocks in the UK.

CLAPBOARD A style of construction using long, thin boards of timber laid horizontally and overlapping to clad exterior walls.

CLASSICAL Relating to the architecture of ancient Greece and Rome, characterized by its column-and-pediment structures, fixed proportions, symmetry, and decorative orders (notably Doric, Ionic, and Corinthian). *See also* order.

CLASSICISM The revival of the forms and techniques associated with Classical architecture. See *Neo-Classical.*

CLERESTORY The high section of a church where the walls rise above the aisle roofs and are set with windows to admit light into the interior. *See also* aisle, church.

CLOISTER A covered walkway around a quadrangle. It is usually attached to a cathedral, monastery, or college.

COB A natural building material made from soil, clay, and straw.

COFFER A square, rectangular, or octagonal sunken panel used in a grid to decorate a ceiling. The most famous example of a coffered ceiling is the interior of the dome of the Pantheon in Rome, where the coffers are used to lighten the weight of the structure.

COLONIAL REVIVAL A design movement in the US and Canada that gained popularity in the 1890s. Inspired by the Georgian architecture of England, buildings are typically two stories and strongly symmetrical, and the detailing is influenced by Classical architecture. Main entrances are usually pronounced, with a large door framed by a portico. *See also* Classical, portico.

COLONNADE A row of columns, often carrying an entablature. *See also* entablature.

COLOSSAL ORDER An order of Classical architecture in which columns extend the height of two or more stories in a building. It is also sometimes called the giant order. *See also* composite order, order.

COLUMN A cylindrical shaft with a base and a capital. Columns usually support a structure, but they can also be free-standing monuments—for example, Trajan's Column in Rome. *See also* capital, order.

COMPOSITE ORDER An order of Classical architecture that combines the volutes found in the Ionic order with acanthus leaves from the Corinthian order. *See also* Classical, Colossal order, Corinthian, Ionic, order.

CHURCH

During the first four centuries of Christianity, worshippers met in private houses. In time, dedicated houses of worship, called churches, developed. A typical early church would be entered from the west, with a nave leading to the chancel, which is an elevated area used by officials, or clergy, who lead the worship. Central to the chancel is the altar, the table used for the Christian rite of Communion. Larger churches may have additional aisles, or transepts, or an ambulatory around the altar. However, there are no set requirements for what constitutes a church. It is simply a meeting place for shared religious observance.

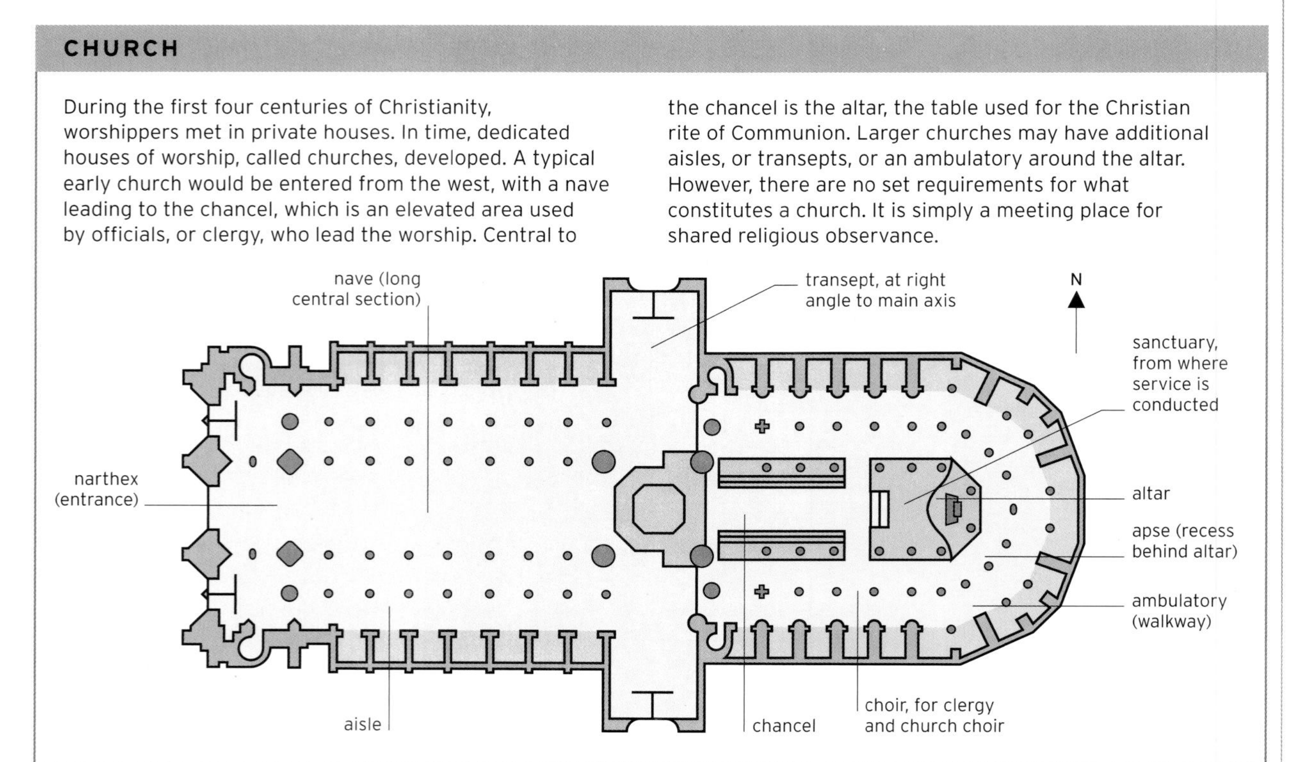

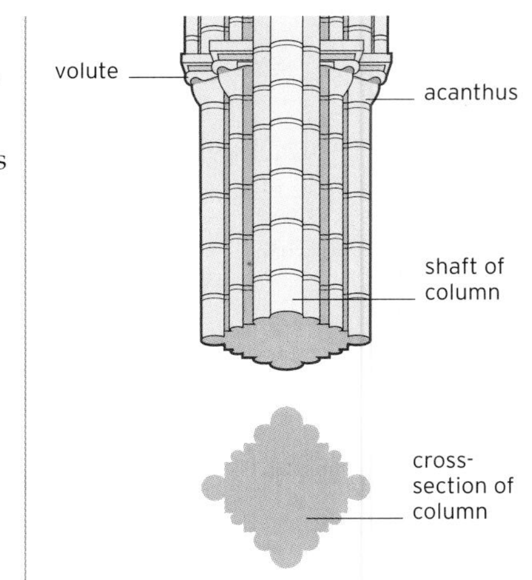

COMPOUND COLUMN ▲ A column made up of a number of slender columns grouped together; this is done for decorative effect. Many compound columns use stones of different types.

COMPUTER-AIDED DESIGN (CAD) The use of computer software packages by architects to aid in the design of buildings and other structures. First introduced in the 1960s, CAD became widespread from the 1990s onward.

CONCRETE A composite material made from cement, water, and an aggregate such as sand or gravel. Relatively cheap, strong, durable, and resistant to fire, water, and decay, concrete can also be molded into almost any shape.

CONFESSIO A niche for relics located near the altar in a church. *See also* altar, church.

CONTEMPORARY Relating to 21st-century architecture. No single style predominates, but what many structures have in common is the use of advanced technology to push the boundaries of what a building can be. Buildings are reaching higher than ever and assuming unconventional nonrectilinear, asymmetrical shapes.

CORBEL A type of bracket that projects from a wall to support an arch, beam, or balcony. Corbels tend

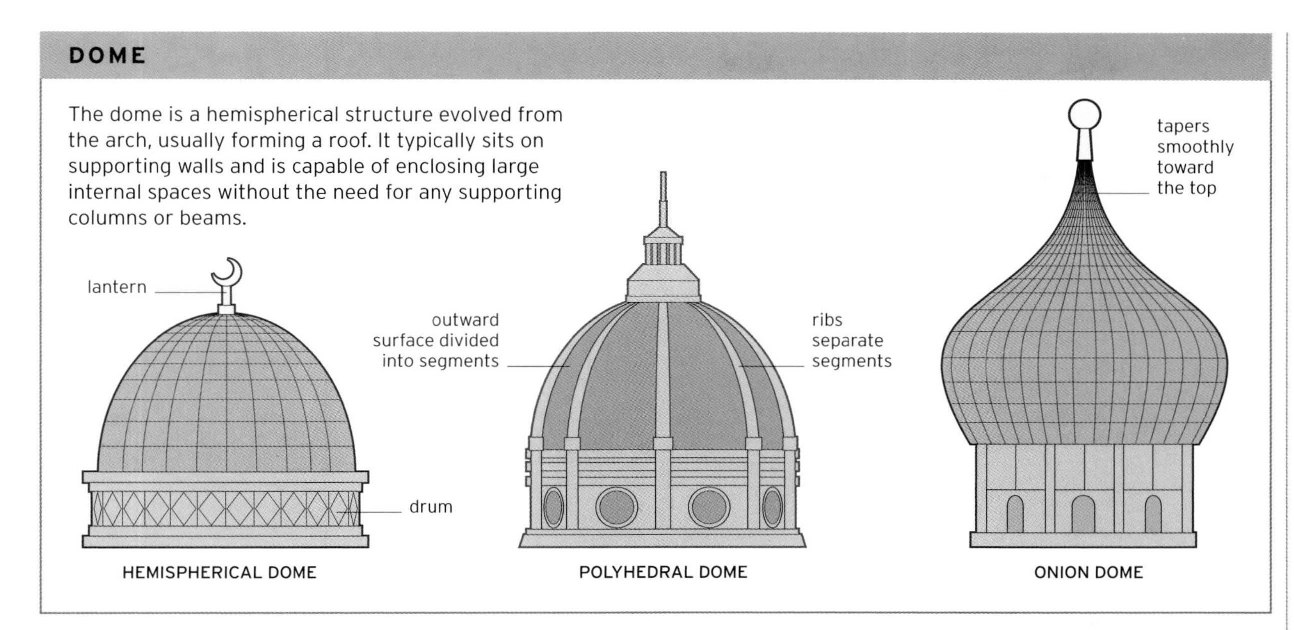

to be embedded deeply into a wall for greater strength and are employed more on external walls, as opposed to brackets, which are often internal.

CORINTHIAN The third of the main Classical orders of architecture. Its main characteristic is an ornate capital carved with acanthus leaves. *See also* acanthus, capital, Classical, Doric, Ionic, order.

CORNICE The horizontal molding that runs along the top of a building.

CORTILE The internal courtyard of a palazzo—often with arcades or colonnades rising many stories—that is open to the sky. *See also* colonnade.

COVING A plain, concave molding used to conceal the joint between a wall and a ceiling.

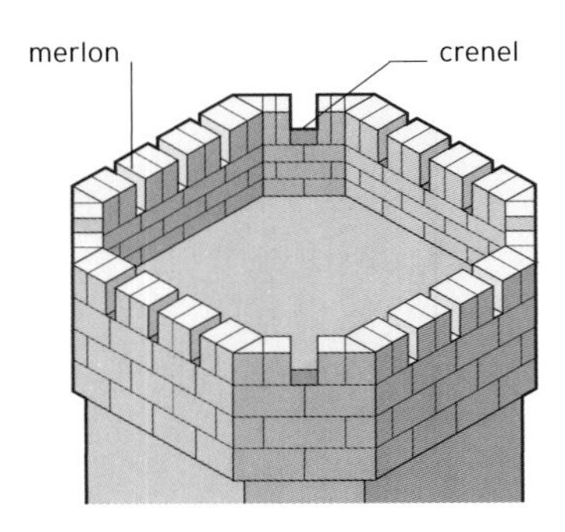

CRENELLATION ▲ The alternating high parts (merlons) and gaps (crenels) that top a fortified wall.

CROCKETS The small, stylized carvings of buds, curled leaves, or flowers that appear at intervals on spires and finials. They are a particular feature of Gothic churches and cathedrals. *See also* finial, Gothic, spire.

CROSSING The junction where the transepts intersect with the main body in a cruciform church. In a cathedral, the tower often rises above the crossing. *See also* church, cruciform, transept.

CRUCIFORM Having the shape of a cross. Western churches have often employed a cruciform floor plan, in which the main nave is intersected by north and south transepts that form the arms of the cross. *See also* church, nave, transept.

CUPOLA A small dome often used to crown a turret, tower, roof, or larger dome.

CURTAIN WALL A non-load-bearing wall attached to a building's structural frame. Almost any modern building of more than five or six stories will employ a curtain wall. This can be formed of thin slabs of stone, veneer, or metal, but most curtain walls are made of glass set in aluminum frames.

DADO The lower part of an internal wall between the skirting board and the dado rail. *See also* chair-rail.

DAIBUTSU Japanese term for a giant Buddha.

DAUB Clay or a similar substance used to coat a surface, especially when mixed with straw and applied to wooden strips (known as laths or wattles) to form a wall, as in wattle-and-daub construction. *See also* wattle.

DECONSTRUCTIVISM A movement in contemporary architecture that seeks to break traditional rules. It questions ideas of harmony, consistency, and symmetry and offers buildings that seem fragmented, unpredictable, and chaotic. A prime example would be Frank Gehry's Guggenheim Museum in Bilbao.

DIOCLETIAN WINDOW A semicircular window divided into three parts by two vertical mullions. It gets its name from windows found in the ancient Baths of Diocletian in Rome. *See also* mullion.

DIPTEROS A building with a double colonnade on all sides. *See also* colonnade.

DOME See panel, left.

DORIC The earliest of the main Classical orders of architecture. It is characterized by a simple and unadorned capital and a sturdy column. *See also* Corinthian, Ionic, order.

DORMER ▼ An upright window projecting out of a sloping roof. Dormers are often later additions to buildings when an attic is converted into a living space requiring natural light.

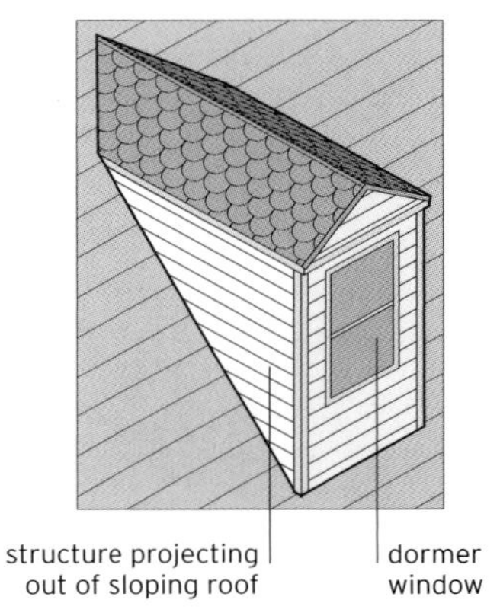

DOUBLE-FRAMED ROOF A roof in which extra longitudinal components, known as purlins, run perpendicular to the rafters and join them together, providing additional stability and strength.

DOVETAILING ▼ A type of carpentry joint in which pieces have shaped pins and tails that fit together securely and strongly without nails or other fasteners.

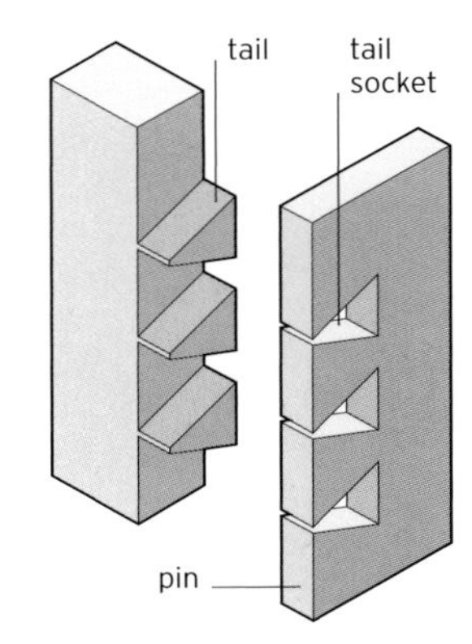

DRUM One of the cylindrical stone blocks that compose most columns. The drum is also the circular or polygonal structure on which a dome sits. *See also* dome.

E

EAVES The edges of a roof that overhang the walls of a building.

ECHAL The enclosure containing the ark in a synagogue, which in turn holds the synagogue's Torah scrolls. It is also written as hekhal, hechal, or heichal. *See also* synagogue.

ECLECTICISM An architectural style that incorporates a mixture of historic styles to create something new and original, as opposed to simply reviving an older style. It flourished in the 19th and 20th centuries.

EGYPTIAN REVIVAL An architectural style that incorporates the forms, motifs, and imagery of ancient Egypt. It became popular following Napoleon's conquest of Egypt in 1798 and briefly again after the discovery of the tomb of Tutankhamun in 1922.

ELEVATION A flat representation of one side of a building, usually the facade. The term is most commonly used in reference to architectural drawings.

ENFILADE A series of openings between rooms and spaces that align along an axis to provide an extended view.

ENGAGED COLUMN A column that is partially embedded in a wall. The columns are load-bearing and also act as buttresses for the wall; they are distinct from pilasters, which are purely decorative. *See also* pilaster.

ENTABLATURE See panel, below.

ENTASIS ▼ The addition of a slight convex curve to a structure for aesthetic purposes. For example, the columns of the Parthenon in Athens curve slightly to counteract the fact that columns with straight sides appear to the eye to be more slender in the middle and have a waist.

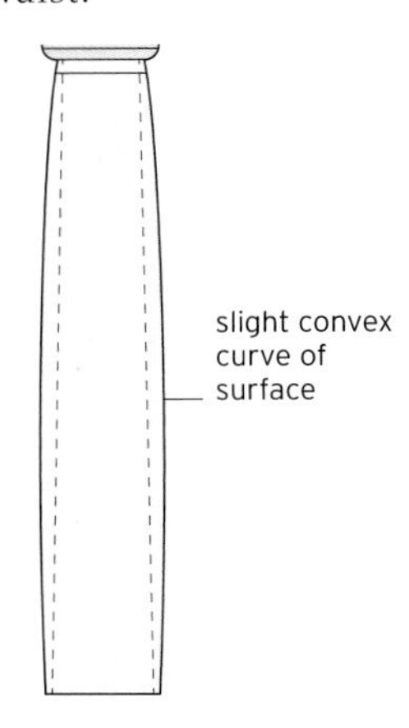

ENVIRONMENTAL DESIGN The practice of considering environmental factors when carrying out urban planning or building design.

ESPLANADE A pedestrian walkway, often on a seafront. In North America, an esplanade is also a strip of raised land dividing a road.

F

FABRIC The load-bearing structure of a building.

FACADE The front or main face of a building.

FAN VAULT A form of highly decorative ceiling vault with ribs radiating out from a central column, like a fan. It is associated with the Gothic style and is mostly found in the ecclesiastical architecture of England. *See also* Gothic, vault.

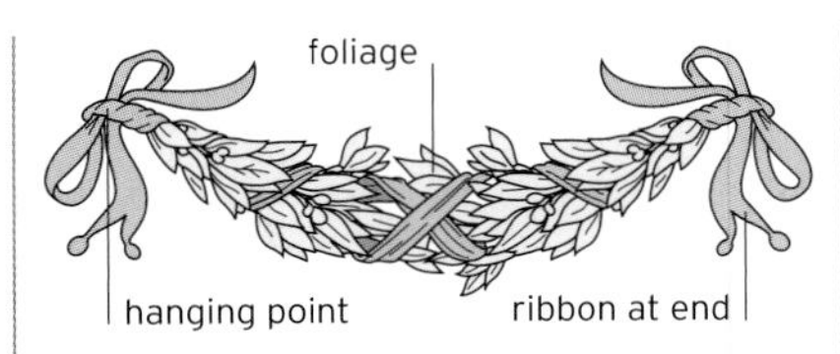

FESTOON ▲ A form of stone-carved ornamentation depicting a wreath or garland hanging from two points.

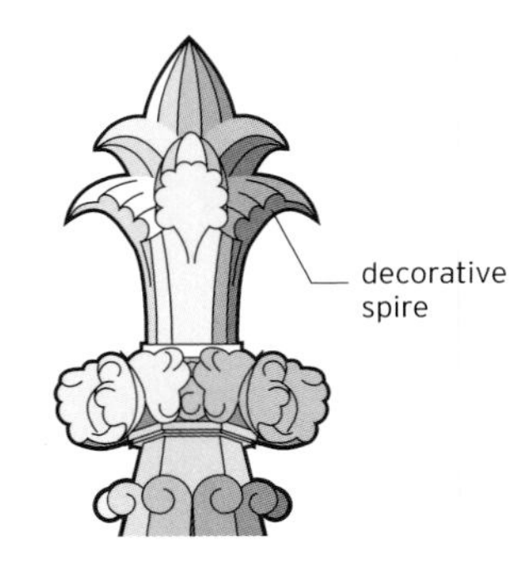

FINIAL ▲ The uppermost part of a pinnacle, spire, minaret, or gable end. This usually has some decorative element, such as a spike or knob.

FLAMBOYANT A style of Gothic architecture popularized in 15th-century France and Spain with an emphasis on decoration. Its main characteristic is the use of a flamelike S-shaped curve in stone tracery. *See also* Gothic, tracery.

FLÈCHE A small spire set on the ridge of a church or cathedral roof.

FLEURON A stone-carved, flower-shaped ornament.

FLUTING The shallow grooves running vertically on a column or pilaster and occasionally on other surfaces. Fluted columns are particularly associated with the architecture of ancient Greece; columns of the Doric order have 20 flutes, while Ionic and Corinthian columns traditionally have 24. *See also* order.

FLYING BUTTRESS ▲ A type of buttress with a free-standing pier from which extends an arched structure that connects with and

ENTABLATURE

The entablature is the name for the combined horizontal elements that sit on top of a column in a Classical structure. It generally comprises a decorative architrave and, above that, a plain frieze, although sometimes the frieze may have sculpted reliefs. The whole structure is topped by a decorative cornice. Frequently in Classical architecture, the entablature will be topped by a pediment.

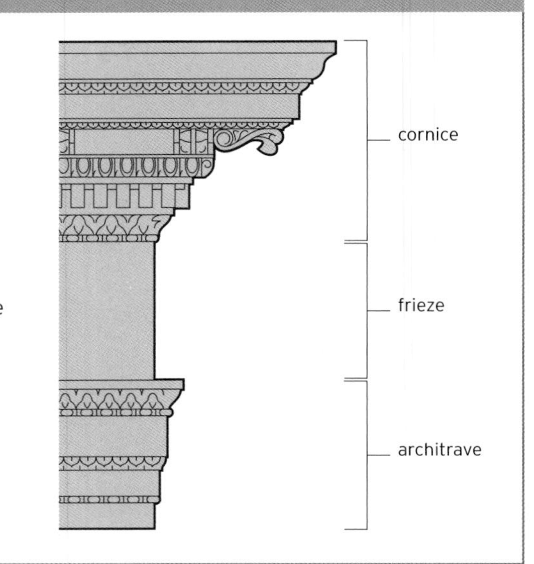

FOIL

Derived from the Latin word *folium*, meaning "leaf," a foil is a decorative pattern that uses combinations of touching circles to create cloverlike patterns. There are three basic types of foliated decoration: trefoil, quatrefoil, and cinquefoil (respectively, three-, four-, and five-leaved). This type of design is typically seen in the shape of small windows or in the patterned tracery within large windows. It is particularly associated with the Gothic style.

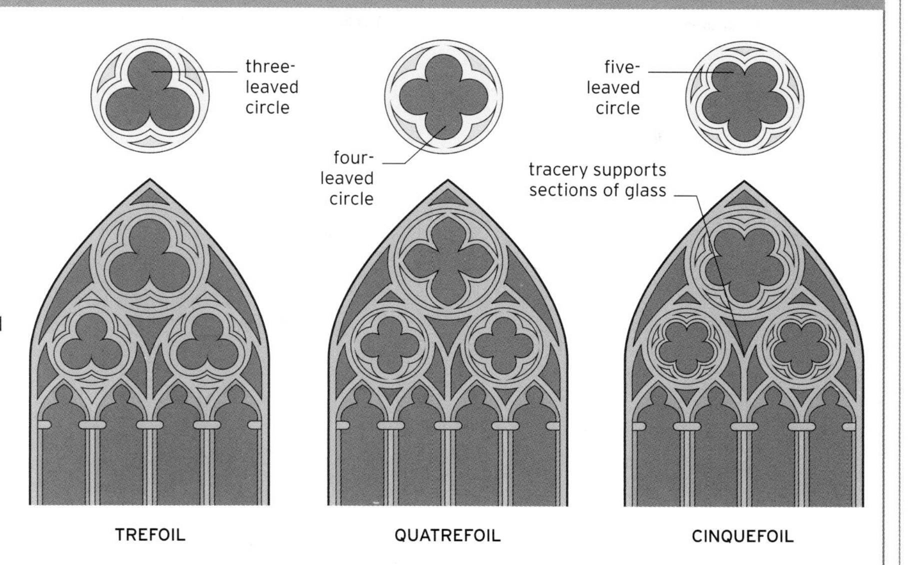

supports the wall of the main building. The forces from the main structure are transmitted across the arch to the pier. *See also* Gothic.

FOIL See panel, above.

FOLLY A generally nonfunctional and whimsical building erected to enhance a landscape.

FORUM An ancient Roman public square or marketplace.

FRAMED BUILDING A building in which the weight is carried by a frame, typically of steel or concrete columns and beams, rather than by load-bearing walls.

FRESCO A type of wall painting executed on freshly laid or wet lime plaster.

FRIEZE In Classical architecture, the wide central part of the entablature, between the architrave (bottom) and cornice (top). The frieze is often decorated, the most famous example being the Parthenon frieze in Athens. In interiors, the frieze is a painted or sculpted section of the wall directly under the cornice or coving. *See also* architrave, Classical, cornice, entablature.

FRONTISPIECE The elements that frame and decorate the main entrance to a building.

FULCRUM The pivot about which a lever turns.

FUNCTIONALISM The principle that buildings should be designed based solely on their purpose and function. This notion took hold following World War I and was part of the greater wave of Modernism. *See also* Modernism.

GABLE ▼ The triangular section of wall formed below a pitched roof.

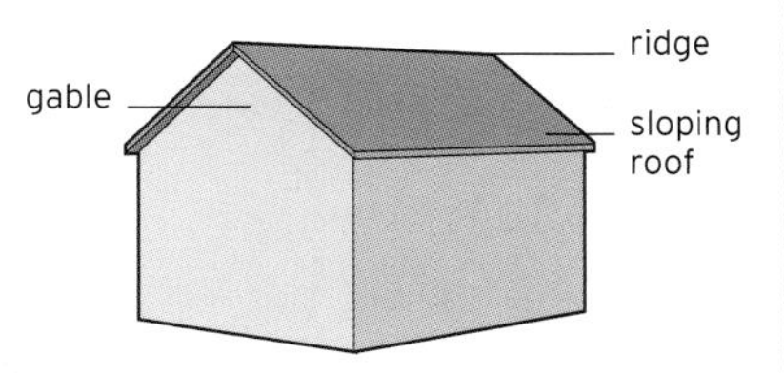

GADROONING A form of sculpted decoration consisting of a series of convex curves, sometimes tapering, sometimes running diagonally. It features on Roman sarcophagi and on metalwork and porcelain.

GALLERY A narrow balcony running the length of the wall overlooking a large interior.

GARBHAGRIHA The innermost sanctum of a Hindu temple, where the idol of the primary deity of the temple is kept.

GARGOYLE A carved imaginary creature, often a grotesque, projecting from the edge of a roof, with a spout for a mouth through which rainwater is directed away from the walls of a building. *See also* Gothic, grotesque.

GAZEBO A small, free-standing pavilion in a garden, constructed to take advantage of a view.

GEOMANCY An age-old belief in the art of reading signs in a landscape to identify beneficial energies. It is similar to the Eastern tradition of feng shui.

GLAZED BRICKS Bricks with a ceramic coating. The coating is often colored, and the bricks are used for decorative effect.

GOPURAM A monumental tower acting as the gateway to a Hindu temple enclosure. *Gopurams* are typically ornate, covered with painted sculptures.

GOTHIC An architectural style that originated in 12th-century France and flourished throughout Europe during the later part of the Middle Ages (5th to 15th centuries). It is characterized by an emphasis on verticality, with walls reduced to a minimum by high-pointed arches and arcades and expanses of stained-glass windows. It also extensively uses flying buttresses, pinnacles, and spires. Exteriors typically feature a lot of statuary and carvings illustrating Biblical stories and themes for the benefit of illiterate worshippers. *See also* flying buttress, gargoyle, Gothic Revival, grotesque, pinnacle, spire.

GOTHIC REVIVAL A movement in the late 18th and 19th centuries to revive medieval Gothic forms. Some buildings in Gothic style built after medieval times are also referred to as Neo-Gothic.

GREEK CROSS A cross with arms of equal length. *See also* Latin cross.

GREEN BUILDING Construction that is environmentally responsible and resource-efficient.

GRIDIRON PLAN A type of city plan in which streets run at right angles to each other. The most famous example is the street layout of Manhattan.

GROIN VAULT A vault made up of two barrel vaults that intersect at right angles. *See also* rib vault, vault.

GROTESQUE A stone-carved fantastical beast or mythical creature. Grotesques are a common feature of Gothic architecture, decorating the exteriors of predominantly religious buildings.

GROUT A mixture of cement, sand, and water used to seal gaps between tiles or stonework or to repair cracks.

GYNAECIUM In ancient Greece, the women's quarters.

HACIENDA A large estate, especially in a Spanish-speaking country. The term is also applied to the type of manor house that would typically be found on a hacienda.

HAMMAM The Arabic name for a Turkish-style bathhouse.

HAMMER BEAM ▼ A short timber beam projecting from a wall on which the rafters that support the roof rest. This arrangement allows for greater spans than the normal arrangement, in which the width of the roof can be no longer than the length of individual timbers.

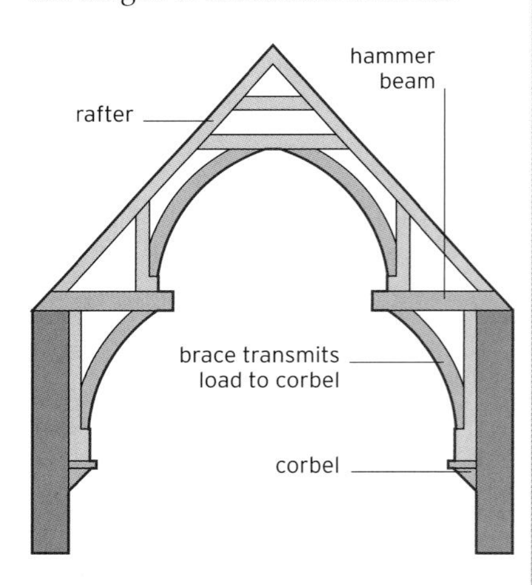

HARMONIC PROPORTIONS A relationship between the dimensions of a building, or part of a building, that conforms to a mathematical formula that is closely bound to the ancient Greek concept of the golden ratio and the Fibonacci series. This describes the proportions that are most pleasing to the eye and echo those found in elements of nature.

HENGE A Neolithic earthwork featuring a circular bank and ditch.

HERM Originally, a sacred stone connected with the cult worship of the Greek god Hermes. Over time, the term came to refer to a pillar, generally square and tapering toward the bottom, suggestive of a male figure. Herms were often topped with a portrait bust.

HIGH-TECH An architectural style that emerged in the 1970s that visually references technology and industry in its outward design. Prime examples are the Georges Pompidou Centre in Paris, which displays the building's mechanical elements of ducts and piping on its outside, and the HSBC Hong Kong headquarters with its framework of columns and lateral supports that appears on the facade.

HIPPODROME In ancient Greece and Rome, a stadium for horse- and chariot-racing.

HISTORICISM The idea of drawing inspiration from recreating historical architectural styles.

HYPOSTYLE An interior space with a roof that rests on pillars or columns.

I

I-BEAM A steel beam with an I- or H-shaped cross-section. The vertical part is known as the web, while the horizontal parts are called flanges. It is also known as a universal beam (UB) or rolled-steel joist (RSJ).

ICONOSTASIS A wall or screen of icons that separates the nave from the sanctuary in an Eastern Christian church. *See also* church, nave.

INDUSTRIAL ARCHITECTURE Buildings serving the needs of industry such as factories, power plants, mills, and warehouses.

INGLENOOK A recess set next to a fireplace where a person could sit to enjoy the warmth.

INTARSIA The fitting together of pieces of intricately cut wood of different types and colors to make often complex images.

IONIC One of the three main orders of ancient Greek architecture. It is characterized by having the narrowest columns, with capitals decorated by volutes. *See also* capitals, Classical, columns, Corinthian, Doric, order.

IWAN A vaulted hall, walled on three sides and open on the fourth. Iwans are typically placed facing the central courtyard of a mosque or *madrasa*.

JALI An ornamental lattice screen, often made of carved stone, commonly found in Hindu temples.

JOIST ▼ A load-bearing structural component that spans between walls to provide support for a floor or ceiling. A joist is a form of beam but one that is typically made of timber, spans short distances, and is associated with domestic architecture.

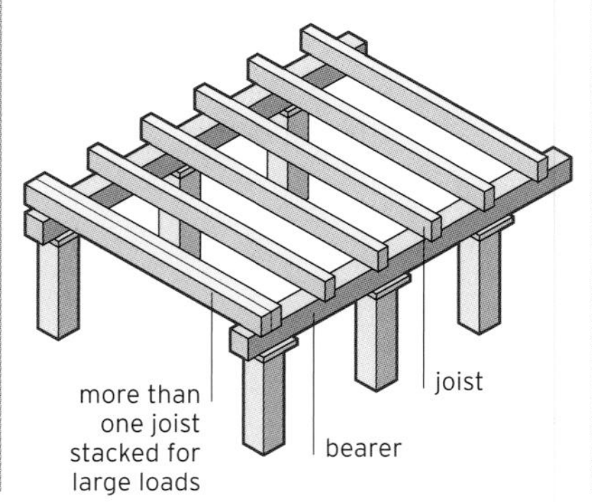

KASUGA-ZUKURI A Japanese Shinto temple style that incorporates Chinese-style roofs and a curved *chigi*; the shrine buildings are decorated with red, gold, and vermillion. A *kasuga-zukuri* has its entrance on the gable end, covered by a veranda. *See also* chigi.

KATSUOGI Short decorative logs placed horizontally and in a row along the ridges of roofs on Japanese Shinto shrines. They often feature in conjunction with *chigi*. They are purely decorative and have no structural function. *See also* chigi.

KEEP A large fortified tower within a castle or fortress. *See also* castle.

KEYSTONE The wedge-shaped stone at the apex of an arch. It is the final stone laid in construction and locks all the other stones in place, allowing the arch to be self-supporting and bear weight. *See also* arch.

KING POST A vertical post acting as a tie between the apex of a triangular truss and the crossbeam (also known as a tie beam). *See also* truss.

KING TRUSS A simple triangular framework on a single plane, used for roof construction. It is composed of a crossbeam with two diagonal rafters that meet to form an equilateral triangle. The crossbeam is secured to the rafters at their apex by a king post. There may also be additional supporting struts at an angle between the king post and the rafters. *See also* king post, truss.

KONDO The main shrine hall at the heart of a Japanese Buddhist temple compound. It may also be referred to in English as a hondo.

L

LANCET ▼ A tall, narrow window with a sharp, pointed arch at its apex.

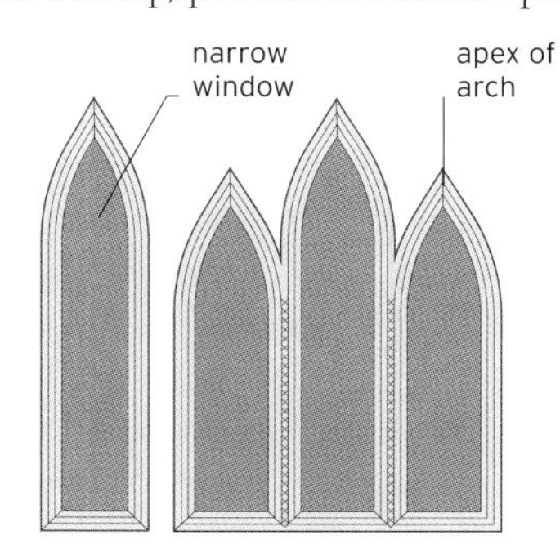

LANTERN An element on a roof or the top of a dome or tower that allows natural light into the room or space below.

LATIN CROSS A cross in which the horizontal element is shorter than the vertical and the two elements intersect high up on the vertical. It resembles the cross on which, in Christian tradition, Jesus Christ was crucified. *See also* Greek cross.

LIME A mineral used in the production of mortar, cement, and concrete. Its use dates back to ancient times.

LINTEL A horizontal block that spans the space between two vertical supports. Lintels are commonly found over doors and windows.

LOGGIA A covered corridor or gallery on the exterior of a building, open to the elements on one side and usually supported on columns or arches.

LONGHOUSE ▼ A long, narrow dwelling, typically wooden-framed, common to many ancient cultures and some contemporary indigenous cultures, notably some of the inhabitants of Borneo.

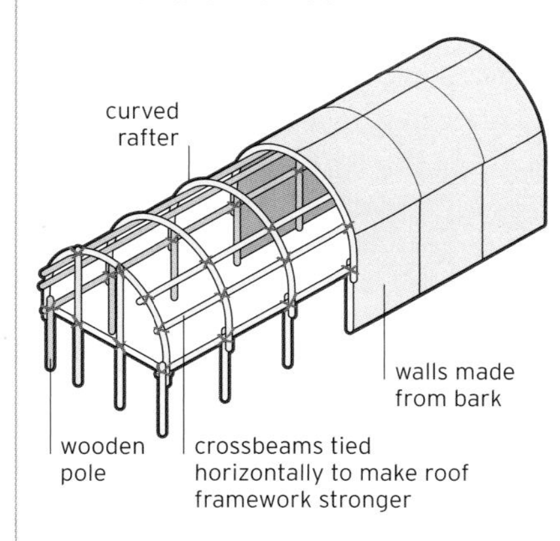

LOUVRE ▼ A shutter with horizontal slats angled to admit light and ventilation but to keep out rain and direct sunshine.

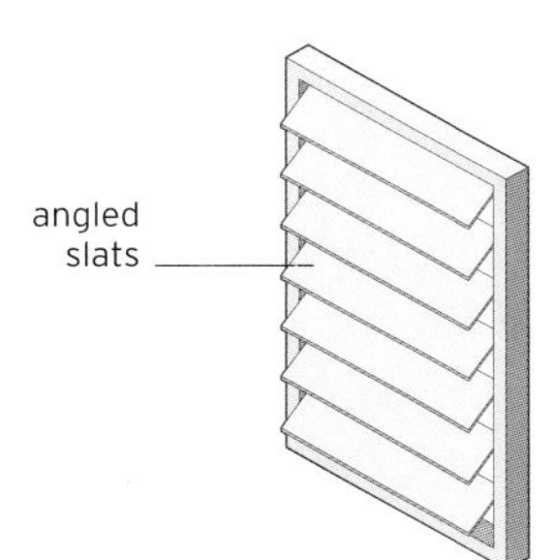

LUNETTE A small, semicircular section of wall, usually over a door or window—for example, on the end wall of a barrel-vaulted room. It may contain a mural, sculpture, or another window.

MADRASA Also spelled *madrasah*, an Islamic theological school or college. Some buildings serve a double function as a mosque-madrasa. There is no particular architectural form associated with a *madrasa*, but historically it can be lavishly decorated.

MANNERISM An architectural style originating in 16th-century Italy as a reaction against the harmony and forms of the High Renaissance. It is characterized by the distortion of proportions and arbitrary arrangements of decorative features. For example, the colossal order has its origins in Mannerism. *See also* Renaissance architecture.

MANSARD A type of roof associated with France that is characterized by having two slopes on each of its sides; the lower slope has steep sides and is typically punctured by dormer windows, while the upper slope is shallower. It is also known as a gambrel roof. *See also* dormer, roof.

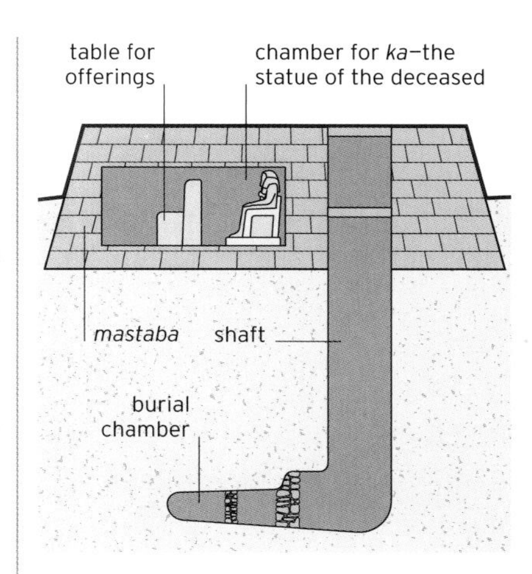

MASTABA ▲ A type of early ancient Egyptian tomb that takes the form of a simple low, flat, rectangular structure with inward-sloping sides.

MEGALITH A large stone erected, mainly in the Neolithic period, as a monument or as part of a monument with other stones. *See also* monolith.

MEZZANINE An intermediate floor in a building that does not extend over the whole space of the floor below and so, in part, overlooks the lower floor.

MIHRAB A small niche or alcove in the wall of a mosque that indicates the direction of prayer. *See also* mosque.

MINARET See panel, below.

MODERNISM An architectural style that emerged in the 1930s, associated with functionality,

MINARET

In Islamic societies, the muezzin calls the neighborhood to prayer five times a day. To allow his voice to carry as far as possible, the call would traditionally be made from the top of a minaret, or tower. The architectural styles of minarets have varied by region and era. Even today, when the call to prayer is broadcast via loudspeakers, mosques maintain the tradition of the minaret.

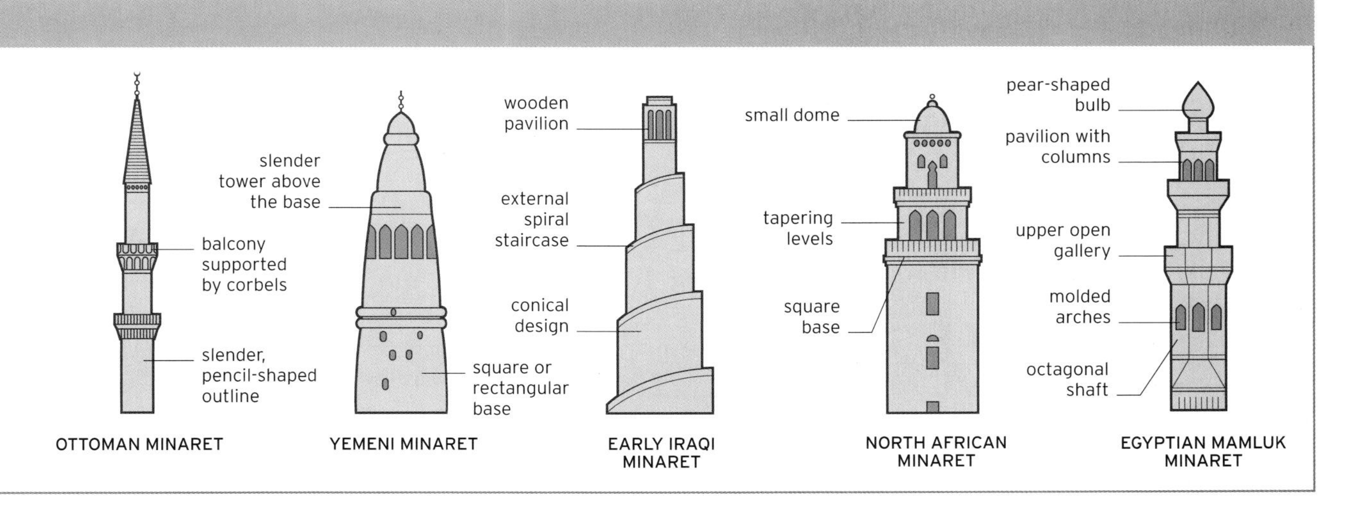

the use of modern materials (notably concrete, steel, and sheet glass), structural innovation, and the elimination of ornament. The Bauhaus school was Modernist, although the movement's most famous exponent was the Swiss-born French architect Le Corbusier.

MODULOR A scale of proportions devised by Le Corbusier to create a harmonic standard for design elements. The system is based on the height of an average man with one arm raised. *See also* Modernism.

MOLDING A strip of plaster, stone, wood, or other material that decorates the top or bottom of a wall or is added around a door or window. Moldings are usually decorated with some form of pattern.

MONOLITH A single great stone that stands as a monument or as part of a monument. There is some overlap with the term megalith, although megalith tends to be used in reference to prehistoric structures. *See also* megalith.

MOSAIC An image made from an assemblage of small, flat pieces of colored stone, glass, or other materials. The small pieces are known as tesserae.

MOSQUE See panel, right.

MOTTE ▼ A raised mound of earth flattened on top for the placement of a defensive keep. Mottes were often encircled by a courtyard (bailey), an arrangement that is known as a motte-and-bailey castle. *See also* bailey, keep.

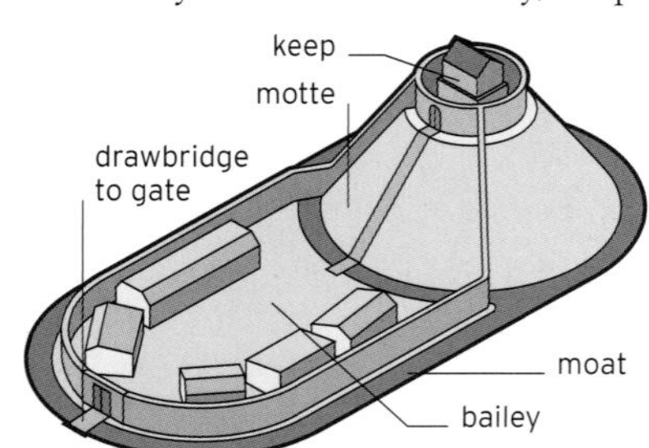

MUGHAL Relating to the Mughal empire, which ruled in 1526–1540 and 1555–1857. At its height, the Mughals ruled most of the Indian subcontinent and built monuments such as the Taj Mahal at Agra and the Red Fort at Delhi.

MULLION The vertical or horizontal element—of stone, wood, or other material—that divides the casements or panes of a window.

MULTIFOIL ARCH A highly ornamental arch with semicircular indentations in the underside of its curved area. It is particular to the Moorish architecture of Arab Andalusia. It is sometimes called a scalloped arch.

MUQARNA A type of vaulting found in Islamic architecture comprising clusters of little niches that decorate the uppermost parts of alcoves and the corners where walls transition into domes. *Muqarnas* are often said to resemble honeycombs or stalactites.

MURAL A painting, mosaic, or other work of art, generally on a large scale, made directly onto a wall.

N

NAGARE-ZUKURI ▼ A Japanese Shinto temple style characterized by an asymmetrical gabled roof projecting outward on one of the nongabled sides to form a portico.

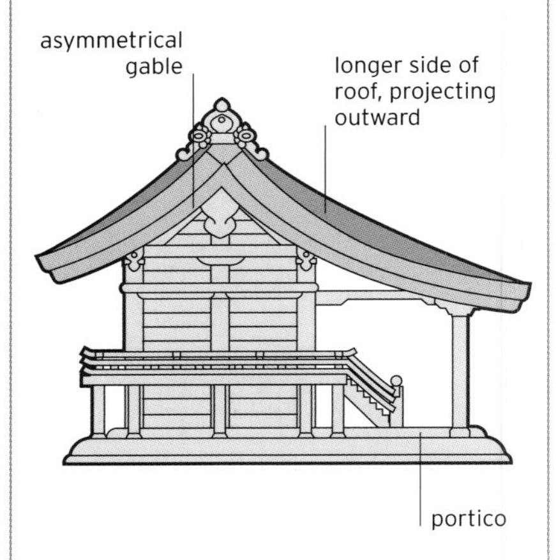

NAOS The inner chamber of a temple in ancient Greek architecture. It usually contained a statue of the deity to whom the temple was dedicated.

NARTHEX The entrance or lobby area of a church, distinct from the church proper and located at the opposite end of the nave to the altar. *See also* altar, church, nave.

NAVE The central longitudinal part of the church, where typically the congregation sits. Any aisles in a church are not part of the nave. *See also* church.

NEO-CLASSICAL Relating to the revival of simple forms associated with Classical architecture, such as columns, and a preference for blank walls, during the 18th and early 19th centuries. It represented a reaction to the excesses of the Baroque and Rococo styles. *See also* Baroque, Classical, Rococo.

NEO-GOTHIC *See* Gothic Revival.

NICHE A decorative recess set into a wall, typically semicircular and topped by a semidome. A niche is usually used for displaying a statue, vase, or similar object.

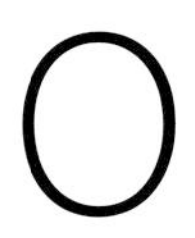

OBELISK A tall, four-sided monument that ends in a pyramid-shaped cap. The form is most often associated with ancient Egypt.

MOSQUE

The mosque is an Islamic place of worship. As with a church, a mosque can be a simple meeting room. All mosques, however, have some form of *mihrab*, which is the niche that indicates the direction of the Kaaba at Mecca and the direction of prayer. Traditionally, a mosque will also feature a *minbar* for the preaching of sermons, an ablutions fountain for ritual washing, and a minaret for the call to prayer. Most traditional mosques in warmer regions also have a large arcaded courtyard to accommodate the expanded congregations that gather for Friday prayers, or *Jummah*.

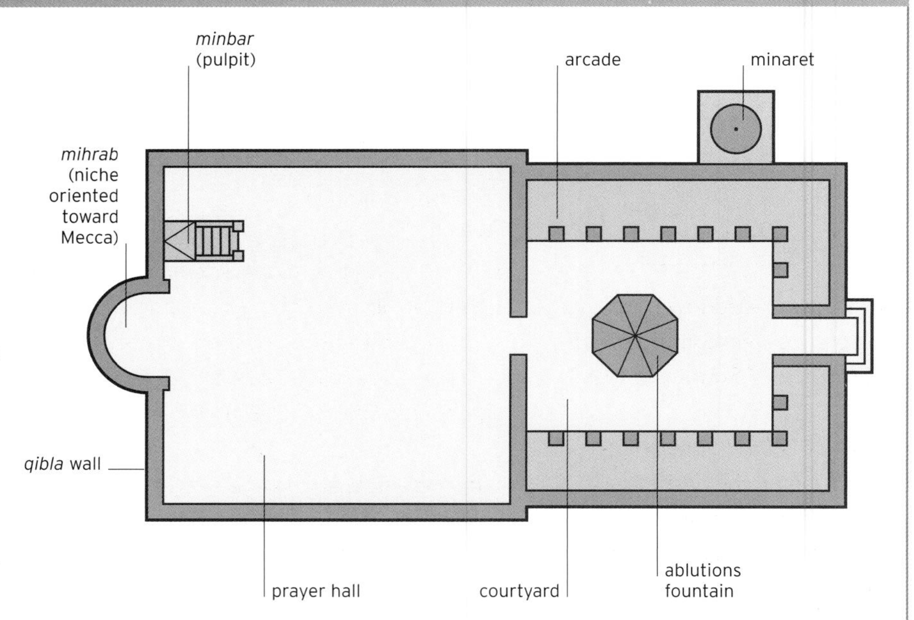

ORDER

Established by the ancient Greeks and Romans, the architectural orders are styles of Classical architecture characterized by their prescribed proportions and detailing. This is most evident in columns. Of the three main orders, the Doric column is short and heavy with plain, rounded capitals. The Ionic column is more slender, and it is topped by a capital decorated with volutes. The Corinthian column is the most slender, and has an ornate capital decorated with two rows of acanthus leaves. Doric columns have 20 flutes and the other two, 24. To these, the Romans added the plainest order–Tuscan–and the most decorative order–Composite.

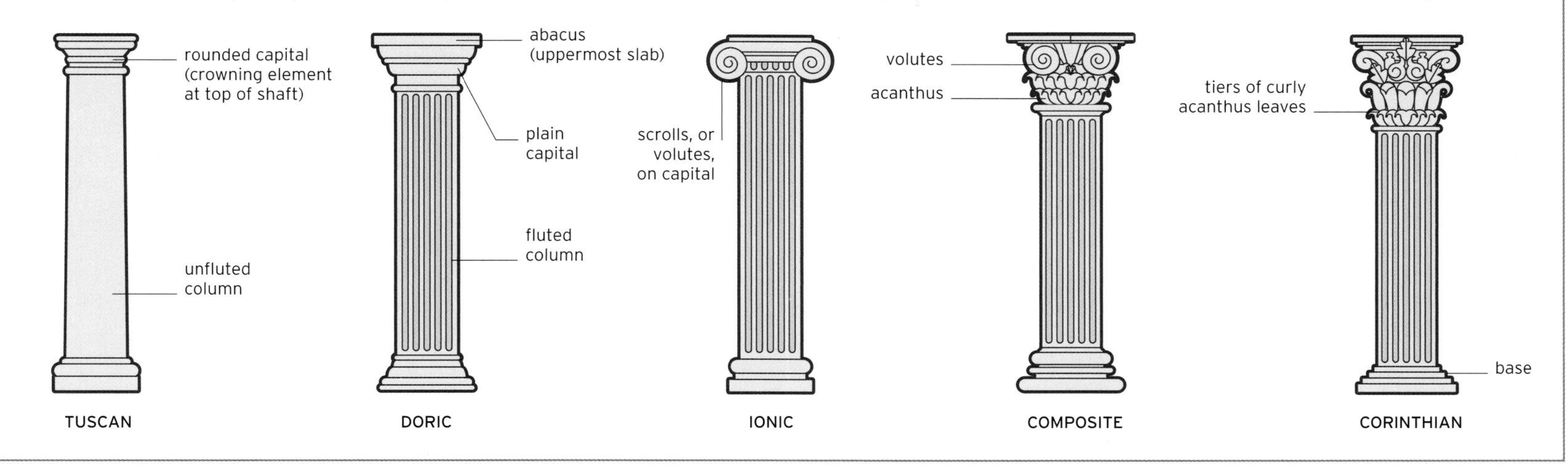

OCULUS A circular opening at the apex of a dome or in a wall.

OGEE An S-shaped curve. For example, an ogee arch is one in which the two sides curve upward at the top of the arch to meet in a point. *See also* arch.

ONION DOME A dome that has a larger diameter than the drum on which it sits so that it curves inward at the bottom. The height is greater than the width, and the dome tapers to a point at the top. *See also* dome.

ORDER See panel, above.

PAGODA A tiered tower with projecting eaves, common to the architecture of Southeast Asia.

PALLADIANISM An architectural style based on the designs and ideas of 16th-century Italian architect Andrea Palladio. Palladio was heavily influenced by the buildings of ancient Rome, and his architecture was based on Classical proportions with plain, typically symmetrical exteriors, often with porticos and a cruciform plan. Palladio's *Four Books of Architecture*, first published in 1570, was highly influential. *See also* Classical, cruciform, portico.

PANOPTICON A prison with cells arranged in a circle so that they can all be seen from a central observation point.

PARAPET A low wall around the edge of a roof, terrace, or balcony.

PARTERRE A garden with beds divided in an ornamental pattern.

PAVILION A subsidiary building that stands separately or is attached to a main building.

PEDIMENT In Classical architecture, the usually triangular gable that sits above the entablature, the whole of which is typically supported by columns. *See also* entablature, gable.

PENDENTIVE A structural device that facilitates the placing of a circular dome over a square room. Pendentives are the triangular segments of a sphere that follow the curve of the dome and then taper downward to a point at the corners of the walls. *See also* squinch.

PERISTYLE In Classical architecture, a colonnaded walkway around an inner courtyard. It is similar to a medieval cloister. *See also* colonnade.

PIANO NOBILE The principal floor of a large house, with higher ceilings and grander rooms than on the other floors. The term came into use in Renaissance Italy, where the "noble floor" was usually on the first floor and expressed on the facade by enlarged windows.

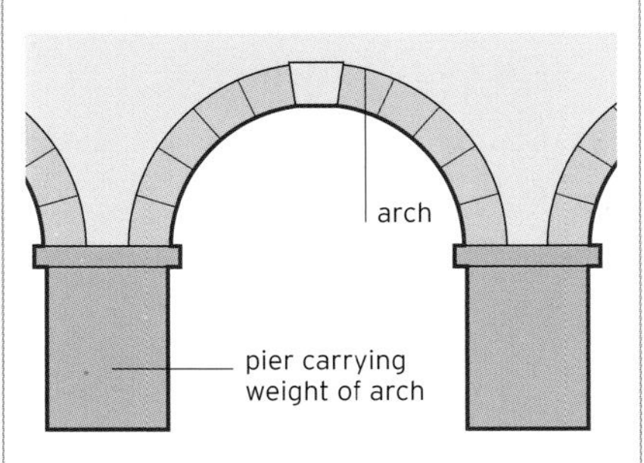

PIER ▲ A vertical supporting element of a structure, such as an arch or bridge.

PILASTER A false column, rectangular in shape and projecting only slightly beyond the wall into which it is built. It is purely decorative.

PILE ▼ A large post of timber, steel, or poured concrete sunk deep into the ground to support a structure above.

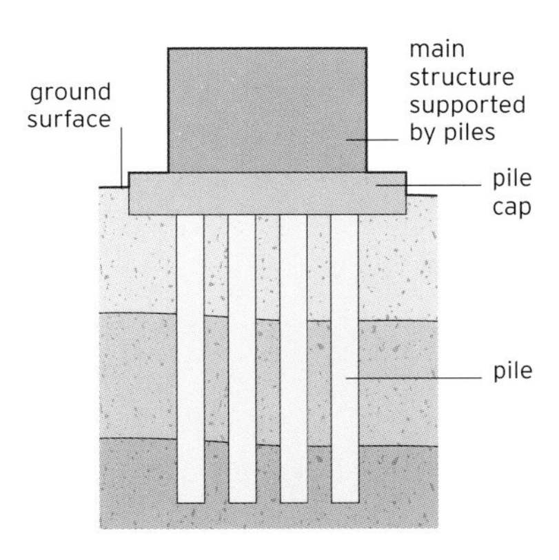

PILLAR An isolated vertical structural element, of any cross-section, that typically has a load-bearing function.

PILOTI A support, such as a column, pillar, or stilt, that lifts a building off the ground—for example, in a structure built above water or a building with an open area beneath the lower floor.

PINNACLE A small spire frequently employed in Gothic architecture to ornament the corners of towers or the tops of buttresses.

PLINTH The base of a column.

PODIUM A platform used to raise a building off the ground.

PORTCULLIS A vertically closing gate typically found in medieval fortifications. It was often constructed of a lattice of wood reinforced with metal.

PORTICO A roofed space, usually supported by columns, forming an entrance and often the central feature of a facade.

PROPYLAEUM A monumental gateway in ancient Greek architecture.

PROSCENIUM In Greek theater, a narrow raised platform on which actors performed. In the Roman era, it referred specifically to the vertical front dropping from the stage to the floor of the orchestra.

PYLON A monumental structure that served as a gateway to an ancient Egyptian temple. Pylons took the form of giant vertical slabs that sloped inward to the top.

QIBLA The direction of Mecca, toward which Muslims face when they perform prayer. In a mosque, the direction is marked by the *mihrab*, and the wall in which the *mihrab* is set is known as the *qibla* wall. *See also* mosque.

QUADRIGA A sculpture depicting a chariot drawn by four horses harnessed abreast that often appears on ancient Roman victory monuments or triumphal arches.

RELIEF CARVING

This is a technique in which designs are carved into a flat piece or panel of wood, stone, or other material so that figures and patterns project from the background. If the design projects only slightly from the background and there is no undercutting of the outline, the work is known as low- or bas-relief. If the design projects more prominently, with parts that may be completely disengaged from the background, approximating sculpture, it is known as high- or alto-relief.

BAS-RELIEF

ALTO-RELIEF

QUEEN POST One of two posts that work together to support a structure. *See also* king post, truss.

QUEEN TRUSS Similar to a king truss, except it employs two posts and can span longer openings than a king truss. *See also* king truss, truss.

QUOIN A decorative corner stone on a wall. Quoins differ in size, color, or texture from the rest of the masonry used in the wall. They are often toothed—that is, set in a pattern of alternating lengths with one short and one long.

RATIONALISM An architectural style practiced in Italy from the 1920s to the 1940s that married Neo-Classicism with Modernism. It was characterized by symmetrical plans, rhythmic colonnades (with columns repeating at regular intervals), the use of columns and arches, and marble slab coating. It was promoted by the fascist regime of Benito Mussolini. *See also* Modernism, Neo-Classicism.

REINFORCED CONCRETE ▼ Concrete strengthened by having steel reinforcing bars embedded within it before it sets. The steel adds tensile strength to complement the natural compressive strength of the concrete.

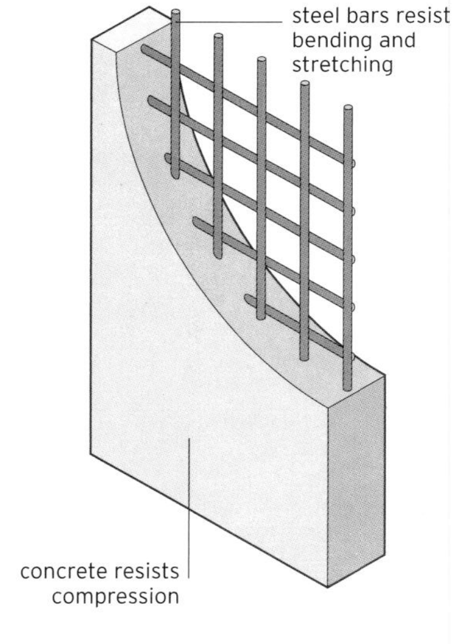

RELIEF CARVING See panel, above.

RELIEVING ARCH An arch built over a lintel to redistribute weight away from the lintel. It is also known as a discharging arch. *See also* lintel.

RENAISSANCE ARCHITECTURE A style that was developed first in Florence but which spread throughout Europe from the early 14th century onward. It revived elements of ancient Greek and Roman architecture and stood in stark contrast to the Gothic architecture that preceded it. It favored the reintroduction of columns, colonnades, Classical orders, arches, and domes.

RETAINING WALL ▼ A wall that holds back a mass of soil, such as in garden terracing.

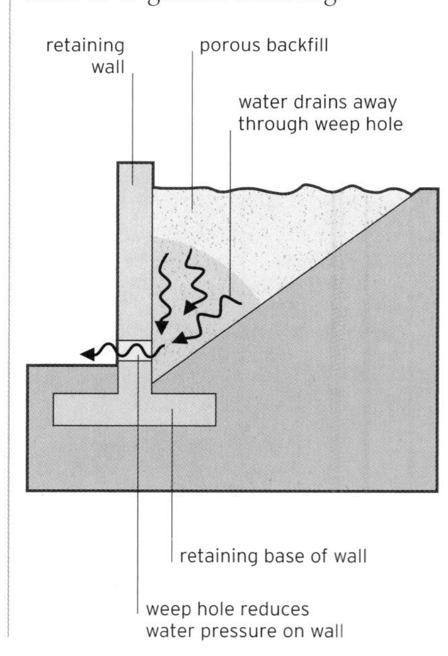

RIB VAULT A type of vault that, like a groin vault, is formed by the intersection of two barrel vaults at right angles. The apexes of a rib vault are marked by stone ribs that divide the surface into curved triangular panels. This style of vaulting developed during the era of Gothic architecture. *See also* Gothic, groin vault, vault.

ROCOCO A highly decorative style of the mid-18th century that spread from France and Italy throughout central Europe. Characterized by an abundance of curves, moldings, and painted stucco, it was intended to inspire awe in the viewer. However, the embellishments were typically confined to the interiors of buildings and the exteriors were usually more restrained. *See also* stucco.

ROMANESQUE An architectural style of the early Middle Ages that combined elements of ancient Roman and Byzantine buildings. It is characterized by thick walls, round arches, the use of barrel vaulting, and a relative simplicity of plan and form. *See also* arch, barrel vault.

ROOD SCREEN An ornate partition of open tracery—made of wood, stone, or wrought iron—between the chancel and nave in a church. The screen is typically topped by a cross or crucifix, which when hung above the entrance to a chancel is known as a rood. *See also* chancel, church, nave.

ROOF See panel, below.

ROSE WINDOW A large, circular, decorative window, often with stained glass. Rose windows are typically found in churches and cathedrals at the end of the nave, behind the altar and in the transepts. *See also* nave, transept.

ROTUNDA A building or a room with a circular floor plan covered by a dome.

ROUNDEL A round tower built as part of a set of defensive walls and meant to accommodate the placement of artillery.

RUSTICATION Finishing stone blocks with a rough surface. The blocks are cut back around the edges to give prominence to the rough area.

S

SARCOPHAGUS A receptacle for a corpse that also serves as a monument and is often displayed above ground. The sarcophagus can take the form of a stone box with carved decoration or, in the case of the ancient Egyptians, it can be shaped from wood and have a more humanoid shape.

SASH WINDOW A window with one or more vertical panels that are opened by sliding up or down.

SCREEN A thin, lightweight, free-standing wall that can be moved around to divide a room.

SEGMENTAL ARCH An arch in which the curve is less than a full 180 degrees. *See also* arch.

SEPULCHRE A burial vault or tomb.

SETBACK A steplike recession in a wall.

SEXPARTITE VAULT A rib vault divided into six bays. *See also* rib vault.

SHAFT The long, narrow, vertical cylinder that forms the body of a column.

SHELL VAULT A thin, self-supporting vault, typically made of reinforced concrete. *See also* reinforced concrete, vault.

SHINGLES Small, thin squares or rectangles of material—usually wood, asphalt, or slate—that are laid as overlapping courses for a roof covering or hung as a weather-proof covering on the sides of buildings. If they are made out of clay, and possibly shaped, they are called tiles.

SKYSCRAPER A habitable high-rise building of more than 40 or 50 stories. The Council on Tall Buildings and Urban Habitat defines a skyscraper as a building that reaches or exceeds 490 ft (150 m) in height.

SMART BUILDING A building that incorporates advanced technology such as intelligent management systems that monitor the energy consumption and carbon footprint of the building and gather other data in order to continually improve performance and user efficiency.

SOFFIT The underside of a construction element such as an arch or architrave, or the board covering the undersurface of the eaves on a roof.

ROOF

The roof is the covering of the top of a building, protecting it against the elements. The form of roofs varies greatly from region to region; for example, simple flat roofs are common in hot climates, while in parts of the world with frequent rainfall, the preference is for gabled and hipped roofs that are pitched to shed water. Mansard roofs prove useful if the building needs to be expanded vertically, due to the almost vertical bottom slope.

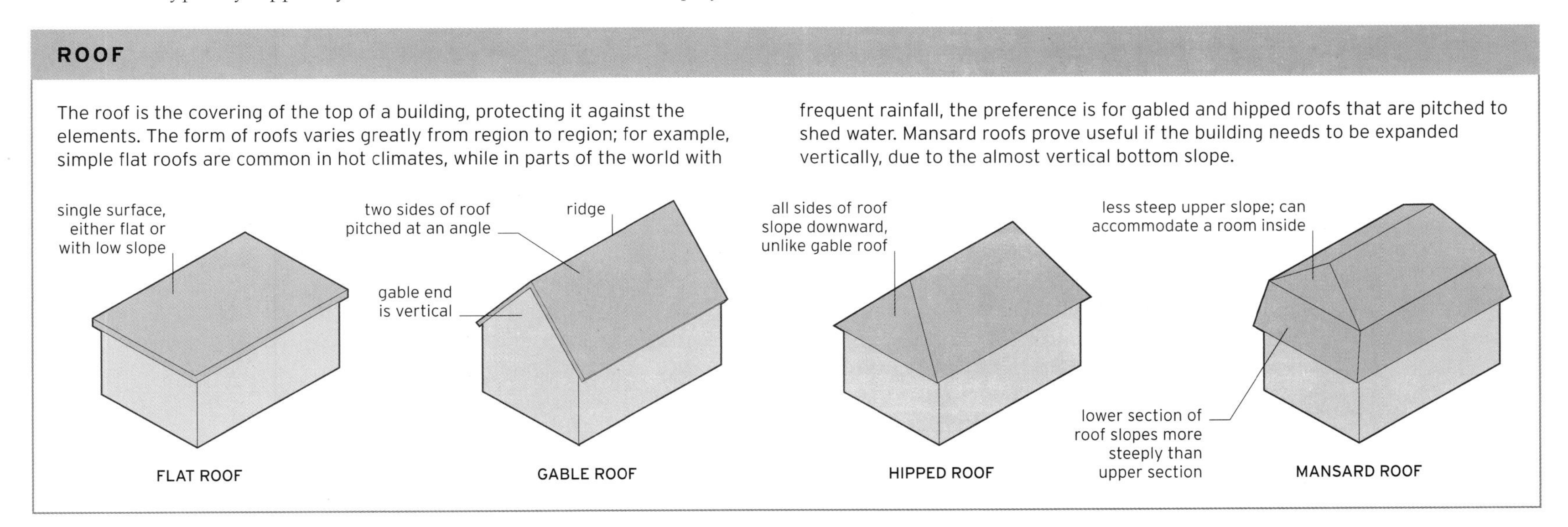

SPIRE

Most commonly associated with European Gothic architecture, the spire is a conical or multifaced tapering structure that provides the termination of a church or cathedral tower. The spire originated in the 12th century as a simple, four-sided pyramidal roof capping a tower, which over time developed into an ever higher, slimmer structure. Spires also became increasingly ornate with the addition of broaches (triangular sections of masonry added to the bottom of the spire faces that did not coincide with the tower sides), gabled dormers, flying buttresses, and additional pinnacles around the base of the spire. The aim was always the same: to symbolize the heavenly aspirations of those who worshipped within the building.

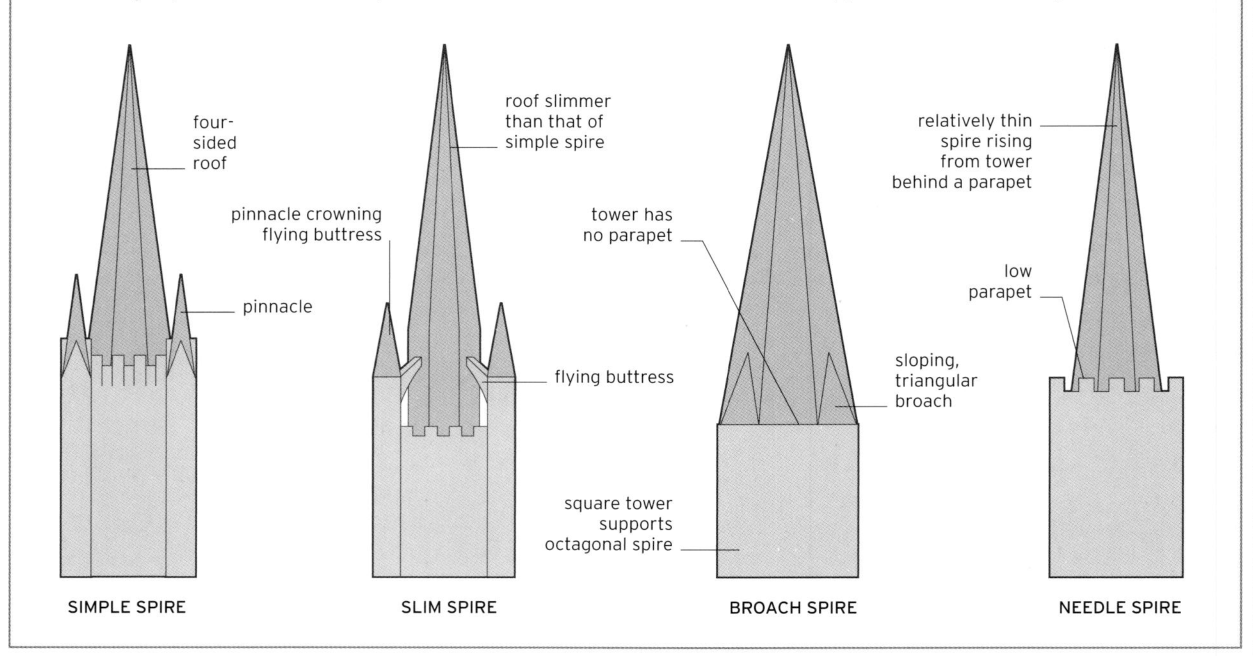

SPANDREL ▼ The space between a curved figure and a rectangular boundary—for example, the roughly triangular areas above and on either side of an arch and below a horizontal line aligned with the arch's apex. In modern times, a spandrel is also the area between the top of a window and the sill of the window above it.

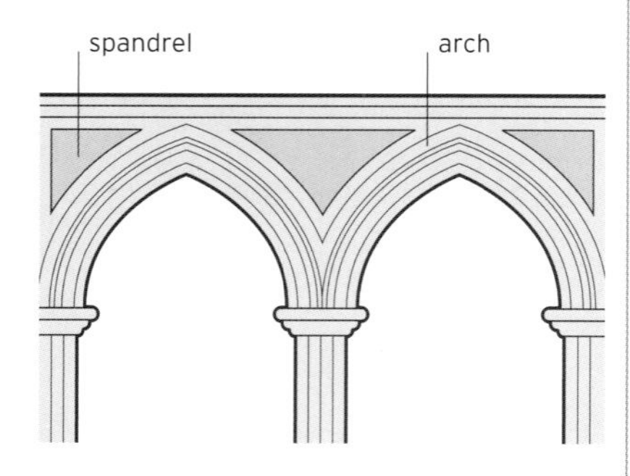

SPIRE See panel, above.

SPRINGING LINE The point at which an arch starts to curve from its vertical supports. *See also* arch.

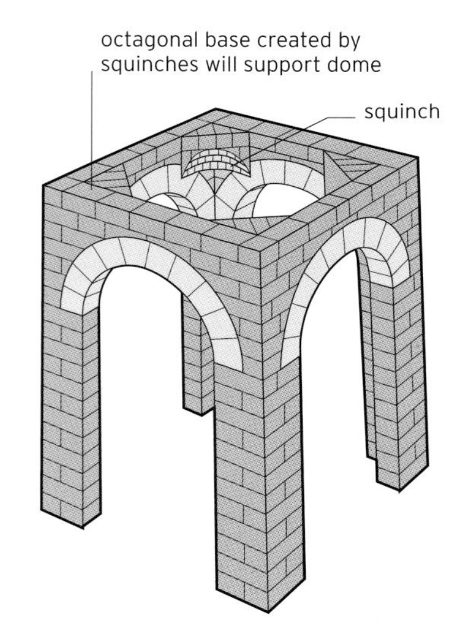

SQUINCH ▲ A corner filler of masonry placed diagonally at the interior angle of a square structure to create an octagonal upper space on which a dome can sit. The squinch evolved into the more sophisticated pendentive. *See also* pendentive.

STAR VAULT A type of rib vault in which secondary ribs connect with the main ribs to form a star pattern. *See also* rib vault.

STAVE A vertical wooden post or plank.

STELA A stone or wooden slab that is taller than it is wide. A stela generally has carved or painted text or ornamentation and serves as a monument, boundary marker, or official notice.

STOA In ancient Greek architecture, a covered walkway with the roof supported by one or more colonnades. Stoas often surrounded marketplaces.

STRAPWORK A form of decoration that makes ribbonlike elements into geometric patterns, usually within a frame. These elements are slightly raised on a flat surface, and strapwork is usually done in wood, metal, or plaster.

STUCCO A type of plaster that is applied wet and carved, often with highly elaborate designs. It is common in the Islamic architecture of North Africa. Baroque and Rococo architecture also make heavy use of carved stucco in decorative interiors. *See also* Baroque, Rococo.

STUPA ▼ A Buddhist shrine for the veneration of saints, as well as the Buddha himself. The design of a stupa is derived from a burial mound and sometimes takes the form of a hemispherical dome topped by a small square structure. With the spread of Buddhism to Tibet, Nepal, and Sri Lanka, the stupa also evolved, in structure and symbolism, into an elongated cone with horizontal rings that typified stages of enlightenment. Stupas are not entered; instead, worship takes the form of walking around the monument in a clockwise direction.

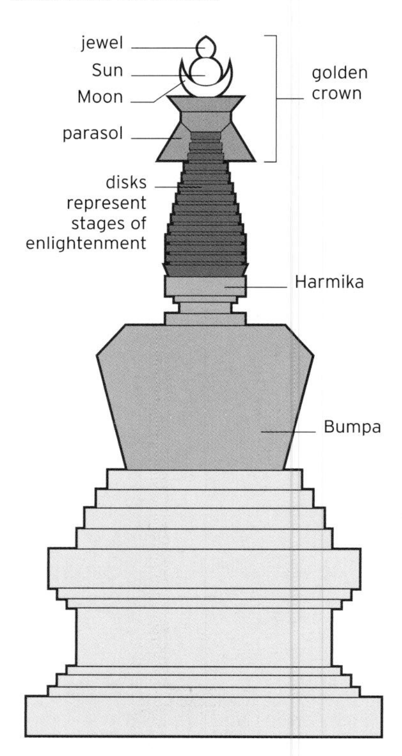

STYLE The appearance of a building and, in particular, the features that identify it as belonging to a specific historical period or approach to design and construction.

STYLOBATE The top step of the stepped platform on which an ancient Greek temple was built.

TRACERY

Tracery is the decorative stone molding found in vaulting, panels, and especially windows. It originated with the pointed Gothic arch window; this usually held two vertical, pointed windows within a single frame, leaving space below the peak of the main arch. This space was often carved with a decorative window, a technique known as plate tracery, after the carved "plate" of stone.

This gave way to bar tracery, in which panes of glass were separated by thin mullions of stone. Tracery developed into complex patterns, such as intersecting (mullions curve to the arch head), reticulated (circles forming ogee patterns), geometrical (simple decoration formed by circular patterns), and curvilinear (based on ogee curves) forms.

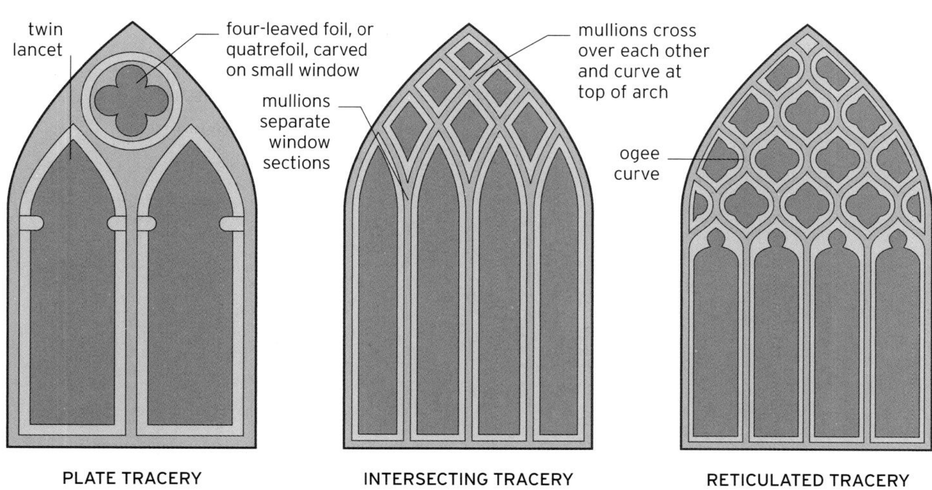

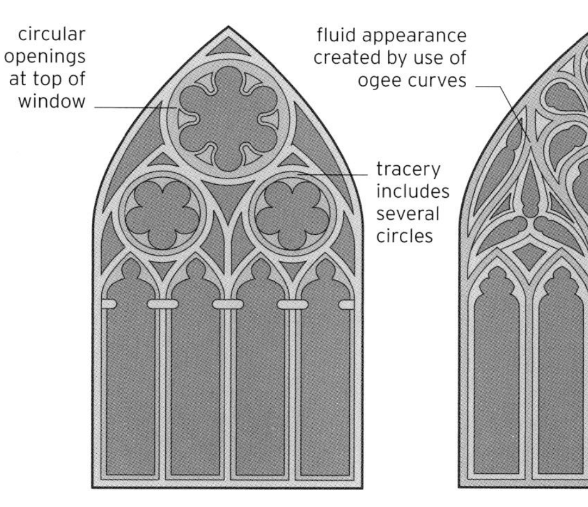

SYNAGOGUE

As with churches and mosques, there are no set requirements for synagogues. They vary greatly in form according to when and where they are built. According to Jewish tradition, worship is possible wherever there is a *minyan*, a quorum of 10 worshippers. However, a traditional synagogue always contains an ark in which the Torah scrolls are kept.

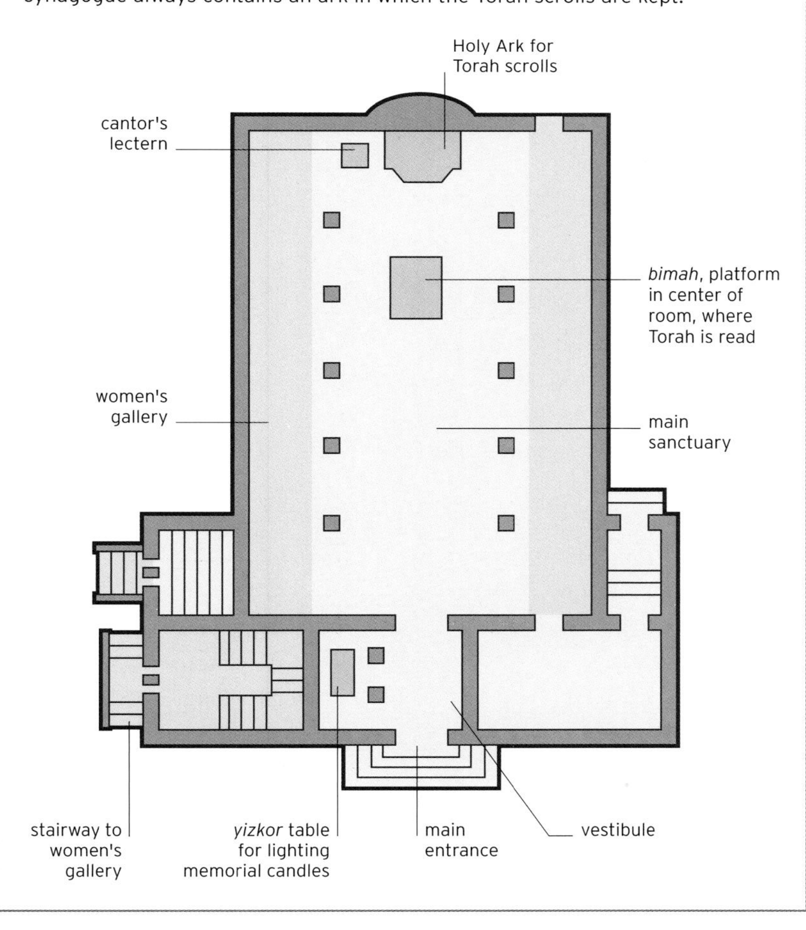

SUBSTRUCTURE The elements below ground, such as foundations and basements, that support a structure above.

SUPERSTRUCTURE The part of a structure that is above ground.

SYNAGOGUE See panel, left.

T

TALUD-TABLERO A stylistic feature of pyramid construction in Pre-Columbian Mesoamerica that alternates sloping surfaces (talud) with vertical surfaces (tablero).

TALUS A sloping face at the base of a fortified wall in a castle. It was designed to prevent siege towers (portable high towers that allowed attackers to shoot arrows into the fortification) from approaching the base of the wall and to present an obstacle to attackers using scaling ladders.

TESSERA An individual tile used in creating a mosaic. *See also* mosaic.

THOLOS An ancient Greek round temple with a ring of columns supporting a conical roof.

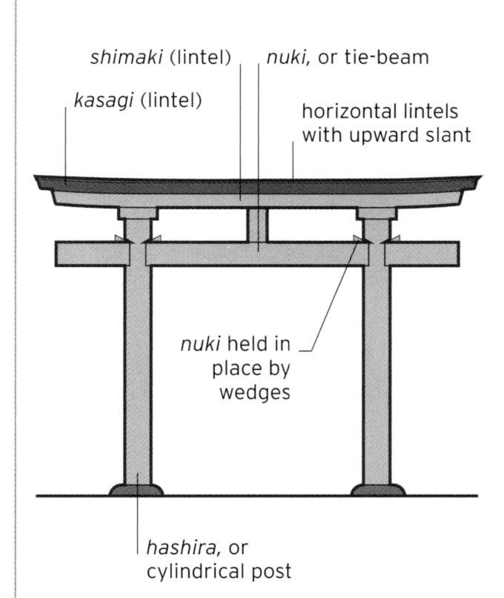

TORII ▲ A gateway at the entrance to the precincts of a Japanese Shinto shrine. It typically consists of two cylindrical posts topped by a horizontal rectangular tie-beam and lintels that extend beyond the posts on either side, with a second horizontal beam below the first. *See also* lintel.

TRACERY See panel, above.

TRANSEPT The parts of a cruciform church that are at right angles to the principal axis, as in the arms of a cross. *See also* church, cruciform.

TRIGLYPH ▼ A design element found on friezes of the Doric order in ancient Greek architecture; it consists of a panel scored with three vertical grooves. *See also* Doric, order.

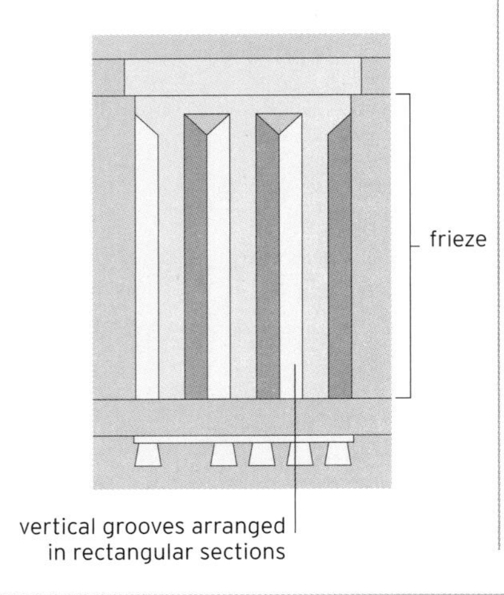

TROMPE L'OEIL A painting that aims to deceive the viewer into thinking that it is a three-dimensional object.

TRUSS See panel, below.

TUMULUS A large mound of earth raised over a burial site. Tumuli are also known as barrows.

TUNNEL VAULT An alternative name for a barrel vault. *See also* barrel vault, vault.

TUSCAN An order of Classical architecture simpler than all others, with an unadorned shaft that sits on an unadorned base and has an unadorned capital. *See also* order.

TYMPANUM In Classical architecture, the triangular or semicircular part of the wall over an entrance or window. It often contains sculpture or other decoration.

U

UNDERCROFT A cellar or storage room.

V

VAULT See panel, right.

VERNACULAR ARCHITECTURE The domestic architecture of a region.

TRUSS

A truss is a flat structural framework of beams, joists, and posts. The simplest truss is a single triangle, but trusses more commonly consist of a combination of triangles. It is a form that offers great rigidity because a triangle cannot be distorted by stress. Individual trusses are typically set in parallel, resting on supporting walls and tied together by purlins and a ridge beam. The two simplest trusses are the king and queen trusses. A queen truss can support a roof spanning a greater distance than a king truss.

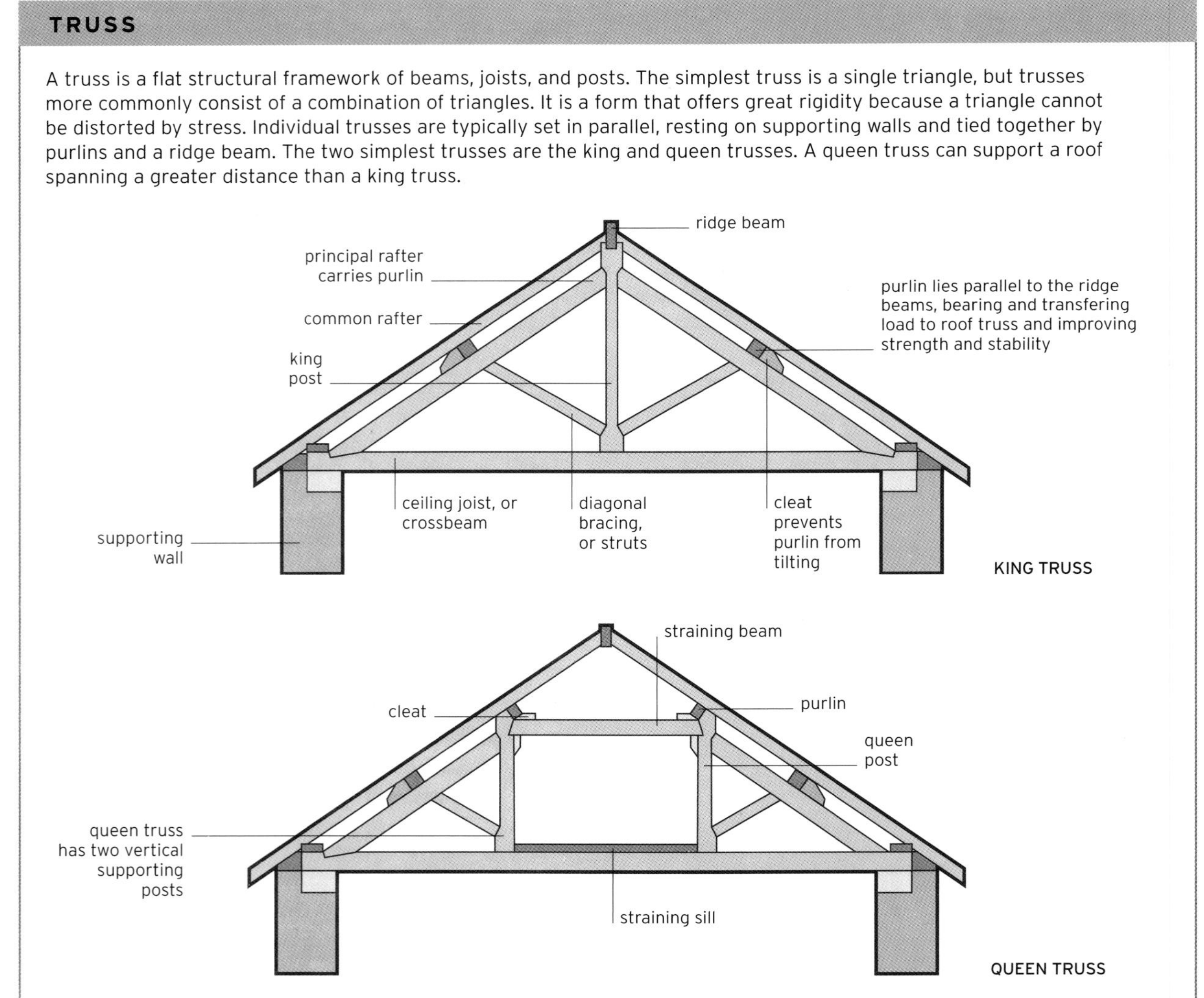

VESTIBULE A small foyer leading into a larger space.

VIADUCT A long bridge composed of many small spans that crosses a valley floor, an area of wetland, or a stretch of water.

VIMANA The pyramidlike tower of a Hindu temple in South India; its equivalent in North India is called a shikhara.

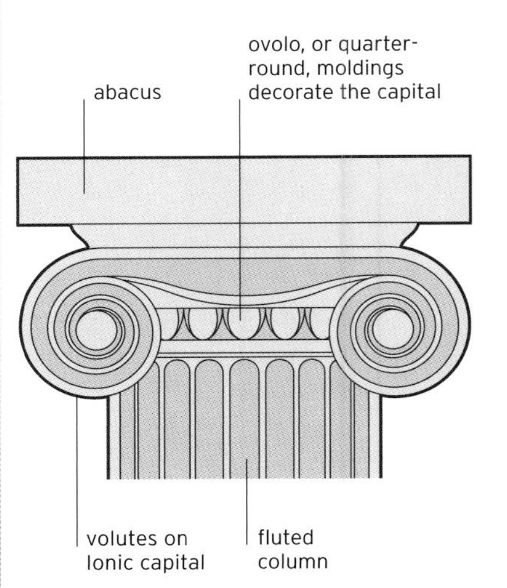

VOLUTE ▲ The scroll-like ornament that decorates the capitals of columns in the Ionic order. *See also* Ionic, order.

VOUSSOIR The wedge-shaped stone used in forming the curve of an arch. *See also* arch.

VYSE A spiral staircase or a staircase that winds around a central column.

W

WAINSCOT The paneled wooden lining of an interior wall. It can also refer specifically to the lower portion of an interior wall if this is finished differently from the rest.

WAT A Buddhist sacred precinct.

VAULT

A vault is an arch extended lengthwise to cover a space. The most basic form is called a barrel vault, and its use dates from as far back as the time of the ancient Egyptians. When two vaults intersect at right angles, the result is a groin vault. Skilled medieval stonemasons evolved this into the rib vault, which is a groin vault with pointed arches, and the fan vault, with multiple ribs radiating from the central groin. The barrel vault came back into common use during the Industrial Revolution with the invention of iron framing and was popular for large spaces such as railway sheds and exhibition halls.

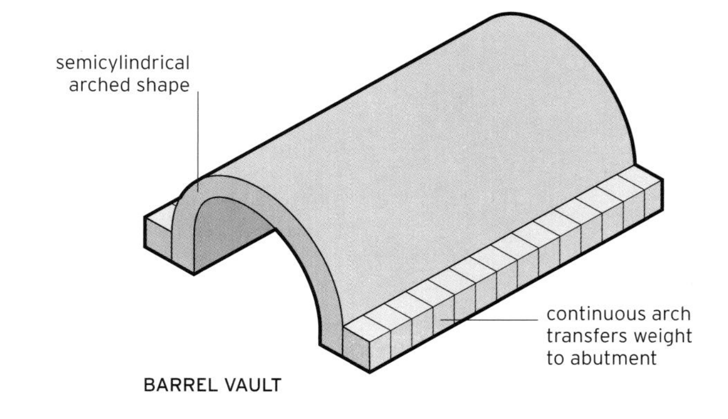

BARREL VAULT

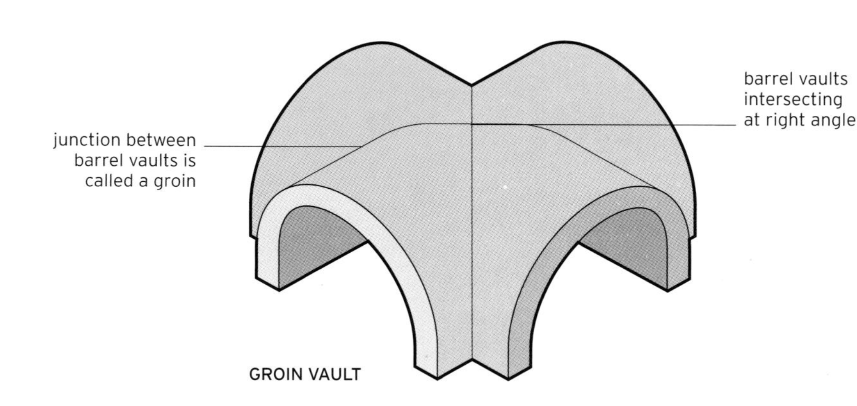

GROIN VAULT

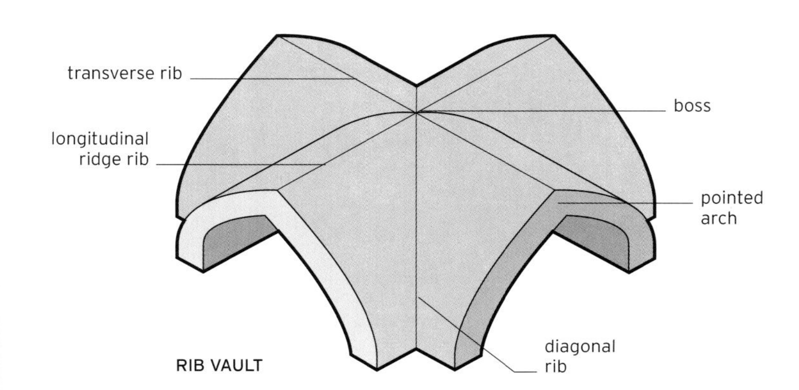

RIB VAULT

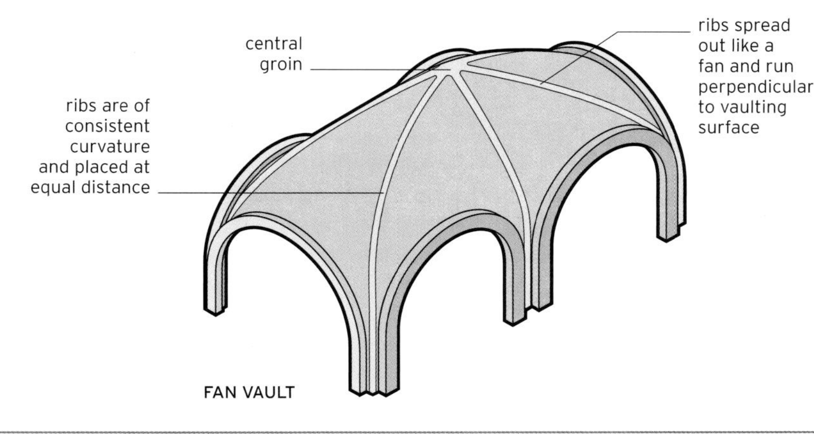

FAN VAULT

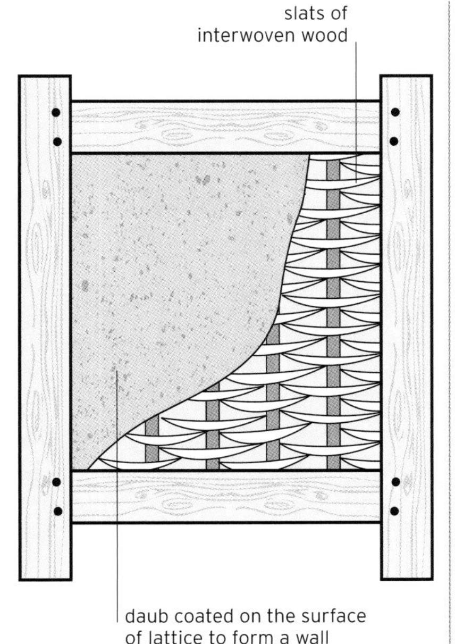

WATTLE ▲ A construction using thin branches or light, thin slats of wood woven into a lattice. It was traditionally used for fences. Combined with daub—a sticky, wet clay—it is also used for the walls of buildings. *See also* daub.

WHEEL WINDOW A round window with a central boss from which spokelike mullions radiate. *See also* mullions.

XYSTUS The ancient Greek term for the covered portico of a gymnasium used by athletes to exercise. In ancient Rome, it meant a garden walkway or terrace. *See also* portico.

Z

ZIGGURAT A characteristic monumental structure of ancient Mesopotamia that takes the form of a series of ascending terraces that decrease in size as one climbs them, like a stepped pyramid.

INDEX

Page numbers in **bold** refer to main entries.

C

D

E

F

G

N

O

P

Q

R

S

T

U

V

W

Z

ACKNOWLEDGMENTS

Dorling Kindersley would like to thank: Phil Wilkinson for helping to compile the list of sites; Claire Gell for additional artwork research; Katie John for proofreading; Duncan Turner for design assistance; Simon Mumford and Casper Morris for assistance with cartography; Steve Crozier for image retouching; Alex Lloyd for illustrations; and Suhita Dharamjit (Senior Jacket Designer), Emma Dawson (Jacket Editor), Harish Aggarwal (Senior DTP Designer), Priyanka Sharma (Jackets Editorial Coordinator), and Saloni Singh (Managing Jackets Editor).

DK India would like to thank Nobina Chakravorty and Meenal Goel for design assistance; and Ashwin Raju Adimari for picture research assistance.

SMITHSONIAN ENTERPRISES

President Carol LeBlanc

Vice President, Consumer and Education Products Brigid Ferraro

Director, Licensed Publishing Jill Corcoran

Senior Manager, Licensed Publishing Ellen Nanney

Product Development Manager Kealy Gordon

The publisher would like to thank the following for their kind permission to reproduce their photographs:

(Key: a-above; b-below/bottom; c-center; f-far; l-left; r-right; t-top)

1 Shutterstock: MBL. **2-3 RNPictures**. **4 AirPano images:** (cla). **Getty Images:** Michael H (cra). **Enrico Pescantini:** (fcra). **5 AirPano images:** (fcra). **Jonathan Danker:** (cra). **Getty Images:** DigitalGlobe / ScapeWare3d (cla). **Bachir Moukarzel:** (fcla). **7 Getty Images:** Wang Qin / Chengdu Economic Daily / VCG. **8-9 AirPano images**. **10 Alamy Stock Photo:** EmmePi Images. **11 Alamy Stock Photo:** Wendy Connett / robertharding (fclb); Paul Strawson (tr); Andrew Roland (clb); Philip Scalia (cb). **Getty Images:** Dennis K. Johnson / Lonely Planet Images / Getty Images Plus (crb). **12 Alamy Stock Photo:** Sarah Akad (cb); Michael Runkel / robertharding (crb). **Dreamstime.com:** Sergio Bertino / Serjedi (clb). **iStockphoto.com:** Teerayuth Mitrsermsarp (cra). **13 Alamy Stock Photo:** Oliver Hoffmann (r). **iStockphoto.com:** ChiccoDodiFC (clb). **14 Getty Images:** Tim Graham / Hulton Archive (cl). **14-15 iStockphoto.com:** R.M. Nunes. **15 123RF.com:** Piotr Piatrouski (cra); sophiejames (crb). **Alamy Stock Photo:** Francois Roux (cr). **16-17 Getty Images:** Michael H. **18 Alamy Stock Photo:** George H.H. Huey (bc). **Getty Images:** Richard A. Cooke / Corbis Documentary / Getty Images Plus (br). **19 Alamy Stock Photo:** Wiliam Perry (bl). **Dreamstime.com:** Ckchiu (cra); Zeynep Ayse Kiyas Aslanturk / Zaka00 (bc); Sean Pavone (br). **20 Alamy Stock Photo:** George H.H. Huey (cb). **Jim Shoemaker:** (t). **21 Alamy Stock Photo:** robertharding (b). **22-23 Alamy Stock Photo:** Tom Till. **23 Getty Images:** Danita Delimont (cb). **24 Alamy Stock Photo:** Felix Lipov (cra). **Nini Jin:** (br). **25 Architect of the Capitol**. **26 Alamy Stock Photo:** Luis Leamus (c). **Getty Images:** Museum of the City of New York (br). **26-27 Dreamstime.com:** Prochasson Frederic. **30 Getty Images:** Detroit Publishing Company / Interim Archives (bc). **NASA:** Bill Ingalls (bl). **31 4Corners:** Richard Taylor. **32-33 iStockphoto.com:** buzbuzzer. **33 Alamy Stock Photo:** Danita Delimont (tc). **Dreamstime.com:** Carole Rigg (br). **iStockphoto.com:** aladin66 (tl). **34-35 Getty Images:** Library of Congress / Corbis Historical. **35 Bridgeman Images:** Private Collection / Avant-Demain (bc). **36-37 Susan Candelario**. **36 iStockphoto.com:** Medioimages / Photodisc (c). **Dan McQuade:** (bl). **37 Alamy Stock Photo:** Melvyn Longhurst (crb). **38-39 Bethany DiTecco**. **39 Getty Images:** Dr. Antonio Comia (br); Ambrose Vurnis (clb). **40-41 Alamy Stock Photo:** Jesse Kraft. **40 Getty Images:** George Rinhart (tr). **42 Alamy Stock Photo:** Dan Highton (cr). **Craig T Fruchtman:** @craigsbeds (l). **43 Alamy Stock Photo:** robertharding (bc). **Dominic Kamp:** (t). **44-45 Danny du Plessis**. **45 Jordan Lloyd, Dynamichrome:** United States Bureau of Reclamation archive image (br). **Getty Images:** Popperfoto (cr). **46-47 iStockphoto.com:** franckreporter. **47 Getty Images:** Underwood Archives (br). **48 Dreamstime.com:** Ivan Cholakov (bl). **48-49 Dreamstime.com:** Littleny. **49 David Leventi:** (br). **50 Rainer Kühn:** © ARS, NY and DACS, London 2019 / © DACS 2019 (t). **Jose Francisco Salgado:** (br). **51 Alamy Stock Photo:** Granger Historical Picture Archive (cla). **Tristan Zhou:** (br). **52 iStockphoto.com:** SeanPavonePhoto (bc). **Louis-Philippe Provost:** (cr). **53 Sanjay Chauhan:** (tl). **Getty Images:** Bettmann (tr). **54-55 © DACS 2019:** © OMA / DACS 2019. **56-57 iStockphoto.com:** jimkruger. **57 Smithsonian Institution, Washington, DC:** Alan Karchmer (tc, tr). **58-59 Enrico Pescantini**. **60 Alamy Stock Photo:** Diego Grandi (tl); Gábor Kovács (tc). **61 Alamy Stock Photo:** Ionut David (bl); Angus McComiskey (br). **Getty Images:** Jason Bleibtreu / Sygma (crb); Marcelo Nacinovic / Moment / Getty Images Plus (tr). **62 Alamy Stock Photo:** Witold Skrypczak (cl). **David Coventry:** (bl). **62-63 Alamy Stock Photo:** Tim Hester. **63 Getty Images:** Diego Lezama (br). **64-65 Getty Images:** Robert Clark / National Geographic Image Collection. **64 Science Photo Library:** David Nunuk (bc). **65 Getty Images:** De Agostini / G. Dagli Orti (cra); Stephan de Prouw (br). **66-67 AirPano images**. **66 SuperStock:** Iberfoto (br). **67 Getty Images:** Brigitte Merle (b). **68 Alamy Stock Photo:** Diego Grandi (bl). **Getty Images:** Manuel Romaris (c). **69 Alamy Stock Photo:** Photogilio (br). **Getty Images:** Jean-Pierre Courau (bl). **Science Photo Library:** John R. Foster (t). **72 Alamy Stock Photo:** Jan Wlodarczyk (b). **Getty Images:** Werner Forman / Universal Images Group (ca). **73 Dreamstime.com:** Saletomic (cra). **Robert Harding Picture Library:** Robert Frerck (bl). **74 Getty Images:** Luis Davilla (cr). **74-75 Robert Harding Picture Library:** Michael Nolan. **75 Alamy Stock Photo:** Nicholas Charlesworth (tl). **76 Dorling Kindersley:** University of Pennsylvania Museum of Archaeology and Anthropology (cra). **76-77 Chabrov Andrey**. **77 iStockphoto.com:** juliandoporai (tr). **78-79 iStockphoto.com:** AlbertoLoyo. **78 Dreamstime.com:** Byelikova (br). **79 Leonardo Cavallini:** (br). **80 Alamy Stock Photo:** Jan A. Csernoch (c). **Getty Images:** James P. Blair (bl); George Rinhart / Corbis (br). **80-81 Photo Courtesy of the Panama Canal Authority**. **82 Shane Hawke:** (br). **82-83 Ricardo Zerrenner**. **84 Alamy Stock Photo:** David R. Frazier Photolibrary, Inc. (br). **84-85 Marcos de Freitas Mattos:** © NIEMEYER, Oscar / DACS 2019 / © DACS 2019. **85 Alamy Stock Photo:** age fotostock (bl). **Getty Images:** Sergio Lopes Viana / Moment (tr). **86 © DACS 2019:** © NIEMEYER, Oscar / DACS 2019. **Getty Images:** Jane Sweeney (cb). **Sokari Higgwe:** (t). **87 Getty Images:** Bloomberg (b); DigitalGlobe (ca). **88-89 Bachir Moukarzel**. **90 Alamy Stock Photo:** funkyfood London - Paul Williams (bl). **Dreamstime.com:** Ivan Bastien (clb); Linda Williams (cra). **91 Alamy Stock Photo:** Raga Jose Fuste / Prisma by Dukas Presseagentur GmbH (ca). **AWL Images:** Mark Sykes (bc). **iStockphoto.com:** ChiccoDodiFC (c). **92-93 Anthony Murphy**. **92 Getty Images:** DEA / G. Dagli Orti (crb). **93 Stephen Emerson:** (br). **Getty Images:** Joe Cornish (cra); DEA / G. Dagli Orti / De Agostini (bl). **94-95 David Stoddart**. **94 Alamy Stock Photo:** Hemis (cl). **Getty Images:** DEA / A. Dagli Orti / De Agostini (bl). **Tommy Tenzo:** (crb). **98-99 Dreamstime.com:** Carafoto. **98 Getty Images:** Westend61 (br). **Photo Scala, Florence:** (clb). **100-101 SuperStock:** imageBROKER. **100 Getty Images:** CM Dixon / Print Collector (bl). **101 iStockphoto.com:** IPumbaImages (br). **102-103 Dave Bowman Photography**. **102 Bachir Moukarzel:** (bl). **104-105 Marco Rovesti**. **104 Getty Images:** Cristian Negroni (br). **105 John Kehayias:** (bc). **106-107 Alamy Stock Photo:** Jorge Tutor. **106 iStockphoto.com:** mrak_hr (bl). **107 Alamy Stock Photo:** Realy Easy Star (bc). **Getty Images:** nimu1956 (br). **108-109 iStockphoto.com:** klug-photo. **109 Getty Images:** Westend61 (cb). **110-111 Getty Images:** Nicolas Cazard / EyeEm. **111 Getty Images:** Westend61 (br). **112-113 iStockphoto.com:** sorincolac. **113 Getty Images:** Gonzalo Azumendi (bc). **Mohammad Reza Domiri Ganji:** (cb). **114-115 Stefan Muel:** Mädchenchor am Aachener Dom. **115 Alamy Stock Photo:** Bildarchiv Monheim GmbH (cl). **Getty Images:** Angelo Hornak / Corbis Historical (br). **iStockphoto.com:** jotily (cla). **116-117 Pixabay:**

Julius_Silver. **116 Alamy Stock Photo:** Dave Stamboulis (cb). **118 iStockphoto.com:** smartin69. **119 Getty Images:** Jason Hawkes. **120-121 iStockphoto.com:** The_Chickenwing. **120 Getty Images:** DEA / G. Nimatallah / De Agostini (crb). **121 Daniela Sbarro:** (clb). **122 Getty Images:** Print Collector / Hulton Fine Art Collection (bl). **122-123 Robert Harding Picture Library:** Christian Kober. **123 iStockphoto.com:** mammuth (tl). **124 Alamy Stock Photo:** funkyfood London - Paul Williams (bc). **Dreamstime.com:** Jonathan Braid (t). **125 Gary Lobdell:** (r). **SuperStock:** Funkystock (clb). **128 Getty Images:** Douglas Pearson (tl); PK . (b). **129 Alamy Stock Photo:** dleiva (bl). **Christian Barrette:** (tl). **130-131 Dreamstime.com:** Pavel V. **131 Oleg Anisimov:** (br). **Bridgeman Images:** (clb). **132 Bjorn Letink:** (cr). **132-133 Viktor Goloborodko**. **133 Getty Images:** DEA / G. Sioen (cla). **134-135 4Corners:** Antonino Bartuccio. **135 iStockphoto.com:** espiegle (cb). **136-137 iStockphoto.com:** Yulia-B. **137 Getty Images:** Jaap Mechielsen (br). **Mochalov Maxim:** (bl). **138 AWL Images:** Emily M. Wilson. **139 Alamy Stock Photo:** Zoonar GmbH (br). **Violeta Meletis:** (crb). **140 Getty Images:** Luis Alvarenga / EyeEm. **141 Getty Images:** DEA / G. Nimatallah (cra); Terence Kong (br). **142-143 Nico Trinkhaus**. **143 akg-images:** Album / Oronoz (br). **iStockphoto.com:** AlKane (bl). **144 Getty Images:** Andrea Thompson Photography. **145 Alamy Stock Photo:** Hercules Milas (cr). **Laurent Dequick:** (bl). **148 Alberto Barrera Rodríguez:** (bl). **148-149 Alamy Stock Photo:** Tamas Karpati. **149 Alamy Stock Photo:** Rolf Richardson (br). **150-151 Getty Images:** Yuliya Baturina. **150 Alamy Stock Photo:** Zoonar GmbH (bl). **Getty Images:** All Canada Photos (bc). **151 Getty Images:** Gavin Hellier (cr). **152-153 Dreamstime.com:** Reidlphoto. **152 AirPano images:** (bc). **154 Getty Images:** DEA / G. Dagli Orti / De Agostini (c); Enrique Díaz / 7cero (b). **155 Chuck Bandel:** (t). **Getty Images:** Sylvain Sonnet (bc). **156 iStockphoto.com:** -AZ-. **157 Thomas Mitchell:** (br). **Dr. Rana Nawab:** (t). **158-159 Stavros Argyropoulos**. **158 AirPano images:** (bc). **Alamy Stock Photo:** Vito Arcomano (c). **160 Getty Images:** Mark Edward Harris. **161 Alamy Stock Photo:** Luciano Mortula (t). **Getty Images:** Sabine Lubenow / LOOK-foto (br). **162-163 Jason Hawkes Aerial Library**. **163 Alamy Stock Photo:** Jo Miyake (br). **166 Alamy Stock Photo:** Paul Dymond (c). **Getty Images:** © Philippe Lejeanvre (bl). **Picfair.com:** Fabien Desmonts (bc). **166-167 Emmanuel Charlat**. **168 Alamy Stock Photo:** Oxford_shot (bl). **168-169 Alamy Stock Photo:** Anton Ivanov. **169 Getty Images:** Jason Hawkes (br). **170-171 Bayerische Verwaltung der Staatlichen Schloesser, Gaerten und Seen:** © Bayerische Schlösserverwaltung, Achim Bunz, München. **170 Getty Images:** Tomekbudujedomek (bl). **171 Getty Images:** Skyworks Places (bl). **Julius Silver:** (tr). **172 Getty Images:** Atlantide Phototravel (bl); Claude Gariepy (cb). **172-173 Depositphotos Inc:** Foto-VDW. **173 Dreamstime.com:** Ccat82 (br). **174 Alamy Stock Photo:** dbimages (b). **Getty Images:** Beatrice Lecuyer-Bibal / Gamma-Rapho (tr). **175 AirPano images:** (t). **Getty Images:** Alberto Suárez (br). **176 Alamy Stock Photo:** Glenn Harper (ca). **Getty Images:** Yvan Travert (bl). **176-177 Birgit Franik**. **177 Alamy Stock Photo:** Falkensteinfoto (br). **178 Alamy Stock Photo:** eye35 stock (tl). **iStockphoto.com:** benedek (cr). **179 iStockphoto.com:** franckreporter. **180 Getty Images:** Bettmann (cb). **180-181 Getty Images:** Agapicture Chang. **182 Getty Images:** George Pickow / Three Lions (c). **182-183 iStockphoto.com:** fotoVoyager. **184 Pol Albarran**. **185 Alamy Stock Photo:** Rob Whitworth (bc). **Robert Harding Picture Library:** Nico Tondini (br). **186 Alaa Othman**. **187 Getty Images:** Steven Blackmon / 500px (br). **National Geographic Creative:** Robert Harding Picture Library (t). **188 Shutterstock:** Jaroslav Moravcik (bl). **Hanaa Turkistani:** 500px.com / hanaaturkistani (t). **189 Alamy Stock Photo:** Manjik photography (ca). **Getty Images:** VWB photos (br). **190 AeroShots:** (tl). **Getty Images:** Artur Debat (tr). **190-191 Cristina Rocca**. **192-193 © DACS 2019:** © F.L.C. / ADAGP, Paris and DACS, London and © ADAGP, Paris and DACS, London 2019. **194 Matjaz Vidmar:** © DACS 2019. **195 Alamy Stock Photo:** Architectural Images (crb); Vichaya Kiatying-Angsulee (clb); Heritage Image Partnership Ltd (cra). **196 SuperStock:** Aurora Photos (bl). **196-197 Steven Blin:** used with permission of Mr. Richard Rogers and Mr. Renzo Piano. **198 AirPano images:** (crb). **Joep de Groot:** (l). **199 Alamy Stock Photo:** Hemis / La Grande Arche © 2019 Johan Otto Von Spreckelsen. **200-201 Getty Images:** Yann Arthus-Bertrand. **201 Alamy Stock Photo:** age fotostock (cb). **202 Getty Images:** Jean-Pierre Lescourret / Foster & Partners / CEVM Eiffage. **203 Prad Patel:** (t). **Unsplash:** Andrea Leopardi (cb). **204-205 Getty Images:** DigitalGlobe / ScapeWare3d. **206 Alamy Stock Photo:** Jack Jackson / robertharding (cra). **Dreamstime.com:** Witr (crb). **Getty Images:** Kitti Boonnitrod / Moment (tr); Philipp Klinger / Moment (br). **207 Alamy Stock Photo:** Robert Preston Photography (br). **Getty Images:** Alberto Manuel Urosa Toledano / Moment Open (bc). **208-209 Alamy Stock Photo:** Dereje Belachew. **208 Getty Images:** MyLoupe / UIG (bl). **209 Getty Images:** Jochen Schlenker / robertharding (br). **210 Getty Images:** Yann Arthus-Bertrand (crb). **Picfair.com:** annmarie (bc). **211 Robert Harding Picture Library:** Richard Ashworth. **214-215 Getty Images:** Tibographie - Thibaud Chosson. **214 Alamy Stock Photo:** Art Kowalsky (bc). **Muhammad Saber:** (br). **215 Getty Images:** DEA / S. Vannini / De Agostini (bc). **216 Bridgeman Images:** © 2019 Museum of Fine Arts, Boston, Massachusetts, USA / Harvard University-Boston Museum of Fine Arts Expedition (cra). **216-217 Alan Mandic**. **217 Getty Images:** Torsten Antoniewski (ca). **iStockphoto.com:** mason01 (cra). **218-219 George Steinmetz**. **219 Alamy Stock Photo:** robertharding (bc). **Getty Images:** DEA / G. Dagli Orti (br). **220 Alamy Stock Photo:** Ariadne Van Zandbergen (cb). **iStockphoto.com:** mtcurado (crb). **220-221 Rachid Hakka**. **222-223 Maurizio Camagna**. **223 Getty Images:** George Steinmetz (bc); Sam Tarling / Corbis (c). **224-225 Shutterstock**. **224 Getty Images:** DEA / W. Buss / De Agostini (br). **225 Getty Images:** José Fuste Raga (br). **226 Alamy Stock Photo:** Gary Cook (br). **Magnum Photos:** George Rodger (t). **227 Getty Images:** DEA / G. Roli / De Agostini (tr); Roger Wood / Corbis / VCG (tl). **228 Robert Harding Picture Library:** Gavin Hellier. **229 Alamy Stock Photo:** Abdellah Azizi (cla). **iStockphoto.com:** narvikk (br). **230-231 Alamy Stock Photo:** Christopher Scott. **230 iStockphoto.com:** evenfh (bl). **231 Alamy Stock Photo:** Black Star (br). **Getty Images:** Raquel Maria Carbonell Pagola / LightRocket (cr). **232-233 Getty Images:** Gavin Hellier. **233 Getty Images:** George Steinmetz (br). **234-235 Alamy Stock Photo:** Fabian Plock. **234 Getty Images:** Shamim Shorif Susom / EyeEm (crb). **235 Dreamstime.com:** Fabian Plock (bl). **Getty Images:** Hans Georg Roth (br). **236-237 Jonathan Danker**. **238 Getty Images:** Yann Arthus-Bertrand (bc). **iStockphoto.com:** real444 (br). **239 Alamy Stock Photo:** Thant Zaw Wai (cra). **Dreamstime.com:** Neophuket (crb). **Getty Images:** John W Banagan / Photographer's Choice / Getty Images Plus (bl). **240 Alamy Stock Photo:** Sezai Sahmay (bl). **Getty Images:** Vincent J. Musi (c). **240-241 SuperStock:** Biosphoto. **242-243 Getty Images:** Ozgur Donmaz. **243 Getty Images:** Thaaer Al-Shewaily (br); Nadeem Khawar (ca). **244-245 Getty Images:** JX K. **244 Depositphotos Inc:** Buurserstraat38 (crb). **245 Getty Images:** George Thalassinos (bl). **246 Getty Images:** MediaProduction (bl). **246-247 SuperStock:** Timothy Allen / Axiom Photographic / Design Pics. **248 Alamy Stock Photo:** ephotocorp (clb). **Dreamstime.com:** Saiko3p (br). **249 Getty Images:** Glen Allison (tr); Christian Kober (crb). **250 Getty Images:** Sylvain Grandadam (tr). **Nima Malek:** (bl). **250-251 Alamy Stock Photo:** Wiktor Szymanowicz. **254 Getty Images:** J. Baylor Roberts / National Geographic Image Collection. **255 Alamy Stock Photo:** age fotostock (cra); Michele Burgess (b). **256-257 Getty Images:** Weerapong Chaipuck. **256 Getty Images:** Rick Wezenaar (c). **Benny Welson:** @junteng99 (cb). **258-259 Alamy Stock Photo:** Fabrizio Troiani. **259 Dreamstime.com:** Lightfieldstudiosprod (crb). **Getty Images:** Geography Photos / UIG (bl). **260 David Blacker:** (bl). **iStockphoto.com:** pidjoe (ca); sandsun (br). **260-261 Getty Images:** Ryan Pyle / Corbis. **262 Alamy Stock Photo:** Alexey Kornylyev (t). **Getty Images:** Artie Photography (Artie Ng) (bc). **263 Muslianshah Masrie:** (b). **Joe Routon:** (ca). **264-265 Dreamstime.com:** Bidouze Stephane. **265 iStockphoto.com:** ugurhan (cb). **266-267 Kensuke Izawa**. **266 Alamy Stock Photo:** Alamy Premium (bc). **267 Dreamstime.com:** Bruno Pagnanelli (br). **Thomas Risse:** (bl). **Linda Tobey:** (bc). **268-269 iStockphoto.com:** Mike Fuchslocher. **269 Robert Harding Picture Library:** Michael Nolan (cb). **270-271 David Dillon**. **270 Getty Images:** Gerard van den Akker / 500px (br). **271 Alamy Stock Photo:** Eric Nathan (cr). **Jafarov Etibar Fikret:** (br). **272 iStockphoto.com:** SeanPavonePhoto. **273 Alamy Stock Photo:** Panther Media GmbH (br). **Getty Images:** KR_nightview / Multi-bits (t). **276-277**

123RF.com: Martin Molcan. **277 AWL Images:** Adam Jones (bc). **Getty Images:** Yongyuan Dai (c). **278-279 Getty Images:** Rob Zhang. **280 Getty Images:** Amith Nag Photography (t); zhouyousifang (cb). **281 4Corners:** Paul Panayiotou (b). **Alamy Stock Photo:** David Pearson (cra). **282 Getty Images:** Malcolm P Chapman (bl). **282-283 Getty Images:** Al-Hassan. **283 Getty Images:** Dedy Wibowo / EyeEm (br). **286-287 Dr. Ali Kordzadeh**. **286 Alamy Stock Photo:** age fotostock (br). **iStockphoto.com:** efesenko (bl); mariusz_prusaczyk (ca). **288-289 Getty Images:** Naveen Khare. **288 Alamy Stock Photo:** imageBROKER (cra). **289 Alamy Stock Photo:** Dinodia Photos (cla). **290-291 AWL Images:** Michele Falzone. **291 Dreamstime.com:** Sundraw (bl). **292 Alamy Stock Photo:** Leonid Andronov (bl). **292-293 Zhang Zhe**. **293 iStockphoto.com:** Hung_Chung_Chih (br). **296 Getty Images:** narvikk (bl). **296-297 Getty Images:** Tetra. **297 Alamy Stock Photo:** Michel & Gabrielle Therin-Weise (br). **298-299 Dreamstime.com:** Aliaksandr Mazurkevich. **299 Alamy Stock Photo:** mauritius images GmbH (cb). **300-301 George Nuich:** 500px.com / georgenuich. **300 Stewart Donn:** (bl). **Lynda McArdle:** (ca). **301 State Library of South Australia:** (br). **302 Dreamstime.com:** Kinek00 (bl). **Getty Images:** Malcolm Chapman (crb). **303 Vimal Konduri:** (cra). **Latitude Image:** Nicolas Chorier (b). **304-305 Alamy Stock Photo:** Avalon / Construction Photography. **305 Alamy Stock Photo:** David Ball (cra); Pablo Valentini (crb). **306-307 Getty Images:** Anuchit Kamsongmueang. **306 Alamy Stock Photo:** Peter Cook-VIEW (clb). **AWL Images:** Marco Bottigelli (br). **Getty Images:** David Greedy (bc). **307 Getty Images:** Nigel Killeen (br). **308-309 Jiti Chadha**. **308 Guo Hao:** (ca). **iStockphoto.com:** lkunl (bl). **309 iStockphoto.com:** Extreme-Photographer / E+ (br). **310 Getty Images:** Andrew Madali (t). **Imre Solt:** (cb). **311 Jonathan Danker:** (crb). **iStockphoto.com:** TwilightShow (bl). **312-313 Alamy Stock Photo:** SIPA Asia / ZUMA Wire. **313 MVRDV:** Ossip van Duivenbode (br). **314-315 AirPano images. 316-317 Getty Images:** Mike Biggs / 500Px Plus. **317 Getty Images:** Bloomberg. **318 Alamy Stock Photo:** Simone Celeste. **319 Getty Images:** David Sailors / Corbis Documentary / Getty Images Plus. **320-321 Alamy Stock Photo:** Megapress. **321 Alamy Stock Photo:** Adwo. **322 iStockphoto.com:** diegograndi. **323 Getty Images:** SambaPhoto / Cassio Vasconcellos. **324 Dreamstime.com:** Georgios Tsichlis. **324-325 Alamy Stock Photo:** Fabian von Poser / imageBROKER. **326 Alamy Stock Photo:** Songquan Deng. **327 Getty Images:** Georgy Rozov / EyeEm. **328-329 Getty Images:** Hans Georg Eiben / Stockbyte. 328 Steve Watson. **330 Alamy Stock Photo:** David Lyons. **331 Getty Images:** NurPhoto. **332 Alamy Stock Photo:** Kellee Kovalsky. **333 Alamy Stock Photo:** Violeta Meletis. **334-335 Alamy Stock Photo:** mauritius images GmbH. **335 Alamy Stock Photo:** Teo Moreno Moreno. **336 Alamy Stock Photo:** Karl F. Schöfmann / imageBROKER. **336-337 Alamy Stock Photo:** Salvatore Conte. **338 Dreamstime.com:** Marcorubino. **339 Alamy Stock Photo:** Stefano Politi Markovina. **340-341 Getty Images:** Norbert Achtelik. **342 Dreamstime.com:** Plotnikov. **343 123RF.com:** joserpizarro. **344-345 iStockphoto.com:** Frank Leung. **345 Alamy Stock Photo:** Zoonar / Konstantin Kalishko. **346-347 Alamy Stock Photo:** Karl Johaentges / Look. **347 Alamy Stock Photo:** Will Dale. **348 Getty Images:** Yang Liu / Corbis Documentary / Getty Images Plus. **348-349 Getty Images:** Chiara Salvadori / Moment Open. **350-351 Getty Images:** Bob Edwards / 500px. **351 Dreamstime.com:** Sjankauskas. **352 Alamy Stock Photo:** Kenneth Taylor. **353 Alamy Stock Photo:** Viennaslide. **354-355 Alamy Stock Photo:** Luxy Images Limited / L Collection. **356-357 Getty Images:** Torsten Andreas Hoffmann / LOOK-foto / LOOK. **357 Dreamstime.com:** Mihai-bogdan Lazar / Bogdan. **358 Alamy Stock Photo:** Reinhard Dirscherl. **359 Getty Images:** DEA / C. Sappa / De Agostini. **360-361 Alamy Stock Photo:** Marcin Dobas. **361 Alamy Stock Photo:** Salvador Aznar. **362 123RF.com:** asafeliason. **363 Alamy Stock Photo:** Albert Knapp. **364-365 Dreamstime.com:** EPhotocorp. **365 Alamy Stock Photo:** Dario Bajurin. **366-367 iStockphoto.com:** saiko3p. **367 Alamy Stock Photo:** Chon Kit Leong. **368 Getty Images:** Ian Griffiths / robertharding. **368-369 Dreamstime.com:** Mahmod5cy. **370-371 Alamy Stock Photo:** Sean Pavone. **371 Alamy Stock Photo:** Leonid Andronov. **372 AWL Images:** Jon Arnold. **373 Getty Images:** DuKai photographer / Moment. **374 iStockphoto.com:** Hung_Chung_Chih. **375 iStockphoto.com:** janssenkruseproductions. **376 Alamy Stock Photo:** Nattee Chalermtiragool. **376-377 iStockphoto.com:** Nikada. **378-379 Getty Images:** DuKai photographer / Moment

Endpaper images: Front: **iStockphoto.com:** fanjianhua; Back: **iStockphoto.com:** fanjianhua

For further information see: www.dkimages.com

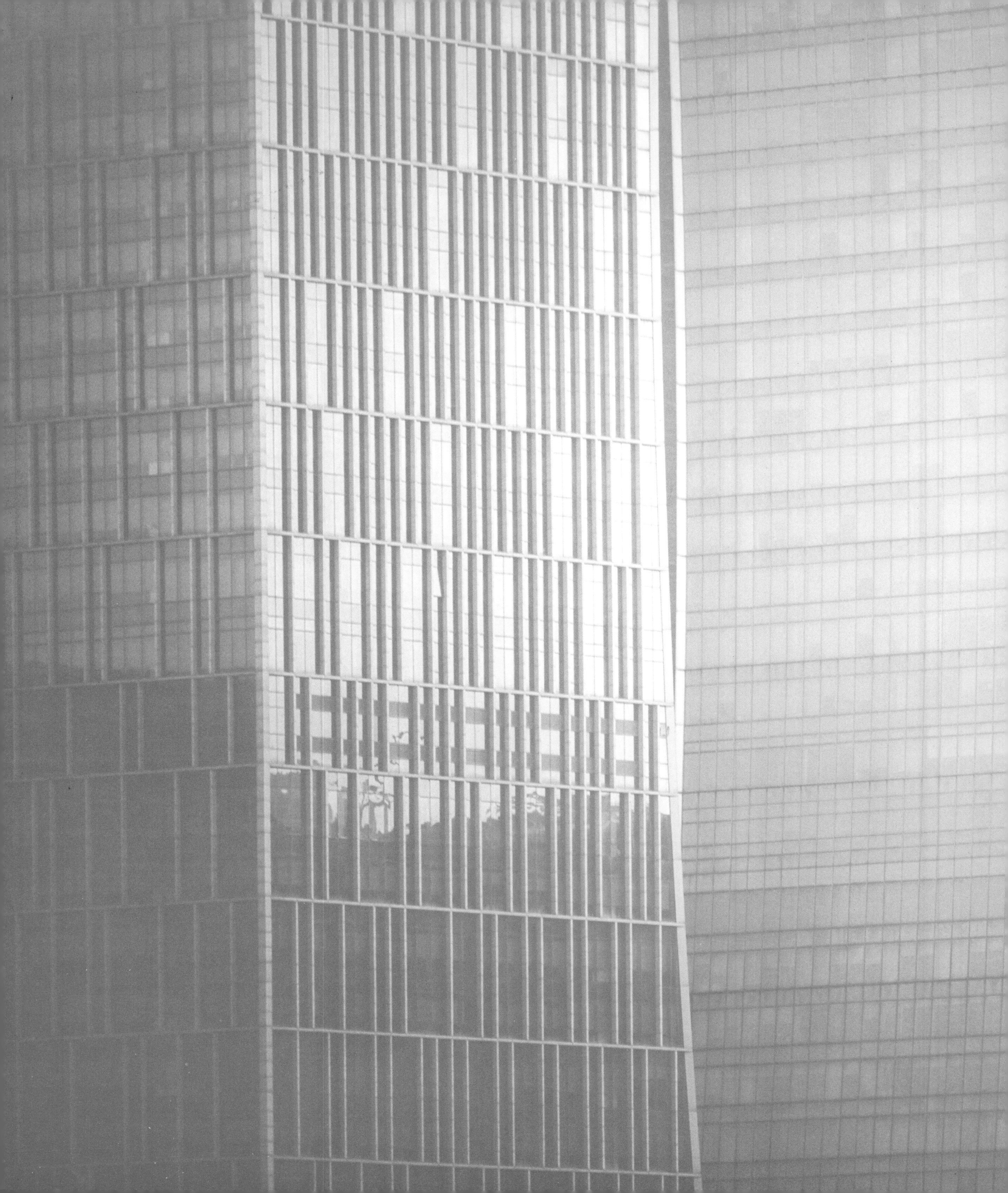